Electro-Coagulation and Electro-Oxidation in Water and Wastewater Treatment

Other Titles of Interest

Green Technologies for Sustainable Water Management, edited by Huu Hao Ngo, Wenshan Guo, Rao Y. Surampalli, and Tian C. Zhang. (ASCE/EWRI 2016). The 28 chapters in this collection describe science-based principles and technological advances behind green technologies that can be effective solutions to pressing problems in sustainable water management. (ISBN 978-0-7844-1442-2)

Computational Fluid Dynamics: Applications in Water, Wastewater and Stormwater Treatment, edited by Xiaofeng Liu and Jie Zhang. (ASCE/EWRI 2019). This book provides an introduction, overview, and specific examples of computational fluid dynamics and their applications in the water, wastewater, and stormwater industry. (ISBN 978-0-7844-1531-3)

Statistical Analysis of Hydrologic Variables: Methods and Applications, edited by Ramesh S. V. Teegavarapu, Jose D. Salas, and Jery R. Stedinger. (ASCE/EWRI 2019). This book provides a compilation of statistical analysis methods used to analyze and assess critical variables in the hydrological cycle. (ISBN 978-0-7844-1517-7)

Sustainable Wastewater Management in Developing Countries: New Paradigms and Case Studies from the Field, by Carsten Laugesen, Ole Fryd, Hans Brix, and Thammarat Koottatep. (ASCE Press 2010). He and his colleagues draw upon their experiences in Malaysia, Thailand, and other countries to inspire innovation and improvement in wastewater treatment and management. (ISBN 978-0-7844-0999-2)

Stormwater Manufactured Treatment Devices: Certification Guidelines, by the Joint Task Committee on Guidelines for Certification of Manufactured Stormwater BMPs; edited by Qizhong (George) Guo. (ASCE/EWRI 2017). This book provides a framework for regulatory agencies to create verification and certification programs to assess compact stormwater treatment systems. (ISBN 978-0-7844-1479-8)

Electro-Coagulation and Electro-Oxidation in Water and Wastewater Treatment

Electro-Coagulation and Electro-Oxidation in Water and Wastewater Treatment Task Committee of the Technical Committee on Hazardous, Toxic, and Radioactive Waste Engineering of the Environmental Council of the Environmental and Water Resources Institute of the American Society of Civil Engineers

Edited by
Patrick Drogui, Ph.D.
R. D. Tyagi, Ph.D.
Rao Y. Surampalli, Ph.D., P.E.
Tian C. Zhang, Ph.D., P.E.
Song Yan, Ph.D.
Xiaolei Zhang, Ph.D.

ASCE

EWRI ASCE ENVIRONMENTAL & WATER RESOURCES INSTITUTE

Published by the American Society of Civil Engineers

Library of Congress Cataloging-in-Publication Data

Names: Environmental Council (Environmental and Water Resources Institute). Electro-Coagulation and Electro-Oxidation in Water and Wastewater Treatment Task Committe, author. | Drogui, Patrick, editor. | Tyagi, R. D., 1952- editor. | Surampalli, Rao Y., editor. | Zhang, Tian C., editor. | Yan, Song (Sanitary engineer), editor. | Zhang, Xiaolei (Sanitary engineer), editor.
Title: Electro-coagulation and electro-oxidation in water and wastewater treatment / Electro-Coagulation and Electro-Oxidation in Water and Wastewater Treatment Task Committe of the Technical Committee on Hazardous, Toxic, and Radioactive Waste Engineering of the Environmental Council of the Environmental and Water Resources Institute of the American Society of Civil Engineers ; editors/authors Patrick Drogui, Ph.D., R. D. Tyagi, Ph.D., Rao Y. Surampalli, Ph.D., P.E., Tian C. Zhang, Ph.D., P.E., Song Yan, Ph.D., Xiaolei Zhang, Ph.D.
Description: Reston, Virginia : American Society of Civil Engineers, [2022] | Includes bibliographical references and index. | Summary: "This book provides a detailed overview of the origins, principle, benefits, impacts, and applications of electro-coagulation and electro-oxidation for water/wastewater treatment"-- Provided by publisher.
Identifiers: LCCN 2021055826 | ISBN 9780784416020 (print) | ISBN 9780784483992 (PDF)
Subjects: LCSH: Water--Purification--Coagulation. | Sewage--Purification--Oxidation. | Water--Purification--Electrochemical treatment. | Sewage--Purification--Electrochemical treatment. | Electrocoagulation. | Electrolytic oxidation.
Classification: LCC TD471 .E58 2022 | DDC 628.1/62--dc23/eng/20220125
LC record available at https://lccn.loc.gov/2021055826

Published by American Society of Civil Engineers
1801 Alexander Bell Drive
Reston, Virginia 20191-4382
www.asce.org/bookstore | ascelibrary.org

***Errata:* Errata, if any, can be found at https://doi.org/10.1061/9780784416020.**

ISBN 978-0-7844-1602-0 (print)
ISBN 978-0-7844-8399-2 (PDF)
ISBN 978-0-7844-8420-3 (ePub)

Manufactured in the United States of America.

27 26 25 24 23 22 1 2 3 4 5

Contents

Preface

Electro-coagulation (EC) and electro-oxidation (EO) are based on the principle of introducing an electrical current to induce a chemical reaction in water, causing the destabilization of most pollutants such as suspended particles, bacteria, viruses, dissolved materials, metals, hydrocarbons, and many organics. The low energy consumption and absence of chemical utilization with the potential recovery and reuse of treated water makes EC and EO truly green technologies for water/wastewater treatment. EC- and EO-based electrochemical water treatments as a green technology have a major advantage in terms of total size flexibility and are applicable in both urban and rural areas.

This book provides state-of-the-art technologies of EC and EO applications in various water and wastewater treatment processes. Chapter 1 introduces general principles and the basics of electro-coagulation. Chapter 2 discusses the origins and principles of the electro-coagulation process. Chapter 3 discusses the origins and principles of the electro-oxidation process. The mathematical modeling of the electro-coagulation (Chapter 4) and electro-oxidation processes (Chapter 5) are also presented respectively. Chapters 6, 8, 10, and 12 provide the fundamental and recent developments of EC-based technologies, including various treatment techniques (Chapter 6), environmental applications of electro-coagulation processes (Chapter 8), comparative studies between electro-coagulation, adsorption and chemical precipitation (Chapter 10), and full-scale application of electro-coagulation processes (Chapter 12). Topics related to EO-based techniques and processes include combined electro-oxidation (Chapter 7), environmental applications of EO processes (Chapter 9), comparative studies between EO and other oxidation processes (Chapter 11), and full-scale applications of EO processes (Chapter 13). Chapter 14 discusses the cost comparison of EC/EO processes with other clean-up technologies, and Chapter 15 presents the challenges and future perspectives of the EC and EO processes.

We hope this book will be of interest to students, scientists, engineers, government officers, process managers, and practicing professionals. As a reference, this book will help readers readily find the information they are looking for.

The editors gratefully acknowledge the hard work and patience of all authors who have contributed to this book. The views or opinions expressed in each chapter of this book are those of the authors and should not be construed as opinions of the organizations they work for.

Patrick Drogui
R. D. Tyagi
Rao Y. Surampalli
Tian C. Zhang
Song Yan
Xiaolei Zhang

Contributing Authors

Bhadana, Bharti, *INRS, Université du Québec, Québec, QC, Canada*

Drogui, P., *INRS, Université du Québec, Québec, QC, Canada*

Ebrahimi, Mitra, *INRS, Université du Québec, Québec, QC, Canada*

Gernaey, Krist V., *Department of Chemical and Biochemical Engineering, Technical University of Denmark, Lyngby, Denmark*

Gholami, Shirkoohi Majid, *INRS, Université du Québec, Québec, QC, Canada*

Karimi, Estahbanati M. R., *INRS, Université du Québec, Québec, QC, Canada*

Khosravanipour, Mostafazadeh Ali, *INRS, Université du Québec, Québec, QC, Canada*

Khajvand, Mahdieh, *INRS, Université du Québec, Québec, QC, Canada*

Kiendrebeogo, Marthe, *INRS, Université du Québec, Québec, QC, Canada*

Kumari, Anita, *INRS, Université du Québec, Québec, QC, Canada*

Kumar, L. R., *INRS, Université du Québec, Québec, QC, Canada*

Kumar, Sushil, *INRS, Université du Québec, Québec, QC, Canada*

Nayernia, Zahra, *School of Chemical and Petroleum Engineering, Shiraz University, Shiraz, Iran*

Ouarda, Yassine, *INRS, Université du Québec, Québec, QC, Canada*

Panidepu, H., *Indian Institute of Technology, New Delhi, India*

Pilli, S., *INRS, Université du Québec, Québec, QC, Canada*

Ram, S. K., *INRS, Université du Québec, Québec, QC, Canada*

Ramin, Pedram, *Department of Chemical and Biochemical Engineering, Technical University of Denmark, Lyngby, Denmark*

Surampalli, R. Y., *Global Institute for Energy, Environment and Sustainability, Kansas, USA*

Tiwari, B., *INRS, Université du Québec, Québec, QC, Canada*

Tyagi, R. D., *INRS, Université du Québec, Québec, QC, Canada*

Vasavi, C., *University of Mumbai, India*

Yan, S., *INRS, Université du Québec, Québec, QC, Canada*

Zhang, Tian C., *University of Nebraska-Lincoln, Lincoln, NE*

Zhang, X. L., *Harbin Institute of Technology (Shenzhen), Shenzhen, Chin*

CHAPTER 1
Introduction

Song Yan, Patrick Drogui, X. L. Zhang, R. D. Tyagi, Rao Y. Surampalli, Tian C. Zhang

1.1 PRINCIPLE AND DEFINITION OF ELECTRO-COAGULATION

Electro-coagulation (EC) is a process that originates from conventional chemical coagulation. EC technology delivers the coagulant in situ by anodic dissolution and produces subsequently iron (or aluminum) hydroxides having a considerable sorption capacity, whereas the simultaneous cathodic reaction allows the removal of pollutants either by deposition on the cathode electrode or by flotation (the evolution of hydrogen at the cathode) (Chen 2004, Mollah et al. 2001). EC is based on the principle of introducing an electrical current to induce a chemical reaction in water, causing destabilization of most pollutants: suspended particles, bacteria, viruses, dissolved materials, metals, hydrocarbons, and many organic compounds. Indeed, during EC treatment of wastewater, several processes (electrochemical, physicochemical, and chemical processes) take place: (1) cathodic and chemical reduction of organic and inorganic matter; (2) chemical interaction between ions of iron (or aluminum), formed during the dissolution of anodes, and anions present in wastewater, resulting in the formation of insoluble compounds; (3) flotation of solid and emulsified impurities by hydrogen gas produced on the cathode; and (4) sorption of ions and molecules of dissolved impurities on the surface of iron (or aluminum) hydroxides having considerable sorption capacity. Likewise, during EC, the resulting effluent is not enriched with anions, and salt content does not increase, in comparison with chemical coagulation (Cenkin and Belevtsev 1985).

The electrodes used can be in varied forms, depending on the flow rate of the effluent and the geometrical form of the reactor. Cylindrical, circular, and rectangular electrodes (plate or grit) can be used (Chen 2004). The simplest electrode in industrial use consists of rectangular (aluminum or iron) electrodes. According to electrode arrangements, two configurations of the electrolytic cell can be distinguished: *monopolar* and *bipolar* electrodes.

Bipolar electrolytic cell consists of parallel pieces of aluminum or mild steel plate electrodes, with only two outermost electrodes physically connected to the power supply. Each electrode, except for those at each end, thus functions as an anode on one face and as a cathode on the other. The inner electrodes function as electrodes owing to their activation induced by the mobility of ions in solution, each ion transporting a fraction of current intensity imposed. The fraction of current transported by each ion (1) is called *transport number* (t_i), which is defined as the ratio between the conductivity of the corresponding ion (1) and the total conductivity of the solution. The transportation of these ions induces the activation of the inner electrodes, and the electric current flows through these electrodes. It is worth noting that, during electrolysis, the mobility of ions in solution is because of the electric field induced by the potential difference existing between the electrodes connected to the power supply. Likewise, the mobility of ions transporting the current intensity also results from the mixing of the electrolyte (mobility by convection). Thus, every two adjacent electrodes and the intervening solution is a single unit. Each reactor unit is electrically in series with the others, and the same current flows through every unit.

By comparison, *monopolar electrolytic cell* consists of parallel pieces of rectangular plate electrodes (mild steel or aluminum), each individually connected to the power supply, with anodes and cathodes being alternated in the electrode pack. In this configuration, each unit operates at the same voltage, with the total current being the sum of the individual unit currents; the activation of the electrodes by the ions transporting the electrical current is negligible because each electrode is individually connected to the power supply.

Both bipolar and monopolar methods of connection are extensively used. The difference between the two configurations can be seen while comparing the potential difference required for a given current flow. In the monopolar configuration system, the electric current imposed is divided among all the electrodes, and the potential difference would be that required by a single cell. This is in contrast to the bipolar configuration, in which a higher potential difference is required because of the higher resistance for the cells connected in series. In the bipolar configuration system, the same electrical current flows through all the electrodes inducing the higher resistance in the electrolytic cell. The bipolar system is preferred, because fewer electrical connections are required and less power dissipation occurs in the external circuit (Laridi et al. 2005).

1.2 ELECTRO-COAGULATION PROCESS: EMERGING TECHNOLOGY FOR WATER AND WASTEWATER TREATMENT

In general, EC process treatment is characterized by simple equipment, ease of operation, short retention time, and negligible equipment for adding chemicals, which would contribute to reducing the operating costs in large-scale applications (Rumeau 1989, Garcia-Segura et al. 2017, Moussa et al. 2017). Besides, electrolytic cells can easily be automated and coupled with other processes, including biological, chemical, and physical processes, to enhance the efficiency of the treatment. The interest of using electrochemical coagulation for water/wastewater treatment can be summarized in three main points: (1) EC treatment has a practical advantage of producing an effluent having a pH close to the neutral value, which is often required for effluent discharge in the receiving water; (2) during EC, the resulting effluent is not enriched with anions, and salt content does not increase, unlike in chemical coagulation [using, for instance, $FeCl_3$ or $Al_2(SO_4)_3$], where the effluent is enriched with chloride ions; and (3) EC allows in situ producing a coagulating agent, which would contribute to reduce the costs related to chemical transportation (no need to add chemical products) in large-scale applications. Likewise, this approach enables to get rid of the constraints linked to the storage of chemical products. EC processes are suitable for decentralized sanitation and water treatment in extreme conditions (e.g., in the nordic regions and remote areas).

1.3 ELECTRO-COAGULATION PROCESS: GREEN AND CLEAN ELECTROCHEMICAL TECHNOLOGY

The definition of a *green chemical* has been given by Anastas and Warner (1998). They have given a broad definition of green chemistry based on 12 principles that relate to several steps, some of which are a chemical should be synthesized in a safe and energy-efficient manner; its toxicity should be minimal, whereas its biodegradation should be optimal; its impact on the environment should be as low as possible. Wastewater treatment technology has undergone many transformations in terms of energy consumption, chemical utilization, and the quality of treated water with the possible reuse potential in the treatment of sewage and industrial wastewater. The low energy consumption, the absence of chemical utilization with the potential recovery and reuse of treated water, and its total size flexibility make EC a true green technology that can be used in both urban and rural areas for wastewater treatment. Likewise, solar energy can be used as an energy source. Solar panels can be used to activate the electrodes inside the electrolytic cells.

1.4 INTEGRATION INTO WATER/WASTEWATER TREATMENT PLANTS OR APPLICATION FOR DECENTRALIZED SANITATION

EC for water treatment purposes was patented over a century ago (Dietrich 1906). Until the last few decades, with the rapid development of the power industry and the increasing complexity of wastewater components, there has been renewed interest in the application of EC. EC is based on in situ electrolytic dissolution of the anode metal material and may be readily automated. Simultaneously with the anodic reaction, gas bubbles are generated at the cathode surface, promoting electroflotation. Usually, Al, Fe, and/or stainless steel (SS) are used as electrode materials (Chen 2004, Mollah et al. 2001). EC treatment is now being accepted as one of the emerging and promising water and wastewater treatment technologies with an advanced, convenient, efficient, and economic electrochemical process that combines the benefits of coagulation, flotation, and electrochemistry.

The EC process has been successfully used to (1) remove metals, oil, and grease from wastewater; (2) harvest protein, fat, and fiber from food processes' waste streams; (3) remove biological oxygen demand (BOD), total suspended solids, total dissolved solids (TDSs), and so on from wastewater before its disposal to the municipal sewerage system; (4) recycle water, allowing closed-loop systems; (5) pretreatment, conditioning, and polishing of drinking water; (6) recondition antifreeze by removing oil, dirt, and metals; (7) pretreatment prior to the application of membrane technologies such as reverse osmosis (RO); (8) preconditioning of the boiler make-up water by removing silica, hardness, TSS, and so on; (9) recondition the boiler blowdown by removing dissolved solids and eliminating the need for the chemical treatment of boilers; and (10) dewater the sewage sludge and stabilize the heavy metals in sewage, lowering freight cost and allowing the sludge to be land-applied (Chen 2004, Garcia-Segura et al. 2017, Moussa et al. 2017, Bharath et al. 2018, Nidheesh and Singh 2017, Song et al. 2017). Table 1-1 presents various pollutants removed by EC in water and wastewater, some of the recent research studies, and the associated references.

A typical municipal sewage treatment plant may include primary treatment to remove solid material, secondary treatment to degrade dissolved and suspended organic material as well as the nutrients (nitrogen and phosphorus), and (sometimes but not always) disinfection to kill pathogens. EC technology is unique and offers many opportunities that are hard to match by other technologies. EC can be used as primary treatment, secondary treatment, tertiary treatment, and with combinations with other treatments. Details on how to integrate EC into different treatment processes are described subsequently.

1.4.1 Electro-Coagulation Used as Primary Physicochemical Treatment

In general, preliminary and primary treatments are applied to remove rough and coarse material from raw wastewater to protect mechanical and electrical parts in the downstream treatment steps. The primary treatment step is essential because it allows for the efficient and prolonged use of the secondary treatment unit. In a traditional wastewater treatment plant, primary treatment consists of temporarily holding the sewage in a quiescent basin (i.e., a primary sedimentaiton tank) where heavy solids can settle to the bottom while oil, grease, and lighter solids float to the surface. The settled and floating materials are removed, and the remaining liquids may be discharged or subjected to secondary treatment.

The classical physicochemical treatment processes that are used for the primary (or advanced) wastewater treatment include adsorption, ultrafiltration, RO, electrodialysis, volatilization, and gas stripping. A common practice involves using chemicals in wastewater treatments, which are sometimes toxic. One of the

Table 1-1. Summary of Pollutants Removed by Electrocoagulation in Water and Wastewater.

Pollutants, wastewater type	*Electrode materials, anode–cathode*	*Optimum conditions*	*Removal efficiency (%)*	*Special marks*	*Level of treatment (primary, secondary, tertiary treatment, or combination)*	*Reference*
Colloids and organics in shale gas fracturing flowback water (SGFFW) during shale gas extraction	Coagulation combined with oxidation might be a promising process for SGFFW; an Al plate as the anode and a carbon-polytetrafluoroethylene gas diffusion electrode as the cathode	Electrocoagulation (EC) and peroxielectrocoagulation (PEC), optimal current density of 50 mA cm^{-2}	The average current efficiency for PEC was 35.4% and 12% higher than that for ozonation and EC, respectively	A high efficiency and space-saving electrochemical system with integrated anodic coagulation and cathodic electro-peroxone for SGFFW treatment	Primary	Kong et al. (2019)

(*Continued*)

Table 1-1. Summary of Pollutants Removed by Electrocoagulation in Water and Wastewater. (Continued)

Pollutants, wastewater type	*Electrode materials, anode–cathode*	*Optimum conditions*	*Removal efficiency (%)*	*Special marks*	*Level of treatment (primary, secondary, tertiary treatment, or combination)*	*Reference*
Textile wastewater	A combined EC, H_2O_2/Fe^{2+}/UV, and activated carbon adsorption; EC) + Fenton (F) or photo-Fenton (PF) + active carbon adsorption (EC + F/PF + AC)	Optimum conditions pH = 4.3, $[Fe^{2+}] = 1.1$ mM, and $[H_2O_2] = 9.7$ mM	EC: dye's color: 94% COD: 56%; TOC: 54%. The EC + PF color: 100%; COD: 76%; TOC: 78%	The total operational costs, including chemical reagents, electrodes, energy consumption, and sludge disposal, were of USD 1.65 m^{-3} and USD 2.3 m^{-3} for EC + F and EC + PF, respectively	Combination	GilPavas et al. (2019)

High organic load and low-volume egg processing industry wastewater	Al–Al and Fe–Fe electrodes	Aluminum electrode in 10 min retention time at a current density of 6.6 mA cm^{-2}	COD (99%), TSS (98.5%), TC (99.99%)	In addition, the aluminum electrode with an EEC of 14.2 kW h m^{-3} and an anode consumption of 0.53 mg cm^{-2} was more cost-efficient than an iron electrode	Primary	Azarian et al. (2018)
Paint manufacturing wastewater (PMW)	Electrode type (Al or Fe)	Initial pH (2–10), current density (5–80 A/m^2), and operating time (0–50 min)	For Fe: COD: 93%, TOC: 88%; for Al electrodes: COD: 94%, TOC: 89% at optimum conditions: (35 A m^{-2}, 15 min, and pH 6.95)	Operating costs: at optimum conditions: Fe: €0.187m^{-3}, Al electrodes: €0.129m^{-3}	Primary	Akyol(2012)

(*Continued*)

Table 1-1. Summary of Pollutants Removed by Electrocoagulation in Water and Wastewater. (Continued)

Pollutants, wastewater type	*Electrode materials, anode–cathode*	*Optimum conditions*	*Removal efficiency (%)*	*Special marks*	*Level of treatment (primary, secondary, tertiary treatment, or combination)*	*Reference*
Hybrid electrocoagulation—nanofiltration process textile wastewater	Real textile wastewater treatment by electrocoagulation (EC), nanofiltration (NF), and a combined EC-NF		COD 64%, color: 94% ultralow rejections of inorganic salts (less than 4%)	EC using an Al electrode was superior to that using other electrodes. Applying EC pretreatment (using Al electrode) effectively controlled NF fouling. The hybridization of EC-NF could augment each other's strengths and mitigate their individual drawbacks	Primary	Tavangar et al. (2019)

Remnant from chemical retting of coconut fiber through electrocoagulation and activated carbon treatment		Current flow (100–2,000 mA), time (15–90 min), and the number of iron electrodes (2, 4, 6). Optimum treatment condition was 6 electrodes, pH 7, 90 min, and 2,000 mA	COD (92.8%) TOC (56%) TDS (99%)	Cost of EC (USD 0.087/L) was found to be much less than that of the traditional method (USD0.384 L^{-1}). Lethal concentration of 135 mg L^{-1} indicates its safer disposability. LC50 value denotes safe disposal	Combination	Jose et al. (2019)
Phenolic wastewater treatment	The Zn anode/stainless-steel cathode pair	Initial pH of the effluent was equal to 3.2, at a current density of 250 A m^{-2}, and the distance between the electrodes of 1.0 cm and 1.5 g L^{-1} of NaCl	Total phenolic (TPh) content: 84.2%, chemical oxygen demand (COD): 40.3% The depuration of a filtered real olive mill effluent without NaCl addition: TPh: 72.3%, COD: 20.9%	An energy consumption of 40 kW·h m^{-3} and 34 kW·h m^{-3} was observed for the treatment of simulated and real wastewater, respectively	Primary	Fajardo et al. (2015)

(Continued)

Table 1-1. Summary of Pollutants Removed by Electrocoagulation in Water and Wastewater. (Continued)

Pollutants, wastewater type	*Electrode materials, anode–cathode*	*Optimum conditions*	*Removal efficiency (%)*	*Special marks*	*Level of treatment (primary, secondary, tertiary treatment, or combination)*	*Reference*
Biodiesel wastewater oil, grease	Aluminum anode and a graphite cathode	pH 6.06, voltage 18.2 V, and reaction time 23.5 min	COD: 55.43%, oil and grease (O&G): 98.42%, SS: 96.59%		Primary	Chavalparit and Ongwandee (2009)
Domestic wastewater	Coupled EC–electro-Fenton	pH: 7.0 Electrode: iron (anode) and carbon vitreous (cathode) Current: 0.34 A dm^{-2} Time: 60 min	COD: 50.07, TSS: 90.40, turbidity: 70.80	0.21 (including electrode cost, energy consumption, and sludge disposal)	Secondary treatment	Daghrir and Drogui (2013)
Landfill leachates	Anode = Al EC and biofiltration (BF) were sequentially used to treat landfill leachate. FA and hydrophilic compounds were removed during the BF	($j = 8$ mA cm^{-2} and $t = 20$ min).	EC treatment: COD: 37 ± 2% of Turbidity: 82 ± 2.7%, true color: 60 ± 13%, Zn: 95 ± 2.6% Iron 95 ± 2.3% Phosphorus: 82 ± 5.5%		Secondary	Dia et al. (2018)

Domestic wastewater	A migration electric–field assisted electrocoagulation (MEAEC) reactor.	Anode (Fe or Al mesh) and the air cathode	Highly efficient phosphate removal with low energy consumption. • With a secondary clarifier effluent, the obtained removals were < 0.1 mg L^{-1} of PO_4^{3-}. • Fe-MEAEC worked better than Al-MEAEC with low PO_4^{3-} concentrations in wastewater	• The Ti plate was more durable than graphite as the charging (inert) electrode.	Post treatment Tertiary	Tian et al. (2018)
Molasses wastewater with high nitrogen content obtained from a baker's yeast industry.	Iron and copper electrodes were used in various forms	3.5 h treatment time and 33 mA cm^{-2} current density	COD: 10 to 54%, ammonium: 70%	Operation cost: ranging from €0.2 to €33 kg^{-1}, COD removed, optimized conditions could be established under moderate molasses wastewater dilution (e.g., 45%).	Post treatment	Tsioptsias et al. (2015)

(*Continued*)

Table 1-1. Summary of Pollutants Removed by Electrocoagulation in Water and Wastewater. (Continued)

Pollutants, wastewater type	*Electrode materials, anode–cathode*	*Optimum conditions*	*Removal efficiency (%)*	*Special marks*	*Level of treatment (primary, secondary, tertiary treatment, or combination)*	*Reference*
	Electrolysis cell in the presence of mild steel (anode) and vitreous carbon (cathode) electrodes	Current density of 4.0 mA cm^{-2} time of 60 min	COD: 67 ± 9, SS: 98 ± 2, turbidity: 55 ± 10, color: 61 ± 9, total phosphorus (P_{tot}): 97 ± 0%, pathogens (fecal coliform): 99 ± 1%		Tertiary	Senghor et al. (2015)
Spiked municipal wastewater (MWW).	Bipolar electrolytic cell made of mild steel electrodes	Current density of 38.2 mA cm^{-2} through 20 min	P_{tot}: 97%, regardless of the initial P_{tot} concentration imposed (5.0–50 mg L^{-1}) in MWW		Tertiary	Tran et al. (2012)

advanced electrochemical technology–based techniques is the EC process. EC not only provides a fast rate of pollutant removal and simplicity of operation but also requires no chemical additive (Azarian et al. 2018, Dia et al. 2018, Nariyan et al. 2017, Song et al. 2017). In addition, EC produces less amounts of sludge (Mouedhen et al. 2008, Linares-Hernández et al. 2008). These beneficial properties render EC more favorable than conventional physicochemical treatment processes.

EC can be used to remove most suspended materials and metal ions from solution and, therefore, has possible uses in treating, mining, electroplating, and other wastewater processes, contaminated groundwater as well as pretreatment of drinking water. For example, EC treatment can be used as a pretreatment before a RO process because EC removes suspended material, reduces calcium scaling, and, in general, reduces the total salt level in water and wastewater (Chen 2004, Garcia-Segura et al. 2017, Moussa et al. 2017, Bharath et al. 2018, Song et al. 2017). Flotation is the conventional primary treatment for oil and grease (O&G) removal prior to biological treatments. Recently, EC has also been successfully applied to remove O&G and chemical oxygen demand (COD) from the oily wastewater that emanates from restaurants and the mechanical and metallurgical industries (Xu and Zhu 2004, Tir and Moulai-Mostefa 2008).

The study by Chavalparit and Ongwandee (2009) adopted EC to treat biodiesel wastewater. The production of biodiesel through a transesterification method produces a large amount of wastewater that contains high levels of COD and O&G. Currently, flotation is the conventional primary treatment for O&G removal prior to biological treatments. The experimental results show that EC could effectively reduce COD, O&G, and SS by 55.43%, 98.42%, and 96.59%, respectively, at the optimum conditions of pH 6.06, applied voltage 18.2 V, and reaction time 23.5 min. The experimental observations were in reasonable agreement with the modeled values (Chavalparit and Ongwandee 2009).

Batch EC experiments were carried out to evaluate the removal of sulfate and COD from petroleum refinery wastewater using three types of electrodes: aluminum, SS, and iron. The effects of current density, electrode arrangement, electrolysis time, initial pH, and temperature were investigated for two wastewater samples with different concentrations of COD and sulfate. The experimental results indicated that the utilization of aluminum, as the anode and cathode, was by far the most efficient arrangement in the reduction of both contaminants. The treatment process was found to be largely affected by the current density and the initial composition of the wastewater. Although EC was found to be most effective at 25 °C and a pH of 8, the influence of these two parameters on the removal rate was not significant. The results demonstrated the technical feasibility of EC as a possible and reliable technique for the pretreatment of heavily contaminated petroleum refinery wastewater (Chavalparit and Ongwandee 2009).

Black liquor is one of the main by-products of the pulp paper industry, which is considered as a pollutant because it contains about 50% of lignin. Lignin is a mixture of polyphenolic compounds with a rather complex chemical structure that resists traditional biological treatment processes because of their nonbiodegradable nature. The presence of lignin in wastewater involves a strong increase in COD and BOD. In addition to lignin, the black liquor contains aliphatic acids, acid greases, resins, and polysaccharides. These organics, mainly dissolved, create a relatively high BOD and COD. The discharge of this effluent in nature without any treatment is responsible of serious damage for the environment and constitutes a threat for human health. EC is an effective, fast, and economic technique for treatment of black liquor resulting from the paper industry (Zaied and Bellakhal 2009). The effect of electrolysis time, current density, type of electrode material, and initial pH was studied in an attempt to achieve a higher removal capacity. Under the optimal experimental conditions (initial pH 7, $t = 50$ min, and $J = 14$ mA cm^{-2}), the treatment of black liquor by EC has led to a removal capacity of 98% of COD, 92% of polyphenols, and 99% of color intensity with a good repeatability (RSD < 3%), making it possible to highlight the industrial interest of this electrochemical process (Zaied and Bellakhal 2009).

In primary treatment processes, chemical treatment addresses suspended solids, O&G, and some heavy metals but may require up to three polymers and multiple pH adjustments for proper treatment. This technology requires the addition of chemicals resulting in expensive, messy, and labor-intensive treatment. This process also requires the addition of compressed air for floatation of coagulated contaminants. In general, filtration is also required as a post-treatment phase for polishing. EC requires no filters, no daily maintenance, and no additives, but remarkably, it removes suspended solids of any size, O&G, and heavy metals (Akyol 2012, Li et al. 2018, Ogando et al. 2019, Pujol Pozo et al. 2019, Sharma and Simsek 2019).

1.4.2 Electro-Coagulation Used as Secondary Treatment

Secondary treatment removes dissolved and suspended biological matter; it is designed to substantially degrade and remove the dissolved and suspended biological content of the sewage that are derived from human waste, food waste, soaps, and detergent. Secondary treatment is typically performed by indigenous, water-borne micro-organisms in a managed habitat. Secondary treatment may require a separation process to remove the micro-organisms from the treated water prior to discharge or tertiary treatment.

Biological processes (BP) are traditionally used for the treatment of landfill leachate. The use of BP is mainly justified by their simplicity, a low treatment cost, and a good performance for the removal of biodegradable organic materials and nitrogen (Kurniawan et al. 2006). However, these technologies face some limits when it comes to dealing with *mature*" leachates with a high proportion of biorefractory compounds (Xiao et al. 2013). To overcome this problem, BP can be coupled with other nonbiological systems to increase the purification efficiency (Mandal et al. 2017). Among these combinations, the EC process could be an interesting approach to support a biological system. The main advantages of EC are a low sludge production in comparison with the chemical coagulation, a low or no chemical reagents requirement, short treatment time, and an easy automation (Drogui et al. 2007). Depending on the treatment needs, the BP can be used as a pretreatment or posttreatment step.

The feasibility of treating landfill leachate by using a combining approach of EC and biofiltration (BF) processes had been investigated (Dia et al. 2018). The pretreatment by the EC process allowed removing $37 \pm 2\%$ of the total COD. The fractionation of the organic matter (before and after EC treatment) indicated that the fraction of COD removed from raw landfill leachate was mainly attributed to the removal of insoluble COD and humic acid (HA). Furthermore, other pollutants such as turbidity, true color, zinc, and phosphorus were successfully removed (removal of $82 \pm 2.7\%$, $60 \pm 13\%$, $95 \pm 2.6\%$, and $82 \pm 5.5\%$, respectively). The subsequent biological treatment allowed removing the ammonia pollution as well as a partial fraction of the dissolved organic matter. More than 99% of the initial NH_4 had been removed, and nitrate was the main nitrogen by-product formed from the nitrification process. Surprisingly, the phosphorus deficiency and the residual aluminum ions in the electrocoagulated landfill leachate had no negative impact on the nitrification. Therefore, it can be concluded that the EC/BF hybrid process is a promising technology for landfill leachate. The complementarity of these two technologies allows an efficient removal of various ranges of pollutants found in landfill leachate effluent. This hybrid process is also economically viable. The operating cost of the process had been estimated at around \$US 1.23 m^{-3}, which is below the operating cost of most alternate processes used for landfill leachate treatment. This cost takes into account the aeration of the biofilter, the renewal of anode material and biofilter media, the energy consumption, and the EC sludge treatment (Dia et al. 2018).

1.4.3 Electro-Coagulation Used as Tertiary Treatment

Tertiary treatment is supplementary to primary and secondary treatment for the purpose of removing the residual organic and inorganic substances and, in some cases, the refractory and dissolved substances to the degree necessary. It provides a final treatment stage to further improve the effluent quality before it is discharged to the receiving environment (sea, river, lake, wetland, ground, etc.). More than one tertiary treatment process may be used at any treatment plant.

Tertiary treatment is sometimes defined as anything more than primary and secondary treatment to allow ejection into a highly sensitive or fragile ecosystem (estuaries, low-flow rivers, coral reefs, etc.). Treated water is sometimes disinfected chemically or physically (e.g., by lagoons and microfiltration) prior to discharge into a stream, river, bay, lagoon, or wetland, or it can be used for the irrigation of a golf course, greenway, or park. If it is sufficiently clean, it can also be used for groundwater recharge or agricultural purposes.

Effluents from municipal wastewater treatment plants (WWTPs) can contain residual phosphorus concentration above the recommended limiting values, and it could have a high environmental impact (e.g., eutrophication as a result of excess nutrients). To produce an effluent suitable for stream discharge, electrochemical techniques (Tran et al. 2012) have been explored at the laboratory pilot scale for total phosphorus (P_{tot}) removal from spiked municipal wastewater (MWW). The results showed that P_{tot} removal from MWW was highly effective on occasions when EC was combined with flocculation. Under optimal conditions, electrolysis was operated at a current density of 38.2 mA cm^{-2} through 20 min of treatment, and then, the electrocoagulated effluent was flocculated using a cationic polymer (10 mg L^{-1}) before filtration. At least, 97% of P_{tot} was removed regardless of the initial P_{tot} concentration imposed (5.0 to 50 mg L^{-1}) in MWW. The optimal conditions determined for P_{tot} removal, including energy consumption, electrode consumption, and metallic sludge disposal, involved a total cost varying from $CAN 0.24 to 0.35 m^{-3} of treated MWW in continuous mode of operation (Tran et al. 2012).

The possibility of treating landfill leachate by using BF as a pretreatment process, followed by EC, was investigated (Dia et al. 2016). The BF process was used as secondary treatment to remove effectively ammonia nitrogen ($N–NH_4$ removal of 94%), BOD (94% removed), turbidity (95% removed), and phosphorus (more than 98% removed). Subsequently, the EC process using magnesium-based anode was used as tertiary treatment. The best performances of COD and color removal from landfill leachate were obtained by applying a current density of 10 mA cm^{-2} through 30 min of treatment. The COD removal reached 53%, whereas 85% of color removal was recorded. It has been proved that the alkalinity had a negative effect on COD removal during EC treatment. COD removal efficiencies of 52%, 41%, and 27% were recorded in the presence of 1.0, 2.0, and 3.0 g L^{-1} of sodium bicarbonate ($NaHCO_3$), respectively. Hydroxide ions produced at the cathode electrode reacted with the bicarbonate ions to form carbonates. The presence of bicarbonates in solution hampered the increase in pH, so that the precipitation of magnesium hydroxides could not take place to effectively remove organic pollutants (Dia et al. 2016). The results (Dia et al. 2016) proved the feasibility of combined BF and EC processes for the effective treatment of mature sanitary landfill leachate.

A process that combines EC and electroperoxidation (EC–EP) was developed to evaluate its performance in treating domestic wastewater (DWW) (Senghor et al. 2015). Electrolysis was performed using a parallelepipedic electrolytic cell (0.5 L) containing one sacrificial anode (mild steel or aluminum) and one cathode (vitreous carbon). The experimental results revealed that a current density of 4.0 mA cm^{-2} was beneficial for DWW treatment. A decrease in COD, SS, turbidity, color, and total phosphorus (P_{tot}) by $67 \pm 9\%$, $98 \pm 2\%$, $55 \pm 10\%$, $61 \pm 9\%$, and $97 \pm 0\%$, respectively, was observed for a treatment time of 60 min in the electrolysis cell in the presence of mild steel (anode) and vitreous carbon (cathode) electrodes. The process was also determined to be effective for removing pathogens ($99 \pm 1\%$ removal) such as fecal coliform (the log inactivation was higher than two units) (Senghor et al. 2015). It provided a feasibility evaluation of the EC–EP process as a tertiary treatment for DWW containing different types of pollutants (organic, inorganic, and microbial pollutants) (Senghor et al. 2015). Finally, the EC–EP process could form the basis of a process (as a tertiary treatment) that is capable of simultaneously removing inorganic, organic, and microbial pollutants from many wastewaters (Senghor et al. 2015).

The EC process was applied as a tertiary treatment of residual organic matter for landfill leachate after a BF process (Dia et al. 2017). Organic matter such as humic acids (HAs), fulvic acids, and hydrophilic compounds (Hyl) was fractionated using DAX-8 resin to estimate the efficiency of EC on each fraction. Initial characterization of the biofiltrated landfill leachate showed that humic substances (HA and FA) constituted nearly 90% of the total organic carbon (TOC). The effects of current densities, type of the anode (aluminum versus iron), and treatment time on the performance of COD removal were investigated. The best COD removal performances were recorded at a current density ranging between 8.0 and 10 mA cm^{-2} during 20 min of treatment time. Under these conditions, 70% and 65% of COD were removed using aluminum and iron electrodes, respectively. The fractionating of organic matter after EC treatment revealed that HA was completely removed using either aluminum or iron anode. However, FA and Hyl fractions were partially removed, with the percentages varying from 57% to 60% and 37% to 46%, respectively. FA and Hyl removal were quite similar using either aluminum or iron anode. Likewise, a significant decrease in 254 nm absorbance was recorded (UV254 removal of 79% to 80%) using either type of the anode. These results prove that EC is a suitable and efficient approach for treating the residual refractory organic matter from a landfill leachate previously treated by a biological system (Dia et al. 2017).

In another study (Laridi et al. 2005), electrochemical techniques have been explored at the laboratory pilot scale for refractory residual organic compound removal in liquid swine manure (LSM) following a BF process. Two types of electrolytic cells (monopolar and bipolar electrode cells) using aluminum and mild steel electrodes were studied. Effectiveness was measured in terms of COD and BOD reduction. The amount of residue sludge produced and energy consumed were also considered. The results showed that the best performances of COD and BOD removal from LSM were obtained using either aluminum bipolar (Al-BP) electrodes or mild steel monopolar (Fe-MP) electrodes operated at current intensities of 0.5 and 2.0 A, respectively, through 30 min of treatment. The COD removal yields varied from 65% to 68%, whereas BOD removal reached 87%. The optimal conditions determined for organic compound removal, including energy consumption and metallic sludge disposal, involved a total cost of only $US 0.24 and 0.29 m^{-3} of treated LSM. The treatment using the Fe-MP system was found to be more economical and practical than the chemical treatment using $FeCl_3$ as a coagulating agent (Laridi et al. 2005). Finally, in view of enhancing dissolved organic carbon (DOC) removal from LSM, nonconsumable electrodes (such as Ti/RuO_2, Ti/IrO_2, Ti/SnO_2, or Pt) can be used, which are capable of oxidizing refractory organic pollutants to water and carbon dioxide, and no further treatment is then required (Laridi et al. 2005).

The efficiency of EC as a tertiary treatment (Zodi et al. 2011) of paper mill wastewaters was investigated by testing the effect of batch EC for 90 min with two parallel iron or aluminum plates at two values of the current density (100 and 150 A m^{-2}). This effluent is subject to very strict discharge limits by local authorities (DREAL1): the limits have been set to 470 mg L^{-1} for COD and 20 μg L^{-1} for arsenic, but they can be lowered in the case of low river water flow rate. DOC removal ranged between 24% and 46%, and COD removal ranged between 32%

and 68%. UV–visible spectroscopy showed a reduction of the aromaticity of the treated effluent. The process was also very efficient for the removal of lignin-based pollution, characterized by the fluorescence of humic substances. Arsenic was selected as an example of a nonorganic micropollutant and was also satisfactorily removed (from 4 to 0.5 $\mu g\ L^{-1}$). The settling characteristics of the sludge obtained after the EC treatment were also evaluated. The sludge aptitude to settling was better with Fe electrodes than that with Al electrodes. The experimental results obtained in the present study indicate that EC treatment can be very effective and capable of improving the quality of paper mill wastewaters downstream of the biological treatment (Zodi et al. 2011).

The efficiency of a bench-top EC unit with aluminum sacrificial electrodes to reduce concentrations of the biological and chemical pollutants from raw and tertiary-treated DWW was studied (Symonds et al. 2015). The EC unit can concomitantly reduce concentrations of phosphate, microbial surrogates representing several major pathogen types (DNA/RNA viral, bacterial, protozoan parasite), as well as several personal care products (PCPs) in DWW. EC treatment resulted in significant reductions ($p < 0.05$, $a = 50.05$) in phosphate, all microbial surrogates, and several personal care products from raw and tertiary-treated domestic wastewater. When wastewater was augmented with microbial surrogates representing bacterial, viral, and protozoan pathogens to measure the extent of reduction, EC treatment resulted in up to 7 $\log_{10}$ reduction of microbial surrogates; the results enhance previous assertions that EC is a promising sustainable wastewater treatment technology for DWW (Symonds et al. 2015).

In addition, several technologies of EC focus on the laboratory and industrial scales (Bharath et al. 2018, Song et al. 2017). Electrochemical units can be combined with other treatment modalities such as biological treatments or physical processes like sedimentation, settling, or filtration (Chen 2004, Garcia-Segura et al. 2017, Moussa et al. 2017). It is possible to have in the domain of EC numerous variants according to the forms of electrodes (flat, cylindrical, fixed beds of spherical balls or rotary electrodes) or according to their way of electric connection (Bharath et al. 2018, Nidheesh and Singh 2017).

Dairy industry wastewater is characterized by high BOD, COD, and other pollution loads. A study (Bazrafshan et al. 2013) was designed to investigate the effects of operating parameters such as applied voltage, number of electrodes, and reaction time on the treatment of a real dairy wastewater in the EC process. Aluminum electrodes were used in the presence of potassium chloride as electrolytes. It was shown that the removal efficiency of COD, BOD, and TSS increased with increasing the applied voltage and the reaction time. The results indicate that EC was efficient and able to achieve 98.84% COD removal, 97.95% BOD_5 removal, 97.75% TSS removal, and >99.9% bacterial indicators at 60 V in 60 min. The experiments demonstrated the effectiveness of EC techniques for the treatment of dairy wastewaters. The results demonstrated the technical feasibility of the EC process using aluminum electrodes as a reliable technique for the removal of pollutants from dairy wastewaters.

EC was also applied to assess as a possible technique for the reduction of sulfate and COD concentration in petroleum refinery wastewater (El-Naas et al. 2009). Wastewater samples, collected from two different locations in the wastewater treatment system of a petroleum refinery, were treated in a batch electrochemical cell using three types of electrode materials: aluminum, SS, and iron. The experimental results showed that EC can achieve percentage removal of up to 93% and 63% for sulfate and COD, respectively. It was shown that the utilization of aluminum, as anode and cathode, was by far the most effective arrangement in the reduction of both sulfate and COD. The performance of the electrochemical reactor was found to be highly influenced by the initial composition of the wastewater and the current density. The EC process was most effective at 25 °C and a pH value of 8. The study demonstrated that EC may prove to be a practical approach for the pretreatment of petroleum refinery wastewater (El-Naas et al. 2009).

The variations of some characteristics of the mixed liquor solution in WWTPs under the effect of the EC technique were studied experimentally (Al-Shannag et al. 2013). Unlike conventional methods, EC was performed using cylindrical perforated iron electrodes to achieve a good distribution of the applied DC field onto the MWW. The effects of EC time, applied voltage gradient, and aeration on the mixed liquor characteristics were considered. The study figured out the dependency of the local pH and oxidation potential reduction values near anode and cathode on these operational conditions. It was found that the EC process reached steady-state conditions in no more than 60 min. At high voltage gradient ($6\ V\ cm^{-1}$), the steady removal efficiencies of COD and nutrients exceeded 89%. Although aeration showed a moderate influence on removal efficiencies and wastewater sludge filterability, it significantly affected the pH and ORP near the electrodes, especially at the low voltage gradient. The results also indicated that the sludge filterability was improved notably by increasing the voltage gradient to have minimal levels of specific resistance to filtration.

In EC treatment, the most important parameter is pH to help coagulating iron and aluminum ions. Some researchers reported that the pH should be adjusted to 7; however, others indicate that a higher pH is preferable (Zaied and Bellakhal 2009). Studies have been made on the EC process, combined with other treatment processes or methods such as oxidation, Fenton, ultrasonic process, TiO_2 photo-assisted, and photocatalytic process, to enhance the efficiency of wastewater treatment; more detailed information is given in later chapters.

1.5 PRINCIPLE AND DEFINITION OF ELECTRO-OXIDATION

Electro-oxidation (EO) is an application of electrochemical properties that can be effectively used to remove many types of organic, inorganic, and microbial pollutants from water. Water or wastewater is treated by direct and/or indirect oxidation inside the electrolysis cell. Insoluble electrodes having a high overvoltage of oxygen are currently used (Deng and Englehardt 2007, Gandini et al. 1998, Rajkumar et al. 2005). At the anode, these electrodes allow oxidizing many types of organic molecules contained in water and wastewater (Morao et al. 2004, Wang et al. 2004). The oxidation of these pollutants can exclusively take place at the anode surface (direct effect of electrolysis) and/or by in situ generation of an oxidant in solution (indirect effect of electrolysis).

1.5.1 Direct Effect of Electro-Oxidation

The direct effect of electrolysis obtained by anodic discharge of water molecules induces the formation of hydroxyl radicals ($HO^{\bullet}$) on the surface of the electrode. Hydroxyl radicals are species that are highly active, capable of oxidizing many types of complex nonoxidizable or hardly oxidizable organic compounds (aromatic, aliphatic, polar or nonpolar compounds, poly-aromatic hydrocarbons, pharmaceuticals, pesticides, etc.) (Pillaud 1987, Pulgarin et al. 1994). Anodic oxidation can lead to a total oxidation of organic pollutants (release of CO_2 and H_2O formation) or partial oxidation by forming simple compounds, nontoxic or less toxic, and biodegradable compounds.

As indicated previously, the effect of the direct effect of electrical current is mainly attributed to hydroxyl radicals generated at the anode surface. At the anode material, there is a competitive reaction between physisorption of radicals (combustion process) and chemisorption of oxygen atom (conversion process). Electrochemical conversion only transforms the nonbiodegradable organic pollutants into biodegradable compounds, whereas electrochemical combustion yields water and carbon dioxide, and no further treatment is then required (Pulgarin et al. 1994). Thus, the anode electrodes having low overvoltage of oxygen (oxygen is easily released on such anodes) favor the conversion process and molecular oxygen formation. When the ultimate objective of the process is to effectively oxidize organic pollutants or persistent organic pollutants from water, it is essential to use anodic electrodes having a high overvoltage of oxygen (oxygen is not easily released on such electrodes) so that adsorbed $HO^{\bullet}$ formation increases on the surface of electrode. Many kinds of electrodes have been found having a high overvoltage of oxygen capable of treating effluents or water containing various organic compounds: Ti/Pt, Ti/Pt-Ir, Ti/IrO_2, Ti/RuO_2, Ti/PbO_2, Ta/PbO_2, Ti/SnO_2, Nb/BDD (boron doped diamond), and so on. The influence of the nature of anodic electrodes has been clearly put into evidence by Comninellis (1994) while treating effluents contaminated by phenolic compounds. The anodic electrodes used are titanium coated with the catalytic films of Pt, RuO_2, IrO_2, PbO_2, and SnO_2. The best performance of pollutant removal was obtained using Ti/SnO_2 with a current efficiency of 55%, whereas 10% to 18% of current efficiencies were recorded for the other electrodes (Ti/Pt, Ti/RuO_2, Ti/IrO_2, and Ti/PbO_2).

The BDD electrodes are one of the anodic electrodes having the highest catalyst activity. Hence, this anode material is interesting and very attractive for the degradation of refractory organic compounds. Beck et al. (2000) revealed the influence of anode materials by comparing the effectiveness of different materials (such as Si/BDD, Ti/SnO_2, Ta/PbO_2, and Pt) while treating an effluent contaminated by phenolic compounds. The results obtained with the different electrode materials are, respectively, as follows: 97%, 80%, 57%, and 37% of phenolic degradation.

The EO of polycyclic aromatic hydrocarbons (PAHs) in solution was investigated (Tran et al. 2009). Two anode materials (Ti/IrO_2 and Ti/SnO_2) were used for the treatment of synthetic creosote oily effluent containing several PAHs. The Ti/SnO_2 electrode was found to be more effective (82% to 84% of PAH removal) in removing PAHs than the Ti/IrO_2 electrode. As reported (Comninellis 1992, Comninellis and Nerini 1995, Feng et al. 2003), tin oxide is one of the noble metal oxides having a better performance for organic compound degradation in comparison with traditional electrodes (Pt, IrO_2, and RuO_2). This is attributed to the highly crystalline nature of tin oxide, which catalyzes the reaction of electrochemical oxidation (Comninellis 1992).

Furthermore, the effectiveness of the electrolytic cell can also be linked to the shape and structure of electrodes. Rectangular, circular, and concentrical electrodes are currently employed, and their use is dependent on the type of rector and hydrodynamic conditions (laminar or turbulent flow) (Chen 2004). The main objective is to obtain the largest possible exchange surface of electrodes with a high mechanical resistance. This high mechanical resistance of electrodes allows ensuring a clearing and carbonate removal from electrodes. Expanded metal electrodes (mesh electrodes) are currently used. This configuration offers a high mechanical resistance and a large exchange surface of electrodes (compared with solid electrode plates) and acts as a good promoter of turbulence, allowing to increase the mass transfer coefficient at the interface of the electrode–electrolyte (Coeuret and Storck. 1993).

1.5.2 Indirect Effect of Electro-Oxidation

In some cases, the direct effect of electrolysis can be ineffective or slow. It is better to electrochemically generate oxidant species that can react with pollutants in solution, and whose action can be extended at the outlet of the electrolyser. The indirect effect can be obtained by oxidation of halides (Cl^-, Br^-, or I^-), by oxidation of sulfate $\left(SO_4^{2-}\right)$, or by reduction of dissolved oxygen. Halides can be oxidized and they generate some oxidants such as hypochlorous acid (HClO), hypobromous acid (HBrO), and hypoideous acid (HIO). Sulfate ions can be oxidized into persulfate ($H_2S_2O_8$), whereas dissolved oxygen can be electrochemically reduced at the cathode into hydrogen peroxide (H_2O_2). The oxidation of halides (mainly chlorides and bromides) is currently used for water disinfection. The molecular forms of disinfectants (i.e., HClO and HBrO) are more bactericidal than the ionic forms (ClO^- and BrO^-). The indirect effect of electrolysis allows destroying bacteria and producing a remnant effect.

Bureau et al. (2012) studied at the laboratory scale an electrochemical process for enhancing dewaterability and stabilizing wastewater sludge from MWW treatment. The electrolytic cell was comprised of titanium coated with ruthenium oxide (Ti/RuO_2) used as an anode and titanium (Ti) used as a cathode. The efficacy of the electrolytic cell for active chlorine (Cl_2, HClO, and ClO^-) production was first evaluated. The specific chlorine production rate recorded in the present study was 26.1 mg min^{-1}-A, with a sodium chloride concentration of 3,000 mg L^{-1}. The electrochemical process was effective in removing indicator pathogens (abatement >4 to 5 log units of total and fecal coliforms) and unpleasant odors. Likewise, the electrolytic treatment induced particles to agglomerate and enhanced the filterability of wastewater sludge during dewatering in the presence of an organic polymer. The increase in dryness (6% to 10% more solids than in untreated sludge) reduced the amount of dewatered sludge produced (up to 30% sludge reduction) (Bureau et al. 2012).

The indirect effect of electrolysis obtained by hydrogen peroxide generated at the cathode has been used as well for water disinfection (Drogui et al. 2001). H_2O_2 is produced by the cathodic reduction of oxygen in an electrolytic cell comprised of titanium coated with ruthenium oxide (Ti/RuO_2) used as the anode material and carbon felt used as cathode material. The contribution to disinfection from the direct and indirect effects of electrolysis was evaluated. A synthetic wastewater contaminated by *Pseudomonas aeruginosa* was treated with a current of 1.5 A. Disinfection was much more effective with the simultaneous direct and indirect effects of electrolysis, which induced more than 3 log difference compared with the sole effect of H_2O_2 (the indirect effect of electrolysis for which 1 log unit was recorded). The direct effect of electrical current allowed destroying protein inclusions within the bacterial membrane, so that bacterial cells could not exchange ions. However, they could be reactivated in a more favorable environment (Cremieux and Freney 1995). The total destruction of bacteria cells requires an oxidant able to cross the bacterial membrane and reach its vital centers, which is done by H_2O_2 generated at the cathode (Garnerone 1979).

The indirect effect of EO is also used to remove dyes from effluents. Zaviska et al. (2009) investigated the performance of an electrolytic cell comprised of two anodes and two cathodes. Four synthetic effluents containing a methyl violet 2B dye (MV2B), an eosin yellowish dye (EOY), a trypan blue dye (TRB), and an acridine orange dye (ACO), respectively, were studied (Zaviska et al. 2009). The best performance for dye removal was obtained using Ti/IrO_2 anodes operated at a current density of 15 mA cm^{-2} during 40 min of treatment in the presence of 3.42 mM of chloride ions. Under these conditions, more than 99% of dye was removed. TRB was the most difficult dye to remove from solution with a value reaction rate constant of 0.12 min^{-1}, compared with 0.20, 0.19, and 0.24 min^{-1} recorded for the MV2B, ACO, and EOY dyes, respectively. More than 99% of these dyes were removed by electrochemical oxidation (Zaviska et al. 2009).

The combination of the direct and indirect effects of electrolysis has been put into evidence while treating effluent contaminated by carbamazepine (CBZ) (Komtchou et al. 2015). CBZ has been found to be a persistent drug in the environment that resists treatment by WWTPs (Calisto et al. 2011, Clara et al. 2004). The effluents used throughout the study were synthetically prepared dissolving CBZ (12 mg L^{-1}) in deionized water. To simulate natural water contaminated by CBZ, HA was also added at a concentration of 5 mg L^{-1}. Sodium sulfate (1.0 g L^{-1} Na_2SO_4) was used as the supporting electrolyte. The EP process consists of exposing CBZ solution to electrolysis treatment using carbon felt at the cathode (for H_2O_2 formation) and Ti/Pt (titanium coated with platinum) at the anode. The AO process was performed using the electrolytic reactor with Ti/Pt at the anode and Ti at the cathode to avoid H_2O_2 formation. A current intensity of 0.2 A was imposed, and after 120 min of electrolysis, the degradation of CBZ was around 36% while applying the EP process compared with 8.5% of CBZ removal recorded for AO. However, when BDD was used as the anode material (instead of Ti/Pt), 50% of CBZ degradation was recorded using the EP process compared with 36% of CBZ removal recorded with the AO process. This oxidant agent contributed to CBZ degradation.

1.6 INTEGRATION OF ELECTRO-OXIDATION INTO WATER/WASTEWATER TREATMENT PLANTS OR APPLICATION FOR DECENTRALIZED SANITATION

Pollution-free electricity, ease of operation, and the reduction of equipment volume in wastewater treatment processes are parameters that should contribute to the industrial development of electrochemical techniques compared with conventional processes (chemical or biological processes). Several electrochemical systems were developed for full-scale applications (at the industrial scale) for the treatment of various organic pollutants (phenolic compounds, HA, pharmaceutical compounds, etc.), inorganic pollutants (toxic metals, phosphate, sulfur, fluoride, etc.), and microbial pollutants (*E. coli*, etc.).

Mendia (1982) studied the possibility of integrating electrochemical systems in WWTPs located in the coastal region of the city of Naples, Italy. The wastewater was subjected to screening and then grit removal before being transferred to the tank where it could mix with seawater (salt water) in appropriate proportions [seawater/wastewater ratio (v/v) is 1:3 and 1:4]. The mixture solution was then transferred into the electrolytic reactor comprised of graphite (anode) and iron (cathode). The electrolytic reactor had a retention time varying from 5 to 10 s and a flow rate of 4.8 $m^3 h^{-1}$. Once electrolyzed, the effluent was transferred in a tank of flocculation/clarification, in which disinfection and sedimentation take place simultaneously. The retention time of flocculation/clarification tank was 60 to 110 min. The supernatant was then discharged to the seawater, whereas the solid fraction (sludge) was dried (via a drying bed) before its final disposal owing to a relatively high content of chlorine (Mendia 1982).

Electrochemical water treatment is a promising alternative for small-scale and remote water systems that lack operational capacity or convenient access to reagents for chemical coagulation and disinfection. In a study (Heffron et al. 2019), the mitigation of viruses was investigated using EC as a pretreatment prior to EO treatment using BDD electrodes. This research is the first to investigate a sequential EC–EO treatment system for virus removal. Bench-scale batch reactors were used to evaluate mitigation of viruses in variable water quality via (1) EO, and

(2) a sequential EC–EO treatment train. The EO of two bacteriophages, MS2 and ΦX174, was inhibited by natural organic matter (NOM) and turbidity, indicating the probable need for pretreatment. However, the EC–EO treatment train was beneficial only in the model surface waters employed. In model ground waters, EC alone was as good or better than the combined EC–EO treatment train. Reduction of human echovirus was significantly lower than one or both bacteriophages in all model waters, although Bacteriophage ΦX174 was a more representative surrogate than MS2 in the presence of NOM and turbidity. Compared with conventional treatment by ferric salt coagulant and free chlorine disinfection, the EC–EO system was less effective in model surface waters but more effective in model ground waters. Sequential EC–EO was beneficial for some applications, although practical considerations may currently outweigh the benefits (Heffron et al. 2019).

Small drinking water systems serve approximately 20% of the US population, but they can struggle to comply with the Total Coliform Rule and the Disinfectant and Disinfection By-product Rule (Lynn 2019). Issues with insufficient funds to effectively treat the water and difficulties with the transportation of required chemicals can affect compliance. Electrochemical processes may offer an alternate approach for small water systems because they have demonstrated some advantages over traditional treatments, such as reduced handling and storage of chemicals and cost-effectiveness. Sequential electrochemical processes have yet to be tested for the treatment of *E. coli* in drinking water. In a study (Lynn 2019), EC and EO were investigated using two model surface waters and two model ground waters to determine the efficacy of sequential EC–EO for mitigating *E. coli*. At a current density of 1.67 mA cm^{-2} for 1 min, bench-scale EO alone achieved four logs mitigation of *E. coli* in the model shallow aquifer. Increasing the EO current density to 6.67 mA cm^{-2} for 1 min provided similar levels of *E. coli* mitigation in the model deep aquifer (characterized by lower initial chloride concentrations compared with the shallow aquifer). Using a current density of 10 mA cm^{-2} for 5 min, EC achieved 1 log or greater *E. coli* mitigation in all model waters. No additional mitigation beyond EC alone was achieved using sequential EC–EO. Reductions in the initial pH of the surface waters to target higher NOM removal did not enhance *E. coli* treatment with EC–EO compared with EC alone. In fact, an average of 64% of NOM was removed irrespective of the change in pH, which likely limited *E. coli* mitigation. Additional reasons for the lack of improvement in *E. coli* treatment may have included the presence of iron following EC or insufficient EO current density. Decreasing the initial water pH did improve *E. coli* mitigation using EO when pretreated by EC compared with the baseline water matrix pH. Total EC residual iron concentration also increased, and it correlated slightly with *E. coli* mitigation. This correlation and oxidation of ferrous iron may indicate that Fenton-like reactions occurred during EO after EC pretreatment (Lynn 2019).

The performance of an electrochemical process was investigated at the laboratory scale for enhancing dewaterability and stabilizing wastewater sludge from municipal WWTPs (Bureau et al. 2012). This process has the advantage of simultaneous disinfection and odor removal from sludge. The equipment consisted of a 12 L cylindrical electrolytic cell containing a $Ti/RuO_2Ti/RuO_2$ anode and Ti cathode. Current intensities varying from 5.0 to 10 A were tested in the absence and presence of the electrolytes $H_2SO_4H_2SO_4$ [10 to 27 kg ton^{-1} of dry matter (tDM)] and NaCl (88 to 354 kg tDM^{-1}). The best performance with respect to sludge treatment and stabilization was obtained when the electrolytic cell was operated for 60 min at 8.0 A, with an energy consumption of 856 kW·h tDM^{-1}, in the presence of 177 kg NaCl tDM^{-1} and 23.3 kg H_2SO_4 tDM^{-1}. The electrochemical treatment induced particles to agglomerate and enhanced the filterability of wastewater sludge during dewatering in the presence of the cationic polymer Percol 789 (2.5 kg tDM^{-1}). The process was efficient in increasing the dryness of sludge. A 6% to 10% increase in total solids in dehydrated sludge was expected. The electrochemical process was also found to be effective in removing indicator pathogens (abatement > 4 to 5 log units of total and fecal coliforms) and unpleasant odors. At the same time, it preserved the fertilizing properties of the sludge by maintaining concentrations of inorganic nutrients (total phosphorus and total Kjeldahl nitrogen) and organic matter (COD) in dewatered sludge (Bureau et al. 2012).

During the electrochemical oxidation of real wastewater, different species present in the effluent may interact, creating complex scenarios and making the prediction of the behavior of the whole system difficult. Different phenomena that occur during the EO process of landfill leachate at a pilot plant scale with BDD anodes were elucidated (Anglada et al. 2009). The total BDD anode area of the pilot plant was 1.05 m^2. The evolution of the concentration of chloride ions, chlorate, and inorganic carbon, as well as the value of pH and redox potential, were found to be inter-related. In turn, the concentration of chloride affected the oxidation of ammonia, which took place through indirect oxidation by active chlorine. Moreover, chloride ions competed with organic matter to be oxidized at the anode. The effect of current density was also investigated. Organic matter and ammonia oxidation were highly influenced by the applied current density value. A change in the mechanism of organic matter oxidation was observed when high current densities were applied. Two mathematical models, previously applied to the oxidation of synthetic wastewaters in the literature, were able to predict the evolution of COD and ammonia for low current density values (Anglada et al. 2009).

Leachate concentrates, an effluent produced from nanofiltration (NF) and/or RO, contain a high amount of salts and dissolved organics, especially refractory organics. Thus, the treatment of leachate concentrates would lead to the consumption of high energy or a large amount of chemicals. A study by Soomro et al. (2020) was made to develop an effective treatment method by using coupled electrochemical methods with the least possible energy consumption. The leachate concentrates were pretreated by EC, with aluminum or iron electrodes as anodes, to reduce the dissolved organic content. EC with Al electrodes was found to be more efficient by consuming 1.25 kW·h m^{-3} energy to remove 70% of TOC. EC effluent was further subjected to a novel simultaneous EO and in situ peroxone process, which used a Ti-based nickel and antimony doped tin dioxide as an anode and a carbon nanotube–coated carbon–polytetrafluoroethylene as a cathode for the oxygen reduction reaction (ORR). Compared with a traditional EO with a cathode for the hydrogen evolution reaction (HER-EO), ORR-EO produced a higher efficiency and an energy consumption of 26.25 kW·h m^{-3}, which was much lower than 35.5 kW·h m^{-3} for HER-EO. The results showed that after ORR-EO, a final TOC of 57.3 mg L^{-1} was obtained. Thus, EC, in tandem with the ORR-EO process, has an excellent capability and economic merit in the area of treating leachate concentrates (Soomro et al. 2020).

An EC, coupled with a novel ORR-EO, was developed. Complete removal of refractory organics in leachate concentrate was achieved with this technique. The Al electrode was found to be more efficient in organics removal during the EC process. ORR-EO produced an in situ peroxone effect with a higher efficiency and reduced energy consumption (Soomro et al. 2020).

The performance of a pilot-scale combined process of a fluidized biofilm process, chemical coagulation, and electrochemical oxidation for textile wastewater treatment was studied (Kim et al. 2002). To enhance biological treatment efficiency, two species of microbes, which can degrade textile wastewater pollutants efficiently, were isolated and applied to the system with supporting media. $FeCl_3 \cdot 6H_2O$, pH 6, and 3.25×10^{-3} mol L^{-1} were determined as the optimal chemical coagulation condition and 25 mM NaCl of electrolyte concentration, 2.1 mA cm^{-2} of current density, and 0.7 L min^{-1} of flow rate were chosen for the most efficient electrochemical oxidation at pilot-scale treatment. The fluidized biofilm process showed 68.8% of COD and 54.5% of color removal efficiency despite using a relatively low MLSS concentration and a short solid retention time (SRT). COD and color removal of 95.4% and 98.5%, respectively, were achieved by the overall combined process. The contribution of the fluidized biofilm process to the overall combined process increased more than 25.7% of COD reduction and 20.5% of color reduction by adopting support media in biological treatment. It can be inferred that the fluidized biofilm process was effective and pollutant loading on post-treatment greatly decreased by this system (Kim et al. 2002). This combined process was highly competitive in comparison with the other similar combined systems. It was successfully employed and it effectively decreased pollutant loading on post-treatment for textile wastewater treatment at the pilot scale (Kim et al. 2002). EO treatment of pollutants in water and wastewater is shown in Table 1-2.

1.6.1 Electro-Oxidation Used as Pretreatment

Wastewaters may require pretreatment before disposal and assurances that the treated waters will not adversely impact BP at wastewater treatment facilities or receiving waters. An electrochemical advanced oxidation process (EAOP) pretreatment for contaminated waters, using a BDD anode, prior to discharge to wastewater treatment facilities (Phillips et al. 2018), was studied. A range of contaminants were studied, including herbicides, pesticides, pharmaceuticals, and flame retardants. The resulting toxicities varied with the supporting electrolytes from 5% to 92%, often increasing, indicating that microbial toxicity, in addition

Table 1.2. Summary of Pollutants Removed by Electro-oxidation in Water and Wastewater.

Pollutants, wastewater type	*Electrode materials, anode-cathode*	*Optimum conditions*	*Removal efficiency (%)*	*Special marks*	*Level of treatment (primary, secondary, and tertiary treatment or combination)*	*Reference*
Industrial wastewater (receives a discharge of 144 different facilities)	Electrochemical oxidation using boron-doped diamond electrodes	More than 2 h per 0.7 L batch	COD by 99.9%		Pretreatment	García-Morales et al. (2013)
Industrial wastewater (receives a discharge of 144 different facilities)	Coupling of electrochemical oxidation and ozonation	1 h	COD: 99.9% along with most color and turbidity in about an hour.		Combination	García-Morales et al. (2013)
Dye wastewater	Individual electro-oxidation	10.25 A dm^{-2}	Individual electro-oxidation process, maximum COD removal of about 80% and color removal of 97.9%		Pretreatment	Senthilkumar et al. (2012)

(Continued)

Table 1.2. Summary of Pollutants Removed by Electro-oxidation in Water and Wastewater. (Continued)

Pollutants, wastewater type	*Electrode materials, anode-cathode*	*Optimum conditions*	*Removal efficiency (%)*	*Special marks*	*Level of treatment (primary, secondary, and tertiary treatment or combination)*	*Reference*
Dye wastewater	Electrochemical and biological oxidation combined processes	Density of 7.6 and 10.2 A dm^{-2}	COD and color removal of 90% and 98.5%		Combination biological treatment followed by EO post-treatment	Senthilkumar et al. (2012)
Petroleum refinery wastewater	Boron-doped diamond anode (BDD), direct electrochemical oxidation by using a ruthenium-mixed metal oxide (Ru-MMO) electrode	The most efficient method was the electrofenton process followed by the electrochemical oxidation using the BDD anode	Phenol removal of 98.74% in 6 min of electrolysis; COD removal of 75.71% was achieved after 9 min of electrolysis in electrofenton.		Combination	Yavuz et al. (2010)
Petroleum refinery wastewater	Direct electrochemical oxidation by using a ruthenium-mixed metal oxide (Ru-MMO) electrode	5 mA cm^{-2}	99.53% phenol and 96.04% COD removal		Pretreatment	Yavuz et al. (2010)

Paper mill bleaching	Pb	2 g/L NaCl, 6.6 mA cm^{-2} 60 min	COD: 97%, color: 100%		El-Ashtoukhy et al. (2009)
landfill leachates	Ti/Pt electrodes	$t = 1$ h, $J = 4$ A dm^{-2}	Removal: COD (85%), TKN (94%), $color_{436}$ (99%)	Combination of a MBR with electrochemical oxidation	Feki et al. (2009)
Textile wastewater	Ti-Pt/b-PbO_2	Electrolyte: mol L^{-1} Na_2SO_4, pH = 7.0, 75 mA cm^{-2} 3 h	COD removal: 90%		Aquino et al. (2014)
Textile industry wastewater	Graphite	Electrolyte: 0.5 mol L^{-1} NaCl, pH = 5.5, 111.11 mA cm^{-2} 2 h	COD removal: 91%, color removal: 96%	Post treatment tertiary	Bhatnagar et al. (2014)
Textile industry wastewater	Graphite rod	pH 1.3 28 mA cm^{-2} 60 min 0.6 A	Removal of COD, total solids, total dissolved solids, and total organic carbon was found to be approximately 68%, 49.2%, 50.7%, and 96.8%, respectively	Post treatment tertiary	Radha et al. (2009)

to parent compound degradation, should be monitored during treatment. These toxicity results are particularly novel because they systematically compare the microbial toxicity effects of a variety of supporting electrolytes, indicating that some electrolytes may not be appropriate in certain applications. Further, these results are the first known report of the use of the nitrification inhibition test for this application. Overall, these results systematically demonstrate that anodic oxidation using the BDD anode is useful for addressing the problem of water contaminated with refractory organic pollutants, while minimizing impacts to WWTPs or receiving waters accepting EAOP-treated effluent. The results indicate that nitrate can be a suitable electrolyte for incident response and, more importantly, serve as a baseline for site-specific EAOP usage (Phillips et al. 2018).

Drain disposal of water from contamination incidents may impact utility operation. An EAOP using a BDD anode was utilized for pretreatment to address microbial toxicity. Electrolytes reported to be inert exhibited microbial toxicity or quenching issues. Microbial toxicities varied throughout treatment, even increasing in some cases. This type of oxidation process may enable drain disposal if operated to minimize toxicity (Phillips et al. 2018).

Phytoremediation of reverse osmosis concentrate (ROC) with microalgae can simultaneously achieve multifunctions of ROC treatment, CO_2 mitigation, and microalgae biolipid production. However, the performances are usually inhibited by high free ammonia nitrogen (*FAN*) concentration and chromaticity of ROC. To offset these negative effects, an integrated technique including EO pretreatment and *Chlorella vulgaris* remediation was proposed (Chang et al. 2020), in which the ROC was first pretreated with EO to decrease *FAN* and chromaticity, and then the oxidized ROC was remediated with microalgae to reclaim nutrients and produce biolipid. The results showed that *FAN* was sharply reduced from 53.0 to 13.9 mg N L^{-1}, and chromaticity was decreased from 1,600 to 100 Pt-Co via EO. A possible reaction mechanism of nutrient removal was discussed in terms of electron mass balance. Explanation on chromaticity decrease was revealed by analyzing the HA conversion path with fluorescence characteristics. During the microalgae remediation process, the nutrient removal rate, microalgae biomass concentration, and lipid yield were effectively enhanced in electro-oxidized ROC. Energy balance analysis indicated that microalgae lipid energy under a current density of 3.25 mA cm^{-2} basically compensated the total input energy despite ROC sterilization. It provided a promising strategy for large-scale ROC treatment and microalgae biolipid production (Chang et al. 2020). FAN and chromaticity were greatly reduced via EO. HA was degraded into a family of soluble microbial by-products. Precipitation of nitrogen and phosphorus occurred based on electron mass balance. Microalgae growth and lipid synthesis were enhanced via EO. Microalgae lipid energy almost compensated the total input energy despite autoclaving (Chang et al. 2020).

A study (Vasanthapalaniappan et al. 2021) focused on a novel method to integrate the EO process with a membrane bioreactor (MBR) to reduce biofouling and increase the biodegradability index. EO was used as a pretreatment with a MBR operating at a current density of 1.5 mA cm^{-2} with a hydraulic retention time at 6 h. The mixed liquor suspended solid concentration was maintained at a constant level of 3,200 mg L^{-1} throughout the experiment for 30 days. The results obtained were promising, with the percentage removal of COD, TOC, total nitrogen, and chlorides being in the range of 97%, 90%, 94%, and 15%, respectively, which was comparatively higher than that of the existing MBR. The integrated EO process was efficient for the complete removal of pollutants from wastewater. In addition, the phytotoxicity test showed a significantly higher quality of treated water compared with that of raw tannery effluent. Hence, the integrated EO process can be used to decrease biofouling with an increased biodegradability index as a replacement for MBR (Vasanthapalaniappan et al. 2021).

Wastewater from soft drink manufacturing, having a high organic load (COD = 4,500 mg L^{-1}) and high alkalinity (2,653.7 mg L^{-1} $CaCO_3$; pH 12), was pretreated with a calcium-modified zeolite to reduce the alkalinity and improve the EO of organic matter (Victoria-Salinas et al. 2019). The $CaCl_2$-modified zeolite (ZSACaCl-72h) was more effective for the treatment of soft drink wastewater than the congener modified with $Ca(OH)_2$, where the former reduced the alkalinity by 86% after 8 h. The EO of soft drink wastewater without zeolite pretreatment was carried out with BDD electrodes under the optimal conditions (current intensity: 3 A; sample pH: 12), with 98% and 94.05% reduction of the COD and TOC, respectively, after 14 h of treatment. Soft drink wastewater pretreated with calcium-modified clinoptilolite was also electro-oxidized using the BDD system. The results showed that the pretreatment was extremely convenient, reducing the treatment time to 6 h compared with the EO of wastewater. At a current intensity of 3 A, the treatment time was 8 h, with 100% reduction of color and COD as well as 97.5% reduction of TOC (Victoria-Salinas et al. 2019).

The efficiency of EO used as the single pretreatment of landfill leachate was examined (Rada et al. 2013). The yields obtained were considered satisfactory, particularly given the simplicity of this technology. Like all processes used to treat refluent water, the applicability of this technique to a specific industrial refluent needs to be supported by feasibility studies to estimate its effectiveness and optimize the project parameters. This could be a future development of the work (Rada et al. 2013).

A study (Bhaskar Raju et al. 2009) aimed to ascertain the efficacy of electrochemical techniques as pretreatment methods to RO. The textile wastewater was initially treated by EC to remove the suspended solids. After the EC, the wastewater was further treated by EO for COD removal. Mild steel as an anode was found to be effective for coagulation of suspended solids. For EO, graphite and $RuO_2/IrO_2/TaO_2$-coated titanium were used as electrodes. The COD was removed to the extent of 90% to 93% using graphite and 54% with the $RuO_2/IrO_2/TaO_2$-coated titanium electrodes. Current efficiencies of 40% and 11% were achieved with graphite and $RuO_2/IrO_2/TaO_2$-coated titanium, respectively (Bhaskar Raju et al. 2009).

The effluent from a WWTP was pretreated with an electrochemical cell using BDD electrodes to control the fouling of an ultrafiltration membrane (Gonzalez-Olmos et al. 2018). The results showed that the electrochemical pretreatment decreased the transmembrane pressure by 36% to 67%, and consequently, the membrane fouling, with increasing applied current densities. The removal of DOC and turbidity by the membrane process was enhanced to 40% and 41%, respectively, using the EO pretreatment. The application of EO as a pretreatment stage points out to be a promising alternative to reduce membrane fouling and improve water quality for water reuse applications.

Disintegration of municipal waste–activated sludge (WAS) is regarded as a prerequisite for the anaerobic digestion process to reduce sludge volume and improve biogas yield. Pretreatment of WAS using thermo-alkaline (TA), H_2O_2 oxidation, electrolysis, and EO (EO) processes were investigated and compared in terms of COD solubilization and biogas production (Feki et al. 2015). At optimum conditions, EO gave the maximum COD solubilization (28%). The effects of pretreatments under the optimum conditions on anaerobic digestion were experienced with biochemical methane potential assay. Significant increases in biogas yield up to 78% and 40% were observed in the EO and TA pretreated samples, respectively, compared with raw sludge yield. These results clearly reveal that the application of EO is a significant alternate method for the improvement of WAS anaerobic digestion (Feki et al. 2015).

1.6.2 Electro-Oxidation Used as Tertiary Treatment

A combination of a high-performance MBR equipped with ultrafiltration and EO with BDD electrodes was used to effectively treat highly contaminated old landfill leachate (Daghrir and Drogui 2013). MBR and EO were optimized for raw and pretreated landfill leachates. For MBR, an organic load rate of 1.2 gCOD L^{-1} day^{-1} and an SRT of 80 days were considered as the optimum operating conditions, in which COD, TOC, NH_4^+, and phosphorus removal reached average efficiencies of 63%, 35%, 98%, and 52%, respectively. The best performance of EO was in a current intensity of 3 A with a treatment of time of 120 min. Furthermore, the energy consumption of EO was decreased from 22 to 16 kW·h m^{-3} for biologically treated and raw landfill leachates, respectively (Zolfaghari et al. 2016).

An integration of MBR with the electrochemical process was investigated (Feki et al. 2009) for the treatment of stabilized landfill leachates collected from Djebel Chekir (Tunisia). The results showed that at optimum conditions for the membrane and with organic loading rates of 1.9 and 2.7 g COD L^{-1} day^{-1}, MBR-treated effluent was still undergoing the process of coloring and contained high COD and ammonia concentrations. To reduce these high pollutant concentrations, an electrochemical oxidation process using Ti/Pt, graphite, and PbO_2 electrodes was tested as effluent post-treatment. At optimal operational conditions ($t = 1$ h, $J = 4$ A dm^{-2}), the final COD and total Kjeldahl nitrogen concentrations (TKN) were 1,000 and 86 mg L^{-1}, respectively. The final treated wastewater COD, TKN, colors, and pH met the discharge standards in the sewer. The combination of MBR with EO can be a technical suitable solution for stabilized landfill leachate treatment with an efficient reduction of different parameters, essentially COD (85%), TKN (94%), and color $_{436}$ (99%) (Feki et al. 2009).

The electrochemical treatment of biotreated landfill leachate was carried out using a Ti/SnO_2-Sb_2O_5-IrO_2 anode and a porous carbon nanotube–containing cathode (Chu et al. 2015). The influences of cathodic potential and external addition of Fe^{2+} on TOC decay evidenced that both electro-Fenton oxidation and anodic oxidation were accounted for pollutant degradation. Cl^- can play an important role in the removal of NH_3–N and TN. The determination of dehydrogenase activity and BOD_5/COD showed that the leachate toxicity became weaker and the biodegradability was enhanced after the electrochemical treatment. These results suggest that the electrochemical process may present a promising alternative for the advanced treatment of biotreated landfill leachate (Chu et al. 2015).

In a study (Zhou et al. 2016), the high-quality BDD electrodes with excellent electrochemical properties were deposited on niobium (Nb) substrates by the hot filament chemical vapor deposition method. The electrochemical oxidation of landfill leachate concentrates from a disc tube RO process over a BDD anode was investigated. As a result, the best conditions obtained were as follows: current density 50 mA cm^{-2}, pH 5.16, and flow velocity 6 L h^{-1}. Under these conditions, 87.5% COD and 74.06% NH_3–N removal were achieved after 6 h treatment, with a specific energy consumption of 223.2 kW·h m^{-3}. These results indicate that EO with a BDD/Nb anode is an effective method for the treatment of landfill leachate concentrates (Zhou et al. 2016).

In a study (Panizza and Martinez-Huitle 2013), the electrocatalytic properties of Ti-Ru-Sn ternary oxide ($TiRuSnO_2$), PbO_2, and BDD anodes have been compared for the EO process of a real landfill leachate from an old municipal solid waste landfill (average values of COD 780 mg dm^{-3} and $NH_4^+ - N$ 266 mg dm^{-3}). The experimental results indicated that after 8 h of electrolysis, the $TiRuSnO_2$ anode yields only 35% COD, 52% color, and 65% ammonium removal. Using PbO_2, ammonium and color were completely removed, but a residual COD (i.e., 115 mg dm^{-3}) was present (Panizza and Martinez-Huitle 2013). On the contrary, BDD enables complete COD, color, and ammonium removal because of the electrogeneration of hydroxyl radicals from water discharge and active chlorine from chloride ion oxidation. BDD also exhibits greater current efficiency along with a significantly lower energy cost than other electrodes. These results indicated that EO with a BDD anode is an effective process for the treatment of landfill leachate (Panizza and Martinez-Huitle 2013).

1.7 SUMMARY

EC and/or EO are more effective treatment processes than conventional methods such as chemical coagulation. The technology is unique and offers many opportunities that are hard to match with other technologies. The undertaken research and experiments proved that EC is a revolutionary technology, significantly underused in wastewater treatment. Based on experimental data, and because no precipitating agents were applied, the EC and/or EO processes proved to be not only feasible and environmentally friendly but also a cost-effective technology. EC and/or EO may be a potential answer to environmental problems dealing with water reuse and rational waste management. EC and/or EO have become one of the affordable wastewater treatment processes around the world by reducing electricity consumption and miniaturization of the needed power supplies. EC and/or EO have a wide variety of wastewater treatment capabilities. This treatment is the process of destabilizing suspended, emulsified, or dissolved contaminants in aqueous medium by introducing a minimal amount of electrical current. It thereby reduces additional costs involved in the process. It has even replaced traditional treatment processes such as filtration and chemical treatment, which have proved to be less-effective and expensive processes.

EC can be used in any step of the treatment process—primary, secondary, and tertiary treatment. Over the decades, it has been noted that EC and/or EO are capable of providing high removal efficiencies of color, COD, BOD, sulfate, phosphorus, TSS, SS, O&G, and other types of pollutants in various wastewaters and water to achieve treatment processes that are more efficient and quicker than traditional coagulation and inexpensive than other methods of treatment such as ultraviolet (UV) and ozone. The removal rates, particularly for difficult-to-treat contaminants, are superior to the results achieved using the traditional wastewater treatment methods. EC has proven effective in treating wastewater and sewage sludge, as well as in sufficient fixation of metals in sewage sludge, to enable land application. Compared with biological treatment, which requires specific conditions, therefore limiting the ability to treat many wastewaters with high toxicity, xenobiotic compounds, and pH, EC can be used to treat multifaceted wastewaters, including industrial, agricultural, domestic, and others. PO_4^{3-} removal increased with EO pretreatment. NO_3^- in the filtered water increased with EO pretreatment. The use of EO pretreatment reduced cleaning costs.

These aforementioned results are based on controlled laboratory analysis. Most of the literature deals with experiments at the laboratory scale using synthetic solutions; thus, each application needs to undergo a proper controlled investigation. Examples of EC alone or in a hybrid process are numerous and attractive in the literature. However, EC remains disregarded in comparison with other treatment processes because no design and scale-up rules are clearly established for it. Therefore, future pilot and full-scale investigations are needed to optimize EC and/or EO treatments for personal care products and energy consumption, elucidating the mechanisms behind microbial reductions and performing life cycle analyses to determine the appropriateness of implementation. Future studies should focus on not only the high efficiency of the treatment but also the modeling of the mechanisms of the EC and/or EO processes to predict the treatment of wastewater. Continual research using this technology will improve its efficiency; new modeling techniques can be used to predict many factors and develop equations that will predict the effectiveness of the EC treatment. The EC technology provides strategic guidelines for further research and development of sustainable water management processes. However, additional test series on continuous operation must be performed to make this concept ready for future large-scale applications.

References

Anastas, P. T., and J. C. Warner. 1998. *Green chemistry: Theory and practice*, 29–56. Oxford University Press, New York.

Akyol, A. 2012. "Treatment of paint manufacturing wastewater by electrocoagulation." *Desalination* 285: 91–99.

Al-Shannag, M., K. Bani-Melhem, Z. Al-Anber, and Z. Al-Qodah. 2013. "Enhancement of COD-nutrients removals and filterability of secondary clarifier municipal wastewater influent using electrocoagulation technique." *Sep. Sci. Technol.* 48 (4): 673–680.

Anglada, A., A. Urtiaga, and I. Ortiz. 2009. "Pilot scale performance of the electro-oxidation of landfill leachate at boron-doped diamond anodes." *Environ. Sci. Technol.* 43 (6): 2035–2040.

Aquino, J. M., R. C. Rocha-Filho, L. A. M. Ruotolo, N. Bocchi, and S. R. Biaggio. 2014. "Electrochemical degradation of a real textile wastewater using β-PbO_2 and DSA® anodes." *Chem. Eng. J.* 251: 138–145.

Azarian, G., A. R. Rahmani, M. Masoudi khoram, Z. Atashzaban, and D. Nematollahi. 2018. "New batch electro-coagulation process for treatment and recovery of high organic load and low volume egg processing industry wastewater." *Process Saf. Environ. Prot.* 119: 96–103.

Bazrafshan, E., H. Moein, F. Kord Mostafapour, and S. Nakhaie. 2013. "Application of electrocoagulation process for dairy wastewater treatment." *J. Chem.* 2013: 640139.

Beck, F., W. Kaiser, and H. Krohn. 2000. "Boron doped diamond (BDD)-layers on titanium substrates as electrodes in applied electrochemistry." *Electrochim. Acta* 45 (28): 4691–4695.

Bharath, M., Krishna, B. M., and Manoj Kumar, B. 2018. "A review of electrocoagulation process for wastewater treatment." *Int. J. ChemTech Res.* 11 (3): 289–302.

Bhaskar Raju, G., M. Thalamadai Karuppiah, S. S. Latha, D. Latha Priya, S. Parvathy, and S. Prabhakar. 2009. "Electrochemical pretreatment of textile effluents and effect of electrode materials on the removal of organics." *Desalination* 249 (1): 167–174.

Bhatnagar, R., H. Joshi, I. D. Mall, and V. C. Srivastava. 2014. "Electrochemical oxidation of textile industry wastewater by graphite electrodes." *J. Environ. Sci. Health Part A* 49 (8): 955–966.

Bureau, M.-A., P. Drogui, B. Sellamuthu, J. F. Blais, and G. Mercier. 2012. "Municipal wastewater sludge stabilization and treatment using electrochemical oxidation technique." *J. Environ. Eng.* 138 (7): 743–751.

Calisto, V., M. R. M. Domingues, and V. I. Esteves. 2011. "Photodegradation of psychiatric pharmaceuticals in aquatic environments—Kinetics and photodegradation products." *Water Res.* 45 (18): 6097–6106.

Cenkin, V., and A. Belevtsev. 1985. "Electrochemical treatment of industrial wastewater." *Effluent Water Treat J.* 25 (7): 243–247.

Chang, H., R. Hu, Y. Zou, X. Quan, N. Zhong, S. Zhao, et al. 2020. "Highly efficient reverse osmosis concentrate remediation by microalgae for biolipid production assisted with electrooxidation." *Water Res.* 174: 115642.

Chavalparit, O., and M. Ongwandee. 2009. "Optimizing electrocoagulation process for the treatment of biodiesel wastewater using response surface methodology." *J. Environ. Sci.* 21 (11): 1491–1496.

Chen, G. 2004. "Electrochemical technologies in wastewater treatment." *Sep. Purif. Technol.* 38 (1): 11–41.

Chu, Y.-Y., M.-H. Zhu, and C. Liu. 2015. "Electrochemical treatment of biotreated landfill leachate using a porous carbon nanotube-containing cathode: Performance, toxicity reduction, and biodegradability enhancement." *Environ. Eng. Sci.* 32 (5): 445–450.

Clara, M., B. Strenn, and N. Kreuzinger. 2004. "Carbamazepine as a possible anthropogenic marker in the aquatic environment: Investigations on the behaviour of carbamazepine in wastewater treatment and during groundwater infiltration." *Water Res.* 38 (4): 947–954.

Coeuret, C., and A. Storck. 1993. *Élément de génie électrochimique. Techniques et Documentation (Lavoisier).* Génie des procédés de l'École de Nancy, Paris, France. ISBN: 978-2-85206-929-9.

Comninellis, C. 1992. "Electrochemical treatment of waste water containing phenol." In *Proc., Electrochemical Engineering and the Environment*, 189–201. Published in: Electrochemical Engineering and the Environment 92; Series: Institution of Chemical Engineers Symposium Series, 127; Conference: Electrochemical Engineering and the Environment 92, Loughborough University of Technology, 7–9 April 1992, Loughborough, England, United Kingdom. ISBN0-85295-280-5.

Comninellis, C. 1994. "Electrocatalysis in the electrochemical conversion/combustion of organic pollutants for waste water treatment." *Electrochim. Acta* 39 (11-12): 1857–1862.

Comninellis, C., and A. Nerini. 1995. "Anodic oxidation of phenol in the presence of NaCl for wastewater treatment." *J. Appl. Electrochem.* 25 (1): 23–28.

Cremieux, A., and J. Freney. 1995. "Bases fondamentales de l'action antimicrobienne des antiseptiques et désinfectants-les mécanismes d'action antimicrobienne." In *Antisepsie et Désinfection*, edited by J. Fleurette, J. Freney, and E. Reverdy, 23–37. Paris: Eska.

Daghrir, R., and P. Drogui. 2013. "Coupled electrocoagulation–electro-Fenton for efficient domestic wastewater treatment." *Environ. Chem. Lett.* 11 (2): 151–156.

Deng, Y., and J. D. Englehardt. 2007. "Electrochemical oxidation for landfill leachate treatment." *Waste Manage.* 27 (3): 380–388.

Dia, O., P. Drogui, G. Buelna, and R. Dubé. 2018. "Hybrid process, electrocoagulation-biofiltration for landfill leachate treatment." *Waste Manage.* 75: 391–399.

Dia, O., P. Drogui, G. Buelna, R. Dubé, and B. S. Ihsen. 2017. "Electrocoagulation of bio-filtrated landfill leachate: Fractionation of organic matter and influence of anode materials." *Chemosphere* 168: 1136–1141.

Dia, O., D. Patrick, G. Buelna, R. Dubé, and B. Salah Ihsen. 2016. "Coupling biofiltration process and electrocoagulation using magnesium-based anode for the treatment of landfill leachate." *J. Environ. Manage.* 181: 477–483.

Dieterich, A. E. 1906. Electric water purifier, United States of America Patent No. 823671 (June 19, 1906).

Drogui, P., J.-F. Blais, and G. Mercier. 2007. "Review of electrochemical technologies for environmental applications." *Recent Pat. Eng.* 1 (3): 257–272.

Drogui, P., S. Elmaleh, M. Rumeau, C. Bernard, and A. Rambaud. 2001. "Oxidising and disinfecting by hydrogen peroxide produced in a two-electrode cell." *Water Res.* 35 (13): 3235–3241.

El-Ashtoukhy, E. S. Z., N. K. Amin, and O. Abdelwahab. 2009. "Treatment of paper mill effluents in a batch-stirred electrochemical tank reactor." *Chem. Eng. J.* 146 (2): 205–210.

El-Naas, M. H., S. Al-Zuhair, A. Al-Lobaney, and S. Makhlouf. 2009. "Assessment of electrocoagulation for the treatment of petroleum refinery wastewater." *J. Environ. Manage.* 91 (1): 180–185.

Fajardo, A. S., R. F. Rodrigues, R. C. Martins, L. M. Castro, and R. M. Quinta-Ferreira. 2015. "Phenolic wastewaters treatment by electrocoagulation process using Zn anode." *Chem. Eng. J.* 275: 331–341.

Feki, E., S. Khoufi, S. Loukil, and S. Sayadi. 2015. "Improvement of anaerobic digestion of waste-activated sludge by using H_2O_2 oxidation, electrolysis, electro-oxidation and thermo-alkaline pretreatments." *Environ. Sci. Pollut. Res.* 22 (19): 14717–14726.

Feki, F., F. Aloui, M. Feki, and S. Sayadi. 2009. "Electrochemical oxidation post-treatment of landfill leachates treated with membrane bioreactor." *Chemosphere* 75 (2): 256–260.

Feng, C., N. Sugiura, S. Shimada, and T. Maekawa. 2003. "Development of a high performance electrochemical wastewater treatment system." *J. Hazard. Mater.* 103 (1–2): 65–78.

Gandini, D., C. Comninellis, N. B. Tahar, and A. Savall. 1998. "Électrodépollution: Traitement électrochimique des eaux résiduaires chargées en matières organiques toxiques." *Actual. Chim.* 10: 68–73.

García-Morales, M. A., G. Roa-Morales, C. Barrera-Díaz, B. Bilyeu, and M. A. Rodrigo. 2013. "Synergy of electrochemical oxidation using boron-doped diamond (BDD) electrodes and ozone (O_3) in industrial wastewater treatment." *Electrochem. Commun.* 27: 34–37.

Garcia-Segura, S., M. M. S. G. Eiband, J. V. de Melo, and C. A. Martínez-Huitle. 2017. "Electrocoagulation and advanced electrocoagulation processes: A general review about the fundamentals, emerging applications and its association with other technologies." *J. Electroanal. Chem.* 801: 267–299.

Garnerone, A. 1979. *Contrbibution à l'étude des mécanismes d'action des agents antiseptiques dans les eaux: Application à la désinfection électrochimique indirecte.* Doctoral thesis, Géotechnique et Génie Civil, Institut National Polytechnique de Grenoble. France.

GilPavas, E., I. Dobrosz-Gómez, and M.-Á. Gómez-García. 2019. "Optimization and toxicity assessment of a combined electrocoagulation, H_2O_2/Fe^{2+}/UV and activated carbon adsorption for textile wastewater treatment." *Sci. Total Environ.* 651: 551–560.

Gonzalez-Olmos, R., A. Penadés, and G. Garcia. 2018. "Electro-oxidation as efficient pretreatment to minimize the membrane fouling in water reuse processes." *J. Membr. Sci.* 552: 124–131.

Heffron, J., D. R. Ryan, and B. K. Mayer. 2019. "Sequential electrocoagulation-electrooxidation for virus mitigation in drinking water." *Water Res.* 160: 435–444.

Jose, S., L. Mishra, S. Debnath, S. Pal, P. K. Munda, and G. Basu. 2019. "Improvement of water quality of remnant from chemical retting of coconut fibre through electrocoagulation and activated carbon treatment." *J. Cleaner Prod.* 210: 630–637.

Kim, T.-H., C. Park, J. Lee, E.-B. Shin, and S. Kim. 2002. "Pilot scale treatment of textile wastewater by combined process (fluidized biofilm process–chemical coagulation–electrochemical oxidation)." *Water Res.* 36 (16): 3979–3988.

Komtchou, S., A. Dirany, P. Drogui, and A. Bermond. 2015. "Removal of carbamazepine from spiked municipal wastewater using electro-Fenton process." *Environ. Sci. Pollut. Res.* 22 (15): 11513–11525.

Kong, F. X., X. F. Lin, G. D. Sun, J. F. Chen, C. M. Guo, and Y. F. Xie. 2019. "Enhanced organic removal for shale gas fracturing flowback water by electrocoagulation and simultaneous electro-peroxone process." *Chemosphere* 218: 252–258.

Kurniawan, T. A., G. Y. Chan, W.-H. Lo, and S. Babel. 2006. "Physico–chemical treatment techniques for wastewater laden with heavy metals." *Chem. Eng. J.* 118 (1–2): 83–98.

Laridi, R., P. Drogui, H. Benmoussa, J.-F. Blais, and J. C. Auclair. 2005. "Removal of refractory organic compounds in liquid swine manure obtained from a biofiltration process using an electrochemical treatment." *J. Environ. Eng.* 131 (9): 1302–1310.

Li, C., G. Feng, C. Song, G. Zhong, P. Tao, T. Wang, et al. 2018. "Improved oil removal ability by the integrated electrocoagulation (EC)-carbon membrane coupling with electrochemical anodic oxidation (CM/EAO) system." *Colloids Surf. A* 559: 305–313.

Linares-Hernández, I., C. Barrera-Díaz, G. Roa-Morales, and F. Ureña-Núñez. 2008. "Remoción de contaminantes biorefractarios en aguas residuales industriales mediante métodos electroquímicos." *Quivera Revista de Estudios Territoriales* 10 (1): 1–15.

Lynn, W. M. 2019. "Impact of electrocoagulation pretreatment on *E. coli* mitigation using electrooxidation." Master's thesis, Department of Civil and Environmental Engineering, Marquette University, USA

Mandal, P., B. K. Dubey, and A. K. Gupta. 2017. "Review on landfill leachate treatment by electrochemical oxidation: Drawbacks, challenges and future scope." *Waste Manage.* 69: 250–273.

Mendia, L. 1982. "Electrochemical processes for wastewater treatment." *Water Sci. Technol.* 14 (1–2): 331–344.

Mollah, M. Y. A., R. Schennach, J. R. Parga, and D. L. Cocke. 2001. "Electrocoagulation (EC)—Science and applications." *J. Hazard. Mater.* 84 (1): 29–41.

Morao, A., A. Lopes, M. P. de Amorim, and I. Gonçalves. 2004. "Degradation of mixtures of phenols using boron doped diamond electrodes for wastewater treatment." *Electrochim. Acta* 49 (9–10): 1587–1595.

Mouedhen, G., M. Feki, M. D. P. Wery, and H. Ayedi. 2008. "Behavior of aluminum electrodes in electrocoagulation process." *J. Hazard. Mater.* 150 (1): 124–135.

Moussa, D. T., M. H. El-Naas, M. Nasser, and M. J. Al-Marri. 2017. "A comprehensive review of electrocoagulation for water treatment: Potentials and challenges." *J. Environ. Manage.* 186: 24–41.

Nariyan, E., A. Aghababaei, and M. Sillanpää. 2017. "Removal of pharmaceutical from water with an electrocoagulation process; effect of various parameters and studies of isotherm and kinetic." *Sep. Purif. Technol.* 188: 266–281.

Nidheesh, P. V., and T. S. A. Singh. 2017. "Arsenic removal by electrocoagulation process: Recent trends and removal mechanism." *Chemosphere* 181: 418–432.

Ogando, F. I. B., C. L. de Aguiar, J. V. N. Viotto, F. J. Heredia, and D. Hernanz. 2019. "Removal of phenolic, turbidity and color in sugarcane juice by electrocoagulation as a sulfur-free process." *Food Res. Int.* 122: 643–652.

Panizza, M., and C. A. Martinez-Huitle. 2013. "Role of electrode materials for the anodic oxidation of a real landfill leachate—Comparison between Ti–Ru–Sn ternary oxide, PbO_2 and boron-doped diamond anode." *Chemosphere* 90 (4): 1455–1460.

Phillips, R. B., R. R. James, and M. L. Magnuson. 2018. "Electrolyte selection and microbial toxicity for electrochemical oxidative water treatment using a boron-doped diamond anode to support site specific contamination incident response." *Chemosphere* 197: 135–141.

Pillaud, J.-L. 1987. "Application de l'oxydation anodique au traitement de certains effluents industriels." Doctoral thesis, Department of Biologie, écologie, évolution, environnement, sciences de la terre et de l'eau, Université de Montpellier II, France.

Pujol Pozo, A. A., E. Bustos Bustos, and F. Monroy-Guzmán. 2019. "Decontamination of radioactive metal surfaces by electrocoagulation." *J. Hazard. Mater.* 361: 357–366.

Pulgarin, C., N. Adler, P. Peringer, and C. Comninellis. 1994. "Electrochemical detoxification of a 1,4-benzoquinone solution in wastewater treatment." *Water Res.* 28 (4): 887–893.

Rada, E. C., I. A. Istrate, M. Ragazzi, G. Andreottola, and V. Torretta. 2013. "Analysis of electro-oxidation suitability for landfill leachate treatment through an experimental study." *Sustainability* 5 (9): 3960–3975.

Radha, K. V., V. Sridevi, and K. Kalaivani. 2009. "Electrochemical oxidation for the treatment of textile industry wastewater." *Bioresour. Technol.* 100 (2): 987–990.

Rajkumar, D., J. Guk Kim, and K. Palanivelu. 2005. "Indirect electrochemical oxidation of phenol in the presence of chloride for wastewater treatment." *Chem. Eng. Technol.* 28 (1): 98–105.

Rumeau, M. 1989. "Méthodesélectrolytiques de traitement des effluents—première partie. Électro-coagulation, flottation, détoxication, chloration." *L'eau et l'industrie*, 38: 47–51.

Senghor, F., P. Drogui, and B. Seyhi. 2015. "A combined electrocoagulation–electroperoxidation process for the tertiary treatment of domestic wastewaters." *Water Air Soil Pollut.* 226 (11): 373.

Senthilkumar, S., C. A. Basha, M. Perumalsamy, and H. J. Prabhu. 2012. "Electrochemical oxidation and aerobic biodegradation with isolated bacterial strains for dye wastewater: Combined and integrated approach." *Electrochim. Acta* 77: 171–178.

Sharma, S., and H. Simsek. 2019. "Treatment of canola-oil refinery effluent using electrochemical methods: A comparison between combined electrocoagulation + electrooxidation and electrochemical peroxidation methods." *Chemosphere* 221: 630–639.

Song, P., Z. Yang, G. Zeng, X. Yang, H. Xu, L. Wang, et al. 2017. "Electrocoagulation treatment of arsenic in wastewaters: A comprehensive review." *Chem. Eng. J.* 317: 707–725.

Soomro, G. S., C. Qu, N. Ren, S. Meng, X. Li, D. Liang, et al. 2020. "Efficient removal of refractory organics in landfill leachate concentrates by electrocoagulation in tandem with simultaneous electro-oxidation and in-situ peroxone." *Environ. Res.* 183: 109249.

Symonds, E. M., M. M. Cook, S. M. McQuaig, R. M. Ulrich, R. O. Schenck, J. O. Lukasik, et al. 2015. "Reduction of nutrients, microbes, and personal care products in domestic wastewater by a benchtop electrocoagulation unit." *Sci. Rep.* 5: 9380.

Tavangar, T., K. Jalali, M. A. Alaei Shahmirzadi, and M. Karimi. 2019. "Toward real textile wastewater treatment: Membrane fouling control and effective fractionation of dyes/inorganic salts using a hybrid electrocoagulation—Nanofiltration process." *Sep. Purif. Technol.* 216: 115–125.

Tian, Y., W. He, D. Liang, W. Yang, B. E. Logan, and N. Ren. 2018. "Effective phosphate removal for advanced water treatment using low energy, migration electric–field assisted electrocoagulation." *Water Res.* 138: 129–136.

Tir, M., and N. Moulai-Mostefa. 2008. "Optimization of oil removal from oily wastewater by electrocoagulation using response surface method." *J. Hazard. Mater.* 158 (1): 107–115.

Tran, L.-H., P. Drogui, G. Mercier, and J.-F. Blais. 2009. "Electrochemical degradation of polycyclic aromatic hydrocarbons in creosote solution using ruthenium oxide on titanium expanded mesh anode." *J. Hazard. Mater.* 164 (2–3): 1118–1129.

Tran, N., P. Drogui, J.-F. Blais, and G. Mercier. 2012. "Phosphorus removal from spiked municipal wastewater using either electrochemical coagulation or chemical coagulation as tertiary treatment." *Sep. Purif. Technol.* 95: 16–25.

Tsioptsias, C., D. Petridis, N. Athanasakis, I. Lemonidis, A. Deligiannis, and P. Samaras. 2015. "Post-treatment of molasses wastewater by electrocoagulation and process optimization through response surface analysis." *J. Environ. Manage.* 164: 104–113.

Vasanthapalaniappan, K., K. Palani, S. S. Saravanabhavan, N. Jonna, M. Pounsamy, K. Natarajan, et al. 2021. "A study on novel coupled membrane bioreactor with electro oxidation for biofouling reduction." *Environ. Eng. Res.* 26 (4): 200039.

Victoria-Salinas, R. E., V. Martínez-Miranda, I. Linares-Hernández, G. Vázquez-Mejía, M. Castañeda-Juárez, and P. T. Almazán-Sánchez. 2019. "Pre-treatment of soft drink wastewater with a calcium-modified zeolite to improve electrooxidation of organic matter." *J. Environ. Sci. Health Part A* 54 (7): 617–627.

Wang, A., J. Qu, H. Liu, and J. Ge. 2004. "Degradation of azo dye Acid Red 14 in aqueous solution by electrokinetic and electrooxidation process." *Chemosphere* 55 (9): 1189–1196.

Xiao, S., J. Peng, Y. Song, D. Zhang, R. Liu, and P. Zeng. 2013. "Degradation of biologically treated landfill leachate by using electrochemical process combined with UV irradiation." *Sep. Purif. Technol.* 117: 24–29.

Xu, X., and X. Zhu. 2004. "Treatment of refectory oily wastewater by electro-coagulation process." *Chemosphere* 56 (10): 889–894.

Yavuz, Y., A. S. Koparal, and ÜB Öğütveren. 2010. "Treatment of petroleum refinery wastewater by electrochemical methods." *Desalination* 258 (1–3): 201–205.

Zaied, M., and N. Bellakhal. 2009. "Electrocoagulation treatment of black liquor from paper industry." *J. Hazard. Mater.* 163 (2–3): 995–1000.

Zaviska, F., P. Drogui, J.-F. Blais, and G. Mercier. 2009. "In situ active chlorine generation for the treatment of dye-containing effluents." *J. Appl. Electrochem.* 39 (12): 2397–2408.

Zhou, B., Z. Yu, Q. Wei, H. Long, Y. Xie, and Y. Wang. 2016. "Electrochemical oxidation of biological pretreated and membrane separated landfill leachate concentrates on boron doped diamond anode." *Appl. Surf. Sci.* 377: 406–415.

Zodi, S., J.-N. Louvet, C. Michon, O. Potier, M.-N. Pons, F. Lapicque, et al. 2011. "Electrocoagulation as a tertiary treatment for paper mill wastewater: Removal of non-biodegradable organic pollution and arsenic." *Sep. Purif. Technol.* 81 (1): 62–68.

Zolfaghari, M., K. Jardak, P. Drogui, S. K. Brar, G. Buelna, and R. Dubé. 2016. "Landfill leachate treatment by sequential membrane bioreactor and electro-oxidation processes." *J. Environ. Manage.* 184: 318–326.

CHAPTER 2

Electro-Coagulation Process: Origins and Principles

Sridhar Pilli, S. Yan, R. D. Tyagi, P. Drogui,
Rao Y. Surampalli, Tian C. Zhang

2.1 INTRODUCTION

The availability of freshwater on the earth's surface is around 3%, of which around two-thirds is unavailable for use because it is tucked away in frozen glaciers. Further, because of industrialization and urbanization, freshwater sources such as rivers, lakes, and aquifers are drying up or becoming too polluted for use. Moreover, climate change is altering the patterns of weather and water on the earth. Thus, preservation of water resources has become one of the major challenges of the twenty-first century, and for this, there is a need for the formulation and implementation of sustainable water and wastewater management strategies.

For treating raw water and wastewater, currently many treatment methods are in use. In conventional treatment processes, coagulation–flocculation, followed by gravity sedimentation, is the most common process for the removal of colloidal particles (size: 10^{-3} to 10^{-6} mm) from raw water and wastewater. Initially, the application of the electro-coagulation (EC) process for treating raw water took place in 1889 at London (Vik et al. 1984). Thereafter, A. E. Dietrich in 1906 filed a patent on the EC process for treating bile water from ships (Moreno-Casillas et al. 2009). In 1909, J. T. Harries received a patent for the electrolysis of wastewater treatment using sacrificial aluminum and iron electrodes (Moreno-Casillas et al. 2009). However, because of high operating cost and the production of sludge of chemical coagulants, all EC plants were abandoned.

Further, in the 1940s, for purifying water, an *electronic coagulator* with dissolved aluminum (from anode) and hydroxyl ion (from cathode) was used (Holt et al. 2005). In 1956, in Britain, to treat river water, an electronic coagulator was used (Moreno-Casillas et al. 2009). During the early 1970s, EC was performed to treat food industry wastewater (Matteson et al. 1995). Furthermore, for the biotechnology process to harvest microorganisms from the cultures and for the removal of proteins and soluble substances from wastewater, EC was performed (Matteson et al. 1995). Following recent advances in instrumentation and rapid industrialization, there is a renewed interest in the application of the EC process because of the enhanced efficiency in the removal of contaminants (greater than filtration or chemical treatment systems). EC is applied for the removal of emulsified oils, textile dyes, fluorine, polymeric waste, organic matter from leachate, turbidity, chemical and mechanical polishing waste, total petroleum hydrocarbons, refractory organics, suspended solids, and heavy metals (Kabdaşlı et al. 2012).

In this chapter, the fundamentals of the EC process treating raw water and wastewater and the associated advantages and disadvantages are detailed. Furthermore, the experimental features, such as current density, anode and cathode material, and the effects of operating parameters on the efficiency of EC processes, are deliberated.

2.2 FUNDAMENTALS OF ELECTRO-COAGULATION FOR WATER AND WASTEWATER TREATMENT

EC means applying electrical current to change the particle surface charge (causing their destabilization) and enhance *coagulation*. In general, EC is an alternate water and wastewater treatment that combines electrochemical processes with conventional chemical coagulation. Coagulation is a chemical process, in which coagulants [e.g., soluble metal salts, $FeCl_3$, $Al_2(SO_4)_3$] are added to form a gelatinous mass to trap (or bridge) the particles, leading to particle agglomeration to form a mass bulky enough to settle (i.e., overcoming the repulsive charge and *destabilizing* the suspension). Detailed studies on chemical coagulation (normally organic polymers with the addition of metal salts) are available when compared with EC. Moreover, chemical coagulation and EC have fundamentally similar destabilization mechanisms. Therefore, it is necessary to understand the fundamentals of colloid destabilization with chemical coagulants.

In water and wastewater, the colloidal particles are in the range of 0.001 to 10 μm. These particles are responsible for turbidity and color (Marriaga-Cabrales and Machuca-Martínez 2014). In general, the coagulation process occurs in three discrete and successive steps (Ghernaout et al. 2011):

1. Coagulant solubilization (coagulants surrounded by water molecules),
2. Particle destabilization, and
3. Interparticle collisions.

The first two steps occur at a fast rate during rapid mixing, whereas interparticle collisions take place during slow mixing.

For colloidal particles, gravitational force is negligible because they are relatively small in *size and mass*; however, the surface property predominates because of a large surface area (Ghernaout et al. 2011). The colloids acquire positive or negative charge because of the disassociation of functional groups or by the preferential

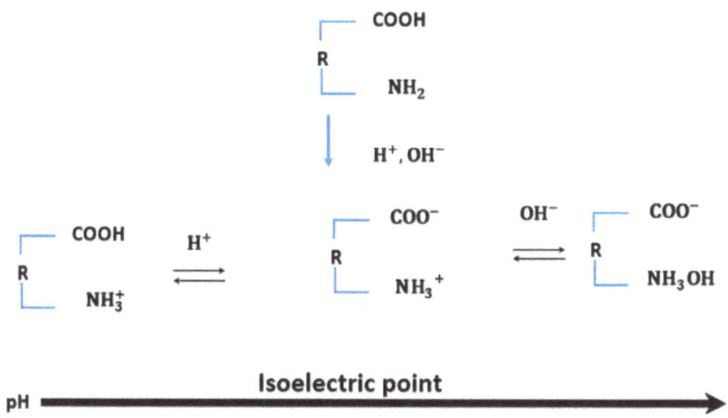

Figure 2-1. Surface charge on an organic colloid as a function of pH.

adsorption of ions from solution (Marriaga-Cabrales and Machuca-Martínez 2014). With a change of pH in the bulk solution, the colloids will have positive or negative charge (Figure 2-1) (Sincero and Sincero 2002). When the pH of the bulk solution is high (or low hydrogen ion concentration), the reactions will shift to the right, and the colloids will attain negative charge (Figure 2-1). For a low pH, the reactions will shift to the left, and the colloids will be positively charged (because of ionization of the amino group) (Figure 2-1). At the isoelectric point (Figure 2-1), the colloidal particles have neither positive nor negative charge, and they are in the neutral state. Table 2-1 explains the importance of properties of the colloid system, colloidal structure, stability, particle destabilization, and the time required for rapid mixing.

In chemical coagulation and EC, the colloidal destabilization principle is similar; however, EC differs in some other aspects, as side reactions occur at the electrodes. Therefore, it is very essential to understand that EC is a process where the sacrificial anode undergoes electrodissolution to produce cations. EC occurs through three discrete and successive steps (Mollah et al. 2004).

- Step 1: Active coagulant precursors form because of the electrolytic oxidation of the sacrificial electrode.
- Step 2: The active coagulant precursors will destabilize contaminants, particulate suspension, and breaking of mixtures.
- Step 3: The destabilized material aggregate to form flocs.

In general, EC occurs in an electrochemical cell consisting of the cathode and sacrificial anode immersed in a solution (water or wastewater to be treated). The anode and cathode connect through a circuit to induce current, as shown in Figure 2-2.

In an EC process, oxidation and reduction reactions take place at the electrode/electrolyte interface (Figure 2-2). The oxidation and reduction processes occur at the anode and cathode, respectively. As the anode is the sacrificial electrode, it releases active coagulant precursors (usually ionic coagulants of aluminium or iron) into the solution. A summary of destabilization of the contaminants,

Table 2-1. Important Parameters and Their Effects on Coagulation.

Description	*Parameter*	*Importance*
Properties of the colloidal system	1. Electrokinetic properties 2. Hydration 3. Brownian movement 4. Tyndall effect	1. This helps in determining the sign and magnitude of the acquired charge on colloidal particles 2. Colloids, which are hydrophilic in nature, possess water-soluble groups on their surface, such as hydroxyl, carboxyl, amino, and sulfonic. These groups exhibit high affinity for hydration and cause a water film surrounding the particles 3. Continuous random movement of colloids in a bulk solution occurs because of the collision with the fast-moving atoms or molecules. This enhances the coagulation 4. The interference with the passage of light is termed the Tyndall effect. Hydrophilic colloids may produce just a diffuse Tyndall cone or not at all. These phenomena are due to the bound water layer surrounding colloids (Ghernaout et al. 2011)
Colloidal structure and stability	The structure and stability of the colloid particle is dependent on its electrokinetic property	The theory of double layer explains the structure and stability of the colloid particle. The double layer consists of three parts: (1) surface charge (charged ions adsorbed on the particle surface), (2) stern layer (is a dense layer of counterions fixed on the surface of the primary a particle), and (3) diffuse layer (a film of the dispersion medium adjacent to the colloid particle). The Nernst potential, Stern potential, and Zeta potential are important to explain the colloidal structure and stability. The Nernst potential is the total potential at the surface of the primary charged particle. The Stern potential is the amount corresponding to the reduction of net charge on the particle by the concentrated counterions within the surface of shear. In detail, the theory of double layer and the stability of colloids is given in Ghernaout et al. (2011), Marriaga-Cabrales and Machuca-Martínez (2014)

Particle destabilization	1. Double-layer compression 2. Adsorption and charge neutralization 3. Entrapment of particles in precipitates 4. Adsorption and bridging between particles	1. The addition of counterions with higher charges decreases or eliminates the net repulsive force. Thus, it allows the particle to approach each other and agglomerate 2. Long-chain organic amines are commonly used coagulants that function by adsorption and electrostatic neutralization. Positively charged organic amines easily attach to the negatively charged colloidal particle, which results in neutralization, and the electrostatic repulsion either decreases or gets eliminated. Thus, destabilization of colloids and agglomeration will occur 3. When coagulants are added in high enough concentration, they react and form metal hydroxide precipitates. Destabilization of the colloidal particles occurs through the formation of bridges among them. Once a polymer is attached to a colloid particle, the remainder of the long-chain molecule extends away into the water. Once the extended portion of polymer gets attached to another colloidal particle, the two particles come together (Ghernaout et al. 2011)
Time required for rapid mixing	Rapid mixing	During rapid mixing, immediate reactions occur with the addition of coagulants, forming active coagulant species. These species induce destabilization. The nature of the coagulants and their coagulation mechanisms are crucial for the accomplishment of rapid mixing (Ghernaout et al. 2011)

particulate suspension, and breaking of mixtures is as follows (Mollah et al. 2004, Naje Ahmed et al. 2017):

1. **Compression**: The ions produced during the oxidation of the sacrificial anode react with charged particles, leading to the compression of the diffuse double layer.
2. **Charge neutralization and coagulation**: Ionic species in the solution (water/wastewater) interact with the counterions produced by the electrochemical dissolution of the sacrificial electrode, leading to charge neutralization. Because of the counterions, the electrostatic interparticle repulsion decreases to the extent that the Van der Waals attraction predominates, thus enhancing coagulation. Thus, colloidal particles attain zero net charge in this situation.
3. **Floc formation**: The flocs formed during coagulation generate a sludge blanket (Figure 2-2) that entraps and bridges colloidal particles remaining in the aqueous phase. The negative surfaces of solid oxides, hydroxides, and oxyhydroxides will adsorb the contaminants present in the solution.

In the electrochemical cell, the most commonly used cathode material is stainless steel or graphite, whereas aluminum or iron is used as an anode metal. However, the choice of the sacrificial anode metal depends on the type/characteristics of wastewater to be treated (Harif et al. 2012). EC has evolved as an alternate, competitive, and effective treatment process that can be used to remove various pollutants ranging from suspended solids (from water and wastewater), heavy metals, petroleum products, colors from dye-containing solution, aquatic humus, fluoride in water, and so on (Moreno-Casillas et al. 2009, Butler et al. 2011).

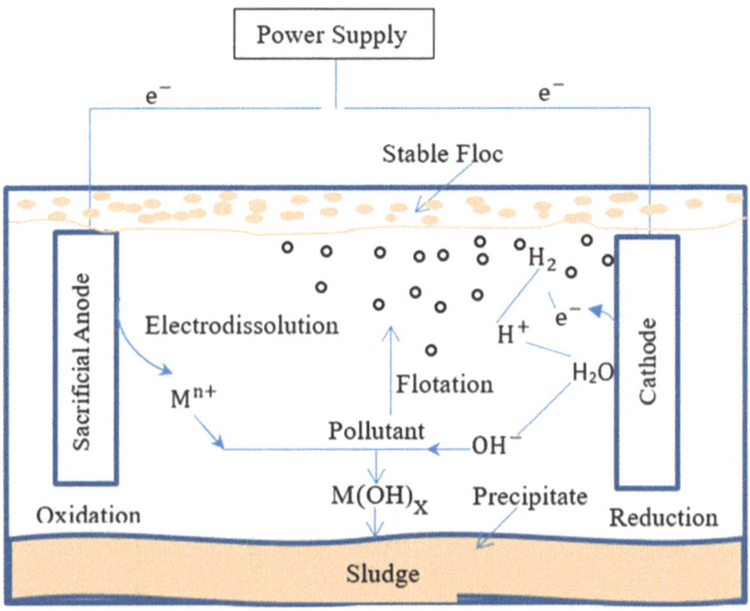

Figure 2-2. Schematic diagram of an EC process.

In this section, the fundamental parameters that impact EC, such as reactor design, electrode configuration, and coagulation agents, are detailed. Further experimental features of EC, such as current density, energetic parameters, the effects of anodic and cathode materials, and operation parameters are summarized in Section 2.3. Moreover, in Section 2.4 of this chapter, the advantages and disadvantages of EC are outlined.

2.2.1 Reactor Design

Wide varieties of EC reactor configurations are available in the literature (Moreno-Casillas et al. 2007, Kabdaşlı et al. 2017). Each arrangement has its own advantages and disadvantages. Primarily, the EC reactor is the alternative for the chemical dosing system; however, the effect of the generation of electrolytic gases during EC is not considered (Kabdaşlı et al. 2017).

Data on the existing literature do not provide a single empirical or systematic approach to the optimization of the EC reactor design and operation (Holt et al. 2005, Kabdaşlı et al. 2017). Mostly, the EC reactor design is still largely empirical and heuristic. The geometry of the EC reactor is of great importance, because it affects the overall performance of EC. The operating parameters of EC are influenced by the geometry of the EC reactor, such as fluid flow regime, floc formation, bubble path, floatation effectiveness, removal yield, and mixing/settling characteristics.

As reported, the EC reactor design (Figure 2-3) is classified based on three major differences (Hakizimana et al. 2017):

1. **Reactor configuration (i.e., batch or continuous systems):** In a batch system, the reactor is filled with a definite volume (water/wastewater to be treated) per treatment cycle. The major disadvantage is associated with both design and operation of the batch reactor. In a continuous system, the liquid (water/wastewater) is continuously fed into the EC reactor to maintain (pseudo) steady-state conditions. The major advantage of a continuous system is that the coagulant requirements are essentially fixed.

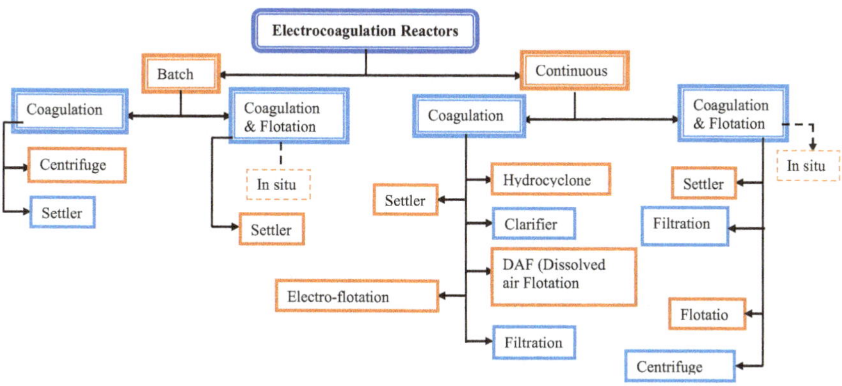

Figure 2-3. Schematic of EC reactor design.

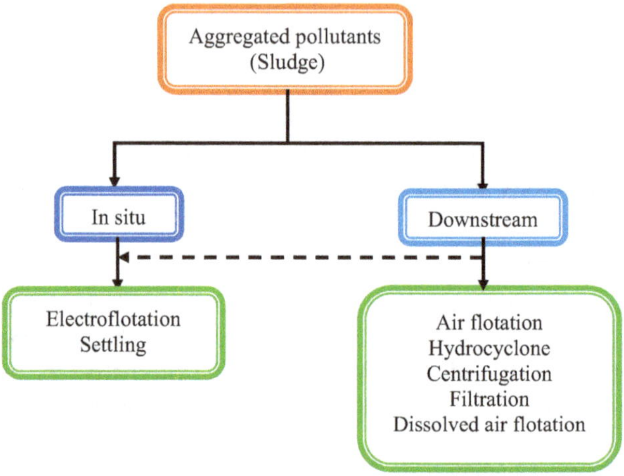

Figure 2-4. Methods to separate aggregated pollutants.

2. **Method to separate aggregated pollutants**: When current passes through the electrochemical cell, the coagulant species generated go through the process of electrodissolution of a sacrificial anode (Figure 2-2). The oxidation and reduction take place at the anode and cathode, respectively, leading to the formation of precipitates. The metal cations released into the aqueous solution go through several equilibrium reactions that correspond to acid/base, complexation, precipitation, and redox reactions in water (Hakizimana et al. 2017). Moreover, the metal cations released from the anode exhibit poor solubility and readily precipitate (i.e., mainly because of the formation of metal hydroxides). For the precipitated contaminant (i.e., sludge), segregation is performed with two physical methods (Figure 2-4).
3. **Design of the electrode geometry:** The most common design in a typical application is the open top vertical plate cell (followed by settler) having planar rectangular electrodes. Mollah et al. (2004) had reviewed EC cells and concluded that there is no significant change in these cells in the last decade. Mainly, vertical anode and cathodes are equally spaced in parallel with any vertical/horizontal length ratio being used. Further, electrode spacing will also affect EC through the reactor working volume, that is, the electrode surface area/volume ratio (A/V). The A/V ratio is a significant scale-up parameter going from lab scale to full scale (Hakizimana et al. 2017). The electrode A/V ratio affects the treatment time and optimum current density, that is, an increase in the A/V ratio will result in a decrease in both treatment time and optimum current density (Kabdaşlı et al. 2017). The typical range of the electrode A/V ratio varies from 15 to 45 $m^2\ m^{-3}$ (Hakizimana et al. 2017). Reducing the gap between the electrodes will lead to the production of gas bubbles, electrochemically bringing about turbulent hydrodynamics. Thus, a high mass transfer and a high reaction rate between the coagulant species and the pollutants will occur (Hakizimana et al. 2017).

2.2.2 Monopolar and Bipolar Configurations

In 1979, Pickett proposed variations in the arrangement of electrodes in the EC (Pickett 1979). The simple arrangement of an electrolytic cell consists of one anode and cathode connected to a direct current. In practical terms, this simple arrangement is not sufficient for the removal of pollutants for solution (Pretorius et al. 1991). Therefore, multiple arrangements are required to enhance the efficiency of pollutant removal. The multiple electrode arrangements in an EC cell include the following: (1) Monopolar electrodes in parallel; (2) monopolar electrodes in series; and (3) bipolar electrodes in series.

1. **Monopolar electrodes in parallel (MP-P):** In this configuration, anodes and cathodes are placed alternatively at the same anodic or cathodic potential, respectively (Mollah et al. 2004, Hakizimana et al. 2017). In a monopolar arrangement, each pair of cathode/anode corresponds to a small electrolytic cell in which voltage is similar. The current of each electrolytic cell is additive for the monopolar electrodes in parallel (Figure 2-5).
2. **Monopolar electrodes in series (MP-S):** In a series configuration, each pair of sacrificial electrodes joins internally without being interconnected with the outer electrodes. As the electrodes are in series, the electric current passing through all the electrodes is the same, whereas the total voltage is the sum of voltage in each individual electrolytic cell (Hakizimana et al. 2017) (Figure 2-6).
3. **Bipolar electrodes in series:** In a bipolar series configuration, two outer electrodes (the monopolar cathode and anode) connect to an electric supply and the sacrificial electrodes are in place between the outer electrodes. The inner electrodes are bipolar, whereas the outer electrodes are monopolar. The inner electrodes (bipolar electrodes) are not interconnected, with each of their sides acting as an anode and a cathode. Thus, the opposite sides of the bipolar electrode are charged with an opposite sign (Figure 2-7); the anodic dissolution take place on the positive side, whereas the negative side is likely to have cathodic reactions.

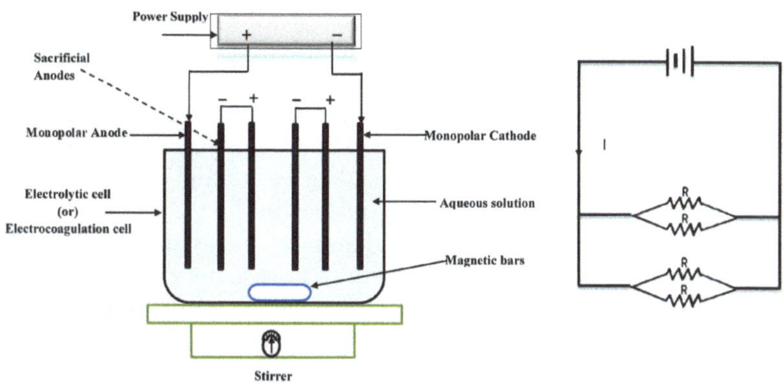

Figure 2-5. Monopolar electrode arrangement in parallel.

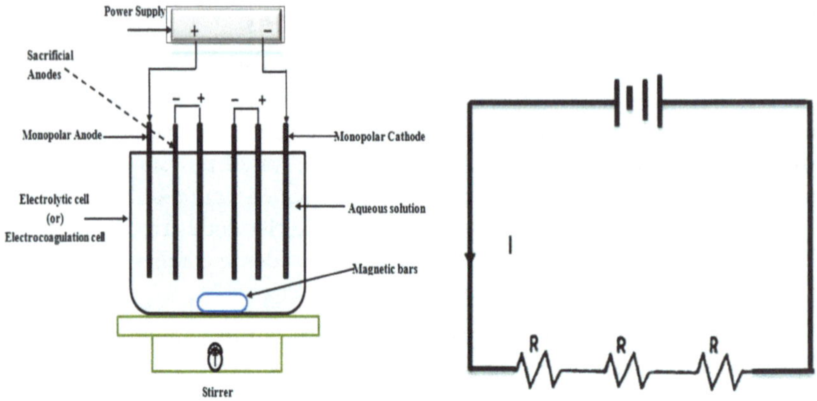

Figure 2-6. Monopolar electrode arrangement in series.

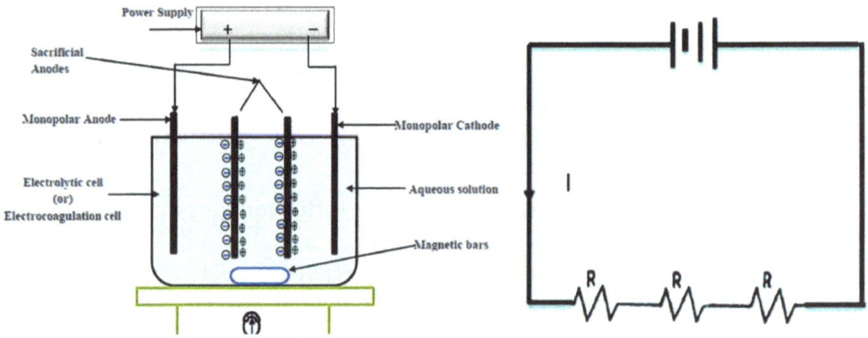

Figure 2-7. Bipolar electrode arrangement in series.

2.2.3 Production of Coagulation Agents

EC is a complex process having various mechanisms that can be of electrochemical (metal dissolution and water reduction, pollutant electro-oxidation or reduction), chemical (acid/base with pH change, hydroxide precipitation, redox reaction in the bulk), and physical (adsorption, coagulation, flotation) functions (Hakizimana et al. 2017). EC involves metal dissolution of the *sacrificial anode* with in situ produced active metal ions acting as coagulant agents, and, thus, it differs from chemical coagulation. The coagulant agents destabilize the contaminants in an aqueous solution. In general, iron and aluminum metals are used as sacrificial anodes, and an iron electrode is superior to aluminum (for each case it can be Al or Fe) in terms of COD removal efficiency and energy consumption (Kobya et al. 2003). In Table 2-2, the chemical reactions occurring at the anode and cathode are presented. In the case of the iron electrode, Chaturvedi (2013) have proposed

Table 2-2. Reactions and Coagulant Agent Production Because of Metal Dissolution in EC

Metal electrodes	Condition	Reactions: Anode	Reactions: Cathode	Overall reaction
Aluminum anode (mechanism 1 or 2)	Alkaline	$Al \rightarrow Al^{3+} + 3e$	$3H_2O + 3e \rightarrow 3/2\ H_2 + 3OH^-$	$2Al + 6H_2O \rightarrow 2Al(OH)_3 + 3H_2$
		$Al^{3+} + 3OH^- \rightarrow Al(OH)_3$		
	Acid	$Al^{3+} + 3H_2O \rightarrow Al(OH)_3 + 3H^+$		
Iron anode (Mechanism1)	Alkaline	$Fe \rightarrow Fe^{2+} + 2e$	$8H^+ + 8e^- \rightarrow 4H_2$	$4Fe + 10H_2O + O_2 \rightarrow 4Fe(OH)_3 + 4H_2$
		$Fe^{2+} + 3OH^- \rightarrow Fe(OH)_2$		
	Acid	$4\ Fe^{2+} + O_2 + 2H_2O \rightarrow 4Fe^{3+} + 4OH^-$		

Source: Cañizares et al. (2005).

two mechanisms. The details of Mechanism 1 are described in Table 2-2, whereas mechanism 2 is described as follows:

At anode

$$Fe \rightarrow Fe^{2+} + 2e^- \tag{2-1}$$

$$Fe^{2+} + 2OH^- \rightarrow Fe(OH)_2 \tag{2-2}$$

At cathode

$$2H_2O + 2e^- \rightarrow H_2 + 2OH^- \tag{2-3}$$

Overall reaction

$$Fe + 2H_2O \rightarrow Fe(OH)_2 + H_2 \tag{2-4}$$

The pH has a strong influence in the chemical dissolution of metal electrodes (Cañizares et al. 2005). The chemical dissolution increases by several times of magnitude at an alkaline pH. At the anode, the pH drops because of the production of a high concentration of protons, whereas at the cathode, the pH increases as the water reduction process results in the formation of hydroxide ions (Cañizares et al. 2005). Furthermore, when aluminum or iron is used, the produced Al^{3+} or Fe^{3+} will immediately undergo spontaneous reactions to yield corresponding hydroxides and polyhydroxides. The ions, Al^{3+} and OH^-, produced during dissolution will further react to form various monomeric species such as $Al(OH)_2^+$, $Al_2(OH)_2^+$, $Al(OH)_4^-$, and polymeric species such as $Al_{16}(OH)_{15}^{3+}$, $Al_{17}(OH)_{17}^{4+}$, $Al_{18}(OH)_{20}^{4+}$, $Al_{13}O_4(OH)_{24}^{7+}$, and $Al_{13}(OH)_{34}^{5+}$. The monomeric and polymeric species will finally transform into $Al(OH)_3$ according to complex precipitation kinetics (Kobya et al. 2003). Similarly, electrogenerated ferric ions may form monomeric ions, $Fe(OH)_3$, and polymeric hydrocomplexes such as $Fe(H_2O)_6^{3+}$, $Fe(H_2O)_5^{2+}$, $Fe(H_2O)_4(OH)_2^+$, $Fe(H_2O)_8(OH)_2^{4+}$, and $Fe(H_2O)_6(OH)_4^{4+}$, depending on the pH of the aqueous media (Chaturvedi 2013). The electrochemical dissolution of metal electrodes can be calculated according to Faraday's second law (Mollah et al. 2004, Kim et al. 2016, Yildiz et al. 2007, Yilmaz et al. 2007, Can et al. 2014):

$$\text{Metal(Al or Fe)}(\text{mg L}^{-1}) = \text{Active metal ions(Al or Fe)generated}(\text{mg s}^{-1})/\text{feed water flow}(\text{L s}^{-1}) \tag{2-5}$$

$$\text{Active metal ions generated}(\text{mg s}^{-1}) = I \times M \times (1{,}000\ \text{mg g}^{-1})/(F * n_e) \tag{2-6}$$

where

I = Current applied in A ($C\ s^{-1}$),
M = Molecular weight of metal (Al or Fe),
F = Faraday constant (96,485 $C\ mol^{-1}$), and
n_e = Number of electrons transferred in the reaction.

The production of Al^{3+} or Fe^{3+} ions and further hydrolyzation activates EC, that is, the precursors (metal cations) produced will react with negatively charged particles and will be carried toward the anode by electrophoretic motion with various destabilization mechanisms (Table 2-2). Moreover, concurrently occurring favorable side reactions, such as pH change and hydrogen bubble formation, will further enhance pollutant removal (Kuokkanen et al. 2013, Sakarinen 2016). The destabilization mechanisms include the following: (1) charge neutralization: the cationic hydrolysis products neutralize the negatively charged colloids in the aqueous solution, and (2) sweep flocculation: the amorphous hydroxide precipitates will act as traps to remove the impurities.

2.3 EXPERIMENTAL FEATURES

The efficiency of pollutant removal depends on the different operational/ experimental features of EC. Parameters such as current density, electrode material, electrolysis time, temperature, pH, conductivity of the solution, supporting electrolyte, and interelectrode distance could affect the efficiency of EC (Valente et al. 2012, Bazrafshan et al. 2013, Liu et al. 2010), which are elaborated as follows.

2.3.1 Current Density and Energetic Parameters

In EC, the amount of electrolytic dissolution of the sacrificial electrode (Al or Fe) and release of ions (Al^{3+} or Fe^{2+}) are dependent on the supply of current (Liu et al. 2010). Naje Ahmed et al. (2017) have noted that current density is an important parameter for analyzing the coagulant dosage rate, bubble formation rate, and the size and development of flocs as they influence the efficiency of the EC process.

The electrolytic dissolution of the sacrificial anode is directly proportional to the current density (Naje Ahmed et al. 2017). When current density increases beyond the optimal value, there is no effect on the contaminant removal efficiency (Barrera-Díaz et al. 2012, Khaled et al. 2019, Naje Ahmed et al. 2017). However, a great chance of electrical energy getting wasted exists while heating up the aqueous solution. Moreover, a large current density will result in a substantial reduction in current efficiency (Chen 2004). To have a high current efficiency in the EC cell, the selection of current density is necessary along with other operating parameters such as pH, temperature, electrolysis time, and flow rate (Chen 2004, Naje Ahmed et al. 2017). The current efficiency depends on the current density as well as the type of electrode used. The relationship between current density (A cm^{-2}) and the amount of substance (g of M cm^{-2}) is established from the Faraday's second law, that is, electrolytic dissolution of the electrode is directly proportional to the current density (Kabdaşlı et al. 2012).

$$Q_m = J \times t \times (M/n) \times F \tag{2-7}$$

where

Q_m = Quantity of electrode material (Al or Fe) dissolved (g of M cm^{-2}),
J = Applied current density (A cm^{-2}),
t = Electrolysis time (s),
M = Relative molar mass of the electrode material,
n = Number of electrons transferred in the reaction at the electrode, and
F = Faraday's constant (96,485 C mol^{-1}).

The electrochemical equivalent mass for Al and Fe is 335.6 and 1,041 mg $(A\ h)^{-1}$, respectively.

2.3.2 Power Supply Type

Mainly DC is required for the EC system. Because the DC used in the EC system, an impermeable oxide layer formation on the cathode and corrosion formation on the anode owing to oxidation will take place. These formations restrict the effective current transport between the electrodes (anode and cathode) and, thus, reduce the efficiency of pollutant removal from the aqueous solution (Khandegar and Saroha 2013, Naje Ahmed et al. 2017). However, a few studies (Kamaraj et al.

2013, Mansour and Hasieb 2012, Vasudevan et al. 2011a, b) are related to the use of alternative current (AC) in the EC system. Vasudevan et al. (2011a) compared the EC and DC for EC systems to remove fluoride from water and concluded that the removal efficiency was similar with both technologies. However, the energy consumption was slightly less for the AC technology than for the DC technology. Further, Vasudevan et al. (2011b) evaluated the effect of AC and DC technologies in the EC system on the removal of cadmium from water. The efficiencies of cadmium removal were 97.5% and 96.2% with an energy consumption of 0.454 and 1.002 kW·h kL^{-1} at a current density of 0.2 A dm^{-2} using AC and DC, respectively. The problem of corrosion formation at the electrodes can be minimized by replacing the DC current with AC current in EC. In addition to the aforementioned studies, Mansour and Hasieb (2012) also investigated the removal of Ni(II) and Co(II) from synthetic drinking water using AC technology, and the results corroborate with the aforementioned findings.

2.3.3 Effect of Anodic and Cathode Materials

Anode and cathode materials determine the electrochemical reactions that take place in EC. Usually, Al and Fe electrodes are used as the electrode material in EC (Hakizimana et al. 2017, Kabdaşlı et al. 2012, Khandegar and Saroha 2013, Naje Ahmed et al. 2017). Mild steel and stainless-steel materials are potential electrodes for EC (Kabdaşlı et al. 2012). The type of electrode material required for EC depends on the chemical properties of electrolytes and pollutant removal. Kobya et al. (2003) stated that the Al electrode seems to be superior to the Fe electrode for treating textile wastewater. Fe is used as an electrode for wastewater treatment because it is relatively cheap. When water contains Ca^{2+} and Mg^{2+} ions, then the recommended cathode material is stainless steel (Chen 2004). Further, according to their report, when Al alloys are used as anode material, it could improve the removal performances as compared with those observed with the pure metal (Vasudevan et al. 2009a, b, 2011c). The Al–zinc alloy was successfully used as an anode to treat water containing phosphates (Vasudevan et al. 2009a), iron (Vasudevan et al. 2009b), arsenate (Vasudevan et al. 2010), chromium (Vasudevan et al. 2011c), and copper (Vasudevan and Lakshmi 2012). The studies showed higher removal efficiencies were observed with Al alloys than those with pure Al. For example, the removal efficiencies of phosphate were 99%, 87%, and 85% with Al alloy, mild steel, and pure Al electrodes, respectively, under similar conditions. Using the Al–copper alloy electrode, the efficiency of the removal of chemical oxygen demand (COD) and total organic carbon of the concentrated oil suspensions is enhanced (Khemis et al. 2005).

In addition, Fe-based alloys are considered as electrodes in the EC system. Fe-based alloy electrodes such as stainless steel are used for the treatment of total suspended solids and turbidity (Bukhari 2008), phthalic acid esters (Kabdaşlı et al. 2009), nitrite (Ghazouani et al. 2015), chromium, color, and turbidity (Mahmad et al. 2016), gray wastewater (Karichappan et al. 2014), sullage wastewater (Santhosh et al. 2015), abamectin pesticide (Ghalwa et al. 2015), heavy metals (Da Mota et al. 2016), strontium (Murthy and Parmar 2011), and so on.

When Al is used as an electrode, it may lead to residual Al. According to the USEPA guidelines, the maximum contamination is 0.05 to 0.2 mg L^{-1}. Therefore, Kamaraj et al. (2013) have used magnesium anode and cathode for the removal of copper from water.

2.3.4 Influence of Operation Parameters

Operational parameters such as pH, conductivity, treatment time, supporting electrolyte, interelectrode distance, alkalinity, and temperature will affect the efficiency of pollutant removal in EC as follows.

pH. The pH of the solution is an important operational parameter that affects the conductivity of the solution, dissolution of the electrodes, speciation of hydroxides, and zeta potential of colloidal species (Naje Ahmed et al. 2017). For a particular pollutant, optimum solution pH is required for maximum pollutant removal. Moreover, the precipitation of a pollutant initiates at a particular pH. The pollutant removal efficiency decreases by either low or high pH values from the optimal value. In an EC process, when the pH of a solution is 4 (acidic), then the effluent pH increases, whereas pH will decrease when the initial pH is above 8 (basic). For a neutral pH (around 6 to 8), the effluent changes only slightly (Kabdaşlı et al. 2012). This phenomenon produces a pH buffering effect during EC, which is unlike traditional chemical coagulation. The pH buffering ability is attributed to the balance between the production and the consumption of hydroxide ions, that is, an increase in pH under acidic conditions is due to the hydrogen evolution at the cathode and the follow-up release of CO_2 from hydrogen bubbling. On the contrary, pH decrease is due to the formation of hydroxide precipitates that release H^+ cations at the anode area and secondary reactions such as water oxidation and chlorine production and its hydrolysis (Hakizimana et al. 2017, Kabdaşlı et al. 2012).

Conductivity/supporting electrolyte. Conductivity of the solution is one of the particularly important parameters in EC because the efficiency of pollutant removal and the operating cost are direct functions of the solution conductivity (Khandegar and Saroha 2013, Naje Ahmed et al. 2017). Moreover, the conductivity of the solution will affect the Faradic yield, cell voltage, and consequently, energy consumption in EC systems (Kabdaşlı et al. 2012). To enhance conductivity, the flow of electric current through the solution is enhanced by the addition of additives such as NaCl or Na_2SO_4. With an increase in conductivity, the current density increases, thereby decreasing the energy consumption (Naje Ahmed et al. 2017). At a constant current density, cell voltage decreases with an increase in conductivity; thus, the total resistance in the solution decreases and, thereby, the consumed electric energy also decreases. Usually, NaCl will enhance the conductivity of water or wastewater. In addition, with an increase in conductivity, chloride ions could significantly reduce the adverse effects of other anions such as HCO_3^- and SO_4^{2-} (Chen 2004, Kabdaşlı et al. 2012, Naje Ahmed et al. 2017). Therefore, the evolution of carbonate and sulfate ions will further lead to the precipitation of Ca^{2+} of Mg^{2+} ions that form an insulating layer on the surface of electrodes. The insulating layer can increase the ohmic resistance of the electrochemical cell (Kabdaşlı et al. 2012). In general, sulfate ions are considered as passivating agents, whereas chloride ions induce a breakdown of the passive layer (Uhlig and Revie 2011). The precipitation reactions are as follows:

$$HCO_3^- + OH^- \rightarrow CO_3^{2-} + H_2O \tag{2-8}$$

$$Ca^{2+} + CO_3^{2-} \rightarrow CaCO_3 \tag{2-9}$$

In water treatment, 20% Cl^- is necessary to ensure normal EC (Chen 2004). If NaCl is added in excessive amounts to the solution, it will induce overconsumption of the Al electrodes because of corrosion pitting (Kabdaşlı et al. 2012), and the electrode dissolution will become irregular. Therefore, NaCl addition should be limited and optimized to ensure normal operation. Further, in EC, in the presence of NaCl, hypochlorite (OCl^-) and hypochlorous acid (HOCl) will be produced. The formation of these oxidants mainly depends on the initial NaCl concentration and the current density applied during electrolysis. OCl^- and HOCl are produced when the NaCl concentration is relatively high in water subjected to treatment.

$$2Cl^- \rightarrow Cl_2 + 2e^- \tag{2-10}$$

$$2H_2O + 2e^- \rightarrow H_2 + 2OH^- \tag{2-11}$$

$$Cl_2 + H_2O \rightarrow HOCl + HCl \tag{2-12}$$

$$HOCl + H_2O \leftrightarrow H_3O^+ + OCl^- \tag{2-13}$$

In a study, NaCl, Na_2SO_4, NH_4Cl, and $(NH_4)_2SO_4$ were used as the supporting electrolytes to evaluate the EC efficiency of unskimmed milk samples; the results showed an increase in electrical power consumption and lower efficiencies with sulfate anions.

Treatment time. Electrolysis/treatment time is also an important parameter that will influence the pollutant removal efficiency (Khandegar and Saroha 2013, Naje Ahmed et al. 2017). Initially, with an increase in treatment time, the pollutant removal efficiency increases, but beyond the optimum electrolysis time, the removal rate becomes constant. Because of the electrolytic dissolution of anode metal hydroxides, production of these hydroxides takes place; that is, with an increase in electrolysis time, the availability of metal hydroxides will increase in solution, leading to floc formation and enhancement of pollutant removal efficiency. If the number of flocs available is sufficient for the removal of pollutants (Khandegar and Saroha 2013) beyond the optimum electrolysis time, the increase in metal hydroxides will not contribute to pollutant removal. For example, Karichappan et al. (2014) investigated the effect of the electrolysis time (5 to 25 min) of EC on the treatment of gray wastewater and concluded that the optimum electrolysis time was 20 min.

In EC, retention time also affects the pollutant removal efficiency. After EC, if enough time (retention time) is provided, coagulated species will settle. With an increase in the retention time, the pollutant removal efficiency will increase. If the retention time is beyond the optimum value, then the removal efficiency will decrease as adsorbed pollutant species will desorb into the solution (Khandegar and Saroha 2013).

Interelectrode distance. Interelectrode distance, that is, the distance between anode and cathode, is an important parameter in the EC system for the removal of a pollutant. The electrostatic field depends on the distance between anode and cathode (Khandegar and Saroha 2013, Naje Ahmed et al. 2017). Moreover, the IR drop (ohmic drop) increases as the distance between the electrodes increases (Hakizimana et al. 2017). As the gap between the electrodes decreases, energy consumption decreases [Equations (2-14) and (2-15)]. Hakizimana et al. (2017) defined electric energy consumption as a function of operation time "*t*":

$$P = \int_0^t U \cdot I \cdot dt \tag{2-14}$$

where

P = Electric energy combustion,
U = Total cell voltage in the reactor,
I = Current, and
T = Operation time.

High current also increases the voltage and ohmic drop between electrodes (anode and cathode). The IR drop resulting from the ohmic resistance of the electrolyte R is as follows (Hakizimana et al. 2017):

$$R = \frac{d}{A}\frac{I}{k} \tag{2-15}$$

where

d = Interelectrode distance (cm),
A = Surface area of the electrode (cm^2),
k = Specific conductivity of the solution ($\mu S\ cm^{-1}$), and
I = Current (A).

As the distance between the electrodes affects the pollutant removal efficiency, optimum distance is required for achieving maximum pollutant removal efficiency. The interelectrode distance and its effect on pollutant removal efficiency is summarized in Table 2-3.

Temperature. Only a few studies on the effect of temperature on EC are available. With increasing the temperature, the pollutant removal efficiency also increases (Yilmaz et al. 2008). A higher solution temperature could lead to an increased mass transfer and the kinetics of particle collisions to a hydroxide polymer (Yilmaz et al. 2008). In a study, with the temperature increasing from 293 to 333 k, boron removal efficiency also increased from 84% to 96%. On the contrary, Katal and Pahlavanzadeh (2011) reported the opposite effect while

Table 2-3. Interelectrode Distance and Pollutant Removal Efficiency.

Distance between anode and cathode	*Description*	*Pollutant removal efficiency*	*Reference*
Low/ minimum	The generated hydroxides that act as flocs and remove pollutants by sedimentation degrade as they collide because of high electrostatic attraction. Moreover, electrochemically generated gas bubbles bring about turbulent hydrodynamics. For a continuous system, the interelectrode distance defines the residence time	Low	Hakizimana et al. (2017), Naje Ahmed et al. (2017)
Optimum	Increasing the interelectrode distance increases the pollutant removal efficiency till an optimum value is reached. Increasing the interelectrode distance, electrostatic attraction decreases and provides more time for the generated metal hydroxides to agglomerate to form flocs, resulting in an increase in pollutant removal efficiency	Maximum	Naje Ahmed et al. (2017)
High (greater than optimum)	As the distance between the interelectrodes increases, the travel time of the ions also increases. Furthermore, a decrease in electrostatic attraction happens, which minimizes the formation of the desired flocs to coagulate the pollutant	Low	Naje Ahmed et al. (2017), Anantha Singh and Ramesh (2013)

treating paper mill wastewater. With an increase in temperature from 293 to 333 k, the efficiency of the removal of color, COD, and phenol decreased by 10% to 20%. High temperature will lead to an increase in the solubility of Al, whereas at lower temperatures, the precipitation of Al is enhanced (Katal and Pahlavanzadeh 2011).

The use of the Al electrode and high solution temperature leads to a shrinkage of large pores of the $Al(OH)_3$ gel, which causes the formation of compact flocs that are more likely to be deposited on the electrode surface (Chen 2004, Naje Ahmed et al. 2017). Thus, temperature increase has both positive and negative effects on pollutant removal efficiency. The effect of temperature on removal efficiency may depend on the pollutant removal mechanisms (Naje Ahmed et al. 2017). Apart from increased mass transfer and the kinetics of particle collisions, high temperature favors the formation of large hydrogen bubbles, enhanced flotation speed, and a reduction in the adhesion of suspended particles.

2.4 ADVANTAGES AND DISADVANTAGES OF ELECTRO-COAGULATION

2.4.1 Advantages

The advantages of EC are as follows:

1. Relatively, the cost of EC is low compared with that of conventional systems.
2. EC can continuously produce coagulant species, requires simple equipment, and is easy to control with the necessary operational freedom to handle most problems encountered on running it (Bazrafshan et al. 2015).
3. Sludge produced from EC tends to be readily settable and easy to de-water compared with the conventional chemical coagulated sludge. The EC-produced sludge mainly contains metallic oxides/hydroxides, which do not have residual charge. In EC, floc formation and a structural evolution pattern are observed under optimal sweep floc conditions; the governing mechanism is diffusion limited cluster aggregation, whereas the mechanism in chemical coagulation is reaction limited cluster aggregation (RLCA).
4. In EC, flocculation increases because of (i) the movement of the smallest charged colloids inside the electric field generated in the electrochemical cell, and (ii) turbulence induced by bubbles.
5. Filtration rate is higher for the flocs produced by EC, as the flocs are bigger in size, contain less bound water, are acid-resistant, and more stable.
6. Comparison of chemical coagulation and EC reveals that the volume of sludge produced in the latter is lower than that in the former by 50% to 70%.
7. EC is more effective and faster for organic matter separation than conventional chemical coagulation. Moreover, EC produces palatable, clear, colorless, and odorless water from wastewater.

8. Total dissolved solids of the treated effluents produced from the EC process is less compared with those from other chemical processes.
9. Compared with the conventional chemical and biological techniques, EC is effective in removing the smallest colloidal particles as they have a greater probability of coagulation by the electric field that gives them motion.
10. In EC, the process of neutralizing excess chemicals and the possibility of secondary pollution because of the addition of a high concentration of chemicals during chemical coagulation cease to exist.
11. Evolution of gas bubbles during electrolysis can carry pollutants to the top of the solution where it can become easily concentrated, facilitating its collection and removal.
12. EC system is very simple and it can be easily established in rural areas even in places where electricity is not available, because a solar panel attached to the EC system may be sufficient to carry out the process.

2.4.2 Disadvantages

1. Regular replacement of sacrificial electrodes is necessary as they dissolve into solution because of oxidation.
2. In some places where electricity is expensive, EC is not possible. Further, sunny days are required to support solar panels for operating the EC system.
3. Cathode passivation (i.e., the formation of an oxide layer on the cathode because of oxidation/consumption of anode) will hinder EC.
4. EC may generate the residual electrode concentration in effluents, a main disadvantage of the system.
5. Because EC is a complex process and no set of configurations is applicable for all its requirements, many parameters need to be adjusted for optimal treatment.
6. Parameters such as electrode materials, electrode design, electrode gap, consistent or alternating polarity, current density, flow configuration, and retention time need to be optimized for a particular pollutant being treated. Moreover, the properties of the solution being treated, including pH, chemical concentrations, temperature, and particle size, also impact the efficiency of EC.
7. Excessive addition of supporting electrolytes will induce overconsumption of the Al electrode because of corrosion pitting, and Al dissolution may become irregular.
8. In some cases, the hydroxides and polyhydroxides (i.e., gelatinous hydroxide) may tend to solubilize.
9. Laboratory results obtained under controlled conditions cannot be extended to full scale (Anantha Singh and Ramesh 2013).

2.5 FUTURE RESEARCH WORK

The EC process has shown that is highly advantageous over traditional chemical coagulation. However, the application of EC in practice is still limited. This is mainly due to the instability of its performance and cost. In light of this, future research should focus on the following aspects:

1. Optimization of parameters to improve the efficiency of the process;
2. Applications of multiple anode–cathode EC systems should be studied as they could improve the efficiency of the process;
3. Combination of other electricity generation methods should be developed to prevent the cessation of EC because of power interruptions; and
4. Necessary development of new materials for anode and cathode to avoid the high concentration of metals in effluents.

2.6 SUMMARY

In this chapter, the fundamentals of the EC process are presented. Parameters such as anode and cathode materials, solution pH, conductivity, treatment time, supporting electrolyte, interelectrode distance, and alkalinity have a great impact on efficiency and effluent quality. A great demand for the application of EC in practice is seen; however, efforts toward the optimization of the process and materials are required to drive this application forward.

References

Anantha Singh, T. S., and S. T. Ramesh. 2013. "New trends in electrocoagulation for the removal of dyes from wastewater: A review." *Environ. Eng. Sci.* 30 (7): 333–349.

Barrera-Díaz, C. E., V. Lugo-Lugo, and B. Bilyeu. 2012. "A review of chemical, electrochemical and biological methods for aqueous Cr(VI) reduction." *J. Hazard. Mater.* 223–224 (Supplement C): 1–12.

Bazrafshan, E., H. Moein, F. Kord Mostafapour, and S. Nakhaie. 2013. "Application of electrocoagulation process for dairy wastewater treatment." *J. Chem.* 2013: 640139.

Bazrafshan, E., L. Mohammadi, A. Ansari-Moghaddam, and A. H. Mahvi. 2015. "Heavy metals removal from aqueous environments by electrocoagulation process—A systematic review." *J. Environ. Health Sci. Eng.* 13: 74.

Bukhari, A. A. 2008. "Investigation of the electro-coagulation treatment process for the removal of total suspended solids and turbidity from municipal wastewater." *Bioresour. Technol.* 99 (5): 914–921.

Butler, E., Y.-T. Hung, R. Y.-L. Yeh, and M. Suleiman Al Ahmad. 2011. "Electrocoagulation in wastewater treatment." *Water* 3 (2): 495–525.

Can, B. Z., R. Boncukcuoglu, A. E. Yilmaz, and B. A. Fil. 2014. "Effect of some operational parameters on the arsenic removal by electrocoagulation using iron electrodes." *J. Environ. Health Sci. Eng.* 12 (1): 95.

Cañizares, P., M. Carmona, J. Lobato, F. Martinez, and M. A. Rodrigo. 2005. "Electrodissolution of aluminum electrodes in electrocoagulation processes." *Ind. Eng. Chem. Res.* 44 (12): 4178–4185.

Chaturvedi, S. I. 2013. "Electro-coagulation: A novel wastewater treatment method." *Int. J. Mod. Eng. Res.* 3 (1): 93–100.

Chen, G. 2004. "Electrochemical technologies in wastewater treatment." *Sep. Purif. Technol.* 38 (1): 11–41.

da Mota, I. d. O., P. d. O. da Mota, J. G. de Oliveira Filho, and L. M. da Silva. 2016. "Removing heavy metals by electrocoagulation using stainless steel mesh electrodes: A study of wastewater from soil treated with metallurgical residues." *Cadernos UniFOA* 11 (31): 47–57.

Ghalwa, A., M. Nasser, and N. Farhat. 2015. "Removal of imidacloprid pesticide by electrocoagulation process using iron and aluminum electrodes." *J. Environ. Anal. Chem.* 2: 1–7.

Ghazouani, M., H. Akrout, and L. Bousselmi. 2015. "Efficiency of electrochemical denitrification using electrolysis cell containing BDD electrode." *Desalin. Water Treat.* 53 (4): 1107–1117.

Ghernaout, D., M. W. Naceur, and B. Ghernaout. 2011. "A review of electrocoagulation as a promising coagulation process for improved organic and inorganic matters removal by electrophoresis and electroflotation." *Desalin. Water Treat.* 28 (1–3): 287–320.

Hakizimana, J. N., B. Gourich, M. Chafi, Y. Stiriba, C. Vial, P. Drogui, et al. 2017. "Electrocoagulation process in water treatment: A review of electrocoagulation modeling approaches." *Desalination* 404 (Supplement C): 1–21.

Harif, T., M. Khai, and A. Adin. 2012. "Electrocoagulation versus chemical coagulation: Coagulation/flocculation mechanisms and resulting floc characteristics." *Water Res.* 46 (10): 3177–3188.

Holt, P. K., G. W. Barton, and C. A. Mitchell. 2005. "The future for electrocoagulation as a localised water treatment technology." *Chemosphere* 59 (3): 355–367.

Kabdaşlı, I., I. Arslan-Alaton, T. Ölmez-Hancı, and O. Tünay. 2012. "Electrocoagulation applications for industrial wastewaters: A critical review." *Environ. Technol. Rev.* 1 (1): 2–45.

Kabdaşlı, I., Z. Atalay, and O. Tünay. 2017. "Effect of solution composition on struvite crystallization." *J. Chem. Technol. Biotechnol.* 92 (12): 2921–2928.

Kabdaşlı, I., A. Keleş, T. Ölmez-Hancı, O. Tünay, and I. Arslan-Alaton. 2009. "Treatment of phthalic acid esters by electrocoagulation with stainless steel electrodes using dimethyl phthalate as a model compound." *J. Hazard. Mater.* 171 (1–3): 932–940.

Kamaraj, R., P. Ganesan, J. Lakshmi, and S. Vasudevan. 2013. "Removal of copper from water by electrocoagulation process—Effect of alternating current (AC) and direct current (DC)." *Environ. Sci. Pollut. Res.* 20 (1): 399–412.

Karichappan, T., S. Venkatachalam, and P. M. Jeganathan. 2014. "Optimization of electrocoagulation process to treat grey wastewater in batch mode using response surface methodology." *J. Environ. Health Sci. Eng.* 12 (1): 29.

Katal, R., and H. Pahlavanzadeh. 2011. "Influence of different combinations of aluminum and iron electrode on electrocoagulation efficiency: Application to the treatment of paper mill wastewater." *Desalination* 265 (1–3): 199–205.

Khaled, B., B. Wided, H. Béchir, E. Elimame, L. Mouna, and T. Zied. 2019. "Investigation of electrocoagulation reactor design parameters effect on the removal of cadmium from synthetic and phosphate industrial wastewater." *Arabian J. Chem.* 12 (8): 1848–1859.

Khandegar, V., and A. K. Saroha. 2013. "Electrocoagulation for the treatment of textile industry effluent—A review." *J. Environ. Manage.* 128 (Supplement C): 949–963.

Khemis, M., G. Tanguy, J. P. Leclerc, G. Valentin, and F. Lapicque. 2005. "Electrocoagulation for the treatment of oil suspensions: Relation between the rates of electrode reactions and the efficiency of waste removal." *Process Saf. Environ. Prot.* 83 (1): 50–57.

Kim, K.-J., K. Baek, S. Ji, Y. Cheong, G. Yim, and A. Jang. 2016. "Study on electrocoagulation parameters (current density, pH, and electrode distance) for removal of fluoride from groundwater." *Environ. Earth Sci.* 75 (1): 45.

Kobya, M., O. T. Can, and M. Bayramoglu. 2003. "Treatment of textile wastewaters by electrocoagulation using iron and aluminum electrodes." *J. Hazard. Mater.* 100 (1–3): 163–178.

Kuokkanen, V., T. Kuokkanen, J. Rämö, and U. Lassi. 2013. "Recent applications of electrocoagulation in treatment of water and wastewater—A review." *Green Sustainable Chem.* 3 (2): 89–121.

Liu, H., X. Zhao, and J. Qu. 2010. "Electrocoagulation in water treatment." In *Electrochemistry for the environment*, edited by C. Comninellis and G. Chen, 245–262. New York: Springer.

Mahmad, M. K. N., M. A. Z. M. R. Rozainy, I. Abustan, and N. Baharun. 2016. "Electrocoagulation process by using aluminium and stainless steel electrodes to treat total chromium, colour and turbidity." *Procedia Chem.* 19 (Supplement C): 681–686.

Mansour, S. E., and I. H. Hasieb. 2012. "Removal of Ni (II) and Co (II) mixtures from synthetic drinking water by electrocoagulation technique using alternating current." *Int. J. Chem. Technol.* 4 (2): 31–44.

Marriaga-Cabrales, N., and F. Machuca-Martínez. 2014. "Fundamentals of electrocoagulation." In *Evaluation of electrochemical reactors as a new way to environmental protection*, edited by M. A. Rodrigo-Rodrigo, C. A. Martínez-Huitle, and J. M. Peralta-Hernández, Research Signpost; Kerala, India. 1–16.

Matteson, J. M., R. L. Dobson, R. W. Glenn, N. S. Kukunoor III, W. H. Waits, and E. J. Clayfield. 1995. "Electrocoagulation and separation of aqueosus suspensions of ultrafine particles." *Colloids Surf. A* 104: 101–109.

Mollah, M. Y. A., P. Morkovsky, J. A. G. Gomes, M. Kesmez, J. Parga, and D. L. Cocke. 2004. "Fundamentals, present and future perspectives of electrocoagulation." *J. Hazard. Mater.* 114 (1–3): 199–210.

Moreno-Casillas, H. A., D. L. Cocke, J. A. G. Gomes, P. Morkovsky, J. R. Parga, and E. Peterson. 2007. "Electrocoagulation mechanism for COD removal." *Sep. Purif. Technol.* 56 (2): 204–211.

Moreno-Casillas, H. A., D. L. Cocke, J. A. G. Gomes, P. Morkovsky, J. R. Parga, E. Peterson, et al. 2009. "Electrochemical reactions for electrocoagulation using iron electrodes." *Ind. Eng. Chem. Res.* 48 (4): 2275–2282.

Murthy, Z. V. P., and S. Parmar. 2011. "Removal of strontium by electrocoagulation using stainless steel and aluminum electrodes." *Desalination* 282: 63–67.

Naje Ahmed, S., S. Chelliapan, Z. Zakaria, A. Ajeel Mohammed, and A. Alaba Peter. 2017. "A review of electrocoagulation technology for the treatment of textile wastewater." *Rev. Chem. Eng.* 33: 263.

Pickett, D. J. 1979. *Electrochemical reactor design*. New York: Elsevier.

Pretorius, W. A., W. G. Johannes, and G. G. Lempert. 1991. "Electrolytic iron flocculant production with a bipolar electrode in series arrangement." *Water SA* 17 (2): 133–138.

Sakarinen, E. 2016. "Humic acid removal by chemical coagulation, electrocoagulation and ultrafiltration." Degree thesis, Plastics Technology, Arcada University of Applied Sciences.

Santhosh, P., D. Revathi, and K. Saravanan. 2015. "Treatment of sullage wastewater by electrocoagulation using stainless steel electrodes." *Int. J. Chem. Sc.* 13 (3): 1173–1186.

Sincero, A. P., and G. A. Sincero. 2002. *Physical-chemical treatment of water and wastewater.* Boca Raton, FL: CRC Press.

Uhlig, H. H., and R. W. Revie. 2011. "Frontmatter." In *Uhlig's Corrosion Handbook*, edited by R. W. Revie. Hoboken, NJ: Wiley.

Valente, G. F. S., R. C. Santos Mendonça, J. A. M. Pereira, and L. B. Felix. 2012. "The efficiency of electrocoagulation in treating wastewater from a dairy industry, part I: Iron electrodes." *J. Environ. Sci. Health Part B* 47 (4): 355–361.

Vasudevan, S., J. Jayaraj, J. Lakshmi, and G. Sozhan. 2009a. "Removal of iron from drinking water by electrocoagulation: Adsorption and kinetics studies." *Korean J. Chem. Eng.* 26 (4): 1058–1064.

Vasudevan, S., B. S. Kannan, J. Lakshmi, S. Mohanraj, and G. Sozhan. 2011a. "Effects of alternating and direct current in electrocoagulation process on the removal of fluoride from water." *J. Chem. Technol. Biotechnol.* 86 (3): 428–436.

Vasudevan, S., and J. Lakshmi. 2012. "Process conditions and kinetics for the removal of copper from water by electrocoagulation." *Environ. Eng. Sci.* 29 (7): 563–572.

Vasudevan, S., J. Lakshmi, J. Jayaraj, and G. Sozhan. 2009b. "Remediation of phosphate-contaminated water by electrocoagulation with aluminium, aluminium alloy and mild steel anodes." *J. Hazard. Mater.* 164 (2–3): 1480–1486.

Vasudevan, S., J. Lakshmi, and G. Sozhan. 2011b. "Effects of alternating and direct current in electrocoagulation process on the removal of cadmium from water." *J. Hazard. Mater.* 192 (1): 26–34.

Vasudevan, S., J. Lakshmi, and G. Sozhan. 2011c. "Studies on the Al–Zn–In-alloy as anode material for the removal of chromium from drinking water in electrocoagulation process." *Desalination* 275 (1): 260–268.

Vasudevan, S., J. Lakshmi, and G. Sozhan. 2010. "Studies on the removal of arsenate by electrochemical coagulation using aluminum alloy anode." *CLEAN—Soil, Air Water* 38 (5-6): 506–515.

Vik, E. A., D. A. Carlson, A. S. Eikum, and E. T. Gjessing. 1984. "Electrocoagulation of potable water." *Water Res.* 18 (11): 1355–1360.

Yildiz, Y. Ş., A. S. Koparal, Ş. İrdemez, and B. Keskinler. 2007. "Electrocoagulation of synthetically prepared waters containing high concentration of NOM using iron cast electrodes." *J. Hazard. Mater.* 139 (2): 373–380.

Yilmaz, A. E., R. Boncukcuoğlu, and M. M. Kocakerim. 2007. "An empirical model for parameters affecting energy consumption in boron removal from boron-containing wastewaters by electrocoagulation." *J. Hazard. Mater.* 144 (1–2): 101–107.

Yilmaz, A. E., R. Boncukcuoğlu, M. M. Kocakerim, M. T. Yilmaz, and C. Paluluoğlu. 2008. "Boron removal from geothermal waters by electrocoagulation." *J. Hazard. Mater.* 153 (1–2): 146–151.

CHAPTER 3

Electro-Oxidation Process: Origins and Principles

Ali Khosravanipour Mostafazadeh,
M. R. Karimi Estahbanati, Patrick Drogui, R. D. Tyagi

3.1 INTRODUCTION

Applied electrochemistry has developed on the path of adopting and following cleaner and more environmentally friendly practices to minimize contamination in the environment by eliminating pollutants. The first collected information about environmental electrochemistry was released around four decades ago. Subsequently, numerous electrochemical methods have emerged by following the green aspects of these methods. By considering the electron as a clean reagent, electrochemical processes may be used as a treatment stage to remove pollutants, such as a pretreatment phase to enhance the biodegradability of contaminants in water/wastewater to adhere to environmental guidelines (Feng et al. 2016).

The anodic oxidation process has been used for a few decades for the treatment of color and certain organic pollutants such as phenol, cyanides, and aniline. For instance, for leachate treatment, electro-oxidation (EO) has yielded relatively satisfactory results. In this, EO processes are used as a pretreatment step to reduce the load of chemical oxygen demand (COD) and ammoniacal nitrogen or as a tertiary treatment for the degradation of recalcitrant organic pollutants (Dia et al. 2016). In general, two types of reactions are there in EO: direct and indirect EO reactions. In direct EO, electron exchange takes place at the anode surface without any contribution from other species. In indirect EO, organic matter is oxidized by the mediation of electroactive substances produced at the electrode surface, which act as mediators for electrons lying between the anode and the organic compounds (Martínez-Huitle and Panizza 2018).

The uncomplicatedness of EO, as exemplified by its simplicity of operation, minimal sludge generation, relatively low hydraulic retention time, and high efficiency in the degradation of macro- and microcontaminants make EO one of the best alternate treatments for nonbiodegradable wastewater such as oily wastewater, landfill leachate, textile dye, pulp and paper, and aquaculture saline water. EO can be applied to a small volume of wastewater with high electrical conductivity, wastewater with a relatively low COD (below 800 mg L^{-1}), or as a post-treatment of the municipal wastewater system with a low electricity intake (4.3 kW · h m^{-3}). Moreover, energy consumption can be reduced by adding some salts such as NaCl and Na_2SO_4. Mainly, anode materials are specifically designed to generate hydroxyl radicals ($OH^{\bullet}$) in preference to oxygen. The oxidation that takes place on the surface of the anode and in the boundary with $OH^{\bullet}$ is direct oxidation. By increasing the current intensity, electro reactions take place between ions and water with $OH^{\bullet}$ or on the surface of the electrode to produce mediator radicals such as hydroxidochlorine (HClO), Marshall's acid ($H_2S_2O_8$), hydrogen peroxide (H_2O_2), and ozone (O_3) (Zolfaghari et al. 2018).

Electrodisinfection is another aspect of EO based on in situ production of dominant reactive oxygen species that have a higher oxidation potential than chlorine derivatives. Free radicals are more than 100 times powerful than chlorine compounds and are extremely reactive and nonselective oxidants. Disinfection may happen through direct oxidation of microorganisms on the surface of the anode because of electron transfer and the generation of hydroxyl radicals [$E°(HO/H_2O) = 2.80$ Volt versus standard hydrogen electrode] or through indirect oxidation by the formation of different active species such as HClO, O_3, $H_2S_2O_8$, H_2O_2, and so on. As a result, the EO process leads to a minimal use of chemical reagents and the production of disinfection by-products (Naji et al. 2018).

This chapter focuses on the origins and principles of the EO process, which is known as one of the most powerful treatment methods in water and wastewater treatment processes. It also discusses the advantages and disadvantages of this process and some of the challenges and perspectives.

3.2 FUNDAMENTALS OF ELECTRO-OXIDATION FOR WATER AND WASTEWATER TREATMENT

One of the first studies on EO was performed by Gaddum in 1924. The mechanism of anodic processes occurring in the electrolysis of organic compounds and their relation to purely chemical oxidation processes was investigated. Gaddum found that anodic processes depend on several factors such as the nature of the organic compound under investigation, the nature of the electrolyte, or particularly, the nature of the anolyte, the nature of the anode (chemical and physical), and the electrode potential, representing the resultant effects of the following factors: temperature, the rate of diffusion of the depolarizer into the anode reaction region, the rate of diffusion of reaction products out of the reaction region, the velocity of discharge of anions, and the velocity of separation of anions in atomic or molecular form (Gaddum 1924). After Gaddum's findings, some researchers performed more studies on the electrolytic oxidation of *para*-toluic acid in alkaline solution in 1927 (Allmand and Puttick 1927). They investigated the oxygen evolution reaction and the effects of current density and overvoltage. Later, Kuwana and French (1964) studied EO of organic compounds into aqueous solutions using a carbon paste electrode. They examined the cyclic voltage scan and studied chronopotentiometry in their investigation. These studies and other research work signaled the start of novel research into EO.

Because the objective of electrochemical treatment is to degrade organic pollutants, it is better to use electrodes with a high oxygen overvoltage. Some of the most used anodes are the oxides of titanium and ruthenium coated with titanium, titanium-coated Sn–Pb–Ru (SPR) ternary electrodes, the boron-doped diamond (DDB), and graphite. The use of these high-oxygen overvoltage electrodes involves the generation of reactive oxygen species, such as the hydroxyl radical, responsible for the degradation of organic matter. These materials also present chemical stability in regard to acidic media and alkaline. The cathode is often made up of simple metallic elements such as stainless steel, copper, or titanium. The choice of the electrode can significantly influence the performance of the treatment. In using different types of anode electrodes, a decreased abatement efficiency of COD is observed in this order: SPR > DSA (oxides of Ru and Ti coated with Ti) > PbO_2/Ti > graphite. Satisfactory abatement rates obtained with the electrode (SPR) are linked to production intermediate oxidants such as HClO that form during indirect oxidation. The SPR shows more ability to generate intermediate oxidants than the other electrodes studied. Anodic reaction oxidizing chloride ions into chlorine gas is followed by its disproportionation in solution and leads to the formation of hypochlorous acid (HclO). However, chlorine can react with organic compounds and form organochlorine compounds, some of which are potentially carcinogenic. The concentration of chloride ions has a positive effect on the reduction of COD. In the presence of ion sulfates, these can also be oxidized to persulfuric acid ($H_2S_2O_8$) on a DDB anode. Persulfuric acid is a strong oxidant that can be used for the oxidation of organic matter in leachate. In short, organic compounds can be directly oxidized at the anode and indirectly oxidized to the solution with hypochlorous acid or persulfuric acid. Like all electrochemical processes, the applied intensity is a major factor in this process. The increase in current density increases the pollution control performance in terms of COD, ammoniacal nitrogen, and color. Corrosion of the graphite anode and the steel electrode has been observed in some studies. These harmful effects related to electrode corrosion highlight the importance of choosing electrodes that have both a high overvoltage in oxygen and good chemical stability. Processing time also influences the effectiveness of treatment. In general, its increase favors the degradation of pollutants. The modest increase in COD abatement in acidic environments is attributed to the low concentration of carbonate ions and bicarbonates found at low pH values. These alkaline species can quickly react with the hydroxyl radicals produced and, thus, limit the effective degradation of organic compounds (Dia et al. 2016).

3.2.1 Electrochemical Reactor Principle and Reaction Mechanism

An electroreactor is a vessel for reactions triggered by electric current to occur. Oxidation occurs at the anode, and reduction occurs at the cathode. The reactor type, configuration, and size for water and wastewater processing depend on the amount and characteristics of the influents and contaminants to be removed. The size of this apparatus differs according to the type of structure, ranging from house water pipelines with dimensions of several centimeters to built-up plants with areas of a few hundred square meters. Oxidation reactions that occur at the anodic part are as follows:

$$\mathrm{Red} \rightarrow \mathrm{Ox} + ze^{-} \tag{3-1}$$

and reduction reactions take place at the cathode like this:

$$\mathrm{Ox} + ze^{-} \rightarrow \mathrm{Red} \tag{3-2}$$

where z is the number of exchanged electrons, and Ox and Red are oxidized and reduced forms, in that order. The potential difference of the two half-cell reactions can be calculated by using the standard electrochemical potentials:

$$E_z = E_{\mathrm{Red}} - E_{\mathrm{ox}} \tag{3-3}$$

To determine whether a reaction is spontaneous, the potential difference and free enthalpy of reaction must be known. To know these, the following should be considered: If E_Z is negative, the reaction is a nonspontaneous electrolysis system such as EO. If E_Z is positive, the reaction is a spontaneous galvanic element such as fuel cells and batteries.

Electrolysis and fuel cell technologies are applied in wastewater systems.

EO reactions normally occur in a few steps (Figure 3-1), with the $OH^{\bullet}$ species at the heart of the EO process, which leads to several consecutive reactions at the

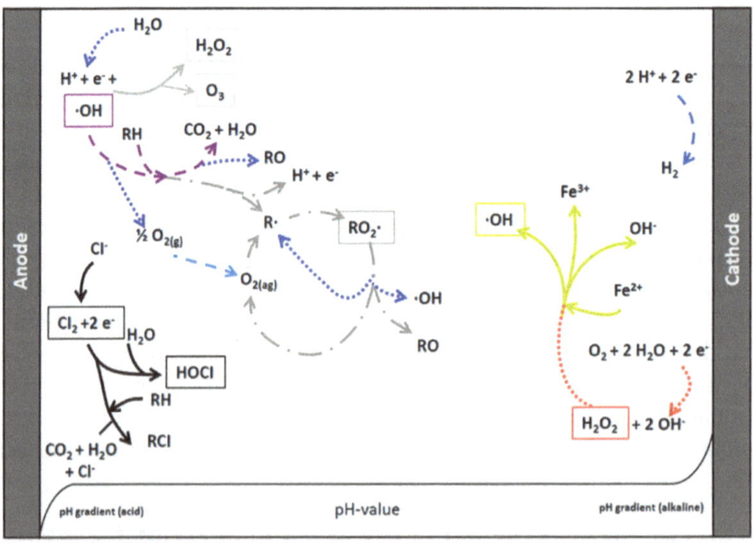

Figure 3-1. Reaction pathways of cathodic and anodic EO.

Source: Adapted from Muddemann et al. (2019) with permission from Wiley.

surface of the electrode and the bulk solution at different pH ranges. The details of direct and indirect EO are explained in Sections 3.3 and 3.4, respectively.

The controlling steps can be mass transfer, electron transfer, or chemical reaction. Barring reaction, if the other steps on the surface of the electrode do not limit the reaction rate, the current intensity will control the rate according to Faraday's law:

$$\frac{n}{t} = \frac{I}{zF} \tag{3-4}$$

Apart from current intensity, mass transfer can affect the reaction rate. If the current is higher than the restrictive value, parasitic reactions take place on account of electron exchange. In the case of water treatment, often oxygen and hydrogen are produced at the anode and cathode, respectively, which can be seen as an advantage in the electroflotation process (Muddemann et al. 2019).

Although the main electroreactions take place at the anode, some electrochemical cell reactions also occur at the cathode for wastewater treatment by performing the reduction reaction or by the formation of hydrogen. Moreover, in acidic and basic solutions, this cathodic H_2O_2 production can directly affect the decontamination process as follows:

$$O_{2(g)} + 2H^+ + 2e^- \rightarrow H_2O_2 \tag{3-5}$$

$$O_{2(g)} + H_2O + 2e^- \rightarrow HO_2^- + OH^- \tag{3-6}$$

Different carbon electrodes can be used as cathodes, such as a mixture of polytetrafluoroethylene (PTFE) with carbon, vitreous carbon, carbon felts, active carbon felt, carbon nanotubes, gauzes, and so on. A carbon-PTFE gas diffusion electrode can generate a higher quantity of H_2O_2 than the other electrode types (Martínez-Huitle and Panizza 2018).

3.2.2 Poisoning Effect

Oxidation at the surface of the anode is hypothetically feasible at low potentials (before oxygen evolution); however, under such circumstances, deactivation on the surface of the electrode occurs because a polymeric layer is formed. This phenomenon (poisoning effect) is governed by the adsorption feature of the surface of the electrode as well as the essence and concentrations of the organic matter in the solution. The anode can be protected by realizing oxidation in the potential range of water discharge because of the contribution of the intermediate species of oxygen evolution (by the following reactions):

$$M + H_2O \rightarrow M(^{\bullet}OH) + H^+ + e^- \tag{3-7}$$

$$M(^{\bullet}OH) + R \rightarrow M + CO_2 + H_2O \tag{3-8}$$

where M is the anode surface, and R is the organic matter. The process efficiency depends on the operating conditions, especially the material of the electrode.

Some anodes having a low oxygen evolution overpotential (IrO_2, RuO_2, or Pt) are called *active electrodes* that tend to realize a partial and selective oxidation of contaminants. Electrodes with a high oxygen evolution overpotential (SnO_2, PbO_2, or BDD) are called *nonactive anodes*, and they are perfect electrodes for combustion and mineralization (complete conversion of organic species into H_2O and CO_2) in wastewater treatment processes. Particularly, BDD is an ideal anode because of its high stability, long life, and broad potential range for water discharge (Martínez-Huitle and Panizza 2018).

3.2.3 By-Products

Although EO is an alternative for on-site wastewater treatment, concerns remain on toxic chemical generation in high concentrations of chloride and organic species. Thus, studies on minimizing toxic by-product formation need to be considered. Although chloride improves the efficiency of this kind of process by the production of reactive chlorine species such as hypochlorous acid, chloramines, and chlorine radicals, oxidation of chloride can result in the formation of some hazardous by-products such as chlorate and perchlorate. Halogenated organic compounds can also be formed by the reaction of chlorines with wastewater containing organic compounds such as carbohydrates, amino acids, and proteins. Some indicators such as trihalomethanes, haloacetic acids, and adsorbable organic chlorine are usually monitored. Therefore, considering the formation of such toxic compounds, more studies on the prevention of harmful by-product production are required (Jasper et al. 2017).

3.3 DIRECT ANODIC OXIDATION

Direct anodic oxidation (or a direct exchange of electrons with the anode) is a method in which the pollutant adsorbs directly on the anode surface without the involvement of any other substance (Feng et al. 2016, Panizza and Cerisola 2009). In this method, the electron is only exchanged between the anode and the pollutant, which causes oxidation of the pollutant (Panizza and Cerisola 2009):

$$R \rightarrow P + e^- \tag{3-9}$$

where R is the adsorbed pollutant, and P is the oxidized pollutant. Direct EO is possible at a very low potential (e.g., less than water discharge potential) and before the oxygen evolution reaction takes place (Feng et al. 2016, Mandal et al. 2017). The reaction rate is usually low but may vary depending on the electrocatalytic activity of different anodes (Mandal et al. 2017, Panizza and Cerisola 2009). Some metallic anodes such as Pt and Pd and metallic oxides such as IrO_2, Ir-TiO_2, RuO_2, and Ru-TiO_2 show high electrocatalytic activity, accelerating the transfer of electrons between the electrode and the pollutant (Mandal et al. 2017, Panizza and Cerisola 2009). In addition to the electrocatalytic activity of the anode, the efficiency of the EO process depends on the pollutant mass transfer rate in water.

The higher the rate of mass transfer in water, the faster the pollutant reaches the electrode surface, and the more likely it is adsorbed and oxidized on the anode surface (Mandal et al. 2017).

The most important advantage of this process is that no additional material is used to oxidize the pollutant, and, therefore, after the reaction takes place, the need to remove excess material does not arise (Panizza and Cerisola 2009). In contrast, the main disadvantage of EO is the poisoning effect, during which the pollutant oxidizes and forms a polymer film on the anode surface, which reduces the catalytic activity of the electrode (Mandal et al. 2017, Panizza and Cerisola 2009). This phenomenon leads to electrode fouling and deactivation, depending on the adsorption properties of the anode surface as well as concentration and the nature of pollutants (Feng et al. 2016, Mandal et al. 2017, Panizza and Cerisola 2009). In the case of using electrodes with weak adsorption properties and inert surfaces (like BDD), the reduction of electrocatalytic activity of the anode is less observed. In contrast, if the concentration of pollutants, especially aromatic substances, is high, electrode fouling is more obvious (Mandal et al. 2017, Panizza and Cerisola 2009). The poisoning effect can be prevented by performing the oxidation process in the potential region of water discharge with the simultaneous oxygen evolution reaction or by indirect oxidation using the production of redox reagents to oxidize the polymer film (Panizza and Cerisola 2009).

Rahmani et al. (2015) investigated the electrochemical oxidation of activated sludge using direct anodic oxidation on graphite and a stainless steel surface. They studied the effect of current density, initial pH, and position of electrodes on COD removal from wastewater. Their results showed that the efficiency of COD removal using this method was 66%. They also reported that the efficiency of removal of COD using the graphite electrode was higher than that using stainless steel. However, if graphite is used at high current densities, the electrode will lose its mechanical strength, leading to increased turbidity of the oxidized activated sludge.

3.4 INDIRECT ELECTROCHEMICAL OXIDATION

One of the solutions to prevent anode deactivation is to use indirect anode oxidation (Feng et al. 2016, Martinez-Huitle and Ferro 2006, Martínez-Huitle and Panizza 2018). Indirect or mediated oxidation means that the pollutant does not exchange electrons directly with the electrode but is oxidized on the electrode surface using electroactive species that mediate the exchange of electrons between the electrode and the pollutant (Martinez-Huitle et al. 2015, Panizza and Cerisola 2009). The advantage of this method over the direct oxidation method is that by-products are not usually produced during the oxidation process (Feng et al. 2016).

Indirect anode oxidation takes place in the potential region where the oxygen evolution reaction occurs because of water oxidation (Mandal et al.

2017). In indirect anodic oxidation, water discharges on the anode surface and generates *active oxygen* that is physically adsorbed (adsorbed hydroxyl radical) or chemisorbed (oxygen in the oxide lattice) (Feng et al. 2016, Johnson et al. 1999, Martinez-Huitle and Ferro 2006, Martínez-Huitle and Panizza 2018). The physical adsorption or chemisorption of active oxygen depends on the type of electrode (Comninellis 1994, Ghanim and Hamza 2018). Equation (3-10) shows the generation of hydroxyl radical on the surface of the electrode by water discharge:

$$M + H_2O \rightarrow M(OH^{\bullet})_{ads} + H^+ + e^- \tag{3-10}$$

where M represents the surface of the electrode. Along with the reaction of hydroxyl radical production, one of the adverse reactions is the oxygen evolution reaction, which inevitably takes place (Mandal et al. 2017, Panizza and Cerisola 2009) as follows:

$$M(OH^{\bullet})_{ads} + H_2O \rightarrow M + O_2 + 3H^+ + 3e^- \tag{3-11}$$

Studies on anodic oxidation showed that electrode material greatly affects the selectivity and efficiency of the process (Comninellis 1994, Comninellis and De Battisti 1996, Panizza and Cerisola 2009, Simond et al. 1997). Electrodes can be divided into two: active and inactive. Active electrodes with a low oxygen evolution overpotential (e.g., IrO_2, RuO_2, and Pt) tend to oxidize pollutants partially and selectively (Comninellis 1994, Drogui et al. 2007, Mandal et al. 2017). The mechanism of EO of pollutants by an active electrode is presented in the following equations:

$$MO_x + H_2O \rightarrow MO_x(OH^{\bullet}) + H^+ + e^- \tag{3-12}$$

$$MO_x(OH^{\bullet}) \rightarrow MO_{x+1} + H^+ + e^- \tag{3-13}$$

$$MO_{x+1} + R \rightarrow MO_x + RO \tag{3-14}$$

where MO_x, R, and x represent the active sites of the anode, pollutant, and stoichiometric coefficient, respectively (Ghanim and Hamza 2018). Accordingly, the pollutant is oxidized by the active surface oxygen that is produced from water.

Inactive electrodes with a high oxygen evolution overpotential (e.g., SnO_2, PbO_2, and BDD) tend to oxidize pollutants completely and convert them to CO_2, which is ideal for the complete EO of pollutants (Comninellis 1994, Drogui et al. 2007, Feng et al. 2016, Grimm et al. 1998, Mandal et al. 2017, Martinez-Huitle and Ferro 2006, Martínez-Huitle and Panizza 2018). Inactive electrodes do not have active sites to adsorb contaminants (Fernandes et al. 2015, Mandal et al. 2017, Panizza and Cerisola 2009). The mechanism of EO of pollutants by an inactive electrode is presented in the following equation:

$$MO_x(OH^{\bullet}) + R \rightarrow MO_x + mCO_2 + nH_2O + H^+ + e^- \tag{3-15}$$

where *m* and *n* are stoichiometric coefficients (Ghanim and Hamza 2018). As can be seen, the pollutant is oxidized indirectly by the hydroxyl radicals that are on the surface of the electrode. Anodic oxidation of pollutants with an electrogenerated hydroxyl radical is relatively rapid and highly temperature-dependent, which leads to an increase in the mass transfer coefficient and a decrease in the activation energy of the oxidation reaction (Aquino et al. 2011, Ghanim and Hamza 2018, Körbahti and Artut 2010).

Oxidation mediators include metal redox pairs such as Ag(II), Ce(IV), Co(III), Fe(III), and Mn(III) as well as strong oxidizing chemicals such as active chlorine, ozone, hydrogen peroxide, persulfate, percarbonate, and perphosphate (Feng et al. 2016, Panizza and Cerisola 2009). Active chlorine species are among the most traditional and most widely used oxidizers in the anodic oxidation process for wastewater treatment. Active chlorine can be in different forms, including gaseous chlorine, hypochlorous acid or hypochlorite ions, and are produced from chlorides that are naturally present in the solution or added into it (Martinez-Huitle and Ferro 2006, Martínez-Huitle and Panizza 2018). The mechanism of hypochlorite generation from chlorine ions is presented in the following equations:

$$2Cl^- \rightarrow Cl_2 + 2e^- \tag{3-16}$$

$$Cl_2 + H_2O \rightarrow HOCl + H^+ + Cl^- \tag{3-17}$$

$$HOCl \rightarrow H^+ + OCl^- \tag{3-18}$$

Accordingly, after the generation of chlorine gas, it hydrolases to hypochlorous acid and then converts to hypochlorite ions. Chlorine intermediates are used to remove various pollutants, but they are particularly suitable for real wastewater treatment with high concentrations of sodium chloride (higher than 5 g dm^{-3}), for example, olive oil, textile, and tannery wastewater (Bonfatti et al. 2000, Comninellis and Nerini 1995, Feng et al. 2016, Martinez-Huitle and Ferro 2006, Martínez-Huitle and Panizza 2018, Tavares et al. 2012).

One of the disadvantages of the indirect oxidation process is the side reaction of oxygen evolution, which reduces the current efficiency because of current consumption, and, thereby, reduces the efficiency of pollutant degradation (Mandal et al. 2017). Another disadvantage with indirect anodic oxidation is that after the process, the produced metal redox pairs or secondary contaminants need to be removed (Panizza and Cerisola 2009).

To achieve high efficiency in indirect anodic oxidation, the following conditions must be met:

- Potential range at which the intermediates are produced should be close to the potential of the oxygen evolution reaction.
- Intermediate production rate should be high.
- Rate of reaction between the intermediate and the contaminant should be higher than that of other competing reactions.
- Adsorption of pollutants on the anode surface should be minimal (Panizza and Cerisola 2009).

Despite successful laboratory and pilot tests, the industrial application of this method is still limited because of the relatively high energy consumption compared with that of other electrochemical methods. Energy consumption can be reduced by modifying the process and performing advanced oxidation, in which anode and cathode are involved in the production of highly oxidizing hydroxyl radicals (Panizza and Cerisola 2009).

COD removal was investigated by indirect anodic oxidation using a Pb/PbO_2 electrode, and a removal rate of 77.5% was achieved, which was higher than that of direct anode oxidation (Rahmani et al. 2015). The real wastewater treatment containing naphthalene using the direct and indirect methods was compared (Panizza et al. 2000). The results showed that direct oxidation was not an effective method because it removed only about 40% of COD from wastewater. However, converting the process into indirect oxidation by the addition of sodium chloride, which causes the production of hypochlorite ions, led to the complete elimination of COD and color.

3.5 CHALLENGES AND FUTURE RESEARCH WORK

One of the drawbacks of conventional treatments is the high elimination of organic chemicals. Advanced EO processes are being used today, but they are still evolving. With these new processes, favorable options for wastewater treatment have emerged, and the use of some of these options has led to efficient organic removal. The application of this removal process on real wastewater treatment has demonstrated an efficient removal of COD and dissolved organic carbon as well as abatement of persistent organic contaminants. Such removal technology holds great promise and potential. This technology is highly useful for decentralized water treatment as well. In this regard, more research is needed on anodic electrode materials to reduce electrode cost. Moreover, controlling the nondesired by-products such as organohalides (Garcia-Segura et al. 2018) needs further studies.

One of the perspectives of the EO process in wastewater treatment is implementing the pilot- and field-scale applications. In comparison with conventional processes, some of the drawbacks of this technology are the potential generation of recalcitrant by-products in some cases, high electricity consumption, costly and high O_2 overpotential electrodes, and a hydrodynamic reactor design. The flow dynamics of the electrolyte can influence the chemical reaction on the surfaces of the electrodes. To do the hydrodynamic analysis of an EO reactor, geometric domain, preprocessing conditions, setup, and postprocessing steps, are required to be considered. Moreover, a numerical calculation during the design of an EO reactor using the software is sometimes required.

The advantages and future research opportunities of EO are as follows: This process can be selective (e.g., focusing on specific targets such as microcontaminants, pathogens, and industrial or domestic wastewaters), flexible for different energy sources or their combinations (e.g., solar and wind energy sources), and versatile for the valorization of different kinds of waste using active anodes. Thus, EO technology holds great promise. Also, it has been shown that it is theoretically feasible to integrate conventional systems with EO (Martínez-Huitle and Panizza 2018).

The challenges involved in the scaling-up of electrochemical advanced oxidation processes (EAOPs) are energy intake and mass transfer restrictions. Electricity consumption is proportionate to the cell voltage, current intensity, and treatment time. Thermodynamic potential, overpotential, and ohmic resistance are the main terms related to applied voltage. In this case, ohmic resistance signifies an efficiency loss and must be reduced as much as possible. Electrical resistance (R_{ohm}) itself is related to the electrolyte type because its conductivity is much lower than that of the other constituents of the electrochemical system.

The conductivity of influents can be improved by adding some salts, although this may increase the operating costs and contaminate the wastewater with inorganic ions. Another alternative is reducing the interelectrode distance even to the order of micrometers. Nevertheless, this configuration may be disadvantaged by significant pressure drops because of the narrow gap between electrodes and the clogging of the network. Mass transfer phenomenon is the main feature in an electroreactor design, which means the charge transport between the electrode and the chemicals in wastewater (heterogeneous process). Hence, the reaction rate of EAOPs can be determined by the rate of the electroactive species reaching the electrode. In some cases, the integration of EO with other electrochemical methods improves the mass transfer between electrodes and species in the solution, such as the preconcentration of the contaminant in a combined electrodialysis/EO reactor for soluble compounds or in an electrocoagulation/EO reactor for suspended solids. Mass transport may similarly be improved by increasing the mass transport coefficient and electrode surface by changing the hydrodynamic conditions in the reactor (Pérez et al. 2017).

EO can be sustained by solar energy and does not need external makeup water, as treated water can be reused for flushing (Jasper et al. 2017). The sustainability of EO can be primarily ensured by powering the system using renewable energies, which can lead to the application of this technology in far-flung areas in decentralized wastewater treatment facilities/plants (Ganiyu et al. 2020). Moreover, EO can be used as a powerful technique to treat wastewater containing microplastics as emerging contaminants (Kiendrebeogo et al. 2021).

3.6 SUMMARY

EO is one of the most effective methods for wastewater treatment and can degrade even nonbiodegradable compounds. Although the scaling-up of this technology is challenging because of high electricity consumption, the efficiency of this process for wastewater treatment application is promising. The literature shows that most nonactive electrodes such as BDD display the best performance and various types of pollutants can be removed by this method. By direct and indirect actions of EO, this treatment method is substantially effective for COD reduction, decontamination, and even disinfection of water, although the formation of toxic by-products needs to be monitored. In the future, the optimal conditions of the process, such as current density, electrolysis time, pH of the medium, and the ionic composition, need to be studied for achieving a better design of the EO system.

References

Allmand, A. J., and A. Puttick. 1927. "The electrolytic oxidation of *para*-toluic acid in alkaline solution." *Trans. Faraday Soc.* 23: 641–650.

Aquino, J. M., G. F. Pereira, R. C. Rocha-Filho, N. Bocchi, and S. R. Biaggio. 2011. "Electrochemical degradation of a real textile effluent using boron-doped diamond or β-PbO_2 as anode." *J. Hazard. Mater.* 192 (3): 1275–1282.

Bonfatti, F., S. Ferro, F. Lavezzo, M. Malacarne, G. Lodi, and A. De Battisti. 2000. "Electrochemical incineration of glucose as a model organic substrate. II. Role of active chlorine mediation." *J. Electrochem. Soc.* 147 (2): 592.

Comninellis, C. 1994. "Electrocatalysis in the electrochemical conversion/combustion of organic pollutants for waste water treatment." *Electrochim. Acta* 39 (11–12): 1857–1862.

Comninellis, C., and A. De Battisti. 1996. "Electrocatalysis in anodic oxidation of organics with simultaneous oxygen evolution." *J. Chim. Phys.* 93: 673–679.

Comninellis, C., and A. Nerini. 1995. "Anodic oxidation of phenol in the presence of NaCl for wastewater treatment." *J. Appl. Electrochem.* 25 (1): 23–28.

Dia, O., P. Drogui, R. Dubé, and G. Buelna. 2016. "Utilisation des procédés électrochimiques et leurs combinaisons avec les procédés biologiques pour le traitement des lixiviats de sites d'enfouissement sanitaires—revue de littérature." *Rev. Sci. Eau* 29 (1): 63–89.

Drogui, P., J.-F. Blais, and G. Mercier. 2007. "Review of electrochemical technologies for environmental applications." *Recent Pat. Eng.* 1 (3): 257–272.

Feng, Y., L. Yang, J. Liu, and B. E. Logan. 2016. "Electrochemical technologies for wastewater treatment and resource reclamation." *Environ. Sci. Water Res. Technol.* 2 (5): 800–831.

Fernandes, A., M. Pacheco, L. Ciríaco, and A. Lopes. 2015. "Review on the electrochemical processes for the treatment of sanitary landfill leachates: Present and future." *Appl. Catal., B* 176–177: 183–200.

Gaddum, L. W. 1924. "A preliminary study of the anodic phenomena occurring in organic electrolysis." Ph.D. thesis, Department of Engineering, Graduate school, University of Missouri.

Ganiyu, S. O., C. A. Martínez-Huitle, and M. A. Rodrigo. 2020. "Renewable energies driven electrochemical wastewater/soil decontamination technologies: A critical review of fundamental concepts and applications." *Appl. Catal. B* 270: 118857.

Garcia-Segura, S., J. D. Ocon, and M. N. Chong. 2018. "Electrochemical oxidation remediation of real wastewater effluents—A review." *Process Saf. Environ. Prot.* 113: 48–67.

Ghanim, A. N., and A. S. Hamza. 2018. "Evaluation of direct anodic oxidation process for the treatment of petroleum refinery wastewater." *J. Environ. Eng.* 144 (7): 04018047.

Grimm, J., D. Bessarabov, and R. Sanderson. 1998. "Review of electro-assisted methods for water purification." *Desalination* 115 (3): 285–294.

Jasper, J. T., Y. Yang, and M. R. Hoffmann. 2017. "Toxic byproduct formation during electrochemical treatment of latrine wastewater." *Environ. Sci. Technol.* 51 (12): 7111–7119.

Johnson, S. K., L. L. Houk, J. Feng, R. Houk, and D. C. Johnson. 1999. "Electrochemical incineration of 4-chlorophenol and the identification of products and intermediates by mass spectrometry." *Environ. Sci. Technol.* 33 (15): 2638–2644.

Kiendrebeogo, M., M. R. Karimi Estahbanati, A. Khosravanipour Mostafazadeh, P. Drogui, and R. D. Tyagi. 2021. "Treatment of microplastics in water by anodic oxidation: A case study for polystyrene." *Environ. Pollut.* 269: 116168.

Körbahti, B. K., and K. Artut. 2010. "Electrochemical oil/water demulsification and purification of bilge water using Pt/Ir electrodes." *Desalination* 258 (1–3): 219–228.

Kuwana, T., and W. G. French. 1964. "Electrooxidation or reduction of organic compounds into aqueous solutions using carbon paste electrode." *Anal. Chem.* 36 (1): 241–242.

Mandal, P., B. K. Dubey, and A. K. Gupta. 2017. "Review on landfill leachate treatment by electrochemical oxidation: Drawbacks, challenges and future scope." *Waste Manage.* 69: 250–273.

Martinez-Huitle, C. A., and S. Ferro. 2006. "Electrochemical oxidation of organic pollutants for the wastewater treatment: Direct and indirect processes." *Chem. Soc. Rev.* 35 (12): 1324–1340.

Martínez-Huitle, C. A., and M. Panizza. 2018. "Electrochemical oxidation of organic pollutants for wastewater treatment." *Curr. Opin. Electrochem.* 11: 62–71.

Martinez-Huitle, C. A., M. A. Rodrigo, I. Sires, and O. Scialdone. 2015. "Single and coupled electrochemical processes and reactors for the abatement of organic water pollutants: A critical review." *Chem. Rev.* 115 (24): 13362–13407.

Muddemann, T., D. Haupt, M. Sievers, and U. Kunz. 2019. "Electrochemical reactors for wastewater treatment." *ChemBioEng Rev.* 6 (5): 142–156.

Naji, T., A. Dirany, A. Carabin, and P. Drogui. 2018. "Large-scale disinfection of real swimming pool water by electro-oxidation." *Environ. Chem. Lett.* 16 (2): 545–551.

Panizza, M., C. Bocca, and G. Cerisola. 2000. "Electrochemical treatment of wastewater containing polyaromatic organic pollutants." *Water Res.* 34 (9): 2601–2605.

Panizza, M., and G. Cerisola. 2009. "Direct and mediated anodic oxidation of organic pollutants." *Chem. Rev.* 109 (12): 6541–6569.

Pérez, J. F., J. Llanos, C. Sáez, C. López, P. Cañizares, and M. A. Rodrigo. 2017. "A microfluidic flow-through electrochemical reactor for wastewater treatment: A proof-of-concept." *Electrochem. Commun.* 82: 85–88.

Rahmani, A. R., K. Godini, D. Nematollahi, and G. Azarian. 2015. "Electrochemical oxidation of activated sludge by using direct and indirect anodic oxidation." *Desalin. Water Treat.* 56 (8): 2234–2245.

Simond, O., V. Schaller, and C. Comninellis. 1997. "Theoretical model for the anodic oxidation of organics on metal oxide electrodes." *Electrochim. Acta* 42 (13–14): 2009–2012.

Tavares, M. G., L. V. da Silva, A. M. S. Solano, J. Tonholo, C. A. Martínez-Huitle, and C. L. Zanta. 2012. "Electrochemical oxidation of methyl red using $Ti/Ru_{0.3}Ti_{0.7}O_2$ and Ti/Pt anodes." *Chem. Eng. J.* 204–206: 141–150.

Zolfaghari, M., P. Drogui, and J. F. Blais. 2018. "Removal of macro-pollutants in oily wastewater obtained from soil remediation plant using electro-oxidation process." *Environ. Sci. Pollut. Res.* 25 (8): 7748–7757.

CHAPTER 4

Mathematical Modeling of Electro-Coagulation Process

S. K. Ram, H. Panidepu, C. Vasavi, P. Drogui, R. D. Tyagi

4.1 INTRODUCTION

To date, a staggering 33% of the world's population still struggles to find pure potable water to use. A major part of this population is drawn from developing and underdeveloped third-world countries. That said, the unavailability of clean drinking water is not just a third-world problem, but, in fact, even developed countries such as the United States and Canada consume polluted water. The reasons for water pollution can be numerous, ranging from accidental spills, anthropological, geological, or mere negligence at various levels of water use, including single households, policymakers, or demotivated capitalists.

Traditionally, polluted water or freshwater obtained from aquifers and other water bodies undergoes a series of physical, chemical, and biological treatments to be certified as purified water for drinking or general use ultimately. Conventional wastewater treatment methods involve primary treatment, secondary biological treatment, tertiary treatment, and, if required, quaternary treatment before discharge or being used as drinking water supply. Wastewater treatment facilities are usually designed and constructed to meet certain treatment needs. However, with increasing population and the dynamically changing personal lifestyle fostered by rapid economic industrialization and modernization of the global society, the existing wastewater treatment facilities may be outdated and poorly functioning. Technological advancements in cosmetics, pharmaceuticals, and many personal care products have led to the release of many emerging contaminants into water eco systems, but these emerging contaminants cannot be treated completely with the existing treatment facilities. The urgent requirement for new water treatment methods has resulted in the introduction of advanced wastewater (or simply water) treatment technologies such as electro-oxidation (EO), electro-coagulation (EC), Fenton oxidation, UV treatment, electroreduction, and ozonation (Chitra and Balasubramanian 2010).

Among different electrochemical techniques, EO and EC are often used for water and wastewater treatment. EC is a technique that utilizes electric current for in situ production of coagulants using sacrificial metal electrodes to adsorb (physical or chemical interactions) pollutant particles, precipitate them out of the solution, and make it easier to separate the remaining particles (often in small flocs after EC treatment) from the bulk liquid by simple filtration or any other solid–liquid separation method. Different reactions take place at anode and cathode in association with the parallel electrocatalytic dissociation of water. Many researchers (Butler et al. 2011, Hakizimana et al. 2017, Holt et al. 2005, Kuokkanen et al. 2013, Verma et al. 2012) have detailed and reviewed the various underlying principles and phenomena of the EC process, which has furthered our understanding of EC.

Within the ambit of engineering and technological applications of EC, it is essential to understand various aspects of the process and their interactions with one another. An understanding of EC from the engineering point of view is required to tune in and optimize the process to get the best-intended outcome from it. In a wastewater treatment scenario, in general, the pollutant levels vary in type and concentration from one day to another; thus, it is essential for an operator to understand the required process adjustments to still achieve the treatment requirements, while sticking to the regulatory norms. To this end, thanks to advanced computational techniques, researchers have been able to model, characterize, and simulate the EC process to unravel the possible ways to optimize and maximize the treatment performance of the process. Different modeling techniques have evolved recently, which can simulate and predict the outcome of EC accurately. This chapter elaborates on the different modeling techniques applied to EC processes using various computational tools. Such types of predictive tools are the need of the hour to rapidly optimize the process parameters and to save a huge amount of time, money, and effort to develop a fully functional industrial-scale process.

4.2 CRITICAL FACTORS TO BE CONSIDERED IN ELECTRO-COAGULATION MODELING

Various important factors affect EC. Various studies present in the literature mainly discuss a subset of these parameters based on the available solicitation from the existing literature. According to the literature (Mollah et al. 2001, 2004), the major factors considered in all modeling studies are the electrode type, the initial pH of the feed solution, the distance between electrodes, temperature, electrolysis time, solution conductivity by supplemented electrolytes, current density, and the initial concentration of the pollutant. The importance of these individual parameters has been discussed in previous chapters and can also be found elsewhere (Butler et al. 2011, Mollah et al. 2001, Verma et al. 2012).

Recently, researchers (Amani-Ghadim et al. 2013, Lakshmanan et al. 2010) attempted to understand the sensitivity of some of the aforementioned factors of EC using the response surface methodology (RSM) techniques. The study focused on the removal of Reactive Red 43 (RR43) dye (used in the textile industry) from synthetic wastewater. It considered four important factors (current density, time, pH, and electrolyte concentration) and devised 31 experiments using a central composite design (CCD). The experimental design resulted in a second-degree (quadratic) polynomial response surface model that could fit the experimental data. The regression equation processed the factor input values to give removal efficiency (RE). The study modeled the quadratic response equation for iron and aluminum electrodes in terms of their individual effects and their mutual interactions. The significance and adequacy analysis was performed by using analysis of variance (ANOVA). The model was validated for its importance using the *F*-test (Fisher's variance ratio test). The study aimed to evaluate the significance of each factor by calculating the significance percentage by Pareto analysis. In Pareto analysis, a percentage effect of each term (P_i) on the response (for instance, the percentage of COD removal) is calculated by using the following equation:

$$P_i = \left(\frac{b_i^2}{\sum b_i^2} \right) \times 100 \tag{4-1}$$

where b_i is regression coefficients in the model equation for the *i*th term. The results indicated that in the EC–Fe process (EC using an iron electrode), current density was the most important parameter and its contribution to the percentage of dye removal was 42.55%, followed by electrolysis time, whose contribution was estimated to be 19.68%. Contrastingly, for the EC–Al process, the pH of the solution was observed to be the most important factor (with a contribution of dye removal estimated to be 68.92%). Among the interaction terms, the pH–time interaction term was very significant. This signified that the EC–Al process is more sensitive to pH, whereas the EC–Fe process was more sensitive to current density. The results of this particular study revealed that the final treatment outcome is dependent on the type of electrode for any given treatment process. Therefore, it is important to perform a similar sensitivity analysis (Pareto analysis) to know which factor should be optimized more precisely.

Many studies available in the literature usually focus on optimizing various factors suggested by the prevailing literature and intuitive experiences such as the time of electrolysis, the pH of feed, and pollutant concentration. Therefore, various tools are available to perform parameter optimization for an electrochemical process, and they have been extensively applied to the EC process.

Aber et al. (2009) performed a similar analysis to identify the weight or relative significance of the independent variables on the final outcome (treatment efficiency) using an artificial neural network (ANN). As anticipated, the results indicated that parameter pH, initial pollutant concentration, voltage, and time of electrolysis were all significant determining factors of an EC process and, thus, could not be overlooked. The literature contains plenty of studies focusing on revealing the most important parameters or interacting parameter couples for different pollutants and wastewater matrices using different modeling techniques (Amani-Ghadim et al. 2013, Kobya et al. 2010, Secula et al. 2014, Thirugnanasambandham et al. 2014a).

4.3 DIFFERENT MODELING TECHNIQUES AVAILABLE FOR ELECTRO-COAGULATION

A mathematical model is a set of equations systematically connected to one another to functionally connect the multiple independent variables of a process to the final outcome (objective function, the final goal expressed in the mathematical equation of a dependent variable) of the process. In most of the EC studies found in the literature so far, the objective function is the removal of one or more than one specific pollutant. A major chunk of these studies is focused on the elimination of colored dyes (heavy metal compounds) from textile industrial effluents (Aber et al. 2009, Alinsafi et al. 2005, Amani-Ghadim et al. 2013, Dalvand et al. 2011, Mirsoleimani-Azizi et al. 2015).

Different modeling techniques that have been applied to EC utilize advanced computational skills and some software tools. The modeling techniques used include ANN, response surface methodology (RSM), adsorption-based models, and mathematical kinetic equation–based models. These techniques are very distinct in terms of their core concept to model the process, but each one of them has been applied successfully to predict the outcomes of the EC process. The models developed for EC can accurately predict the treatment performance for a wide range of operational parameters. These models can be successfully used to optimize the operational parameters. Using these models, researchers have figured out the relative significance of each and every parameter and their sensitivity toward the outcome of the process. Furthermore, a well-defined model can help in multiobjective optimization to reduce energy consumption and can be easily integrated into techno-economic analysis to find the most feasible form of the process. These robust models can not only speed up the process development but also help in reactor design, mechanistic principles' study, and scale-up.

The most prominent modeling techniques applied for simulating electrochemical processes are RSM and ANN. The RSM technique is largely based on a set of experiments designed by a software with a given predefined range of values for each individual variable. Despite being the most straightforward modeling technique practiced by researchers so far, RSM has certain drawbacks that need to be addressed. The primary idea behind applying the statistical optimization technique is to ease the optimization method by moving toward the optimized solution with a minimum number of experiments. The central composite design (CCD) method is often used in RSM to perform the experimental design. The number of experiments in a CCD experimental design is still very high because of different combinations of parameters and levels of their values. For ease of use, researchers have limited themselves to three levels of parameter values, and, in general, CCD experimental design is performed for 4 to 5 parameters simultaneously. In real-life situations, the number of parameters can be too many to handle with RSM optimization techniques (Cassettari et al. 2013).

The optimization modeling response terms utilized in most of the literature is a quadratic equation giving pairwise interaction terms for any two chosen parameters, although realistically, there can be more than two interacting parameters that can have some simultaneous effects. This kind of approximation can be performed only with cubic or higher-order polynomial response equations. Further, experimental designs such as CCD or Box–Behnken Design (BBD) require a prior knowledge of the system boundaries. In fact, once the design has been made, the method is not flexible to accommodate new data levels during the analysis and experiments.

ANN, on the contrary, resolves most of the shortcomings posed by RSM methodology. These optimization techniques are based on deep learning algorithms that can practically handle any kind of interacting system. Although ANN has proven to be a very handy tool for optimization and prediction of complex systems with multiple parameters, it still has many inevitable challenges associated with itself. To develop a neural architecture for ANN , it is essential to have an abundant amount of training data a priori available to the model. The required amount of data set can range from hundreds to thousands of experimental results. To resolve this problem, researchers have often used other statistical modeling tools to predict and generate these data sets. The fundamental problem with ANN is that, using this technique, one cannot develop a predictive model for systems containing multiple stochastic events whose outcome and interactions are based not on correlated laws but purely on random distribution of probabilities without any bias (Navarro and Bennun 2014).

Furthermore, ANN has practical challenges like long computational time because ANN development requires rigorous training. The computational time can increase significantly with an increase in the number of feed parameters. An advanced computational facility is necessary to work and feasibly develop these predictive models. Operating and analyzing ANN data requires high computational skills. Because ANN is not older than 50 years, the availability of trained and skilled ANN researchers is scarce. In the subsequent sections, important elements of various modeling techniques and their workflow are discussed.

4.4 MATHEMATICAL MODELING OF ELECTRO-COAGULATION USING ARTIFICIAL NEURAL NETWORKS

ANN is a promising technique to model complex systems. It does not require a mathematical description of each individual event, but the success of ANN to work is highly dependent on the availability of a large number of experimental data sets (Daneshvar et al. 2006). ANN is inspired from a complex connection present in the human brain with billions of neurons interconnected together communicating with one another with electrical impulses to perform the most complex tasks every second such as visual pattern recognition, decision-making, estimation, and mechanical actions via motor neurons. The ability of a human brain to accurately function a given task is dependent on the history of prior exposure (experience) of the brain. Extrapolating the same logic to the machine world, data scientists have developed an artificial intelligence called ANN based on that fundamental premise. Like human brains, they are also comprised of interconnected nodes (and neurons). The complete network is structured (Mirsoleimani-Azizi et al. 2015) in three layers: (1) input layer, (2) hidden processing layer, and (3) final output layer. The interconnection between these nodes is in the form of mathematical equations giving output from one node as input to another node, which, in turn, relays this information after processing the input by the mathematical definition of the node.

The topology of ANN is decided by the number of layers it consists, the number of nodes it has, and the transfer function (the mathematical functions or rule relaying information). The first step in the development of ANN is to decide the number of independent inputs and outputs. The independent variables in EC are usually the time of electrolysis, the pH of feed wastewater, current density, and so on. (depending on the researcher's choice). Further, the model execution requires a choice of transfer functions, training algorithm, and validation using test data. The validation or accuracy of the model is determined by various parameters such as root mean square error (RMSE) or regression coefficient (R^2). The details of each element of ANN are discussed in the next section.

4.5 IMPORTANT ELEMENTS OF ELECTRO-COAGULATION MODELING BY ARTIFICIAL NEURAL NETWORK

In the last decade, because of the advancement of computational capacities through the availability of advanced high-speed computers, new achievements have been made in ANN designs. Although ANN has been successfully applied to a plethora of advanced wastewater treatment technologies, very limited number of studies are available in EC, which could be possibly due to the lack of computational training available to researchers, among many other reasons. In this section, we detail the various critical elements of ANN design, which are essential for developing a robust model for any complex process.

4.5.1 Topology of Artificial Neural Networks

A typical neural network is composed of three interconnected parallel layers as previously described, namely, an input layer, a hidden layer, and an output layer (Figure 4-1). These layers consist of neurons that are connected to one another with mathematical correlations. The strength of these correlations is determined by the weight associated with them. The mathematical correlations are called *transfer functions*. The input layer consists of as many nodes as the number of independent variables we desire to put in for the model. The hidden layer acts as a feature detector, and there can be multiple layers of hidden layers. The number of neurons in the hidden layer is decided based on the desired accuracy of the model predictions. The input for the hidden layer and the output layer is a weighted summation of all the inputs from the preceding layers. The weighted sums are

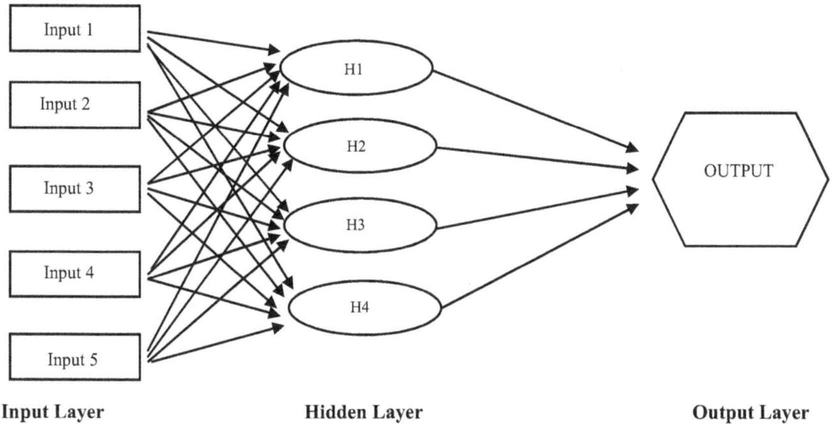

Figure 4-1. Typical structure of a neural network.

transformed using activation functions and fed to the output layer, where they undergo another transformation. Ultimately, the topology of the ANN is decided by the number of layers, number of nodes, and nature of transfer functions. The topology of an ANN is the most important factor that determines its ability and computational strengths.

As of today, not many studies are available on EC dealing with the importance of various ANN topologies simulating the EC process. Aber et al. (2009), while studying Cr(VI) removal by EC, tried to optimize ANN topology for the EC process. The study tested different topologies of ANN by varying the number of hidden nodes from 2 to 20. Each topology was repeated three times.

$$\mathrm{MSE} = \frac{1}{N} * \leftarrow \sum_{i=1}^{i=N} * \left(y_{i.\mathrm{pred}} - y_{i.\mathrm{exp}}\right)^2 \tag{4-2}$$

where

MSE = Mean square error,
N = Number of observations, and
Y_i = Value of the parameter.

The study concluded that the MSE function for the ANN was observed to be minimum for 10 nodes in the hidden layer. MSE [Equation (4-2)] is a very common parameter to qualify the performance of an ANN. Such an evaluation is critical because the number of nodes in a layer eventually decides the computational time required by the software (or computer) to train, program, and perform for a given data set. As the complexity of the problem and the data set size increase, the importance of the topology of ANN becomes evident.

4.5.2 Learning Process of a Model

Learning or training of an ANN is not clearly defined in the literature, which makes it quite subjective according to the perception of a researcher. The learning by the model is a very dynamic process that changes the parameters of the model continuously. Learning is of three types, namely (1) supervised, (2) reinforced, and (3) unsupervised.

In a supervised learning process, a model is given an input from the environment and asked to predict the output. The output is verified or compared by known values, and the model is made aware of the feedback on the difference between the true value and the predicted value (error function) to guide it to correct the weights of the synaptic connections (correlations). However, in a reinforced learning process, no such feedback is provided to the model, and the model has to use the trial-and-error method to decide the direction of adjustment of its weight ratios to finally achieve best-fit outputs. In an unsupervised learning process, no true value indication is provided, and the model itself has to recognize the pattern of the output changes with the change in environment variables and to give similar output for any new input. The readers are advised to refer other sources (Forsyth 1988, Tipping 2004) for more details on machine learning processes.

4.5.3 Training Algorithm

Training a neural network model is the second step of model development. In this step, the interconnections (correlation or transfer functions) of the neural networks, weight ratios, and biases are iteratively updated using various protocols. In practice, three algorithms to improve and amend the weight values exist: (1) backpropagation algorithm, (2) quasi-Newton algorithm, and (3) Levenberg–Marquardt algorithm.

Standard back-propagation includes a gradient descent algorithm, like the Widrow-Hoff learning rule for the multiple-layer networks (Khataee and Kasiri 2010). It is also known as the *gradient decent algorithm* because the network weights move along the negative of the gradient of the performance function. In this algorithm (the simplest form of its application), the weights are updated in the direction in which the performances decrease most rapidly (or the negative of the gradient direction). This can be given by the following equation:

$$x_{k+1} = x_k - \alpha_k g_k \tag{4-3}$$

where

x_k = Weight,
g_k = Gradient, and
α_k = Learning rate.

In this method, it is assumed that changing the weight in the negative of the gradient direction convergence with the true values will be achieved faster.

The quasi-Newton algorithm is another Newton's alternate method for rapid optimization. The basic method is explained by the following formula:

$$x_{k+1} = x_k - A_k^{-1} g_k \tag{4-4}$$

where A_k^{-1} is the Hessian matrix of the performance index of the current values of the weights and biases. This algorithm is the fastest algorithm for convergence, but it is computationally expensive and complex. Alternatively, to avoid the calculation of the Hessian matrix for every iteration (which takes the major fraction of computational resources), a quasi-Newton (or secant) method is available for use. More details about Newtonian algorithms are available elsewhere (Demuth and Beale 1993).

To avoid the necessity of calculating the complex Hessian matrix (H), the Levenberg–Marquardt algorithm was designed to achieve second-order training speed. This method approximates the Hessian matrix (derivatives of the performance function) with Jacobian matrices (J) according to the following equation:

$$H = J^T J \tag{4-5}$$

and the gradient function (g) is

$$g = J^T e \tag{4-6}$$

where J^T is the transpose of the Jacobian matrix, and J is the Jacobian matrix of the first derivatives of the network's error function. In computational terms, Jacobians are easy to calculate as compared to Hessian matrixes. Thereafter, the weights are approximated [Equation (4-7)] with a Newton-like Equation (4-3) substituted with Equations (4-5) and (4-6)

$$X_{k+1} = X_k - [J^T J + \mu I]^{-1} J^T e \tag{4-7}$$

Newton's method is the fastest method of convergence. Therefore, in this method, the aim is to convert this format as quickly as possible to the form of a Newton equation [Equation (4-3)]. Thus, at each step, the value of μ is decreased gradually, tending to zero. A detailed understanding of the training algorithm is beyond the scope of this chapter, but readers are encouraged to refer to other computational modeling resources (Bishop 1995).

4.5.4 Optimization of Neural Network Model

In general, four key steps are available to train the neural network:

1. Assembly of the training data,
2. Creation of the network object,
3. Training of the network, and
4. Simulation of the responses to new inputs.

In practice, a network with biases, a sigmoid layer, and a linear output can simulate any system. Feed-forward networks often contain one or more hidden (sigmoid) neuron layers connected linearly to the output layer. The number of nodes in the hidden layer is optimized by varying the nodes between 2 and 20 and minimizing MSE [Equation (4-2)]. As previously mentioned, networking training is performed by one of the training algorithms. The goal of training is to find the set of weight values for minimum error to achieve values as close to the target value as possible. Various researchers have used the sigmoid transfer function in the hidden layer for which all the inputs are scaled down (normalized) using Equation 4-8:

$$x_i = 0.8\left(\frac{X_i - X_{\min}}{X_{\max} - X_{\min}}\right) + 0.1 \tag{4-8}$$

where X is the value of the variable that can take the maximum, minimum, and ith value. Each topology (network configuration of nodes) is repeated several times with randomly initialized weights until the network's performance stabilizes.

Model fitting evaluation. To evaluate the performance and fitting of the model, the output of the model is scaled back to its original scale and compared with the true values (expected experimental outcomes). Graphically, two lines are drawn. One is $Y = X$ (slope $= 1$; intercept 0, 0), and the other is the best fit line obtained from the predicted values ($Y = aX + b$). If the fit is perfect, the two lines should superimpose or nearly superimpose. Another parameter is the correlation coefficient (R-value) between outputs and the targets. If this number is equal to 1, then it is said to have a perfect fit.

Significance of input parameters. It is also important for practical reasons to have a proper inference of the model data. The relative significance of the input variable is one such parameter that hints at the sensitivity of the model outcome toward the given input parameter. This can be quantified by the neural net weight attributed to the final outcome via various transfer functions within the hidden neural network to the output layer. Garson (1998) proposed a method to evaluate the neural net weight by the following correlation:

$$I_j = \frac{\left\{\sum_{m=1}^{m=N_h}\left(\left(\frac{\left|W_{jm}^{ih}\right|}{\sum_{k=1}^{N_i}\left|W_{km}^{ih}\right|}\right) \times \left|W_{mn}^{ho}\right|\right)\right\}}{\left\{\sum_{k=1}^{N_i}\left\{\sum_{m=1}^{m=N_h}\left(\left(\frac{\left|W_{km}^{ih}\right|}{\sum_{k=1}^{N_i}\left|W_{km}^{ih}\right|}\right) \times \left|W_{mn}^{ho}\right|\right)\right)\right\}} \tag{4-9}$$

where

I_j = Relative importance of the Jth input variable on the output variable;
N_i and N_h = Number of inputs and hidden neurons respectively;
Ws = Connection weights; and
i, h, and o = Input, hidden, and output layers neurons, respectively.

In EC, not many such studies were found that calculated the relative importance index of input variables. Aber et al. (2009) performed a study that utilized this approach to identify the relative significance of their input parameters on the outcome. The study used ANN modeling analysis in its EC work to remove Cr(VI). The authors estimated that current density, time of electrolysis, initial Cr concentration, and electrolyte concentration have index values of 16.51, 32.36,

23.82, and 27.03, respectively, which meant that all the variables have an equally strong effect on the outcome, and none of them can be disregarded.

4.6 ESSENTIAL ELEMENTS OF STATISTICAL MODELING BY RESPONSE SURFACE METHODOLOGY

Traditionally, researchers have performed optimization of multivariable systems using one variable at a time method. These methods require a large numbes of experiments that become impractical to perform. Further, these methods assumed orthogonal independence (having no mathematical correlation) between two parameters.

RSM is a well-known mathematical and statistical modeling technique that has been applied by several researchers to optimize and model their process (systems). RSM optimization has the ability to correlate the independent input parameters to one or more different outcomes (or responses) by a polynomial correlation. RSM overcomes the problem faced by one variable at a time approach by proposing systematic matrices of experimental designs and, thereafter, modeling the final output in terms of polynomial fit equations. The method applies several different design approaches such as BBD and CCD. In general, researchers have applied quadratic equations that enable them to understand the interaction between two independent factors together. Polynomials of an order higher than 2 can reveal interactions among three or more parameters together, but these models turn out to be more complex to compute and interpret. Multiple steps are involved in the application of the RSM modeling technique to any system.

4.6.1 Choosing Independent Variables

The very first step in RSM modeling is choosing the number of independent variables. A process may have more than two variables, but it depends on the insights of a researcher to choose to simulate only the major factors that affect the process outcome significantly. A sound knowledge of the process principle is important to decide the process parameters chosen to simulate in RSM. The number of parameters directly affects the number of experiments required for a given experimental design.

4.6.2 Experimental Design

The simplest form of a model that can be used in RSM is a linear function depicted in the following equation:

$$y = \beta_o \sum_{i=1}^{k} \beta_i x_i + \epsilon \tag{4-10}$$

where

k = Number of variables,

β_o = Constant,

β_i = Coefficient of the parameter,
x_i = Variables, and
ε = Constant to accommodate error.

The linear models cannot present any curvature in responses; therefore, to evaluate a curvature, a second-degree model should be used, and it can also account for interaction terms. Therefore, a quadratic term is often used to facilitate the calculation of maxima or minima and saddle points according to the following equation:

$$y = \beta_o + \sum_{i=1}^{k} \beta_i x_i + \sum_{i=1}^{k} \beta_{ii} x_i^2 + \sum_{i \leqslant i \leqslant j}^{k} \beta_{ij} x_i x_j + \epsilon \quad (4\text{-}11)$$

where β_{ij} is the interaction term, and β_{ii} is the quadratic term. To calculate the values of the parameter, the experimental design needs to run at least three factor levels. The experimental designs that have been extensively used by researchers include the following: (1) BBD; (2) CCD; and (3) Doehlert design. Details on these designs and their features can be found in other relevant literature (Bezerra et al. 2008).

4.6.3 Statistical Treatment of Data

After the experimental plan is designed as suggested by the model, experiments need to be performed in the laboratory to obtain the responses for each set of parameters. After the experimental data are available, the model fits the data into the design equation. The model calculates the values of all the equation coefficients using the method of least square (MLS). Once the model has values of all the coefficients, the model generates the response surfaces (often visualized as 3D surfaces).

4.6.4 Fitting of the Model

After the model has been developed, fitting is attempted by the software tools. In cases when the model is not completely fitting or 100% superimposed, a measure of fitting is performed. The measure of fit of the model is done by the analysis of variance (ANOVA). In ANOVA, the dispersion of the data set variation is studied for all the observed values and its replicates. The square of the deviation is given as Equation (4-12):

$$d_i^2 = (y_{ij} - y_m)^2 \quad (4\text{-}12)$$

where
d_i = Deviation,
y_{ij} = Replicated observation,
y_i = Observations, and
Y_m = Median.

The variation in the model values can be attributed to many factors such as an error caused by regression and residual error. The residual error is further comprised of pure error and error caused by the lack of fit. All these errors are summed up together for all the observations, and when divided by the degree of freedom, it gives the median of the square. For instance, the sum of error caused by regression (SS_{reg}) is calculated as Equation (4-13):

$$SS_{reg} = \sum_{i}^{m}\sum_{j}^{n}(y_i - y)^2 \tag{4-13}$$

with $p-1$ degrees of freedom, the mean sum of error (MS_{reg}) = $SS_{reg}/(p-1)$, where $p$is the number of coefficients of the model, n is the total number of observations, and m is the number of levels used in the investigation.

The significance of regression is calculated as a ratio of median of regression with median of residuals, also termed the F ratio. For the model to statistically fit, the F ratio value should be higher than the values presented in the standard tabulated values of F.

Another parameter often examined to evaluate the fitting of a model is the *lack-of-fit* (LOF) test. The lack-of-fit test is defined as the ratio of median of sum of (MS_{lof}) lack of fit to median of sum of pure error (MS_{pe}). If this ratio is found higher than the standard tabulated F values, then the model lacks a fit and can be improved further.

4.6.5 Finding Optimal Conditions

The surface generated by the RSM model can be used to find optimal solutions. Mathematically, the maximum, minimum, or saddle points refer to the optimal solution in the surface. A two-variable quadratic model looks like Equation (4-14):

$$y = b_o + b_1x_1 + b_2x_2 + b_{11}x_1^2 + b_{22}x_2^2 + b_{12}x_1x_2 \tag{4-14}$$

Differentiating Equation (4-14) with respect to X_1 and X_2 will give us a set of first-order differential equations that can be solved to find X_1 and X_2 values. The visualization of the solution is an n-dimensional surface in an $n+1$-dimensional space. For the equation it implies, it is a 2D surface in a 3D space. A visual inspection of these surfaces (solutions' representation) can give us the optimal solutions. The basic objective of statistical modeling is to find optimal solutions (a set of parameter values) to maximize the treatment performance of EC in terms of pollutant removal.

Researchers performing experimentation in synthetic media with one single deliberately added pollutant have usually had a sole objective of maximizing the removal of that pollutant, but in real applications of EC, the system has to be designed to maximize the removal of multiple pollutants (COD, color, TSS, TOC, metal concentration, etc.) by multiobjective functions. Further, because EC requires electrical energy, the focus has always been on minimizing energy consumption. Other different objectives are possible, such as maximizing treatment capacity, reactor sizing, and so on. Therefore, it can be seen that in real-life implications, we can have multiple objectives to achieve in the same process.

The modeling techniques often apply a strategy called *multiobjective optimization* to achieve the desired overall compounded outcome.

4.7 MULTIOBJECTIVE OPTIMIZATION MODELS

Multiobjective optimization is a concept in which a model optimization is performed keeping two objectives in mind together, that is, maximum pollutant removal and minimization of energy consumption. It is certainly different from classical modeling approaches, because the difference lies in the computation methods so as to arrive at the optimal network topology, training optimization, and final optimal solutions. In other words, the difference in modeling may arise because the two objectives might be antagonistic in nature.

Bhatti et al. (2011a) developed a model for the removal of copper ions from synthetic copper solution by the EC method. The process was experimentally optimized by the RSM optimization technique using a CCD to develop the experimental strategies. A total of 21 CCD experiments were performed to generate the data set, and design expert software was used to analyze and represent the RSM data set graphically. Using RSM, a characteristic equation was developed to represent the output (copper removal %) in terms of input factors. In the study, ANN was also developed using four input factors, namely, copper concentration, pH, voltage, and time of electrolysis. Copper removal efficiency and energy consumption were defined as two final outcome objectives. The network developed was a classical three-layer network with a top layer, a hidden layer, and an output layer. The network was trained with Bayesian regularization–based MATLAB function called TRAINBR to choose an optimum number of neurons in the hidden layer. After optimization, five neurons in the hidden layer were chosen. To optimize the weight factors, in error back-propagation, 202 iterations of the network were performed, resulting in a regression fit of 0.982.

The complexity arises on the premise that when two or more objectives are given to a model, the optimal solution for one objective may not be the optimal solution for the second objective. The model has a dilemma to calculate the optimal solution without compromising the two objectives significantly. Thus, no optimal solution exists per se, but a set of this solution is available, that is equally good for both objectives, and it is known as the *Pareto optimal solution*. A set of these solutions is together called *Pareto front*, which consists of solutions such that there is no solution better than these solutions present in the Pareto solution front. The genetic algorithm (a special class of algorithm) was used to create the Pareto front. These algorithms ("gamultiobj") are available in computational tools like MATLAB (V 2.3).

Finally, in the study by Bhatti et al. (2011a), both RSM and ANN methods were correlated well with experimental data for both the objectives ($R = 0.994$ copper removal and 0.949 energy consumption). The final conclusion of the study was deduced by dividing the objective 1 (copper removal) values with objective

2 (energy consumption) values present in Pareto optimal solution sets to get the final objective of copper removal per unit energy consumed. The result indicated that removal efficiency per watt-hour was minimum (25.37) for 15 mg/L copper concentration at a pH of 7.6, whereas it was maximum (40.63) at a pH of 6.0 among the Pareto solution set.

In a multiobjective modeling, it is very important to understand the interaction terms among parameters so as to know how they affect the final outcome objective(s). The study further concluded that pH has the highest negative impact on copper removal, whereas voltage and treatment time have a positive impact on copper removal (Objective 1). However, if we speak of the second objective (energy consumption), voltage and time have a negative consequence. Both voltage and time increase energy consumption, but voltage has a higher contribution to make than time; therefore, one can increase treatment time but reduce the voltage applied. If the treatment time needs to be increased, it means to achieve the same treatment goals (the rate of water treatment volumetric capacity), and one might need more reactors or a larger volume of reactors for this purpose. Thus, we can discern the strength of modeling approaches in multifaceted modeling in which we can address the bottlenecks of the process in case of variations in the parameter values.

4.8 RECENT MODELING STUDIES USING ARTIFICIAL NEURAL NETWORKS

ANNs have been extensively used in the recent decade. The textile industry is one of the most water-polluting industries. A major part of research is focused on the decolorization of the heavy metal dye contaminants from textile industry effluents (Aber et al. 2009, Bhatti et al. 2011b, Daneshvar et al. 2006, Elemen et al. 2012, Sinha et al. 2012, Turan et al. 2011).

Daneshvar et al. (2006) studied the removal of C.I. Basic Yellow-28 dye by the EC process. The study optimized the conditions for dye removal by performing various experiments. Experimental data were used to develop ANN and validate the model. The study concluded that an ANN with 10 neurons in the hidden layer with sigmoidal transfer functions can successfully predict the dye removal efficiency. The model fit correlation (R^2) was 0.974 when tested against the validation data set. In a similar study by Murugan et al. (2009), a number of neurons in the hidden layer were optimized to simulate an ANN for the EC of textile effluents. In this study, the ANN was also a typical three-layer network. It concluded that using the minimized R^2 error value, the optimum number of neurons found was 10 with three such hidden layers.

A similar study was performed by Aber et al. (2009) to examine the removal of Cr(VI) from synthetic and real wastewater. The study developed an ANN and optimized the number of neurons in a hidden layer. The study found that the mean square error (MSE) was minimum with 10 nodes in the hidden layer. They further calculated the relative significance of input parameters and found that time, current density, NaCl concentration, and initial Cr concentration were all statistically significant .

Bhatti et al. (2011b), in their study, developed an ANN for Cr(VI) removal with four neurons in the hidden layer with a sigmoidal transfer function. The model was trained by normalized data obtained from CCD with 60 iterations. The MSE of the optimized model was found to be 0.0242. The RSM model was optimized using the ANN model, and the optimized RSM model could reduce the energy consumption by 44.8%. The predicted quadratic model for Cr(VI) removal and energy consumption were having a coefficient of determination (R^2) equal to 0.975 and 0.990, respectively. Unlike other studies, Mirsoleimani-Azizi et al. (2015) studied the removal of endosulfan (a pesticide). In this study, the network topology was optimized for a number of neurons in a hidden layer. The results indicated that a hidden layer with eight neurons gave a successful fit to the train data sets, resulting in a 5-8-1 topology configuration for ANN. A correlation coefficient (R^2) of 0.976 indicated that the model could easily fit and predict the validation data set.

Nourouzi et al. (2011) studied the removal of *Reactive Black 5* dye by subsequent EC and flocculation using EC. The study optimized the process by two modeling techniques, RSM and ANN. The study compared the predictive performance of these two techniques. The input parameters studied were time, current, conductivity, and flocculant dosage. The maximum number of iterations was 1,000, and four experimental data sets out of 40 experiments performed were used for checking the performance of the model, whereas 36 data sets were used to train the model. The transfer function used in all the nodes was sigmoidal. The network had one hidden layer with 10 nodes in the layer. The network was trained with the *Quickprop* training algorithm. Model fitting was evaluated by using the R^2 fitting value, and the best model with $R^2 = 0.9968$ was chosen. During evaluation and validation of the model, it was concluded that the model fits significantly with $R^2 = 0.9764$, which was higher than that of the RSM model (0.9446). The study concluded that by using an additional flocculation step, dye removal efficiency could be significantly improved. Similarly, ANN modeling techniques have been shown to successfully simulate the treatment of various other pollutants such as endosulfan, oil tanning effluent, and acid azo dyes (Lakshmi and Sivashanmugam 2013, Mirsoleimani-Azizi et al. 2015, Piuleac et al. 2013, Taheri et al. 2015).

Modeling of EC by ANN has been performed only by a limited number of researchers. The reason for this small number of studies is that ANN has been in existence for only about 50 to 60 years, and its application in EC systems has been quite recent. However, the main reason for the fewer studies in ANN modeling is the lack of proper statistical and computational training that ANN development and data analysis require. Furthermore, ANN simulations are quite computationally expensive and require high-end performance computational machines. In spite of these limitations, ANN still serves as a powerful tool to generate predictive models and analyze the relative significance of operational parameters. It is just a matter of time that with advanced computers and adequate researcher training, ANN will become a mainstream modeling technique for researchers to work on various electrochemical processes.

4.9 RECENT MODELING STUDIES IN ELECTRO-COAGULATION USING RESPONSE SURFACE METHODOLOGY

RSM is among the most preferred modeling techniques by various researchers. According to the recent literature reviews, researchers have tried to model various EC systems (Table 4-1) (Hakizimana et al. 2017). Alinsafi et al. (2005) studied EC of active textile dyes and implemented RSM to optimize the removal of dye pollutants. The parameters considered in this study were pH, time, and current intensity. The objective of modeling in this study was to develop equations to explain the influence of input parameters on the final output of the process. The model developed by RSM was a second-order equation, Equation (4-15):

$$y = b_0 + b_1x_1 + b_2x_2 + b_{12}x_1x_2 + b_{13}x_1x_3 + b_{23}x_2x_3 + b_{123}x_1x_2x_3 \quad (4\text{-}15)$$

where

$Y =$ Objective response,
$xs =$ Independent variables, and
$bs =$ Coefficients.

The model calculated the significance of each coefficient and the results indicated that the coefficients b_0, b_1, and b_3 are significant at a level of less than 5%, whereas the interaction coefficient b_{13} is significant at a level of less than 10%. This result means that the reaction time (b_1), current density (b_3), and interaction among them are most influential factors. The study made a contrary observation that pH has a significant level at 85%, which means that pH is not an important parameter. According to the ANOVA analysis, the model fits 92% with the experimental data. After the preliminary inferences, the study further optimized the two significant parameters, current density and time of electrolysis, by a two-factor RSM model. Using 2D contour plot, it was deduced that COD removal is maximum in the range of 90 to 120 min, with a current density of 12 mA/cm^2. Although increasing the current density and time of electrolysis together results in an increase in decolorization up to 98%, the model target was to achieve two goals simultaneously (i.e., COD removal and decolorization). To execute the simultaneous optimization, the individual desirability of each target is assigned a value (d_i), and a compound desirability (D) was calculated for a different scenario via Equation (4-16):

$$D = (d_1 * d_2 * d_3 \ldots d_m)^{1/m} \quad (4\text{-}16)$$

where D is the compounded desirability, and d_i are individual objective desirability (like decolorization). By using the model, it was found that the optimum conditions were 105 min and 12 mA/cm^2 giving 35% COD removal and 92% decolorization,

Table 4-1. Recent Studies on Mathematical and Statistical Models for EC.

Sr. No.	Pollutant	Matrix	Modeling method	Mode	Electrodes: Cathode	Electrodes: Anode	Optimized solution of various parameters: pH	Pollutant conc.	Current density	Soln. conductivity	Time	Electrode distance/internal resistance	Removal efficiency (additional note)	References
1	C.I. Basic Yellow 28	Real + synthetic WW	ANN	CNTS	Fe (ST-32)	Fe (ST-32)	5–8	50 mg/L	80 A m^{-2}	10 mS cm^{-1}	7 min	15 mm	90%–95%	Daneshvar et al. (2006)
2	Cr(VI)	Real + synthetic WW	ANN	Batch	SS304-Steel	Fe (ST 37-2) and Al (HE18)	5–8	20 mg/L	50 A m^{-2}	16.27 mS cm^{-1}	30 min	15 mm	97%; Add electrolyte (NaCl) to reduce IR-10 mg/L	Aber et al. (2009)
3	Arsenic (As)		RSM Adsorption isotherm	Batch	Steel	Al	4–10	50–150 ppm	0.5–1.5 A dm^{-2}	NA	10–50	15 mm	Langmuir Isotherm	(Balasubramanian et al. 2009)
4	Fluoride		RSM	Batch	Al	Al (98.67% pure)	7	25 mg/L	0.011 A cm^{-2}	NA (NaCl supplemented)	25 min	30 mm	90% removal	(Behbahani et al. 2011, Mollah et al. 2004)
5	Copper		RSM (CCD) and ANN	Batch	Al	Al (91.45% pure)	7.6	15 mg/L	NA	NA	10.3 min	15 mm	Copper removal and energy minimization	(Bhatti et al. 2011a)
6	Cr(VI)	Synthetic	RSM + ANN	Batch	Al	Al (91.45% pure)	5.0	100 mg/L	11.0 V	NA	18.6 min	15 mm	50% removal	(Bhatti et al. 2011b)
7	Chlorophyll	Real lagoon effluent	ANN	CNTS	Steel	Aluminum	7-8.8	NA	3 A dm^{-3}	NA	15-97 L/h	NA	Types of neural networks studied	(Curteanu et al. 2011)
8	As(III)	Synthetic ground water	Mathematical model	Batch	steel	Fe(0)	7.0	0.5 mg/L	NA	NA	NA	75 mm	A mathematical model to simulate kinetics	(Li et al. 2012, Mollah et al. 2004)

9	CI reactive Red 43 (dye)	Synthetic WW	RSM (CCD)	Batch	SS 304	Fe (ST 37) or Al (HE 18)	8.54 4.14	50 mg/L	35.07 A m^{-2} 39.14 Am^{-2}	NaCl as supporting electrolyte	23.75 min 12 min	20 mm	98.9% (EC-Al) 98.88% (EC-Fe)	(Amani-Ghadim et al. 2013)
10	Malathion	Synthetic solution	RSM	Batch	Al	Al	6.0	100 mg/L	1 A	2.5 g/L NaCl	60 min	20 mm	95% pesticide removal	(Behloul et al. 2013)
11	As, NO_3^-, PO_4^{3-}	Synthetic WW	Mathematical model	Batch	Al/Fe	Al/Fe	NA	As, nitrate phosphate (20, 25, and 27 mg dm^{-3})	NA	1 g/dm^3 NaCl 3 g/dm^3 Na_2SO_4 0.5 g/dm^3 Na_2SO_4		9 mm	NA	(Lacasa et al. 2013)
12	Tetracycline	Synthetic solution	Adsorption model	Batch	Al	Al	4–10	10 mg-TC/L	0.8 A	1.25 g/L KCl	2 min	20 mm	96.5% removal	(Ouaissa et al. 2014)
13	Wastewater treatment COD	Real waste-water	RSM (BBD)	Batch	Fe	Fe	7	NA	20 mA/cm^2	1 g/L NaCl	20 min	50 mm	94.75% COD 99.87% TS 97.15% FC	(Karichappan et al. 2014)
14	Chicken WW	Real waste-water	RSM (BBD)	Batch	Fe	Fe	6	30% dilution	14 mA/cm^2	1.075 g/L	15 min	10 mm	95% COD removal	(Thirugnana-sambandham et al. 2014a)
15	Endosulfan	Synthetic solution	ANN	Batch	Al	Al	4	30 mg/L	6.2 mA/cm^{-2}	NaCl	60 min	10 mm	84.57% removal	(Mirsoleimani-azizi et al. 2015)
16	Egg WW-Protein recovery	Real Egg waste-water	RSM (BBD)	CNTS	Fe-SS	Fe-SS	6	NA	25 mA/cm^2	NA	HRT 27 min	45 mm	89% protein recovery	(Thirugnana-sambandham et al. 2016)
17	Bagasse P&P WW	Real-WW	RSM (BBD)	CNTS	Fe-SS	Fe-SS	7	NA	25 mA/cm^2	1 g/L NaCl	40 mL/min	45 mm	84% COD 91% TSS removal	(Thirugnana-sambandham et al. 2015)
18	Theoretical	Synthetic water	CFD	CNTS	Al	Al	NA	NA	NA	NA	NA	10 mm	Design equation validated	(Lu et al. 2017)
19	Denitrification	Synthetic water	Mathematical model	Batch	Al	Al	6–10	50 mg/L	—	—	—	10 mm	Successful demon-stration denitrifica-tion	(Yehya et al. 2014)

respectively. A similar study was performed by Amani-Ghadim et al. (2013) for azo dye (reactive red 43 or RR43) removal. They utilized the concept of RSM to compare the model development for two different electrodes, iron and aluminum. Two quadratic models were developed using experimental design obtained from CCD. The models were tested for the lack of fit, and the significance of the models was evaluated by using the Fisher's F-test. The statistical validation of the two models was confirmed by low p-values. The results indicated that the R^2 value for the iron electrode model (EC–Fe) was 0.981, whereas that for aluminum (EC–Al) was 0.934. The optimized model predicted 99.8% and 99.99% removal with EC-Fe and EC–Al, respectively. After validation experiments, the experimental results strongly supported the model as experimental removal was 98.90% and 98.88%, respectively.

The true potential of the RSM modeling technique lies in the fact that it can address the evaluation of the interaction between parameters and solve the problem of achieving two simultaneous optimizations mathematically. The study performed by Bhatti et al. (2011b) modeled EC for the removal of Cr(VI) using the RSM technique. The developed model was in the form of a quadratic equation [Equation (4-17)]:

$$\text{Cr(VI) removal(\%)} = 21.42 + 4.236\ \text{V} + 4.149t - 0.393\ \text{V} * t - 0.091t^2 \quad (4\text{-}17)$$

where time is in minutes, and volt is in voltage. The study concluded that the model fits well with the coefficient of regression (R^2) up to a value of 0.976 and with a goodness-of-fit value of 1.86% (where up to 5% is considered excellent). The model has an acceptable signal-to-noise (S/N) ratio of 34.17. The time and voltage have an antagonistic effect as suggested by the negative coefficient. The optimization of the model recommended 12.8 V and 19.2 min as optimum values to give 51.9% removal. When the validation experiment was performed, the removal was found to be 52%. Taking energy consumption into account, another quadratic equation was developed to predict energy consumption, and the two model equations were solved to find the optimum solution. The model suggested that 11 V and 18.6 min are the optimum parameters for maximum Cr(VI) removal (50%) and least energy consumption (8.66 W h). The validation experiment conducted gave 49.8% removal of chromium with 8.60 W h energy consumption, which strongly supports the predictive model.

Although the RSM technique has been instrumental in developing and optimizing predictive models for dye and heavy metal compounds in synthetic media, recently, various researchers have also applied RSM techniques to real wastewater matrixes such as chicken processing wastewater (Thirugnanasambandham et al. 2014a), gray wastewater (Karichappan et al. 2014), bagasse-based pulp and paper industry wastewater (Thirugnanasambandham et al. 2015), and egg processing industry wastewater (Thirugnanasambandham et al. 2016). Karichappan et al. (2014) applied the RSM technique to develop a predictive model to predict COD, total solids (TS), and fecal coliform (FC) in a gray wastewater treatment system. Three different quadratic models were designed using BBD with initial pH, current density, electrode distance, and time of electrolysis as the model parameters. Twenty-nine experiments were performed to develop and fit the final model. The adequacy of the model was verified by plotting the experimental versus the predicted plot close to the $y = x$ line. Further, ANOVA analysis was performed, and it was found that Fisher's F value and p-values demonstrated the high statistical significance of the model coefficients. The model with individual parameter and interaction coefficient terms was visualized in 3D plots to study the effect of each parameter. To finally obtain the optimum process parameter, the three objectives, TS, COD, and FC, were combined by Derringer's desired function methodology optimization [explained previously in Equation (4-15)]. According to the optimization process, the final values of parameters were a pH of 7, a current density of 20 mA/cm^2, an electrode distance of 50 mm, and a process time of 20 min. The validation experiments confirmed the robustness of the final experimental outcome with TS, COD, and FC removal percentages of 99.87%, 95.47%, and 97.15%, respectively, which were close to the predicted values of 98.45%, 94.75%, and 96.34%. In subsequent years, the same research group performed on various other real wastewater matrixes using the same strategy and successfully demonstrated that even complex real wastewater matrixes can be reliably modeled for the EC process using the RSM technique. The details of important findings are presented in Table 4-1.

The RSM modeling techniques has been the most favored technique for its simplicity and easily available tools to perform the experimental design data analysis. Bibliographic analysis suggests that lately researchers are focusing on real wastewater matrixes, but very few are studying surface or groundwater for potable use (Behloul et al. 2013, Karichappan et al. 2013, Kobya et al. 2014, Lakshmi and Sivashanmugam 2013, Orssatto et al. 2017, Shankar et al. 2013, Zodi et al. 2010).

Groundwater or surface water for potable use requires very mild treatment conditions. Iron and aluminum are the two most preferred or almost the only electrodes used in EC techniques. The reason for choosing Fe and Al is their cost-effectiveness and their potential to reactively produce multivalent cations in multiple coordinated forms (ionic or hydroxides). The nature of the wastewater matrix affects the efficiency of treatment. For example, as shown in Table 4-2, in the case of bakery yeast wastewater, distillery wastewater, pistachio wastewater, and landfill leachate, the efficiency of COD removal by the most optimized conditions were 41% (Arulmurugan et al. 2007), 52.2% (Gengec et al. 2012), 57.4% (Un et al. 2014), and 60.5% (Moulai-Mostefa et al. 2013), respectively. Some complex wastewater matrixes exist, which show more than 90% treatment efficiency. For instance, Bagasse effluent wastewater, metal cutting wastewater, rice mill wastewater, and Cheese whey wastewater have treatment efficiencies of 98%, 88.43%, 97%, and 86.4%, respectively.

RSM has been a very useful modeling tool to develop robust predictive models for several different wastewater matrixes. RSM can easily simulate the EC process for a wide range of treatment scenarios. Using RSM, we can get optimized solutions for a process in the fastest possible way with the smallest number of experiments. RSM multiobjective optimization can be easily performed to find

Table 4-2. Recent Studies on Response Surface Modeling for EC.

	Pollutant	Matrix	Mode	Electrodes: Cathode	Electrodes: Anode	RSM-Optimized solution of various parameters: Volume (mL)	pH	Pollutant conc. (ppm)	Current densitymA cm^{-2}	Electrolyte concentration	Time	Electrode distance/ Internal Resistance	Fit quality R^2	Efficiency of treatment (additional note)	References
1	Bismarck Brown	Synthetic	Batch	Stainless steel	Mild steel	E	4.5	1,080	6	200 ppm of KCl	10 min	10 mm	0.875	75% COD	(Arulmurugan et al. 2007)
2	COD	Distillery real WW		Al	Al	1,500	6	15,600	120 A/m²	—	150 min	50 mm	0.97	52.2% COD	(Ponselvan et al. 2008)
3	Turbidity, COD, Cr, Zn, Ca	Tannery Real WW	Batch	Fe	Fe	8,000	7.5-9.0	—	68	—	30 min	40 mm	0.96	98% Turbidity 80% COD 98% Cr 90% Zn 62% Ca	(Espinoza-Quinones et al. 2009)
4	Acid Blue 113 dye	Synthetic	Batch	Al	Mild steel	200	6	1,000 ppm	5	400 ppm	—	25 mm	0.984	88% COD	(Murugan et al. 2009)
5	Chromium	Synthetic WW	Batch	Al	Al	5,000	4.23	200	9.14 Volt	—	10 min	10 mm	0.9994	92.83% removal	(Zaroual et al. 2009)
6	COD, TOC, Turbidity,	Metal cutting WW	Batch	Al/Fe	Al/Fe	4,000	5.01	20.124 mg/LCOD 2.65 mg/L TOC Turbidity 13.25 NTU	66.39	—	20.60 min	45 mm	0.927 COD 0.924 TOC, 0.968 turbidity	88.43% COD 79.1%TOC, 99.8% Turbidity	(Kobya et al. 2010)
7	COD	Real WW	Continuous	Al	Al	2,200	7	1,260 mg/L	100 A/m²	—	150 mL/min	20 mm	0.89	70% COD removal	(Zodi et al. 2010)
8	Fluoride	Synthetic photovoltaic WW	Continuous	Al	Al	1,000	—	20-25 mg/L	30 Volts	100 mg/L NaCl	60-80 minute	10 mm	—	60% removed	(Drouiche et al. 2012)
9	COD and TOC	Baker's yeast WW	batch	Al	Al	—	4	—	50	—	22 min.	5 mm	>0.95	41% COD, 39% TOC removal	(Gengec et al. 2012)
10	Napthalene Sulfonate (KA)	Synthetic soln.	batch	SS316	SS316	3,000	7,10	600 mg/L	30	1 g/L NaCl	150 min.	3 mm	0.9464	98% KA removal	(Olmez-Hanci et al. 2012)

11	COD, TSS	Rice mill WW	batch	SS	SS	3,000	7	2,200 mg/L	15	1 g/L NaCl	70 mL/min	50 mm	>0.9	97% COD 89% TSS	(Karichappan et al. 2013)
12	As	synthetic	Batch	Fe	Fe	650	—	184.1	9.2 A/m^2	—	18.3 min	13 mm	0.965	98.6% -removal	(Kobya et al. 2013)
13	Oil tanning effluent	Real WW	Batch	Fe	Fe	250	6.74	—	20	1 g/L NaCl	15 min	15 mm	0.997	85% COD	Lakshmi and Sivashan-mugam (2013)
14	Pretreat-ment	Drinking water	Batch	Al	Al	—	7.0	2-5 NTU 550 ppm hardness	0.3	1 g/L NaCl	5 min	15 mm	0.923	0.22 NTU final	Moulai-Mostefa et al. (2013)
15	Arsenic	Groundwa-ter	Continu-ous	Fe	Fe	16,800	—	133 μg/L	7.03 A/m^2	Air mixed, 1 g/L NaCl	2.5 min (0.25 lpm)	2 mm	0.9507	Final 10 μg/L	García-Lara et al. (2014)
16	Treatment (COD)	Pistachio WW	Batch	Al	Al	750	6	23,250 mg/L	317 A/m^2	—	29 min	20 mm	0.9996	57.4% COD removal	Güçlü (2014)
17	Treatment (COD)	Landfill leach-ate	Batch	Al	Al	250	8	—	20	—	60 min	—	>0.95	60.5% COD removal	Kabuk et al. (2014)
18	Treatment (COD)	Textile dye efflu-ents	Batch	Al/Fe	Al/Fe	4,000	5.5/5.6	—	63.2 / 52.5 A/m^2	—	30.4 min 33.9 min	45 mm	>0.92	82% COD removal/ 69% COD Removal	Kobya et al. (2014)
19	Treatment (COD)	Cheese Whey WW	Continu-ous	Fe	Fe	1,000	4.54	15,500 mg/L	60	—	HRT 20 min	10 mm	0.85	86.4% COD removal	Un et al. (2014)
20	Treatment (COD)	Treated bagasse efflu-ents	Batch	Fe	Fe	1,300	6.5	1,574 mg/L (30% dilution)	8	0.74 g/L NaCl	15 min	40 mm	0.986	98% COD removal	Thirugnana-samband-ham et al. (2014b)
21	As and F	Synthetic	Batch	Al	Al	1,400	7	As 550 μg/L F 12 mg/L	10	714.3 mg/L NaCl	95 min	10 mm	0.998 0.984	98.64% As 84.8% F	Thakur and Mondal (2016)
22	Treatment (COD)	Slaughter-house WW	Continu-ous	Al	Al	800	6.4	4,730 mg/L COD	21.6	—	HRT 25 min	78 mm	>0.9	81.01% COD removal	Orssatto et al. (2017)

the most effective solution space for any EC system. RSM will continue to be one of the preferred modeling techniques in days to come.

Although RSM and ANN provide the most reliable and robust predictive models for EC systems, researchers have also tried other mathematical modeling approaches such as phenomenological models, models based on electrochemical kinetics, adsorption, computational fluid dynamics, flocculation, floatation, and settling. These models provide more insight into the process and the mechanism of action for removal of pollutants during EC. These models are relatively simple and need less computation power in solving the equations. These models are also quite useful for faster predictions or data set development. These techniques can be combined with RSM and ANN techniques for more reliability.

4.10 KINETICS OF ELECTRO-COAGULATION

In phenomenological terms, EC is a process of two simultaneous events: (1) the generation of hydroxylated metal (Fe/Al) compound flocs, and (2) the adsorption of pollutants over these flocs. Many researchers attempted to model the kinetics of EC using mathematical equations. According to a study (Mameri et al. 1998), the kinetics of defluorination follows an exponential path, making the rate of fluoride removal an event of first-order kinetics. Similarly, researchers (Balasubramanian et al. 2009, Chitra and Balasubramanian 2010, Ouaissa et al. 2014) have attempted to mathematically describe the event of pollutant removal by using n-order kinetic equations

$$\frac{dC}{dt} = KC^n \tag{4-18}$$

where

C = Initial pollutant concentration,
K = Reaction rate constant, and
n = Order of the reaction.

Balasubramanian et al. (2009) reported that the rate of arsenic removal is a pseudo first-order equation (the adsorbent concentration assumed constant), as shown in Equation (4-19):

$$\log\frac{C}{C_i} = -kt \tag{4-19}$$

The rate constant can be calculated graphically by plotting log (C/C_i) versus time. Various researchers have also used empirical equations to model the abatement of pollutant. Yehya et al. (2014) while studying fluoride removal performed a multiple regression analysis using statistical package for the social sciences (SPSS) to develop an empirical model to predict the rate constant. The results of the study indicated that the experimental rate constants and the predicted rate constant were in strong agreement. Using the predicted rate constant (empirical equation), the authors successfully presented a model equation to calculate the percentage of fluoride removal:

$$E_t = 1 - e^{-10^{-5}\left[5.9(I/V) - 37.1(C_o) - 82.1(d) + 2746.4\right]t} \tag{4-20}$$

where

Et = Defluoridation efficiency at any time (%),
C0 = Initial fluoride concentration (mg/L),
I/V = Current concentration (A/m3),
d = Distance between electrodes (mm), and
T = Electrolysis time (min).

Chitra and Balasubramanian (2010) proposed a pseudo first-order kinetic model for pollutant removal:

$$q_t = \frac{V(\mathrm{COD}_e - \mathrm{COD}_t)}{m} \tag{4-21}$$

$$\frac{dq_t}{dt} = k_1(q_e - q_t) \tag{4-22}$$

where q_t is the amount of pollutant removed at any time t. Subscript e refers to the equilibrium value of the pollutant.

Sometimes, the systems do not follow first-order kinetics but follow second-order kinetics, which can be expressed as

$$\frac{dq_t}{dt} = k_2(q_e - q_t)^2 \tag{4-23}$$

Solving Equations (4-22) and (4-23) using the normal rules of integral calculus would give solutions in the form of Equations (4-24) and (4-25) for the first-order and second-order kinetics, respectively:

$$\log(q_e - q_t) = \log q_e - \frac{k_1}{2.303}t \tag{4-24}$$

$$\frac{t}{q_t} = \frac{1}{k_2 q_e^2} + \frac{1}{q_e}t \tag{4-25}$$

The values for the rate constants can be easily calculated graphically. For the first-order equation, a graph of log $(q_e - q_t)$ vs time can give the value of the rate constant. Similarly, for the second-order equation [Equation (4-25)], a graph between t/q_t and t can be solved for the value of rate constants. To achieve maximum removal of pollutants, researchers have aimed to maximize the rate constants by varying the different operational parameters.

Other researchers have applied models developed for different adsorption phenomena to EC because of the striking similarities in the principles and mechanisms of pollutant removal in EC. The Elovich model, originally developed for describing the kinetics of heterogeneous chemisorption of gases onto solid surfaces, has been applied to EC by Ouaissa et al. (2014) for studying the removal of tetracycline, a chemical drug (antibiotic) compound. Similar to previously described kinetic models, the linearized form of this model looks similar to the following equation:

$$q_t = \left(\frac{1}{\beta}\right) \cdot \ln\frac{\alpha}{\beta} + \frac{1}{\beta}\ln t \tag{4-26}$$

Ouaissa et al. (2014) applied different kinetic models (first-order, second-order, Elovich model, etc.) to describe the removal of tetracycline. The results indicated that among all models, second-order kinetic equations fit best for the removal of tetracycline by EC.

4.11 MISCELLANEOUS MATHEMATICAL MODELS FOR ELECTRO-COAGULATION

4.11.1 Adsorption Models

Adsorption models have been widely used to describe the EC process. Different adsorption isotherms exist that can be applied to EC. The various isotherm models are (1) Langmuir isotherm, (2) Freundlich isotherm, and (3) Langmuir–Freundlich isotherm. The characteristic equation for these isotherm models is given in Equations (4-27) to (4-29):

Langmuir isotherm

$$q_e = q_{\max}\left(\frac{K_L C_e}{1 + K_L C_e}\right) \tag{4-27}$$

Freundlich isotherm

$$q_e = K_f C_e^{1/p} \tag{4-28}$$

Langmuir–Freundlich isotherm

$$q_e = q_{\max}\left(\frac{K_{\mathrm{LF}} C_e^n}{1 + K_{\mathrm{LF}} C_e^n}\right) \tag{4-29}$$

where q_e and $q_{\max}$ are the equilibrium and maximum amount of pollutants adsorbed per unit mass of adsorbent, respectively. K_L, K_f, and K_{L-F} are the constants for Langmuir, Freundlich, and Langmuir–Freundlich isotherm, respectively. A new method called *variable order kinetics* (VOK) that combines Faraday's law of electrolysis and adsorption models together (Hu et al. 2007) has been used by various researchers to model the removal of pollutants like fluoride. The kinetic form of the model is given in Equation (4-30):

$$-\frac{dC_t}{dt} = \varnothing_m q_e \frac{dM_{\text{tot}}}{dt} \tag{4-30}$$

where $\varnothing_m$ is the efficiency of complexation of fluoride (pollutant) with aluminum (electrode). M_{tot} is the amount of aluminum produced by electrolysis, which can be given by Faraday's law [(Equation (4-31)]:

$$\frac{dM_{\text{tot}}}{dt} = \varnothing \frac{I}{ZFV} \tag{4-31}$$

The VOK model can be easily substituted into adsorption isotherms [Equations (4-27) to (4-29), substitution of q_e] to give a combined model equation [Equations (4-32) to (4-34)]:

Langmuir-VOK

$$\frac{dC_t}{dt} = \varnothing_m \varnothing \frac{I}{ZFV} q_{\max} \left(\frac{K_L C_e}{1 + K_L C_e} \right) \tag{4-32}$$

Freundlich-VOK

$$\frac{dC_t}{dt} = \varnothing_m \varnothing \frac{I}{ZFV} K_f C_e^{1/p} \tag{4-33}$$

Langmuir–Freundlich-VOK

$$\frac{dC_t}{dt} = \varnothing_m \varnothing \frac{I}{ZFV} q_{\max} \left(\frac{K_{LF} C_e^n}{1 + K_{LF} C_e^n} \right) \tag{4-34}$$

Many other versions of adsorption isotherms have been developed by researchers to account for different effects like adsorbent–adsorbent interactions. The Temkin isotherm (Temkin and Pyzhev 1940) is one such model developed to account for adsorbent/adsorbate interaction on the isotherm, and the following form is suggested:

$$q_e = B \ln A + B \ln C_e \tag{4-35}$$

where $B = RT/b$, and A and B are Temkin's constant. A plot of q_e versus ln C_e can be used to determine the values of these constants. A study performed by Chitra and Balasubramanian (2010) compared three major adsorption isotherms to describe the treatment of textile industry effluents. Among the Langmuir, Freundlich, and Temkin isotherms, the Langmuir isotherm best fits the EC data for treating textile effluents.

Hu et al. (2007) studied deflouridation using these combined models and found that the Langmuir-VOK model described the EC process satisfactorily. Adsorption isotherms were also investigated by Balasubramanian et al. (2009)

for arsenic removal. Their study concluded that the standard Langmuir isotherm fits better than the Freundlich isotherm for aluminum and steel electrodes. A study performed by Chou et al. (2010) also compared different adsorption models for the treatment of oxide chemical mechanical polishing wastewater from a semiconductor industry. The results indicated that among Freundlich and Langmuir isotherms, the Freundlich isotherm could better fit the experimental data.

4.11.2 Computational Fluid Dynamics and Electro-Coagulation

Researchers have also applied the concepts of computational fluid dynamics (CFD) to characterize the continuous EC process. Such studies can help simplify the oversimplified assumptions of various kinetic models (like uniform mixing) to mechanistically understand the process. CFD can help understand the effects of mixing, rheology, mass transfer, and other hydrodynamic properties on EC.

The simulation of different electrochemical events is performed by various mathematical equations for mass, momentum, energy, electrical potential, and current spatial distribution. The flow characteristics of an EC flow cell can be laminar or turbulent. The mathematical solutions for mass balance (in Cartesian coordinates) are given by the following continuity equation [(Equation (4-36)]:

$$\frac{\partial \rho}{\partial t} + \rho \frac{\partial u_i}{\partial x_i} = 0 \tag{4-36}$$

where ρ is the density of the fluid, and u_i is the velocity vector. The momentum equation is normally described by the general Reynolds-Averaged–Navier–Stokes (RANS) equation:

$$\frac{\partial \rho u_i}{\partial t} + \frac{\partial \rho u_i u_j}{\partial x_i} = \frac{\partial p}{\partial x_i} + \frac{\partial}{\partial x_i}\left(\mu \frac{\partial u_i}{\partial x_j}\right) + \frac{\partial}{\partial x_j}\left(-\rho u_i' u_j'\right) + \rho g_i \tag{4-37}$$

where

u' = Turbulent velocity,
p = Pressure, and
μ = Kinematic viscosity.

Similarly, for different species of ions or any chemical entity, the conservation of mass is given by Equation (4-38):

$$\frac{\partial C_i}{\partial t} = -\nabla N_i + R_i \tag{4-38}$$

where

C_i = Average concentration;
N_i = Flux of chemical species because of convection of fluid flow; and
R = Reaction rate term.

N_i can be calculated by applying the Nernst–Planck equation [Equation (4-39)]:

$$N_i = C_i u - Z_i u_{m,i} F C_i \nabla \varnothing - D_i \nabla C_i \tag{4-39}$$

where
Z_i = Charge number,
u = Velocity,
$u_{m,i}$ = Mobility of species I, and
D_i = Diffusion coefficient.

Lu et al. (2017) modeled the distribution of various ionic species produced during the EC process in an EC flow cell (30 cm long with a continuous flow of inlet and outlet streams). Their study concluded that Al^{3+} and OH^- concentrations were invariable along the direction of streamline flow. A high concentration of ions was found in the vicinity of the electrodes. A similar distribution was observed for aluminum hydroxide species. The study further investigated the spatial distributions of H^+ and OH^- and concluded that in the EC process, three pH fronts could be observed, namely, acid front (dominated by H^+), base front (dominated by OH^-), and buffering front. The presence of these fronts will decide the overall pH of the area (pH front), which, in turn, can affect the nature of charge on the chemical species. The amount of hydronium and hydroxyl ions is directly proportional to current density, but the shape and size of the pH front are found to be independent of current density. A major part of $Al(OH)_3$ is generated in the buffering front of the EC channel. Thus, the study clearly indicated the spatial distribution, and the amount of the hydroxide flocs of aluminum is controlled by mass transfer and hydrolysis reactions in the EC channel.

Current distribution is another major factor that can influence the EC efficiency and energy usage. In CFD studies, current and potential distributions are categorized (Ferziger and Peric 2012) as three types: (1) primary, (2) secondary, and (3) tertiary. Primary distribution does not take account of the charge transfer on electrodes because of negligible overpotential and the presence of concentration gradients. Secondary distribution considers the overpotential of activation where charge transfer takes place on the electrode surface but still ignores the overpotential because of the concentration gradients. In tertiary distribution, when there is a concentration gradient near electrodes because of the application of high current above the limiting value, there is a superimposition of the overpotential of concentration on the overpotential of activation.

The uniformity of current distribution is often expressed as a function of Wagner dimensionless number [given in Equations (4-40) and –(4-41)].

$$W_a = \frac{K}{L}\frac{\partial n}{\partial i} \tag{4-40}$$

$$\nabla^2 \varnothing = 0 \tag{4-41}$$

where

$\partial n/\partial i$ = Slope of anodic polarization,
K = Electrolyte conductivity, and
L = Length of the system.

CFD has been successfully used to study hydrodynamics phenomena in an EC cell, such as channeling, internal recirculation, and dead zones. CFD studies can help improve the reactor design to obtain optimized operational parameters and enhance the efficiency of the system. Vázquez et al. (2012) studied the distribution of current in an electrochemical cell. The study showed that current and potential are not uniformly distributed in a conventional cell. The edge of electrodes has a comparatively high electric density, which can produce local zone effects resulting in lower and higher current densities in different zones. The theoretical studies of current density distribution can help in designing an efficient EC reactor. Their study concluded that an improvement in current density distribution in EC can result in an improvement in energy usage, and different models can be used to minimize energy consumption. Although CFD has great potential to perform integral theoretical–experimental studies to improve the EC cell design and scale up. The drawback with CFD is that the computational part requires high skills and training to execute and analyze data. The computational times required for simulation necessitate the use of powerful computing devices (computers). Simultaneous development of such computers will solve this part of the problem in future.

4.11.3 Mathematical Model for Electro-Coagulation Using Reaction Kinetics

Modeling EC speciation of pollutants and coagulating agents has become highly necessary (Canizares et al. 2008a, b). From this perspective, two major processes that take place during EC are (1) direct precipitation of ions, and (2) enmeshment of inorganic ions in the flocs of Fe/Al. According to various studies, the pollutant molecules are espectated as follows:

$$AP_0 = AP_1 + AP_2 + AP_3 + AP_4 \tag{4-42}$$

where AP (anionic pollutant) 1, 2, 3, 4, and 0 are refractory, direct precipitation removable, enmeshment removable, residual removable anions, and total anions, respectively. This implies that, by EC, the removed ions will comprise AP_2 and AP_3 (AP_e eliminated). The event of direct precipitation is considered as an equilibrium reaction given by Equation (4-43):

$$AP_4 + nM_3 \leftrightarrow AP_2 + M_4 k_1 \tag{4-43}$$

where n is the coefficient of direct precipitation, and it varies with a charge of the species present in the solution. K_1 is the equilibrium constant. Because $AP_2 = n^* M_4$, K_1 can be calculated as $k_1 = AP_2/M_3 {}^* AP_4$. Similarly, enmeshment reaction can be considered as another equilibrium reaction [Equation (4-44)]:

$$AP_4 + nM_1 \leftrightarrow AP_3 \quad \text{where } AP_3 = mM_1 \tag{4-44}$$

Solving for AP_4, we can get Equation (4-45):

$$AP_4 = AP_0 - AP_1 - mM_1 \tag{4-45}$$

and

$$AP_4 = \frac{AP_0 - AP_1 - mM_1}{1 + k_1 \cdot M_3} \tag{4-46}$$

and

$$M_3 = \frac{M_0/\alpha - M_1 - M_2}{1 + (k_1 AP_4/n)} \tag{4-47}$$

This phenomenological model has been used by Lacasa et al. (2013) for the removal of arsenate using EC. The study tested the fitting suitability of the model and found that the model can be used reproducibly with a regression coefficient (R^2) > 0.9. The study further used this model to simulate the treatment of a broad range of As concentration by EC and minimized the cost of treatment. Similarly, Canizares et al. (2008) developed the model for the removal of textile dye and treated the effluents from the textile industry. The study used the model that had been developed previously with Kaolin suspension (Canizares et al. 2008). Using the speciation of aluminum and pollutants (in terms of turbidity), the model uses pseudo equilibrium equations to characterize the coagulation and separation events in the EC system. The models fit the experimental validation data suitably with an overall regression coefficient (R^2) of 0.92. Li et al. (2012) developed a chemical kinetic model for arsenic removal using various chemical reactions that were envisaged to happen during the EC process [Equations (4-48) to (4-53)]:

Oxidation of Fe(II):

$$\text{Fe(IV)} + \text{Fe(II)} \rightarrow 2\text{Fe(III)}k_1 \tag{4-48}$$

Reduction of Fe(IV) using A (III):

$$\text{Fe(IV)} + \text{As(III)} \rightarrow \text{Fe(III)} + \text{As(IV)}k_2 \tag{4-49}$$

Fe(IV) is used in both the aforementioned processes. Thus, to determine the fraction of Fe(IV) for each purpose, the model proposes that the fraction can be calculated using the following equation:

$$R_{\text{Fe(IV)}\rightarrow\text{As(III)}} = \frac{1}{1 + k_1[\text{Fe(II)}]/k_2[\text{As(III)}]} \tag{4-50}$$

Using the aforementioned equations, the model calculates the rate of As(III) oxidation as

$$\begin{aligned}\frac{d[\text{As(III)}]_{\text{oxidation}}}{dt} &= R * \frac{d[\text{Fe(IV)}]_{\text{oxidation}}}{dt} \\ &= \frac{\beta}{1 + (k_1[\text{Fe(II)}]/k_2[\text{As(III)}])} \frac{d[\text{Fe(II)}]_{\text{oxidation}}}{dt}\end{aligned} \tag{4-51}$$

The rate of oxidation of Fe(II) is given by $d[\mathrm{Fe(II)}]_{\mathrm{oxidation}}/dt = -k_{\mathrm{app}}[\mathrm{Fe(II)}][\mathrm{O_2}]$. The apparent rate constant (K_{app}) is largely influenced by Fe(II) speciation with the corresponding pH, and it is equal to the weighted sum of the oxidation rates of all possible aqueous Fe (II) species. The adsorption of arsenate is given by

$$[\mathrm{As(III)}] = \frac{q_{\max}[\mathrm{Fe(III)}]K_{\mathrm{As(III)}}[\mathrm{As(III)}]}{1 + K_{\mathrm{As(III)}}[\mathrm{As(III)}] + K_{\mathrm{As(V)}}[\mathrm{As(V)}] + K_P[\mathrm{P}] + K_{\mathrm{Si}}[\mathrm{Si}]} \tag{4-52}$$

where

$q_{\max}$ = Adsorption capacity;

Fe (III) = Concentration of HFO; and

$K_{\mathrm{AS(III)}}$, $K_{\mathrm{AS(V)}}$, K_P, and K_{Si} = Adsorption equilibria constants for the competitive adsorption of AS(III), As(V), P, and Si, respectively.

By solving the equations as proposed in the model, the final concentration of As(III) after treatment can be easily calculated, that is, $\mathrm{As(III)}_{\mathrm{final}}$ = initial–oxidized–Adsorbed. Li et al. (2012) performed several experiments to obtain the values of model parameters and then validated them using their experimental results. Their results show a reasonably good agreement between predicted and experimental values. The model can successfully provide minimal iron dosage needed for adequate treatment and, thus, can help efforts aimed at cost minimization.

4.11.4 Electro-Coagulation Modeling Using Flotation and Settling Phenomena

Settling and floatation are two competitive physical phenomena that occur in an EC system. Settling is favored by metal hydroxide flocs that entrap pollutants, whereas floatation is mainly caused by physical interaction between gaseous by-products (hydrogen or oxygen) of electrolysis reactions. Holt et al. (2005) described a model to explain these two phenomena. The study assumed that the removal of pollutant in the bulk solution follows a first-order kinetics:

$$C_{\mathrm{pollutant}} \xrightarrow{k_1} \mathrm{Surface} \quad \text{and} \quad C_{\mathrm{pollutant}} \xrightarrow{k_2} \mathrm{Base} \tag{4-53}$$

Therefore, the rate of pollutant (r_{poll}) removal will be

$$r_{\mathrm{poll}} = -\frac{dC_{\mathrm{poll}}}{dt} = k_1 C_{\mathrm{poll}} + k_2 C_{\mathrm{poll}} = (k_1 + k_2) C_{\mathrm{poll}} \tag{4-54}$$

The rate constants (k_1 and k_2) can be determined from experimental data analysis. In the bulk liquid, once the flock is heavier than the aqueous media, they will start settling down. The settling down of the sludge flock has been modeled by various researchers. These models consist of various variants such as power model, exponential model, Cho's experimental model, third-order model, fourth-order model, and Cho's complex exponential models. Although they have been applied sporadically for providing an estimate, these models are not very precise as they ignore the fact that the system faces a lot of hindrance from rising gaseous bubbles and distortions because of electrical fields. Details of these models can be found in the works of Hakizimana et al. (2017).

4.11.5 Electro-Coagulation Modeling Using Flocculation

Among various processes happening in an EC cell, floc formation and aggregation are among the major ones. The flocculation of particles occurs by the collision of two or more particles by virtue of Brownian motion, convective flow of bulk liquid, and gravitational field. The rate of collision between any two particles can be modeled as a function of collision frequency, particle concentration, and particle size (Thomas et al. 1999). Some reports propose highly simplified models in an idealized reactor system, but they are rarely applicable to real-life systems. Therefore, new and improved models are required to model flocculation in the EC system. The fractal theory can quantify disordered systems, and it is based on power law behavior (Bushell et al. 2002) and mathematically expressed by Equation (4-55):

$$M \alpha r^{D_f} \tag{4-55}$$

where

M = Mass,
r = Radius, and
D_f = Mass fractal dimensions.

The fractal dimensions' value ranges from 1 to 3. The value of D_f varies with the dimensions of the particle, and it is measured by scattering, image analysis, and sedimentation velocity. The fractal theory has not been used in EC, except in a few studies. Argaman and Kaufman (Ofir et al. 2007) studied the overall kinetic model of flocculation by integrating the floc break-up rate in turbulent mixing. Further research can help in understanding the development of flocs in EC and their geometrical characteristics for optimized settling and shortening the settling separation times.

4.12 CONCLUDING REMARKS

In this chapter, we discussed various mathematical and statistical tools that have been applied to simulate and develop predictive models for EC. The modeling in EC is performed by various techniques such as ANN and RSM and by using phenomenological models such as adsorption, flocculation, floatation, computational fluid dynamics, and electrochemical reaction kinetics. All these models have been applied to the EC system and they yield reliable predictions, although RSM and ANN appear to be the most preferred and most robust techniques for model development. Other modeling techniques (phenomenological models) are rapid models that can be developed, but they display lower predictive strength and confidence as compared with ANN and

RSM. In the models based on phenomena, often oversimplification of the process for ease of calculations happens, which makes the models far removed from reality. Although these models can make faster predictions, they cannot be trusted for high scale-up and complex reactor geometry design. To develop the most accurate models, it is advisable to subscribe to the most realistic assumptions (from CFD and other phenomenological models) and apply them together to modify and treat the data obtained from ANN or RSM to achieve more accuracy and precision in data prediction. Modeling techniques from other disciplines like fluid dynamics, solid–liquid systems, and electrochemistry can be unified together to develop one series of compounded modeling techniques to accommodate the effects that are overlooked by the limitations posed by one single model. Further advancement in computational tools and the available high-skill training can help improve modeling techniques and possibly lead to an increase in the number of researchers applying these modeling techniques to EC.

References

Aber, S., A. R. Amani-Ghadim, and V. Mirzajani. 2009. "Removal of Cr(VI) from polluted solutions by electrocoagulation: Modeling of experimental results using artificial neural network." *J. Hazard. Mater.* 171 (1–3): 484–490.

Alinsafi, A., M. Khemis, M. N. Pons, J. P. Leclerc, A. Yaacoubi, A. Benhammou, et al. 2005. "Electro-coagulation of reactive textile dyes and textile wastewater." *Chem. Eng. Process.* 44 (4): 461–470.

Amani-Ghadim, A. R., S. Aber, A. Olad, and H. Ashassi-Sorkhabi. 2013. "Optimization of electrocoagulation process for removal of an azo dye using response surface methodology and investigation on the occurrence of destructive side reactions." *Chem. Eng. Process.* 64: 68–78.

Arulmurugan, A., C. Kumaran, T. Ramamurthy, and B. Subramanian. 2007. "Degradation of textile effluent by electro coagulation technique." *Bull. Electrochem.* 23: 247–252.

Balasubramanian, N., T. Kojima, and C. Srinivasakannan. 2009. "Arsenic removal through electrocoagulation: Kinetic and statistical modeling." *Chem. Eng. J.* 155 (1–2): 76–82.

Behbahani, M., M. R. A. Moghaddam, and M. Arami. 2011. "Techno-economical evaluation of fluoride removal by electrocoagulation process: Optimization through response surface methodology." *Desalination* 271 (1–3): 209–218.

Behloul, M., H. Grib, N. Drouiche, N. Abdi, H. Lounici, and N. Mameri. 2013. "Removal of malathion pesticide from polluted solutions by electrocoagulation: Modeling of experimental results using response surface methodology." *Sep. Sci. Technol.* 48 (4): 664–672.

Bezerra, M. A., R. E. Santelli, E. P. Oliveira, L. S. Villar, and L. A. Escaleira. 2008. "Response surface methodology (RSM) as a tool for optimization in analytical chemistry." *Talanta* 76 (5): 965–977.

Bhatti, M. S., D. Kapoor, R. K. Kalia, A. S. Reddy, and A. K. Thukral. 2011a. "RSM and ANN modeling for electrocoagulation of copper from simulated wastewater: Multi objective optimization using genetic algorithm approach." *Desalination* 274 (1–3): 74–80.

Bhatti, M. S., A. S. Reddy, R. K. Kalia, and A. K. Thukral. 2011b. "Modeling and optimization of voltage and treatment time for electrocoagulation removal of hexavalent chromium." *Desalination* 269 (1–3): 157–162.

Bishop, C. M. 1995. *Neural networks for pattern recognition*. Oxford, UK: Oxford University Press.

Bushell, G., Y. Yan, D. Woodfield, J. Raper, and R. Amal. 2002. "On techniques for the measurement of the mass fractal dimension of aggregates." *Adv. Colloid Interface Sci.* 95 (1): 1–50.

Butler, E., Y.-T. Hung, R. Y.-L. Yeh, and M. S. Al Ahmad. 2011. "Electrocoagulation in wastewater treatment." *Water* 3 (4): 495–525.

Canizares, P., F. Martinez, M. Rodrigo, C. Jimenez, C. Saez, and J. Lobato. 2008a. "Modelling of wastewater electrocoagulation processes. Part I: General description and application to kaolin-polluted wastewaters." *Sep. Purif. Technol.* 60 (2): 155–161.

Canizares, P., F. Martinez, M. Rodrigo, C. Jimenez, C. Saez, and J. Lobato. 2008b. "Modelling of wastewater electrocoagulation processes. Part II: Application to dye-polluted wastewaters and oil-in-water emulsions." *Sep. Purif. Technol.* 60 (2): 147–154.

Cassettari, L., R. Mosca, R. Revetria, and F. Rolando. 2013. "Effectiveness and limits of response surface methodology in application to discrete and stochastic simulation of manufacturing plants." *Appl. Math. Sci.* 7 (83): 4137–4172.

Chitra, K., and N. Balasubramanian. 2010. "Modeling electrocoagulation through adsorption kinetics." *J. Model. Simul. Syst.* 1 (2): 124–130.

Chou, W. L., C. T. Wang, W. C. Chang, and S. Y. Chang. 2010. "Adsorption treatment of oxide chemical mechanical polishing wastewater from a semiconductor manufacturing plant by electrocoagulation." *J. Hazard. Mater.* 180 (1–3): 217–224.

Curteanu, S., C. G. Piuleac, K. Godini, and G. Azaryan. 2011. "Modeling of electrolysis process in wastewater treatment using different types of neural networks." *Chem. Eng. J.* 172 (1): 267–276.

Dalvand, A., M. Gholami, A. Joneidi, and N. M. Mahmoodi. 2011. "Dye removal, energy consumption and operating cost of electrocoagulation of textile wastewater as a clean process." *CLEAN—Soil, Air, Water* 39 (7): 665–672.

Daneshvar, N., A. R. Khataee, and N. Djafarzadeh. 2006. "The use of artificial neural networks (ANN) for modeling of decolorization of textile dye solution containing C. I. Basic Yellow 28 by electrocoagulation process." *J. Hazard. Mater.* 137 (3): 1788–1795.

Demuth, H., and M. Beale. 1993. *Neural network toolbox for use with Matlab–User's guide version 3.0*. Natick, MA: MathWorks.

Drouiche, N., S. Aoudj, H. Lounici, H. Mahmoudi, N. Ghaffour, and M. F. A. Goosen. 2012. "Development of an empirical model for fluoride removal from photovoltaic wastewater by electrocoagulation process." *Desalin. Water Treat.* 29 (1–3): 96–102.

Elemen, S., E. P. Akçakoca Kumbasar, and S. Yapar. 2012. "Modeling the adsorption of textile dye on organoclay using an artificial neural network." *Dyes Pigm.* 95 (1): 102–111.

Espinoza-Quinones, F. R., M. M. T. Fornari, A. N. Modenes, S. M. Palacio, D. E. Trigueros, F. H. Borba, et al. 2009. "Electrocoagulation efficiency of the tannery effluent treatment using aluminium electrodes." *Water Sci. Technol.* 60 (8): 2173–2185.

Ferziger, J. H., and M. Peric. 2012. *Computational methods for fluid dynamics*. Berlin: Springer.

Forsyth, R. 1988. *Machine learning: Principles and techniques*. London: Chapman & Hall.

García-Lara, A. M., C. Montero-Ocampo, F. Equihua-Guillen, J. E. Camporredondo-Saucedo, R. Servin-Castaneda, and C. R. Muñiz-Valdes. 2014. "Arsenic removal from natural groundwater by electrocoagulation using response surface methodology." *J. Chem.* 2014: 857625.

Garson, G. D. 1998. *Neural networks: An introductory guide for social scientists*. London: Sage.

Gengec, E., M. Kobya, E. Demirbas, A. Akyol, and K. Oktor. 2012. "Optimization of baker's yeast wastewater using response surface methodology by electrocoagulation." *Desalination* 286: 200–209.

Güçlü, D. 2014. "Optimization of electrocoagulation of pistachio processing wastewaters using the response surface methodology." *Desalin. Water Treat.* 54 (12): 3338–3347.

Hakizimana, J. N., B. Gourich, M. Chafi, Y. Stiriba, C. Vial, P. Drogui, et al. 2017. "Electrocoagulation process in water treatment: A review of electrocoagulation modeling approaches." *Desalination* 404: 1–21.

Holt, P. K., G. W. Barton, and C. A. Mitchell. 2005. "The future for electrocoagulation as a localised water treatment technology." *Chemosphere* 59 (3): 355–367.

Hu, C.-Y., S.-L. Lo, and W.-H. Kuan. 2007. "Simulation the kinetics of fluoride removal by electrocoagulation (EC) process using aluminum electrodes." *J. Hazard. Mater.* 145 (1): 180–185.

Kabuk, H. A., F. İlhan, Y. Avsar, U. Kurt, O. Apaydin, and M. T. Gonullu. 2014. "Investigation of leachate treatment with electrocoagulation and optimization by response surface methodology." *CLEAN—Soil, Air, Water* 42 (5): 571–577.

Karichappan, T., S. Venkatachalam, and P. M. Jeganathan. 2014. "Optimization of electrocoagulation process to treat grey wastewater in batch mode using response surface methodology." *J. Environ. Health Sci. Eng.* 12 (1): 29.

Karichappan, T., S. Venkatachalam, P. M. Jeganathan, and K. Sengodan. 2013. "Treatment of rice mill wastewater using continuous electrocoagulation technique: Optimization and modelling." *J. Korean Chem. Soc.* 57 (6): 761–768.

Khataee, A. R., and M. B. Kasiri. 2010. "Artificial neural networks modeling of contaminated water treatment processes by homogeneous and heterogeneous nanocatalysis." *J. Mol. Catal. A: Chem.* 331 (1–2): 86–100.

Kobya, M., E. Demirbas, M. Bayramoglu, and M. T. Sensoy. 2010. "Optimization of electrocoagulation process for the treatment of metal cutting wastewaters with response surface methodology." *Water Air Soil Pollut.* 215 (1–4): 399–410.

Kobya, M., E. Demirbas, U. Gebologlu, M. S. Oncel, and Y. Yildirim. 2013. "Optimization of arsenic removal from drinking water by electrocoagulation batch process using response surface methodology." *Desalin. Water Treat.* 51 (34–36): 6676–6687.

Kobya, M., E. Gengec, M. T. Sensoy, and E. Demirbas. 2014. "Treatment of textile dyeing wastewater by electrocoagulation using Fe and Al electrodes: Optimisation of operating parameters using central composite design." *Color. Technol.* 130 (3): 226–235.

Kuokkanen, V., T. Kuokkanen, J. Rämö, and U. Lassi. 2013. "Recent applications of electrocoagulation in treatment of water and wastewater—A review." *Green Sustainable Chem.* 3 (2): 89–121.

Lacasa, E., P. Cañizares, C. Sáez, F. Martínez, and M. A. Rodrigo. 2013. "Modelling and cost evaluation of electro-coagulation processes for the removal of anions from water." *Sep. Purif. Technol.* 107: 219–227.

Lakshmanan, D., D. A. Clifford, and G. Samanta. 2010. "Comparative study of arsenic removal by iron using electrocoagulation and chemical coagulation." *Water Res.* 44 (19): 5641–5652.

Lakshmi, M., and P. Sivashanmugam. 2013. "Treatment of oil tanning effluent by electrocoagulation: Influence of ultrasound and hybrid electrode on COD removal." *Sep. Purif. Technol.* 116: 378–384.

Li, L., C. M. van Genuchten, S. E. Addy, J. Yao, N. Gao, and A. J. Gadgil. 2012. "Modeling As(III) oxidation and removal with iron electrocoagulation in groundwater." *Environ. Sci. Technol.* 46 (21): 12038–12045.

Lu, J., Z. Wang, X. Ma, Q. Tang, and Y. Li. 2017. "Modeling of the electrocoagulation process: A study on the mass transfer of electrolysis and hydrolysis products." *Chem. Eng. Sci.* 165: 165–176.

Mameri, N., A. R. Yeddou, H. Lounici, D. Belhocine, H. Grib, and B. Bariou. 1998. "Defluoridation of septentrional Sahara water of North Africa by electrocoagulation process using bipolar aluminium electrodes." *Water Res.* 32 (5): 1604–1612.

Mirsoleimani-Azizi, S. M., A. A. Amooey, S. Ghasemi, and S. Salkhordeh-Panbechouleh. 2015. "Modeling the removal of endosulfan from aqueous solution by electrocoagulation process using artificial neural network (ANN)." *Ind. Eng. Chem. Res.* 54 (40): 9844–9849.

Mollah, M. Y., P. Morkovsky, J. A. Gomes, M. Kesmez, J. Parga, and D. L. Cocke. 2004. "Fundamentals, present and future perspectives of electrocoagulation." *J. Hazard. Mater.* 114 (1–3): 199–210.

Mollah, M. Y. A., R. Schennach, J. R. Parga, and D. L. Cocke. 2001. "Electrocoagulation(EC)-science and applications." *J. Hazard. Mater.* 84 (1): 29–41.

Moulai-Mostefa, N., S. Ladjelat, H. Kermet-Said, M. Krea, and M. Tir. 2013. "Optimization of operational parameters in the pretreatment of surface water by electrocoagulation using a response surface method." *Desalin. Water Treat.* 52 (13–15): 2382–2387.

Murugan, A. A., T. Ramamurthy, B. Subramanian, C. S. Kannan, and M. Ganesan. 2009. "Electrocoagulation of textile effluent: RSM and ANN modeling." *Int. J. Chem. Reactor Eng.* 7 (1): A83.

Navarro, H., and L. Bennun. 2014. "Descriptive examples of the limitations of artificial neural networks applied to the analysis of independent stochastic data." *Int. J. Comput. Eng. Technol.* 5 (5): 40–42.

Nourouzi, M. M., T. G. Chuah, and T. S. Choong. 2011. "Optimisation of reactive dye removal by sequential electrocoagulation-flocculation method: Comparing ANN and RSM prediction." *Water Sci. Technol.* 63 (5): 984–994.

Ofir, E., Y. Oren, and A. Adin. 2007. "Modified equilibrium-solubility domains and a kinetic model of iron oxide and hydroxide colloids for electroflocculation." *Desalination* 204 (1–3): 79–86.

Olmez-Hanci, T., Z. Kartal, and I. Arslan-Alaton. 2012. "Electrocoagulation of commercial naphthalene sulfonates: Process optimization and assessment of implementation potential." *J. Environ. Manage.* 99: 44–51.

Orssatto, F., M. H. Ferreira Tavares, F. M. da Silva, E. Eyng, B. Farias Biassi, and L. Fleck. 2017. "Optimization of the pretreatment of wastewater from a slaughterhouse and packing plant through electrocoagulation in a batch reactor." *Environ. Technol.* 38 (19): 2465–2475.

Ouaissa, Y. A., M. Chabani, A. Amrane, and A. Bensmaili. 2014. "Removal of tetracycline by electrocoagulation: Kinetic and isotherm modeling through adsorption." *J. Environ. Chem. Eng.* 2 (1): 177–184.

Piuleac, C. G., S. Curteanu, M. A. Rodrigo, C. Sáez, and F. J. Fernández. 2013. "Optimization methodology based on neural networks and genetic algorithms applied to electro-coagulation processes." *Cent. Eur. J. Chem.* 11: 1213–1224.

Ponselvan, F. I. A., M. Kumar, J. R. Malviya, V. C. Srivastava, and I. D. Mall. 2008. "Electrocoagulation studies on treatment of biodigester effluent using aluminum electrodes." *Water Air Soil Pollut.* 199 (1–4): 371–379.

Secula, M. S., C. S. Stan, C. Cojocaru, B. Cagnon, and I. Cretescu. 2014. "Multi-objective optimization of indigo carmine removal by an electrocoagulation/GAC coupling process in a batch reactor." *Sep. Sci. Technol.* 49 (6): 924–938.

Shankar, R., L. Singh, P. Mondal, and S. Chand. 2013. "Removal of COD, TOC, and color from pulp and paper industry wastewater through electrocoagulation." *Desalin. Water Treat.* 52 (40–42): 7711–7722.

Sinha, K., P. D. Saha, and S. Datta. 2012. "Response surface optimization and artificial neural network modeling of microwave assisted natural dye extraction from pomegranate rind." *Ind. Crops Prod.* 37 (1): 408–414.

Taheri, M., M. R. A. Moghaddam, and M. Arami. 2015. "Improvement of the /Taguchi/ design optimization using artificial intelligence in three acid azo dyes removal by electrocoagulation." *Environ. Prog. Sustainable Energy* 34 (6): 1568–1575.

Temkin, M., and V. Pyzhev. 1940. "Kinetics of ammonia synthesis on promoted iron catalysts." *Acta Physiochim. URSS* 12 (3): 217–222.

Thakur, L. S., and P. Mondal. 2016. "Techno-economic evaluation of simultaneous arsenic and fluoride removal from synthetic groundwater by electrocoagulation process: Optimization through response surface methodology." *Desalin. Water Treat.* 57 (59): 28847–28863.

Thirugnanasambandham, K., V. Sivakumar, and J. P. Maran. 2014a. "Efficiency of electrocoagulation method to treat chicken processing industry wastewater—Modeling and optimization." *J. Taiwan Inst. Chem. Eng.* 45 (5): 2427–2435.

Thirugnanasambandham, K., V. Sivakumar, and J. P. Maran. 2015. "Evaluation of an electrocoagulation process for the treatment of bagasse-based pulp and paper industry wastewater." *Environ. Prog. Sustainable Energy* 34 (2): 411–419.

Thirugnanasambandham, K., V. Sivakumar, and M. Prakash. 2014b. "Optimization of electrocoagulation process to treat biologically pretreated bagasse effluent." *J. Serb. Chem. Soc.* 79 (5): 613–626.

Thirugnanasambandham, K., V. Sivakumar, and K. Shine. 2016. "Studies on treatment of egg processing industry wastewater using electrocoagulation method: Optimization using response surface methodology." *Desalin. Water Treat.* 57 (46): 21721–21729.

Thomas, D., S. Judd, and N. Fawcett. 1999. "Flocculation modelling: A review." *Water Res.* 33 (7): 1579–1592.

Tipping, M. E. 2004. "Bayesian inference: An introduction to principles and practice in machine learning." In Vol. 3176 *of Advanced Lectures on Machine Learning. Lecture Notes in Computer Science*, edited by O. Bousquet, U. von Luxburg, and G. Rätsch, 41–62. Berlin: Springer.

Turan, N. G., B. Mesci, and O. Ozgonenel. 2011. "The use of artificial neural networks (ANN) for modeling of adsorption of Cu(II) from industrial leachate by pumice." *Chem. Eng. J.* 171 (3): 1091–1097.

Un, T. U., A. Kandemir, N. Erginel, and S. E. Ocal. 2014. "Continuous electrocoagulation of cheese whey wastewater: An application of response surface methodology." *J. Environ. Manage.* 146: 245–250.

Vázquez, A., I. Rodríguez, and I. Lázaro. 2012. "Primary potential and current density distribution analysis: A first approach for designing electrocoagulation reactors." *Chem. Eng. J.* 179: 253–261.

Verma, A. K., R. R. Dash, and P. Bhunia. 2012. "A review on chemical coagulation/flocculation technologies for removal of colour from textile wastewaters." *J. Environ. Manage.* 93 (1): 154–168.

Yehya, T., M. Chafi, W. Balla, C. Vial, A. Essadki, and B. Gourich. 2014. "Experimental analysis and modeling of denitrification using electrocoagulation process." *Sep. Purif. Technol.* 132: 644–654.

Zaroual, Z., H. Chaair, A. H. Essadki, K. El Ass, and M. Azzi. 2009. "Optimizing the removal of trivalent chromium by electrocoagulation using experimental design." *Chem. Eng. J.* 148 (2–3): 488–495.

Zodi, S., O. Potier, F. Lapicque, and J.-P. Leclerc. 2010. "Treatment of the industrial wastewaters by electrocoagulation: Optimization of coupled electrochemical and sedimentation processes." *Desalination* 261 (1–2): 186–190.

CHAPTER 5

Mathematical Modeling of the Electro-Oxidation Process

Majid Gholami Shirkoohi, M. R. Karimi Estahbanati, Zahra Nayernia, Pedram Ramin, Krist V. Gernaey, Patrick Drogui, R. D. Tyagi

5.1 INTRODUCTION

The electro-oxidation (EO) process has received much attention in recent years because it can be used as an alternative to traditional wastewater treatment methods with more flexibility. The performance of EO processes is usually affected by several parameters such as amperage, electrical conductivity, pH, water matrix, pollution concentration, type of electrode, and reactor volume. In such a complex condition, the modeling approach can be one of the best ways to avoid spending too much time and cost on various tests to analyze this process (Wang et al. 2016, Yao et al. 2019). This approach provides the tools to perform analyses that are difficult to perform by the experimental approach, like analysis of the interaction effects of operating parameters. In this chapter, the fundamentals and different studies on phenomenological and empirical modeling of EO processes are reviewed.

5.2 MODELING TECHNIQUES AVAILABLE FOR ELECTRO-OXIDATION

The developed models for EO can be categorized into phenomenological and empirical models. The phenomenological models are mathematically developed based on physical or chemical concepts of the process. In contrast to empirical models, the phenomena behind the process are important and are mathematically formulated to obtain the phenomenological model. To develop a phenomenological model, mass and heat balances should be formulated and combined with appropriate constitutive equations (e.g., transport kinetics) to describe the evolution of the system. It is, thus, important to understand the behavior and interaction of the model components in different environments.

The empirical models take advantage of mathematical and statistical methods to predict the performance of an EO process using a series of experimental data. An accurate empirical model is developed, thanks to the distribution of experimental data at the entire range of the operating parameters that helps in the identification of a model that is valid in the entire range of available data. Therefore, the phenomena behind the EO process are not important for the development of empirical models. Response surface methodology (RSM) and artificial neural networks (ANNs) are the most frequently used empirical models for the analysis of the EO processes.

RSM is a combination of mathematical and statistical methods and is used for designing experiments, modeling, and optimizing different processes. The purpose of RSM is to optimize the response variables (outputs) that are influenced by independent variables (inputs). In this method, the relationship between the response variables and the independent variables is polynomial (first or second order). The values of the polynomial coefficients are determined to have the best fitting of the response surface to the experimental data (Anderson and Whitcomb 2016a, b; Behera et al. 2018; Montgomery 2017; Morshedi and Akbarian 2014).

The ANN method is inspired by the biological nervous system and is usually preferred over other methods because of its highly accurate prediction ability (Estahbanati et al. 2017). This method consists of several layers that perform mathematical operations on the input data. The input layer receives the information that needs to be processed and sends it to the hidden layers. After transformation, the hidden layer transfers the information to the output layer. The coefficients of the ANN layers are calculated to have the best fitting to the input–output experimental data. This method is highly recommended for complex problems that require a lot of simplifying assumptions, because there is no need to introduce phenomena involved in the process as mathematical equations (Saravanathamizhan et al. 2015).

In this chapter, the details and fundamentals of different modeling approaches, including phenomenological models, RSM, and ANNs, used for the EO process are provided. Also, kinetic analysis of the EO process is reviewed.

5.3 PHENOMENOLOGICAL MODELING

Phenomenological models can be used to predict the behavior of an electrochemical cell based on the information about the phenomena that govern the underlying processes. Therefore, the assumptions on which these models are built should be consistent with the processes occurring in a physical system. In phenomenological models (or first-principles models), the dynamic evolution of the system can be described according to physical laws by the formulation of mass and heat balances combined with a suitable set of constitutive equations. To set up such mathematical formulations for mimicking the dynamics of the real system, it is important to understand in detail how model components (organic chemicals) behave in different environments and how they interact with one another.

In an electrochemical cell, different levels of interactions exist among ionic species. Ions can interact with one another and also with the solvent. Moreover, ions interact with the boundary of the electrodes and also with other boundaries present in the system. A number of steps are involved in interchanging the charge between ionic species in the electrolyte and the electrons from the electrode: (1) Ionic species move toward and/or outward the electrode surface. This migration is governed by mass transfer mechanisms. (2) Ionic species in the vicinity of the electrode absorb or release electrons, through charge transfer, to form products, and (3) these products move away from the electrode surface and new ionic species move toward the electrode surface (Adesokan 2015). Although Step 2 describes the electrochemical reaction and transformation of reactants to products, Steps 1 and 3 describe the movement of chemicals inside the electrolyte solution because of the existence of various forces.

5.3.1 Electrochemical Kinetics

Considering the generic reactant, S_o, and the product, S_R, an electrode reaction can be considered [Equation (5-1)]:

$$S_o + ne^- \underset{k_a}{\overset{k_c}{\rightleftarrows}} S_R \tag{5-1}$$

in which n is the number of participating electrons, and k_c and k_a are the reduction and oxidation constants, respectively. Using Faraday's law, the rate of an electrode reaction as a function of current, I (A) can be formulated [Equation (5-2)]:

$$r = \frac{dN}{dt} = \frac{I}{nF} \tag{5-2}$$

where N is the number of moles, and F is Faraday's constant (A s mol^{-1}). In the case of a constant current density j (A m^{-2}), the total current that passes through the electrode surface, A (m^2), can be calculated as $I = j\,A$. The net electrical current during the electrochemical reactions is then given as $I = I_a - I_c$, where I_a and I_c are the anodic and cathodic currents caused by oxidation and reduction at the electrode and electrolyte interface, respectively. Combining these with Equation (5-2) and the law of mass action, the total current density can then be calculated as

$$j = \frac{I}{A} = j_a - j_c = \frac{nF}{A}\left(\frac{dN_a}{dt} - \frac{dN_c}{dt}\right) = \frac{nF}{A}(k_a S_R - k_c S_O) \tag{5-3}$$

where S_O and S_R are the surface concentration of the reactants and products, respectively. Similar to other chemical reactions, the rates of electrochemical reactions are described with the Arrhenius equation [Equation (5-4)]:

$$k_a = k \exp\left(\frac{-\Delta E_R}{RT}\right), \quad k_c = k \exp\left(\frac{-\Delta E_O}{RT}\right) \tag{5-4}$$

where

k = Reaction rate constant;
R = Universal gas constant (J K^{-1} mol^{-1});
T = Absolute temperature (K); and
ΔE_O and ΔE_R = Activation energy for oxidation and reduction, respectively (J).

In electrochemical reactions, alteration of electrode potentials affects the activation energy. If the activation energy for the oxidation reaction is lowered by a factor β [Equation (5-5)], correspondingly in the reduction reaction, the activation energy is increased by a factor $1 - \beta$ [Equation (5-6)]:

$$\Delta E_O = \Delta E_{O,\text{ref}} + \beta nF\text{ø} \tag{5-5}$$

$$\Delta E_R = \Delta E_{R,\text{ref}} - (1-\beta)nF\text{ø} \tag{5-6}$$

where $\Delta E_{O,\text{ref}}$ and $\Delta E_{R,\text{ref}}$ are the reference activation energy, and ø is the electrode potential difference (V). An overpotential can occur if this electrode potential deviates from its equilibrium potential, ø_{eq}. In this situation, the overpotential controls the transfer of electrons to the ionic species through the electrode/electrolyte interface. By substituting Equations (5-4) to (5-6) into Equation (5-3) with a few manipulations, the total current density can be obtained according to the Butler–Volmer equation:

$$j = j_0\left[\exp\left(\frac{\beta nF}{RT}\eta\right) - \exp\left(-\frac{(1-\beta)nF}{RT}\eta\right)\right] \tag{5-7}$$

in which j_0 is the exchange current density and $\eta = \text{ø} - \text{ø}_{\text{eq}}$. For large overpotentials, the second exponential term is negligible as compared with the other one. This can simplify the calculation of j:

$$\log j = \log j_0 + \frac{\beta nF}{2.3\,RT}\eta = \log j_0 + b\eta \tag{5-8}$$

This equation is a straight line in a log plot and is known as the Tafel equation, where b is the slope of the line. This equation describes the kinetics of current density as a function of only two parameters, exchange current density and Tafel slope.

The kinetic expression in Equation (5-2) refers to the direct electrochemical processes. However, in an electrochemical cell, other chemical processes also occur, such as the formation of hydroxyl radicals as an oxidation product at the electrode surface. The kinetics of these processes are modeled as typical chemical processes depending on the concentration of reactants. A second-order rate expression can be used to describe these chemical processes:

$$r' = kS_oS \tag{5-9}$$

where S is the concentration of the organic compound, and k is the kinetic constant. If the species are highly reactive, the reaction can be rapid or even instantaneous. In this case, S_o (resulting in pseudo first-order kinetics) or both S_o and S (resulting in zero-order kinetics) can be omitted from Equation (5-9).

5.3.2 Mass Transfer in an Electrochemical Cell

In the previous section, it was shown that the current in an electrochemical cell is influenced by the reaction rate constants and applied potential. However, before any current can be recorded, electrode reactions must occur across the electrode/electrolyte interface. For a sufficient net reaction rate, it is necessary that the reactants and products are transported to and from the electrode surface. In an electrochemical reaction, where the reaction is restricted to the electrode surface, the mass transport mechanism of ionic species is characterized by convection, diffusion, and electric migration.

Convection. Convective mass transfer happens if a force is applied on the electrolyte solution. In such a case, the entire electrolyte solution moves; as a result, ionic species move toward or away from the electrode surface. This type of mass transfer for ionic species S_i moving with an average speed of v (m s^{-1}) can be formulated as shown in Equation (5-10):

$$N_{\text{conv}} = v\, S_i \tag{5-10}$$

For an incompressible flow and only considering convective and viscosity forces, the velocity field v can be calculated using the continuity and Navier–Stokes equation:

$$\nabla v = 0 \tag{5-11}$$

$$\rho\left(\frac{dv}{dt} + v\nabla v\right) = -\nabla p + \mu_D \nabla^2 v \tag{5-12}$$

where

ρ = Fluid density (mol m^{-3}),
p = Pressure (Pa), and
μ_D = Dynamic viscosity (Pa s).

Diffusion. Mass transfer caused by diffusion is a consequence of concentration gradients without the involvement of any external physical forces. As a result of diffusion, ionic species diffuse from a region of high concentration to a lower concentration with a Brownian motion. Because electrochemical reactions occur at the surface of the electrode, there is always a higher concentration of reactants in the bulk electrolyte as compared to that in the electrode surface. The concentration gradient is in the opposite direction for products. The diffusive flux for ionic species j is described according to Fick's first law [Equation (5-13)]:

$$N_{\text{diff}} = -D_i \nabla S_i \tag{5-13}$$

in which D_j is the ionic diffusivity of species j (m^2 s^{-1}). The negative sign indicates that a positive gradient causes the diffusive flux in a negative direction.

Migration. Migration occurs because of the potential difference between electrodes and the presence of charged ionic species in the electrolyte. Because of electrostatic forces created by the application of external potential differences, charged species are attracted or rippled near the electrodes. The migration flux can be formulated as shown in Equation (5-14):

$$N_{\text{mig}} = -z_i u_i S_i \nabla \emptyset \tag{5-14}$$

where

$\emptyset$ = Local potential difference in the electrolyte solution (V),
$\nabla\emptyset$ = Electric field,
z_j = Charge, and
u_j = Mobility of the ionic species (m^2 s^{-1} V^{-1}).

5.3.3 Total Ionic Flux in a Bulk Electrolyte

The total flux, N_j (mol m^{-2} S^{-1}), by the contribution of convection, diffusion, and migration, can be formulated according to the Nernst–Einstein equation [Equation (5-15)]:

$$N_j = -D_i \nabla S_i - z_i u_i S_i \nabla \emptyset + v\, S_i \tag{5-15}$$

In an electrolyte solution, the total net current is a result of the movement of all ionic species. Hence, the total net current i can be calculated by considering the overall charge and flux of ionic species:

$$i = F \sum_i z_i N_i \tag{5-16}$$

By substituting Equation (5-16) into Equation (5-14), one obtains Equation (5-17):

$$i = -F\left(\sum_i z_i^2 u_i S_i\right)\nabla\emptyset - F\sum_i z_i D_i \nabla S_i + F\left(\sum_i z_i S_i\right)v \tag{5-17}$$

This equation implies that the total net current in an electrolyte solution because of the presence of ionic species is determined by the potential and concentration gradients as well as the velocity of the electrolyte solution. In most electrochemical systems, electroneutrality conditions can be assumed [Equation (5-18)]:

$$\sum_i z_i S_i = 0 \tag{5-18}$$

This means that the concentrations of the ionic species present must be electrically neutral overall. This consideration eliminates the last term in Equation (5-17), which is the contribution of the convective transport in an electrolyte.

Nevertheless, mass transport by convection still contributes by maintaining a uniform concentration distribution and creating a constant conductivity, except close to the electrode surface where ionic species may be consumed at a high rate. In an electrochemical system with no concentration gradient, the current density becomes Equation (5-19):

$$i = -F\left(\sum_i z_i^2 u_i S_i\right)\nabla\phi \tag{5-19}$$

which indicates that with an assumed constant composition in an electrochemical cell, the electric current is controlled by migration.

5.3.4 Model Selection

As previously described, the concentration of every compound in an electrochemical cell depends on time and space, that is, their distance from the electrode surface. Describing the profile of compounds under such conditions involves a number of partial differential equations, which are often difficult to solve and involve many model parameters. This detailed formulation to describe the removal of organic compounds in wastewater treatment processes is not necessary, because the goal of the model is often to predict the effluent wastewater quality and not to describe the chemical concentration at every point of the cell. Therefore, it is common to simplify the highly distributed system of equations, which is easy to solve and will not require so many parameters. The simplification follows a set of assumptions that need to be carefully evaluated. Two common simplified systems are a lumped system and a semidistributed system.

Lumped system. In this system, only the overall performance of the electrochemical cell is described without the detailed description of what happens inside the cell. It is assumed that the electrochemical cell is a mixed tank reactor where the concentration of chemicals inside the cell depends only on time and not on the position. In this homogeneous system, the mass transfer fluxes, as previously described, are not considered. This assumption, therefore, allows for the simplification of partial differential equations to a set of ordinary differential equations, which are much easier to solve. In a continuous electrochemical cell with inlet and outlet streams, the mass balance for the model species S_i (mol m^{-3}) can be defined as follows [Equation (5-20)]:

$$V\frac{dS_i}{dt} = Q_{\text{in}}S_{i,\text{in}} - Q_{\text{out}}S_{i,\text{out}} + \sum_{j=1}^{p} v_i^j r_j V \tag{5-20}$$

where

V = Reactor volume (m^3),
Q = Flow rate (m^3 s^{-1}),
v_i^i = Stoichiometric coefficient of species i in the process j, and
r_j = Reaction rate (mol s^{-1}).

The subindexes in and out refer to the inlet and the outlet conditions, respectively. If the residence time of species in the electrochemical cell is sufficiently low and the species are not very active, the first two terms in Equation (5-20) can be ignored.

Semidistributed system. In reality, at the reactor level, the concentration gradients and potential differences exist in a three-dimensional space. However, the magnitude of the gradients is often the largest in one direction compared with that in the other two directions. It can be expected that the concentration of a reactant species is lower close to the electrode surface than that to the electrolyte solution, and the opposite is expected to be true for the reaction products. One can, therefore, only consider the direction toward the electrode surface and ignore the other two dimensions. If the flow direction in the reactor is not perpendicular to the electrode surfaces; this assumption can be valid only if the residence time in the reactor is small. This simplifies the complex partial differential equations to a set of one-dimensional partial differential equations where the concentrations of species are changing with time and only across one dimension.

An additional simplification is to divide the electrochemical cell into several interconnected zones where in each zone the concentration of species is assumed to be dependent only on time and not on space. Therefore, each zone is treated as a stirred tank reactor (similar to a lumped approach), which allows converting the one-dimensional partial differential equations to ordinary differential equations. In this way, high reaction rates occurring close to the anode and cathode electrodes are separated from the slow reaction rates in the bulk. For simplicity, the electrochemical reactor can be divided into three zones: one zone close to the anode electrode, one zone close to the cathode electrode, and one zone in between for the bulk solution (Cañizares et al. 2004a, b). These zones and the processes involved in each zone are shown in Figure 5-1. It is assumed that the anode behaves only as an electron sink (nonactive material). The processes follow a sequence: (1) the compounds are exchanged between the bulk zone and the anodic zone via mass transfer; (2) organic compounds are then oxidized on the electrode surface; the oxidation occurs once or in multiple stages until a final oxidant, usually carbon dioxide, is formed; (3) meanwhile, new oxidants (hydroxyl radicals) can be formed by decomposition of water molecules; (4) unstable hydroxyl radicals can lead to formation of other oxidants (hydrogen peroxide, ozone, chlorine, etc.) and cause mediated oxidation of organic compounds; (5) hydroxyl radicals can also promote oxygen formation; (6) at the same time, these oxidants can transfer to the bulk zone; and (7) cause the oxidation of organic compounds. Because most removal of organic compounds occurs in the anodic zone, the reduction reactions in the cathodic zone are assumed to be less important and only hydrogen evolution and reduction of organic species are considered.

If the concentration of a species is low in the reactor, the migration and diffusion fluxes can be negligible as compared to convective mass transfer. Hence, the mass transfer processes between zones are driven by concentration differences between the two zones and the mass transfer coefficient, k (m s^{-1}). For a batch system, without input and output, the following equations can be defined for the

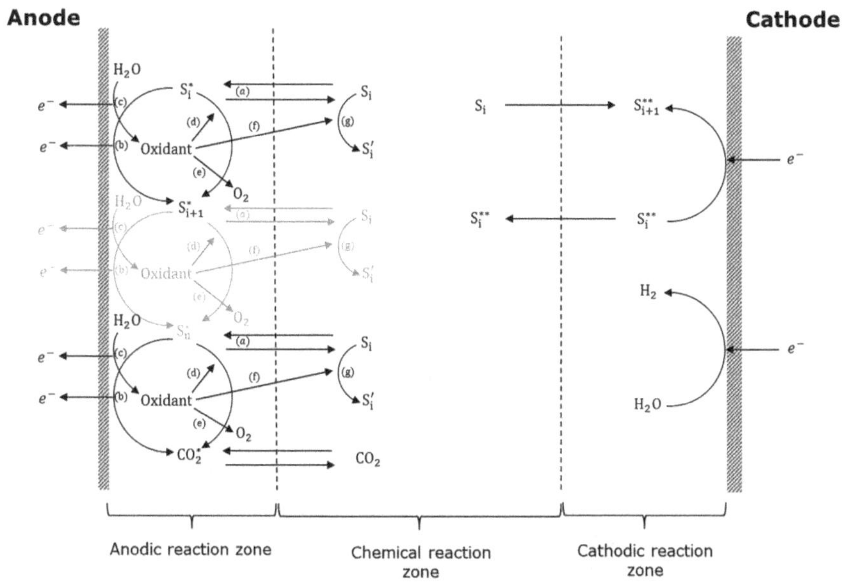

Figure 5-1. Electrochemical processes in an electrochemical cell, divided into three zones for the purposes of semidistributed modeling.

Source: Adapted with permission from Cañizares et al. (2004a). Copyright 2004, American Chemical Society.

anodic zone (index a), the cathodic zone (index c), and the bulk zone (Index c) [Equations (5-21) to (5-23)]:

$$V_a \frac{dS_{i,a}}{dt} = kA(S_{i,b} - S_{i,a}) + \sum_{j=1}^{p} v_i^j \frac{I}{F} \alpha_j^{\text{anode}} \tag{5-21}$$

$$V_c \frac{dS_{i,c}}{dt} = kA(S_{i,b} - S_{i,c}) + \sum_{j=1}^{p} v_i^j \frac{I}{F} \alpha_j^{\text{cathode}} \tag{5-22}$$

$$V_b \frac{dS_{i,b}}{dt} = kA(S_{i,a} - S_{i,b}) + kA(S_{i,c} - S_{i,b}) + \sum_{j=1}^{p} v_i^j r_j V_b \tag{5-23}$$

Because the total current applied should be shared among all the processes at the electrode surface, only a fraction of the current, indicated with α_j^{anode} and $\alpha_j^{\text{cathode}}$, corresponds to process *j* for the anodic and cathodic reactions, respectively. It can be assumed that the difference between the cell potential (ΔV_{work}) and the oxidation/reduction potential (ΔV_j) for each process determines the distribution of electrons [Equation (5-24)]:

$$\alpha_j^{\text{electrode}} = \frac{\Delta V_{\text{work}} - \Delta V_j}{\sum_j \left(V_{\text{work}} - \Delta V_j\right)} \tag{5-24}$$

The calculation of this fraction may need to be updated during the process, for example, if one of the compounds is completely oxidized or reduced.

5.3.5 Selection of Model Variables

Previously, both a lumped model and a semidistributed model were described for *i* number of model species involved in *j* number of processes. The complexity of these models depends on the number of species included in the model. In a multivariable model, all the significant species in an electrochemical cell are included. This, however, requires further knowledge on reaction pathways to account for subsequent formations and transformations. For example, modeling degradation of phenol also involves maleic acid and oxalic acid as transformation products (Cañizares et al. 2004a). In EO of wastewater, the objective is, however, to evaluate the overall treatment performance of the electrochemical cell and not the transformation of individual species. Chemical oxygen demand (COD) and total organic carbon (TOC) are the aggregated parameters, which are commonly considered as the model variables (Dominguez-Ramos et al. 2008, Panizza et al. 2001). In practice, the number of species should be decided based on the modeling purpose and the desired model complexity.

5.4 MODELING BASED ON THE DESIGN OF EXPERIMENTS AND RESPONSE SURFACE METHODOLOGY

Reliable experimental data are required to train the developed models and find the model parameters. These experiments can be conducted based on a design of experiment (DoE) approach to have a uniform distribution at the entire range of operating parameters. The DoE approach can be utilized for either statistical or phenomenological models (Feilizadeh et al. 2015, Karimi Estahbanati et al. 2019); however, mainly RSM models take advantage of this approach. In DoE, two types of variables are there, which are known as (1) independent variable or design variable, and (2) dependent variable or response variable. Identifying the design variables that have the most impact on the EO process and calculating the response variables are the main goals of the DoE. In general, a model consists of one or a set of mathematical equations that comprise dependent variables, independent variables, and constant parameters. Different phenomena should usually be modeled based on different DoE approaches for which experiments must be individually designed. The most common DoE methods that were used for EO processes are factorial design (FD), central composite design (CCD), Box–Behnken design (BBD), Taguchi's design (TD), and Doehlert design (DD). These designs are briefly discussed in this section.

5.4.1 Factorial Design

FD comprises full factorial and fractional FDs. In full FD, all factors are considered at all levels, which increases the number, time, and cost of experiments. Fractional

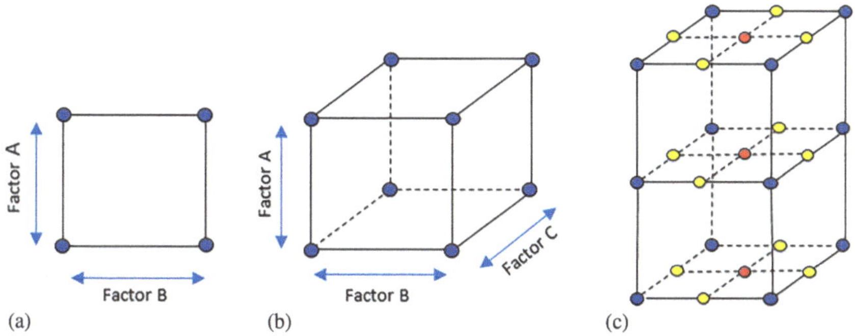

Figure 5-2. Cube plot of a full FD structure with (a) two levels and two factors, (b) two levels and three factors, (c) three levels and three factors.

FD is a method consisting of a subset of the experiments in a full FD. In this method, the upper and lower bounds of the interval for each design variable must be known. The number of experiments for determining the response variables varies depending on the number of points that are considered between the upper and the lower bounds. For a two-level full FD, if the number of independent variables is k, the lowest number of experiments (by considering just the upper and the lower bound of the interval) is equal to 2^k. This method is not a surface method but can be used as an initial estimate of the number of experiments required for modeling of EO processes. Unlike the two-level FD, a three-level FD is in the RSM method category (Anderson and Whitcomb 2016b, Behera et al. 2018). The cube plot of the FD method is represented in Figure 5-2.

5.4.2 Central Composite Design

The CCD method is an RSM-based approach that allows an estimation of the model parameters of a quadratic model. The basis of this method is similar to that of FD, but axial and center points are considered. This method is considered as an alternative to the three-level FD method (which requires 3^k experiments) for quadratic models because it requires fewer experiments (Anderson and Whitcomb 2016b, Behera et al. 2018, Morshedi and Akbarian 2014). The cube plot of the CCD method is depicted in Figure 5-3.

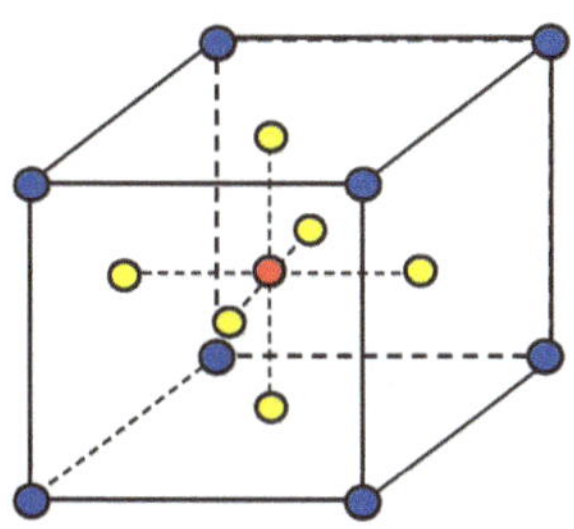

Figure 5-3. Cube plot of the three-factor CCD method.

5.4.3 Box–Behnken Design

The BBD is a second-order method based on three-level fractional FD. This method is one of the most frequently used RSM methods as it requires a slightly smaller number of experiments than other RSM methods like CCD and considerably less experiments than a three-level full FD method. BBD does not consider all the parameters at their highest or lowest value; therefore, it can be used as a way to prevent experiments being performed under extreme conditions and to avoid inappropriate results. The cube plot of the BBD method of DoE is presented in Figure 5-4.

5.4.4 Taguchi's Design

In the TD method, which is not a part of the RSM approach, the parameter space is studied on the basis of fractional factorial arrays called *orthogonal arrays*. This approach argues that there is no need to explicitly consider the interaction between two design variables. This method designs the desired system in the form of a table, which reduces the number of experiments compared with the full FD. One of the benefits of TD is that one is capable of examining discrete variables. However, the method suffers from a nonconsideration of the interactions between the variables (Kozik et al. 2019, Nandhini et al. 2014, Yousefi et al. 2018).

5.4.5 Doehlert Design

DD is another RSM-based DoE method. DoE using the DD is a method that considers the test domain to be spherical and emphasizes the uniformity in space filling. The matrix of DD is neither orthogonal nor rotatable but can be effective enough for DoE. The number of levels of variables in this method does not necessarily have to be equal. For example, one variable can have five levels and the other variable three levels. In DD, five levels are usually considered for the factor that has the greatest impact on the response.

The efficiency of a DoE method is obtained by dividing the estimated model coefficients by the number of conducted experiments. Therefore, by considering the efficiency of different methods, it can be concluded that the efficiency of DD and BBD is higher than that of CCD. The advantages of DD include sequential design, block utilization, and the lack of model fit detection (Ferreira et al. 2004). The illustration of DD with two and three factors is presented in Figure 5-5.

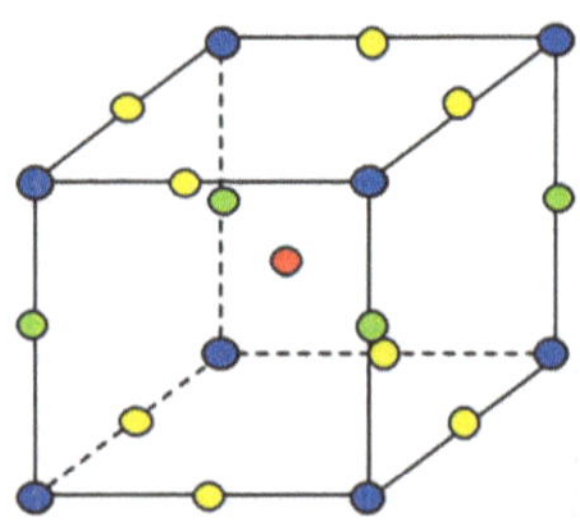

Figure 5-4. Cube plot of the three-factor BBD method.

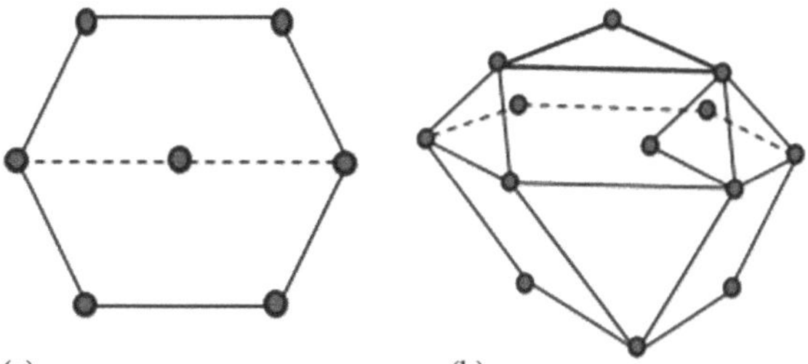

Figure 5-5. Illustration of DD with (a) two factors, (b) three factors.

5.4.6 Modeling Studies Using Response Surface Methodology

The RSM approach is one of the most common methods for modeling EO processes. In this section, the most significant research results on the modeling of EO processes using the RSM approach are explained based on their DoE approach, that is, CCD, BBD, and DD. Analysis of variance (ANOVA) and coefficient of determination (R^2) are mainly used to evaluate and compare the performances of different models. Table 5-1 summarizes different works on RSM modeling of EO processes. As can be seen, in these models, removal efficiency, energy consumption, COD removal, TOC removal, and color removal were selected as the dependent variables. On the contrary, pollution concentration, time, pH, current intensity, electrolyte concentration, temperature, conductivity, and anode material were chosen as the independent variables. In most of the developed RSM models, an R^2 value higher than 0.9 was obtained, which shows a best fitting of the model to the experimental data.

Most of the studies used CCD and BBD for DoE; however, DD was also used in a few works. Liu et al. (2015) studied the EO process for the degradation of bromoamine acid (BAA) using DD. The current density, flow rate, sulfate concentration, and initial BAA concentration were selected as the independent variables, and the BAA degradation efficiency was chosen as the dependent variable. The modeling results showed that the BAA concentration had the greatest impact on the BAA degradation. Azevedo et al. (2014) studied the linear sweep voltammetry for the degradation of 1,2-benzopyrone (BP). They optimized the process using DD and then measured the importance of each variable using a full FD. The results obtained from the DD showed that the optimum pH value of the process was 12.3.

Central composite design. As can be seen from Table 5-1, the CCD method is the most common DoE approach for RSM modeling of the EO processes. One advantage of the RSM method is the capability of performing multiobjective optimizations. Accordingly, the calculation of the constant model parameters is performed for the purpose of simultaneous optimization of multiple responses. García-Gómez et al. (2016) designed the experiments for optimization of EO removal of phenol from wastewater by the CCD method. In this study, they investigated individual effects as well as interaction effects of three different variables, current intensity, process time,

Table 5-1. Comparison of Different Research Works on RSM Modeling of EO Processes.

Model type	*Dependent variable*	*Independent variable*	*Water matrix/ pollutant*	*Model performance*	*Reference*
CCD	Phenol removal, energy consumption	Current intensity, electrolysis time, recirculation flow rate	Biorefractory wastewater	R^2 (0.9261, 0.9075), ANOVA	García-Gómez et al. (2016)
CCD	COD removal, energy consumption, dye removal	Current density, initial pH, electrolysis time	Textile wastewater containing Reactive Black 5 and another effluent	R^2 (0.9187, 0.9023, 0.9103), R^2_{adj} (0.8455, 0.8144, 0.8295), ANOVA	Bansal et al. (2013)
CCD	Diazinon insecticide removal	Current density, electrolysis time, pH, diazinon concentration	Aqueous solution containing diazinon	R^2 (0.9900), R^2_{adj} (0.9858), ANOVA	Moteshaker et al. (2020)
CCD	Dye removal	pH, dye concentration, voltage, treatment time	Color wastewater (CI Reactive Blue 49)	R^2 (0.9509), R^2_{adj} (0.9052), ANOVA	Radi et al. (2018)
CCD	COD removal, dye removal	Electrolyte concentration, current density, reaction time	Semiaerobic landfill leachate	R^2 (0.9958, 0.9868), R^2_{adj} (0.9937, 0.9821), ANOVA	Mohajeri et al. (2010)
CCD	% Degradation of 4-AA, % mineralization	Current density, pH, 4-AA concentration, electrolysis time	Acidic aqueous solution	R^2 (0.956, 0.959), R^2_{adj} (0.905, 0.913)	de Melo da Silva et al. (2018)
CCD	Color removal, COD removal	pH, NaCl concentration, electrolysis time	Textile wastewater containing Reactive Orange 107	R^2 (0.8227, 0.9265), R^2_{adj} (0.6631, 0.8604), ANOVA	Rajkumar and Muthukumar (2012)

Factorial design (FD) and CCD	COD removal, total organic carbon (TOC) removal, dye removal	Current density, time, seasonal change	Highly contaminated old landfill leachate	R^2 (0.983, 0.917, 0.795), ANOVA	Zolfaghari et al. (2016)
FD and CCD	Chlortetracycline degradation, energy consumption	Current intensity, electrolysis time, temperature, type of anode	Water contaminated by antibiotics	R^2 (0.863, 0.969), ANOVA	Zaviska et al. (2013)
FD and CCD	COD removal, TOC removal	Current intensity, treatment time, electrolyte concentration, pollutant concentration	Water containing ethylene glycol	R^2 (0.93, 0.92), ANOVA	Jardak et al. (2017)
FD and CCD	Carbamazepine removal, energy consumption	Current intensity, electrolysis time, anode type, recycling flow rate	Pharmaceutical wastewater with carbamazepine	R^2 (0.9533, 0.9669), ANOVA	Guitaya et al. (2017)
FD and CCD	Carbamazepine removal, energy consumption	Current intensity, electrolysis time, recirculation flow rate, anode material	Wastewater containing carbamazepine	R^2 (0.9605, 0.9991), ANOVA	García-Gómez et al. (2014)
Face CCD	COD removal, TOC removal	Reaction time, pH, salt concentration, voltage	Saline wastewater	R^2 (0.97, 0.94), R^2_{adj} (0.95, 0.84), ANOVA	Darvishmotevalli et al. (2019)
Box–Behnken design (BBD)	% Degradation of color, % COD removal, energy consumption	Initial dye concentration, current density, conductivity	Wastewater containing Acid Yellow 23	R^2 (0.9721, 0.9819, 0.9914), R^2_{adj} (0.9219, 0.9493, 0.9760), ANOVA	GilPavas et al. (2016)

(Continued)

Table 5-1. Comparison of Different Research Works on RSM Modeling of EO Processes. (Continued)

Model type	*Dependent variable*	*Independent variable*	*Water matrix/ pollutant*	*Model performance*	*Reference*
BBD	O/C ratio, tensile strength of carbon fiber	Reaction time, concentration of acidic solution, temperature	Carbon fiber	R^2 (0.9427, 0.9822), ANOVA	Andideh and Esfandeh (2016)
BBD	% COD removal	Current density, initial pH, electrolysis time	Pharmaceutical wastewater	R^2 (0.984), ANOVA	Deshpande et al. (2012)
BBD	Bisphenol-A removal, energy consumption	Current intensity, electrolysis time, electrolyte (sodium sulfate) concentration	Municipal wastewater	R^2 (0.9903, 0.9996), ANOVA	Zaviska et al. (2012)
BBD	COD removal, SCOD removal, TOC removal, DOC removal	pH/H_2O_2 dosage, current density, operation time	Sugar beet wastewater	R^2 (0.9976, 0.9984, 0.9992, 0.9973), R^2_{adj} (0.993, 0.9955, 0.9978, 0.9972), ANOVA	Sharma and Simsek (2020)
Taguchi and fractional FD	COD removal, dye removal	pH, current density, dye concentration, electrolysis time	Synthetic wastewater	R^2 (0.9940, 0.9566), R^2_{adj} (0.9931, 0.941), ANOVA	Yousefi et al. (2018)
Taguchi and CCD	Dye removal, COD removal, TOC removal, suspended solid removal	Potassium ferrate concentration, time, pH	Tannery wastewater	R^2 (0.7565), R^2_{adj} (0.5941), MS (3.4812)	Kozik et al. (2019)

and recirculation flow rate, on energy consumption and phenol removal efficiency as responses. An analysis of the effects of independent variables on the response showed that time had a significant effect on phenol removal efficiency. They also found that an extensive increase in the current, which leads to a higher hydroxyl radical generation, does not guarantee better phenol removal efficiency. Bansal et al. (2013) studied the electrochemical treatment of textile wastewater using an RSM model. They considered initial pH, current density, and electrolysis time as independent variables and used CCD to perform the experiments and optimize the responses to minimize energy consumption and maximize dye and COD removal. By analyzing the model results, they found that as the pH increased from 4 to 7, the COD removal increased, but at a higher pH, the COD removal remained almost constant. They also observed that with increasing current density and reaction time up to almost 100 min, the COD and dye removal efficiencies increased for the entire tested pH range. They argued that as time and current density increase, the amount of generated Al^{3+} will also increase, resulting in the production of more aluminum hydroxide and the elimination of more pollutants. Mohajeri et al. (2010) investigated the applicability of CCD to optimize conditions in the electrochemical oxidation of landfill leachate. They maximized color and COD removal as responses by finding the optimum considering current density, reaction time, and electrolyte concentration as independent variables. With respect to ANOVA and the R^2 value ($R^2 = 0.9953$ for COD removal and $R^2 = 0.9868$ for dye removal), it was found that the developed RSM model fitted very well with the experimental data that were designed by the CCD method.

Simple DoE methods like FD and TD can be used before RSM modeling to identify the most important independent variables affecting the response. This approach contributes to reducing the number of required experiments by first selecting the most effective parameters and suggesting them for RSM analysis. This approach was used in the modeling of the combination of a membrane bioreactor with ultrafiltration and EO for highly contaminated and old landfill leachate treatment where experiments were designed by the CCD method (Zolfaghari et al. 2016). In this study, FD and CCD models were used to respectively find the influence of the parameters and determine the optimum conditions of the process. COD and TOC removal efficiency were considered as response variables, which were influenced by the selected independent variables such as current intensity, treatment time, and seasonal changes. Using the FD method, it was found that treatment time had a significant impact on the COD removal efficiency, whereas for TOC removal, seasonal change was the most effective variable. It was also found that the current intensity had no significant effect on the performance of the EO process. In another work, Kozik et al. (2019) used TD and CCD methods to study the effect of using an industrial product containing potassium ferrate to purify tannery wastewater. Independent variables in both methods were potassium ferrate concentration, time, and pH, whose effects were investigated on COD removal as response. They compared the advantages and disadvantages of these methods and concluded that TD could be a good preliminary method for the elimination of less-effective parameters.

Box–Behnken design. As can be seen in Table 5-1, BBD is the second most used DoE method for the modeling of EO processes. A comparison of the models that employed the CCD and BBD methods for DoE shows that the experimental data that were designed using the latter method resulted in a better fit of the model, as a slightly higher R^2 value was obtained. Deshpande et al. (2012) used BBD to understand the effects of operating parameters such as current density, electrolysis time, and initial pH on the EO treatment efficiency of pharmaceutical wastewater. They concluded that the current density and electrolysis time had the most significant effect on COD removal efficiency. However, there was a limit in the current density as an excessive increase could undermine the electrode stability. By studying different cases, they found that the threshold for increasing the current density depends on the electrode material during which an operational current density of 80 A m^{-2} gave the best results in terms of pollutant removal and electrode stability. They eventually tested the optimum condition obtained from modeling and achieved a COD removal efficiency of 30.2%.

BBD was successfully applied for multiobjective optimization of the EO process. GilPavas et al. (2016) optimized an EO process to remove and degrade dye from wastewater using the BBD method. They evaluated the effect of current density, conductivity, and initial dye concentration (as independent variables) on the degradation percentage of color, COD, and energy consumption (as responses). They found that the initial concentration of dye had the greatest effect on color and COD removal. The lower the initial dye concentration, the faster the rate of dye degradation and the higher the energy consumption. At the optimized number of independent variables, about 99% of dye and 76% of COD were removed. Zaviska et al. (2012) studied the EO of bisphenol-A (BPA) by electrodes consisting of different materials and optimized the process using BBD. They evaluated the effect of independent variables such as current density, reaction time, and sodium sulfate concentration on BPA decomposition efficiency and energy consumption as responses. They tested the process at the optimum condition that was obtained from the model and secured more than 90% removal efficiency.

5.5 MATHEMATICAL MODELING OF ELECTRO-OXIDATION USING ARTIFICIAL NEURAL NETWORKS

Apart from phenomenological and conventional empirical modeling approaches normally used for electrochemical processes for water and wastewater treatment, as previously described , ANNs have emerged as an alternate modeling approach. ANNs do not need heat and mass transport phenomena along with a detailed knowledge of the reaction kinetics. Also, in this approach, it is not required to select the structure of the model a priori, which is otherwise often a challenging task, as one needs to choose the most suitable model from numerous available model candidates, especially for nonlinear processes.

In general, because of the complex relationships between input parameters and outputs of the electrochemical processes for water and wastewater treatment, there can be highly complicated nonlinear systems. Therefore, it is difficult to use phenomenological or empirical models to model, simulate, and optimize the processes. In these situations, when conventional methods for modeling and optimization purposes are not suitable, artificial intelligence tools such as ANNs are an interesting alternate method. They can overcome the classical modeling difficulties and have the following advantages: The possibility of applying them even on multiple input–multiple output complex nonlinear processes, the ability to be constructed solely from historic process input–output data (experimental dataset), and excellent generalization ability when properly trained.

Aleksander and Morton (1990) developed one of the famous descriptions of a neural network: A neural network is a massively parallel distributed processor made up of simple processing units, which has a natural propensity for storing experiential knowledge and making it available for use. It resembles the brain in two respects:

1. Knowledge is acquired by the network from its environment through a learning process.
2. Interneuron connection strengths, known as synaptic weights, are used to store the acquired knowledge.

5.5.1 Artificial Neural Network's Architectures

The main architectures of the ANNs considering how the different neurons are dispositioned and connected to one another, as well as the composition of layers, can be divided as follows: (i) single-layer feedforward networks, (ii) multilayer feedforward networks, and (iii) recurrent networks. Single-layer feedforward networks have just one input layer that projects onto an output layer of neurons, or in other words, the information always flows in a single direction from the input layer to the output layer, and not vice versa. These networks are usually applied to linear filtering and pattern classification problems (Da Silva et al. 2017). The recurrent architecture consists of some neurons in a layer where their output signal would be used as an input signal for other neurons. This feedback feature results in a nonlinear dynamic behavior that can be applied on time-variant systems, relevant for time series prediction, system optimization, and process control (Jain and Medsker 1999). The presence of this feedback structure has a significant impact on the learning capability of the network, hence producing current outputs by taking the previous outputs into consideration.

5.5.2 Multilayer Feedforward Networks and Their Learning Process

Multilayer feedforward ANNs are different from the single-layer feedforward networks in a way that one or more hidden layers with multiple neurons are present. Each neuron in these layers in these types of networks is connected to another neuron in the adjacent layer by connections called *weights* (synaptic weights). This parallel distributed information processing structure provides a condition that nonlinear regression is applied to input/output mappings.

The process in which the connection weights of the network are adjusted and learned by using examples so that the ANN can perform a particular task is called *learning* (training). The objective of the learning process is to minimize an error function by searching for a set of connection weights to produce such outputs that are close to the desired targets (output of any given input) (Hamed et al. 2004). The learning process is basically divided into three categories, namely, supervised learning, unsupervised learning, and reinforcement learning. In supervised learning, an external teacher is involved, who compares the actual output of the neural network with the desired target. In unsupervised learning, which is mostly used for pattern recognition or clustering, there is no performance evaluation of the system by a supervisor or any predetermined correct or incorrect answer. Thus, the need for additional input information or knowledge of the targets is not felt, and the neural network organizes itself by using the correlations among input data to identify groups of similar input patterns. Reinforcement learning, one form of supervised learning, is the type of learning that requires only the fact that the result is good (correct) or bad (incorrect) and, thus, requires less information (Baughman and Liu 1995, Hammerstrom 1993). Applications of ANNs in EO for water and wastewater treatment involve supervised learning. Detailed information about ANNs and their fundamentals can be found in the previous chapter and literature (Da Silva et al. 2017, Haykin 1998).

Table 5-2 summarizes the application of the ANN modeling approaches of electrochemical oxidation processes and their performance for water and wastewater treatment processes. As can be seen in Table 5-2, different configurations of networks have been utilized, including a single hidden layer, multiple hidden layers, and an ensemble modeling approach (stacked neural networks). In one of the studies, Manokaran et al. (2014) applied a feedforward back-propagation model to predict the degradation of a distillery effluent by the EO process. Flow rate, current density, and supporting electrolyte concentration were selected as input parameters, and COD removal efficiency was chosen as the output parameter. In total, 200 data sets were employed for training (150 samples), validation (30), and testing (20) using the scaled conjugate gradient algorithm for training and a sigmoidal transfer function for the hidden layers. The average absolute relative error (AARE) and average root-mean-squared error (RMSE) were utilized to evaluate the performance of the neural networks that were proposed on a trial and error basis. The performance of four different configurations of three-layer networks (one hidden layer) and four different configurations of four-layer networks (two hidden layers) were compared, and they showed that the four-layer BP neural network with the 3:3:3:1 configuration displayed the best performance: RMSE = 0.8633, AARE = 3.4613, $R = 0.9987$, and the COD removal efficiency could be successfully predicted.

Although many authors utilized single ANNs for their modeling processes, others tried other approaches with the so-called ensemble or stacked modeling.

Table 5.2. Application of ANN Modeling Approaches on Electrochemical Oxidation Processes.

Model type	*Dependent variable*	*Independent variable*	*Water matrix/ pollutant*	*Model performance*	*Reference*
ANN; 3:7:1; one hidden layer	COD removal, energy consumption	Current density, time, salt concentration	COD, specialty chemical manufacturer effluent	$R = 0.9977$	Ahamed Basha et al. (2010)
ANN; 3:9:1; one hidden layer	Dye removal	Current density, time, salt concentration	Malachite green dye	$R = 0.9987$	Soloman et al. (2010)
ANN; 3:8:3; one hidden layer	Dye degradation, color removal, energy consumption	pH, current, and reaction time	CBSOL LE red wool dye	$R = 0.995$	Sangal et al. (2015)
ANN; 5:14:1; one hidden layer	Removal efficiency	Nature and concentration of the supporting electrolyte, initial pH, current intensity, reaction time	Oxytetracycline	$R = 0.99$	Belkacem et al. (2017)
ANN; 3:3:3:1; two hidden layers	COD removal	Flow rate, current density, supporting electrolyte concentration	COD, distillery effluent	$R = 0.9987$	Manokaran et al. (2014)
ANN; 5:6:14:1 and 5:8:13:1; two hidden layers	Cell voltage, chlorine current efficiency	Hydrochloric acid concentration, acid flow rate, acid temperature, oxygen flow rate, and applied current density	HCl electrolysis	$R^2 = 0.897$	Ashrafizadeh et al. (2009)

(*Continued*)

Table 5.2. Application of ANN Modeling Approaches on Electrochemical Oxidation Processes. (Continued)

Model type	*Dependent variable*	*Independent variable*	*Water matrix/ pollutant*	*Model performance*	*Reference*
ANN; 5:6:14:1 and 5:8:13:1; two hidden layers	Cell voltage, chlorine current efficiency	Hydrochloric acid concentration, acid flow rate, acid temperature, oxygen flow rate, and applied current density	HCl electrolysis	$R^2 = 0.9525$	Abbasian and Sattari (2013)
ANN; stacked neural networks	COD	Pollutant concentration, pH, temperature, current density, current charge	Phenolic compounds	4.92% Error	Piuleac et al. (2010)
Hybrid and stacked neural networks	COD	Temperature, initial COD, pH, current density, charge, types of chlorine phenol compound, type of nitrophenol compounds	Phenolic compounds	$R^2 = 0.998$	Piuleac et al. (2012)
Multiple ANN topologies and support vector machines	COD, total coliform (TC), fecal coliform (FC), electroconductivity (EC), total dissolved solids (TDS)	Current density, time, electrode type, pH, COD, TC, FC, EC, TDS	Activated sludge of sewage	$R^2 > 0.977$	Curteanu et al. (2014)

The idea of stacked neural networks is based on the premise that improved predictions can be obtained by using multiple networks, instead of simply just one single network, which could be hopefully the optimal network as is usually done (Torres-Sospedra et al. 2006). Therefore, more accurate predictions can be provided by combining different neural network models, each capturing certain aspects of the process, and aggregating their information. Piuleac et al. (2010) performed the idea of stacked neural network modeling on the electrolysis process of wastes polluted with phenolic compounds, including phenol, 4-chlorophenol, 2,4-dichlorophenol, 2,4,6-trichlorophenol, 4-nitrophenol, and 2,4-dinitrophenol. In their work, various types of ANNs were aggregated in a stack whose output response was a weighted sum of the individual network outputs. Also, another approach was considered, which utilizes the concept of a modeling strategy based on considering six subsets of experimental data corresponding to each type of phenol pollutant followed by an assembly of the different individual neural networks that are each designed for the prediction of a specific pollutant compound. A comparison between the different methodologies indicated that in the case with an optimal MLP 7:30:25:1 neural network, a higher error can be observed. By utilizing the stacked neural networks and the assembly of neural networks, smaller validation errors of 5.8% and 4%, respectively, could be obtained. In their work, combined individual neural networks with split data sets for each phenol pollutant type was the most practical strategy. This network was tested with a real wastewater consisting of the effluent of a fine-chemicals plant that combines an aqueous solution of solvents with a high concentration of aromatic compounds. The results showed an average error of around 4.92% between experimental COD values and predicted ones, which gave an excellent representation for using neural networks in the case of a wastewater treatment.

5.5.3 Optimization Techniques Linked to Artificial Neural Networks

Evolutionary algorithms (EAs), and in particular, genetic algorithms (GA) and particle swarm optimization (PSO), have received growing attention in recent years among optimization techniques. GAs with good global searching ability and flexibility, ease of operation, and without the need for gradient information of the objective (fitness) functions, have become powerful techniques for optimization problems (Curteanu and Cazacu 2007, Ding et al. 2011). GAs have been utilized in two ways with ANNs. First, they have been applied in optimization procedures where the objective is to determine the optimum conditions for achieving the best value of the system output and where the trained neural network is used as the objective (fitness) function of the GA. Second, whereas the BP is the most widely used training algorithm, it can get trapped in suboptimal solutions (local optima) for systems containing complex nonlinear relationships. In such cases, GAs can be used in the training process to avoid local optima by searching in several regions simultaneously. A GA starts with a primary population of candidate solutions, and a fitness value is calculated for each solution. Through the algorithm, three stochastic operators are applied to each population, which are analogous to chromosomes in a biological context. Selection is choosing the solutions with the highest fitness value to create an intermediate population. The next population is the result of crossover or mutation. By crossover, the selected members are mated in pairs and recombined through a genetic manipulation of chromosomes to generate two new solutions (offspring). Mutation acts as an assurance against lost genetic material and consists of replacing some of the chromosome's genes with new genes. The generation of new populations and calculation of the fitness value for each population is repeated over and over again in an iterative way. When a specific termination criterion is met, for example, when there is no more change in the population from one iteration to the next or when a satisfactory fitness value is identified, this process ends (Ansari and Bakar 2014, Niculescu 2003, Ridha et al. 2008).

The PSO algorithm, first introduced by Kennedy and Eberhart (1995), is based on social behavior simulation of a flock of birds, called a *swarm*, searching for food. PSO is also a stochastic population–based optimization approach in which particles, a swarm of potential solutions, fly in the problem space to find better regions and finally the optimal solution, while cooperating and competing with other potential solutions (Chen et al. 2010). In PSO, a particle is analogous to a chromosome (population member) in GA and represents a candidate solution to the problem being studied (Eberhart and Shi 1998). The condition of each particle is changed by the impact of three factors: (1) its own inertia, (2) the personal most optimal position, and (3) the swarm's most optimal position (Juneja and Nagar 2016). In the d-dimensional search space of the problem, particle i of the swarm can be represented by $X_i = (x_{i1}, x_{i2}, \ldots, x_{id})$. The velocity of this particle and the best previous position, which is the position giving the best fitness value, are represented as $V_i = (v_{i1}, v_{i2}, \ldots, v_{id})$ and $P_i = (p_{i1}, p_{i2}, \ldots, p_{id})$. Also, the globally best position, the position of the best individual, is noted as $G = (g_1, g_2, \ldots, g_d)$ (Talebi et al. 2010). The velocity and position of the particles are updated as follows:

$$V_i^{j+1} = \omega \times V_i^j + c_1 \times \text{rand}_1 \times (P_i^j - X_i^j) + c_2 \times \text{rand}_2 \times (G_i^j - X_i^j) \quad (5\text{-}25)$$

$$X_i^{j+1} = X_i^j + V_i^{j+1} \quad (5\text{-}26)$$

where

V_i^{j+1} and X_i^{j+1} = Updated velocity and position vector of particle I,
ω = Momentum or inertia weight factor,
c_1 and c_2 = Learning factors, and
rand_1 and rand_2 = Random numbers between (0, 1) (Moghaddam et al. 2012).

Detailed information about the GA and PSO algorithms can be found in the literature (Juneja and Nagar 2016, Whitley 1994).

These EAs have been linked to ANNs to either optimize the hyperparameters of the neural networks or find the best operational conditions for the process. To find the optimal neural network, usually trial and error is utilized, searching between the ANNs' hyperparameters (e.g., the number of hidden layers, number of hidden neurons, transfer functions, training algorithms, etc.). As an alternative to the trial and error method, EAs are utilized to find the best-configured ANN, which results in the highest network performance possible. Viana et al. (2018) used the PSO algorithm to optimize their neural network model parameters, including the hidden neuron number, transfer function, and learning rate. A feedforward back-propagation 4:8:3 network with logsig transfer functions at the hidden layer and a purelin transfer function at the output layer was used for the modeling. Their neural network could successfully predict color removal, COD removal, and energy consumption for the textile dye Reactive Black 5 degradation with the performance of $R^2 = 0.982$, MSE (mean squared error) $= 0.0146$ for the test set.

As mentioned, EAs can be linked to ANNs to find the optimal process conditions. The ANN-GA approach was utilized by Picos and Peralta-Hernández (2018) for the prediction of discoloration of a dye by an EO process. GA optimization was linked to their optimal ANN to find the best operational conditions, where EO can yield a maximum discoloration at the lowest current density, flow rate, experimental time, and at the highest dye concentration. They experimentally validated their ANN-GA result with an experimental reaction time of 110 min, a flow rate of 12 Lps, a current density of 27.34 mA cm^{-2}, and a dye concentration of about 230 mg L^{-1}, where about 95% discoloration could be obtained.

5.5.4 Comparison of Artificial Neural Networks and Response Surface Methodology

The RSM has also been used in parallel to ANNs. Detailed information about RSM is provided in Section 5.4. The input data for ANNs can be derived from experimental design approaches, and the modeling performance of these two approaches can be compared in this way. Sangal et al. (2015) developed a three-layer ANN model to predict the removal of CBSOL LE red wool dye from wastewater by EO. The input parameters such as pH, current, and time were studied to forecast the three outputs, dye degradation, color removal, and energy consumption. The optimal ANN with a 3:8:3 configuration could estimate the outputs with correlation coefficients of 0.995, 0.996, 0.992, and 0.995 for training, validation, testing, and all data sets combined, respectively. It was reported that the proposed ANN can accurately stimulate the outputs from a given set of inputs. Furthermore, they conducted a regression analysis by statistical methods and optimization using the simulated outputs of the neural network. Their CCD model could relate the stimulated data from ANN to predicted values with the R^2 values of 0.9908, 0.9980, and 0.9659 for dye degradation, color removal, and energy consumption, respectively. Although the ANNs' performance was slightly better than that of the RSM methodology, based on their conclusions, a good correlation between the simulated data from ANNs and the predicted values by CCD was obtained.

5.6 KINETIC ANALYSIS OF ELECTRO-OXIDATION

The primary purpose of kinetic models is to optimize the operating conditions for achieving the best response. This approach is very interesting from a practical point of view; however, aside from obtaining the optimum operating conditions, assessing the mechanism and process kinetics is of importance from a scientific point of view. A kinetic analysis of the EO process helps in its comprehensive understanding and diagnosis.

A first-order rate equation is usually suitable for the kinetic modeling of the EO process. García-Gómez et al. (2016) analyzed the rate of phenol removal and showed that a first-order reaction rate could well describe the EO wastewater treatment process. Jardak et al. (2017) analyzed the ethylene glycol degradation by the EO process and concluded that the reaction rate followed the first-order kinetic model. Bansal et al. (2013) investigated the kinetics of EO of dye in wastewater and showed that it could be modeled by a first-order kinetic equation. They obtained the rate constant at two different temperatures and found that the reaction rate constant increased at a higher temperature, which led to a better dye and COD removal. Borras et al. (2007) kinetically studied the influence of concentration and temperature on EO of *p*-methoxyphenol and *p*-nitrophenols. They concluded that the initial oxidation rate follows the Langmuir–Hinshelwood mechanism. They also obtained the reaction rate constant by considering the kinetics of the degradation reaction as first order.

Some studies developed more advanced kinetic models by considering more than one step in the EO process. Deshpande et al. (2012) kinetically studied the EO of pharmaceutical wastewater and showed that it performs in two steps: (1) complete oxidation of organic materials to stable intermediates, and (2) oxidation of intermediates to final products. They then modeled the process using a first-order rate equation to describe both steps. Wang et al. (2016) presented an intrinsic kinetic model using a novel three-stage reaction theory to describe the EO process in a packed bed electroreactor for the removal of organic pollutants. They showed that the kinetics of the EO process can be better simulated using the new method and the results are more accurate than that of the first-order kinetic model.

5.7 CHALLENGES AND FUTURE RESEARCH WORK

The electrochemical oxidation process for water and wastewater treatment has gained more attention in recent years based on its capability to degrade pollutants both with direct (hydroxyl radicals, $OH^{\bullet}$) and indirect oxidation (mediators in solution such as $HClO$, H_2O_2). However, to make the EO process comparable with other conventional methods for water and wastewater treatment, proper design and optimization of this process has been under investigation in recent years. Although various studies employed phenomenological and empirical modeling approaches in this regard, there are still some areas for further research. Phenomenological multivariable models still require further knowledge on reaction pathways to account for subsequent formations and transformations. This will, of course, increase the complexity of developing these models.

On the contrary, empirical modeling approaches such as RSM and ANN do not require a detailed knowledge about the phenomena of the process, which makes them interesting alternate methods. Although RSM modeling provides the tool to analyze the interacting effect of operating parameters, a few studies were found on this aspect. A detailed analysis can be performed on this subject to evaluate the effect of operating parameters on the optimum value of different operating parameters in the EO process.

The performance of the ANN model as an AI technique strongly depends on the hyperparameters of the neural networks. The selection of optimum network hyperparameters such as the number of hidden layers, number of neurons in hidden layers, learning rate, and transfer functions are major tough tasks in ANN modeling, and the usual way to overcome such limitations is by the trial and error method. To reduce uncertainties related to the trial and error method, EAs can be linked to the ANN modeling approach to find the optimal network hyperparameters.

5.8 CONCLUSION

This chapter presented different approaches for modeling electrochemical oxidation processes. Phenomenological modeling based on physical and chemical concepts can mimic the dynamics of the EO process. However, electron transfer and the different ions involved in the reactions and the species involved make this type of modeling complex. Nevertheless, the complexity of this approach can be reduced by selecting suitable aggregated parameters. Alternate empirical methods such as RSM and artificial intelligence techniques have gained growing attention because knowledge about the fundamentals and the detailed phenomena of the process is not a prerequisite for such methods. RSM with DoEs has the ability to reveal the relationship and interaction between the independent and the dependent variables with a minimum number of experiments. The review of different studies showed their good performance for EO modeling and optimization. The drawback of RSM is that the model generated for fitting the experimental data is, in general, a polynomial equation. In these conditions where the nonlinearities of the EO process cannot be described by conventional empirical approaches, artificial intelligence techniques have emerged as interesting alternatives. Based on the information provided, it can be concluded that artificial intelligence techniques, such as ANNs, because of their high ability to simulate the complex, nonlinear input–output systems, demonstrate their potential for modeling, performance prediction, and optimization of the EO process used for water and wastewater treatment processes. To achieve process optimization, GAs and PSO algorithms are mostly linked to ANNs for optimizing the outputs of the ANN models or for finding the optimal neural network structures.

References

Abbasian, M., and A. Sattari. 2013. "Cell voltage and chlorine current efficiency of aqueous HCl electrolysis: Artificial neural network modeling." *Asia-Pac. J. Chem. Eng.* 8 (1): 178–188.

Adesokan, B. J. 2015. "Numerical modeling of microelectrochemical systems." Ph.D. thesis, Department of Applied Mathematics and Computer Science, Technical University of Denmark.

Ahamed Basha, C., P. A. Soloman, M. Velan, L. R. Miranda, N. Balasubramanian, and R. Siva. 2010. "Electrochemical degradation of specialty chemical industry effluent." *J. Hazard. Mater.* 176 (1): 154–164.

Aleksander, I., and H. Morton. 1990. *An introduction to neural computing.* New York: Van Nostrand Reinhold.

Anderson, M. J., and P. J. Whitcomb. 2016a. *DOE simplified: Practical tools for effective experimentation.* Boca Raton, FL: Productivity Press.

Anderson, M. J., and P. J. Whitcomb. 2016b. *RSM simplified: Optimizing processes using response surface methods for design of experiments.* Boca Raton, FL: Productivity Press.

Andideh, M., and M. Esfandeh. 2016. "Statistical optimization of treatment conditions for the electrochemical oxidation of PAN-based carbon fiber by response surface methodology: Application to carbon fiber/epoxy composite." *Compos. Sci. Technol.* 134: 132–143.

Ansari, A., and A. A. Bakar. 2014. "A comparative study of three artificial intelligence techniques: Genetic algorithm, neural network, and fuzzy logic, on scheduling problem." In *Proc., 4th Int. Conf. Artificial Intelligence with Applications in Engineering and Technology,* Kota Kinabalu, Malaysia, IEEE, 31–36.

Ashrafizadeh, S. N., F. Mohammadi, A. Sattari, and N. Shojaikaveh. 2009. "Prediction of cell voltage and chlorine current efficiency of aqueous HCl electrolysis utilizing an oxygen reducing cathode based on artificial neural network." Vienna, Austria, ECS Transactions 25 (33): 27–42.

Azevedo, L. F. M., M. G. Trevisan, J. S. Garcia, and A. M. S. Lucho. 2014. "Electrochemical oxidation, adsorption and quantification of 1,2-benzopyrone employing a glassy carbon electrode." *J. Braz. Chem. Soc.* 25 (3): 389–392.

Bansal, S., J. P. Kushwaha, and V. K. Sangal. 2013. "Electrochemical treatment of Reactive Black 5 textile wastewater: Optimization, kinetics, and disposal study." *Water Environ. Res.* 85 (12): 2294–2306.

Baughman, D. R., and Y. A. Liu. 1995. *Neural networks in bioprocessing and chemical engineering.* Cambridge, MA: Academic Press.

Behera, S. K., H. Meena, S. Chakraborty, and B. C. Meikap. 2018. "Application of response surface methodology (RSM) for optimization of leaching parameters for ash reduction from low-grade coal." *Int. J. Min. Sci. Technol.* 28 (4): 621–629.

Belkacem, S., S. Bouafia, and M. Chabani. 2017. "Study of oxytetracycline degradation by means of anodic oxidation process using platinized titanium (Ti/Pt) anode and modeling by artificial neural networks." *Process Saf. Environ. Prot.* 111: 170–179.

Borras, C., C. Berzoy, J. Mostany, J. C. Herrera, and B. R. Scharifker. 2007. "A comparison of the electrooxidation kinetics of *p*-methoxyphenol and *p*-nitrophenol on Sb-doped SnO_2 surfaces: Concentration and temperature effects." *Appl. Catal., B* 72 (1–2): 98–104.

Cañizares, P., J. Garcia-Gomez, J. Lobato, and M. A. Rodrigo. 2004a. "Modeling of wastewater electro-oxidation processes—Part I. General description and application to inactive electrodes." *Ind. Eng. Chem. Res.* 43 (9): 1915–1922.

Cañizares, P., J. Garcia-Gomez, J. Lobato, and M. A. Rodrigo. 2004b. "Modeling of wastewater electro-oxidation processes—Part II. Application to active electrodes." *Ind. Eng. Chem. Res.* 43 (9): 1923–1931.

Chen, W., X. Yu, H. Wang, and B.-G. Wen. 2010. "PSO-GA on endmember extraction for hyperspectral imagery." In *Proc., Int. Conf. Computer Application and System Modeling.*

Curteanu, S., and M. Cazacu. 2007. "Neural networks and genetic algorithms used for modeling and optimization of the siloxane–siloxane copolymers synthesis." *J. Macromol. Sci. Part A Pure Appl. Chem.* 45 (1): 23–36.

Curteanu, S., K. Godini, C. G. Piuleac, G. Azarian, A. R. Rahmani, and C. Butnariu. 2014. "Electro-oxidation method applied for activated sludge treatment: Experiment and simulation based on supervised machine learning methods." *Ind. Eng. Chem. Res.* 53 (12): 4902–4912.

Darvishmotevalli, M., A. Zarei, M. Moradnia, M. Noorisepehr, and H. Mohammadi. 2019. "Optimization of saline wastewater treatment using electrochemical oxidation process: Prediction by RSM method." *MethodsX* 6: 1101–1113.

Da Silva, I. N., D. Hernane Spatti, R. Andrade Flauzino, L. H. B. Liboni, and S. F. Dos Reis Alves. 2017. *Artificial neural networks: A practical course.* Cham, Switzerland: Springer.

de Melo da Silva, L., F. Gozzi, I. Sires, E. Brillas, S. C. de Oliveira, and A. J. Machulek. 2018. "Degradation of 4-aminoantipyrine by electro-oxidation with a boron-doped diamond anode: Optimization by central composite design, oxidation products and toxicity." *Sci. Total Environ.* 631–632: 1079–1088.

Deshpande, A. M., Ramakant, and Satyanarayan, S. 2012. "Treatment of pharmaceutical wastewater by electrochemical method: Optimization of operating parameters by response surface methodology." *J. Hazard. Toxic Radioact. Waste* 16 (4): 316–326.

Ding, S., C. Su, and J. Yu. 2011. "An optimizing BP neural network algorithm based on genetic algorithm." *Artif. Intell. Rev.* 36 (2): 153–162.

Dominguez-Ramos, A., R. Aldaco, and A. Irabien. 2008. "Electrochemical oxidation of lignosulfonate: Total organic carbon oxidation kinetics." *Ind. Eng. Chem. Res.* 47 (24): 9848–9853.

Eberhart, R. C., and Y. Shi. 1998. "Comparison between genetic algorithms and particle swarm optimization." In *Vol. 1447 of Evolutionary Programming VII, Lecture Notes in Computer Science*, edited by V. W. Porto, N. Saravanan, D. Waagen, and A. Eiben, 611–616. Berlin: Springer.

Estahbanati, M. R. K., M. Feilizadeh, and M. C. Iliuta. 2017. "Photocatalytic valorization of glycerol to hydrogen: Optimization of operating parameters by artificial neural network." *Appl. Catal., B* 209: 483–492.

Feilizadeh, M., I. Alemzadeh, A. Delparish, M. R. Karimi Estahbanati, M. Soleimani, Y. Jangjou, et al. 2015. "Optimization of operating parameters for efficient photocatalytic inactivation of *Escherichia coli* based on a statistical design of experiments." *Water Sci. Technol.* 71 (6): 823–831.

Ferreira, S. L., W. N. Dos Santos, C. M. Quintella, B. B. Neto, and J. M. Bosque-Sendra. 2004. "Doehlert matrix: A chemometric tool for analytical chemistry—Review." *Talanta* 63 (4): 1061–1067.

García-Gómez, C., P. Drogui, F. Zaviska, B. Seyhi, P. Gortáres-Moroyoqui, G. Buelna, et al. 2014. "Experimental design methodology applied to electrochemical oxidation of carbamazepine using Ti/PbO_2 and Ti/BDD electrodes." *J. Electroanal. Chem.* 732: 1–10.

García-Gómez, C., J. A. Vidales-Contreras, J. Nápoles-Armenta, and P. Gortáres-Moroyoqui. 2016. "Optimization of phenol removal using Ti/PbO_2 anode with response surface methodology." *J. Environ. Eng.* 142 (4): 04016004.

Gilpavas, E., I. Dobrosz-Gómez, and M. Á Gómez-García. 2016. "Electrochemical degradation of acid yellow 23 by anodic oxidation—Optimization of operating parameters." *J. Environ. Eng.* 142 (11): 04016052.

Guitaya, L., A. Azaïs, F. Zaviska, P. Drogui, J.-F. Blais, and B. Gourich. 2017. "Electrochemical oxidation as treatment for contaminated wastewaters by carbamazepine: Process optimization through response surface methodology." *Water Air Soil Pollut.* 228 (10): 384.

Hamed, M. M., M. G. Khalafallah, and E. A. Hassanien. 2004. "Prediction of wastewater treatment plant performance using artificial neural networks." *Environ. Modell. Software* 19 (10): 919–928.

Hammerstrom, D. 1993. "Working with neural networks." *IEEE Spectr.* 30 (7): 46–53.

Haykin, S. 1998. *Neural networks. A comprehensive foundation.* Hoboken, NJ: Prentice-Hall.

Jain, L. C., and L. R. Medsker. 1999. *Recurrent neural networks: Design and applications.* Boca Raton, FL: CRC Press.

Jardak, K., A. Dirany, P. Drogui, and M. A. El Khakani. 2017. "Statistical optimization of electrochemical oxidation of ethylene glycol using response surface methodology." *Process Saf. Environ. Prot.* 105: 12–20.

Juneja, M., and S. K. Nagar. 2016. "Particle swarm optimization algorithm and its parameters: A review." In *Proc., Int. Conf. Control, Computing Communication and Materials*, Allahbad, India, IEEE, 1–5.

Karimi Estahbanati, M. R., M. Feilizadeh, and M. C. Iliuta. 2019. "An intrinsic kinetic model for liquid phase photocatalytic hydrogen production." *AlChE J.* 65 (11): e16724.

Kennedy, J., and R. Eberhart. 1995. "Particle swarm optimization." In *Vol. 1944 of Proc., Int. Conf. Neural Networks, 1942–1948.* Perth, Australia: IEEE.

Kozik, V., K. Barbusinski, M. Thomas, A. Sroda, J. Jampilek, A. Sochanik, et al. 2019. "Taguchi method and response surface methodology in the treatment of highly contaminated tannery wastewater using commercial potassium ferrate." *Materials* 12 (22): 3784.

Liu, L., B. Li, Z. He, C. Zhang, and D. Fu. 2015. "Degradation of bromoamine acid by BDD technology—Use of Doehlert design for optimizing the reaction conditions." *Sep. Purif. Technol.* 146: 15–23.

Manokaran, P., R. Saravanathamizhan, C. A. Basha, and T. Kannadasan. 2014. "Feed-forward back-propagation neural network for the electro-oxidation of distillery effluent." *Chem. Eng. Commun.* 201 (10): 1404–1416.

Moghaddam, A. A., A. Seifi, and T. Niknam. 2012. "Multi-operation management of a typical micro-grids using particle swarm optimization: A comparative study." *Renewable Sustainable Energy Rev.* 16 (2): 1268–1281.

Mohajeri, S., H. A. Aziz, M. H. Isa, M. A. Zahed, M. J. Bashir, and M. N. Adlan. 2010. "Application of the central composite design for condition optimization for semi-aerobic landfill leachate treatment using electrochemical oxidation." *Water Sci. Technol.* 61 (5): 1257–1266.

Montgomery, D. C. 2017. *Design and analysis of experiments.* Hoboken, NJ: Wiley.

Morshedi, A., and M. Akbarian. 2014. "Application of response surface methodology: Design of experiments and optimization: A mini review." *J. Fund Appl. Life Sci.* 54 (S4): 2434–2439.

Moteshaker, P. M., S. Saadi, S., and S. E. Rokni. 2020. "Electrochemical removal of diazinon insecticide in aqueous solution by Pb/β-PbO_2 anode. Effect of parameters and optimization using response surface methodology." *Water Environ. Res.* 92 (7): 975–986.

Nandhini, M., B. Suchithra, R. Saravanathamizhan, and G. P. Dhakshinamoorthy. 2014. "Optimization of parameters for dye removal by electro-oxidation using Taguchi design." *J. Electrochem. Sci. Eng.* 4 (4): 227–234.

Niculescu, S. P. 2003. "Artificial neural networks and genetic algorithms in QSAR." *J. Mol. Struct.-Theochem.* 622 (1): 71–83.

Panizza, M., P. A. Michaud, G. Cerisola, and C. Comninellis. 2001. "Electrochemical treatment of wastewaters containing organic pollutants on boron-doped diamond electrodes: Prediction of specific energy consumption and required electrode area." *Electrochem. Commun.* 3 (7): 336–339.

Picos, A., and J. M. Peralta-Hernández. 2018. "Genetic algorithm and artificial neural network model for prediction of discoloration dye from an electro-oxidation process in a press-type reactor." *Water Sci. Technol.* 78 (4): 925–935.

Piuleac, C., C. Saez, P. Cañizares, S. Curteanu, and M. Rodrigo. 2012. "Hybrid model of a wastewater-treatment electrolytic process." *Int. J. Electrochem. Sci.* 7: 771–775.

Piuleac, C. G., M. A. Rodrigo, P. Cañizares, S. Curteanu, and C. Sáez. 2010. "Ten steps modeling of electrolysis processes by using neural networks." *Environ. Modell. Software* 25 (1): 74–81.

Radi, M. A., N. Nasirizadeh, M. Mirjalili, and M. Rohani Moghadam. 2018. "Ultrasound-assisted electrochemical decolorization of anthraquinone dye C.I Reactive Blue 49, its optimization and synergic effect: A comparative study." *Int. J. Environ. Sci. Technol.* 16 (5): 2455–2464.

Rajkumar, K., and M. Muthukumar. 2012. "Optimization of electro-oxidation process for the treatment of Reactive Orange 107 using response surface methodology." *Environ. Sci. Pollut. Res. Int.* 19 (1): 148–160.

Ridha, H., M. Njah, and M. Chtourou. 2008. "Multilayer perceptron training using an evolutionary algorithm." *Int. J. Modell. Identif. Control* 5 (4): 305–312.

Sangal, V., P. Kaur, and J. Kushwaha. 2015. "Modeling and evaluation of electro-oxidation of dye wastewater using artificial neural networks." *RSC Adv.* 5 (44): 34663–34771.

Saravanathamizhan, R., K. Harsha Vardhan, D. Gnana Prakash, and N. Balasubramanian. 2015. "RSM and ANN modeling for electro-oxidation of simulated wastewater using CSTER." *Desalin. Water Treat.* 55 (6): 1445–1452.

Sharma, S., and H. Simsek. 2020. "Sugar beet industry process wastewater treatment using electrochemical methods and optimization of parameters using response surface methodology." *Chemosphere* 238: 124669.

Soloman, P., C. Basha, V. Manickam, and B. Subramanian. 2010. "Electro oxidation of malachite green and modeling using ANN." *Chem. Biochem. Eng. Q.* 24 (4): 445–452.

Talebi, A., M. A. Molaei, and M. J. Sheikh. 2010. "Performance investigation and comparison of two evolutionary algorithms in portfolio optimization: Genetic and particle swarm optimization." In *Proc., 2nd IEEE Int. Conf. Information and Financial Engineering*, Chongqing China, IEEE, 430–437.

Torres-Sospedra, J., C. Hernández-Espinosa, and M. Fernández-Redondo. 2006. "Combining MF networks: A comparison among statistical methods and stacked generalization." In *Vol. 4087 of Artificial neural networks in pattern recognition*, edited by F. Schwenker and S. Marinai, 210–220. Heidelberg, Germany: Springer.

Viana, D. F., G. R. Salazar-Banda, and M. S. Leite. 2018. "Electrochemical degradation of Reactive Black 5 with surface response and artificial neural networks optimization models." *Sep. Sci. Technol.* 53 (16): 2647–2661.

Wang, L., B. Wu, P. Li, B. Zhang, N. Balasubramanian, and Y. Zhao. 2016. "Kinetics for electro-oxidation of organic pollutants by using a packed-bed electrode reactor (PBER)." *Chem. Eng. J.* 284: 240–246.

Whitley, D. 1994. "A genetic algorithm tutorial." *Stat. Comput.* 4 (2): 65–85.

Yao, J., Y. Mei, G. Xia, Y. Lu, D. Xu, N. Sun, et al. 2019. "Process optimization of electrochemical oxidation of ammonia to nitrogen for actual dyeing wastewater treatment." *Int. J. Environ. Res. Public Health* 16 (16): 2931.

Yousefi, Z., A. Zafarzadeh, and A. Ghezel. 2018. "Application of Taguchi's experimental design method for optimization of Acid Red 18 removal by electrochemical oxidation process." *Environ. Health Eng. Manage.* 5 (4): 241–248.

Zaviska, F., P. Drogui, J.-F. Blais, and G. Mercier. 2012. "Electrochemical treatment of bisphenol-A using response surface methodology." *J. Appl. Electrochem.* 42 (2): 95–109.

Zaviska, F., P. Drogui, J.-F. Blais, and G. Mercier. 2013. "Electrochemical oxidation of chlortetracycline using Ti/IrO_2 and Ti/PbO_2 anode electrodes: Application of experimental design methodology." *J. Environ. Eng.* 139 (6): 810–821.

Zolfaghari, M., K. Jardak, P. Drogui, S. K. Brar, G. Buelna, and R. Dube. 2016. "Landfill leachate treatment by sequential membrane bioreactor and electro-oxidation processes." *J. Environ. Manage.* 184 (Pt 2): 318–326.

CHAPTER 6

Combined Electro-Coagulation Processes

Ali Khosravanipour Mostafazadeh,
Patrick Drogui, R. D. Tyagi

6.1 INTRODUCTION

When wastewater moves through the applied electric field in the electro-coagulation (EC) process, different phenomena such as ionization, electrolysis and hydrolysis, electrophoresis, and so on may occur. Ultrasound (US) and low pH (e.g., pH < 3) can accelerate the formation of the free radical and hydroxyl radical. However, there is no evidence of advanced oxidation processes (AOPs) during the EC process (Ghernaout 2013). Therefore, the AOP and EC processes must be combined to have these phenomena to enhance the efficiency of wastewater treatment. Currently, electrochemical AOPs have been recognized as an effective method for wastewater treatment. These processes are environmentally friendly methods that are capable of electrochemically producing hydroxyl radical [$E(^{\bullet}OH/H_2O) = 2.80$ Volts versus standard hydrogen electrode], which is a powerful oxidant capable of oxidizing a broad range of organic compounds (Daghrir and Drogui 2013a).

In general, the advantages of AOPs are that they can operate at room temperature and pressure; have high oxidation ability and rapid reaction rate; remove contaminants, including organic and inorganic, with minimum cost; and have the ability to integrate with other treatment processes such as EC (Aziz et al. 2016). However, the main disadvantage of AOPs is their high operating and investment costs. Therefore, the coupling of AOPs with other techniques such as EC can improve efficiency and reduce costs (Ibarra-Taquez et al. 2017). Electrochemical approaches have been applied in recent years for the treatment of organic substances with high toxicity and low biological degradability. For instance, EC, electro-oxidation (EO), and electro-photo-oxidation (EPO) have been used for treating wastewater containing heavy metals, foodstuff, oil industries, textiles, solutions containing ultrafine particles, nitrate, phenolic compounds, arsenic, and refractory organics such as lignin and ethylenediaminetetraacetic acid (EDTA) (Orkun and Kuleyin 2012).

Over last years, various AOPs have been used to treat wastewater. Chemical (Fe^{2+}/H_2O_2)/(O_3/H_2O_2), photochemical ($UV/Fe^{2+}/H_2O_2$), sonochemical (US/Fenton, US/photocatalytic), photocatalytic (TiO_2 Photocatalyst), photo-Fenton, and photochemical (O_3/UV, H_2O_2/UV, $O_3/UV/H_2O_2$) AOP techniques have been investigated. These processes are environmentally friendly, do not need the addition of more chemicals, and are able to produce in situ hydroxyl radical groups. Coupling AOPs and ozonation with EC proved that they are effective processes for the treatment and disinfection of wastewater by removing organic and inorganic contaminants. Technical and economical factors must be considered in these hybrid systems, such as process efficiency, uncomplicatedness, flexibility, investment, operating costs, energy and chemical intake, and sludge disposal. In a study of a combined treatment (e.g., via ozonation, EC, ozone-EC, peroxi-EC, photo-EC, and peroxi-photo-EC) of distillery industrial effluents for investigating the color and chemical oxygen demand (COD) removal by considering the effective parameters such as ozone flow rate, pH, current density, and H_2O_2 concentration, the highest efficiency was achieved by the use of the ozone-EC process, although it consumed the highest energy (Aziz et al. 2016). In another study, the treatment of pharmaceutical effluents was investigated in terms of COD removal by various combined physicochemical processes (EC, photo-EC, peroxi-EC, and peroxi-photo-EC). They examined the different crucial parameters such as pH, current density, applied voltage, operation time, and amount of hydrogen peroxide. The combined EC processes can be classified as follows (based on the purification performances): peroxi-EC > peroxi-photo-EC > photo-EC > EC. However, the energy consumption of the processes showed the following order: photo-EC > peroxi-photo-EC > EC > peroxi-EC (Farhadi et al. 2012).

Sludge separation, filtration, flotation, and sedimentation are necessary processes that must be done along with EC. However, other treatment technologies, including the integration of other methods for the treatment of wastewater (WW), result in larger removal efficiency compared with single EC, which is a safe and effective approach for WW treatment (Naje et al. 2017).

In a sequential treatment of ice diary wastewater (EC-Fenton/ozone oxidation), only 40% removal efficiency was recorded by the sole EC process. However, by applying post-AOP, the total removal of 70% COD was achieved. The removal efficiency could be achieved by optimizing AOP (Torres-Sánchez et al. 2014). Another study for dairy wastewater treatment using coupled EC-Fenton oxidation demonstrated 98%, 99%, 84.6%, and 99.4% removal efficiencies for COD, biochemical oxygen demand (BOD), total dissolved solids (TDS), and color, respectively (Shivayogimath and Vinayak Rao 2016).

Overall, Figure 6-1 shows the combined EC and AOP in one pot. EC has simultaneously two effects, coagulation and catalysis, because of the formation of $Fe(OH)_n$ ($n = 2$ or 3) coagulant and Fe^{3+}/Fe^{2+} ions. The main catalytic AOPs are heterogeneous catalysis, including TiO_2 photocatalysis and electrocatalysis with high oxygen-overvoltage anodes, and homogeneous catalysis, such as ozonation, Fenton, and photo-Fenton (chemically and/or electrochemically). The combined treatment of wastewater containing Tartrazine (synthetic dye) using a two-step process, that is, EC (Fe/steel)/photo-electro-Fenton (PEF) with a boron-doped diamond (BDD)/air-diffusion cathode showed that this process is able to overcome

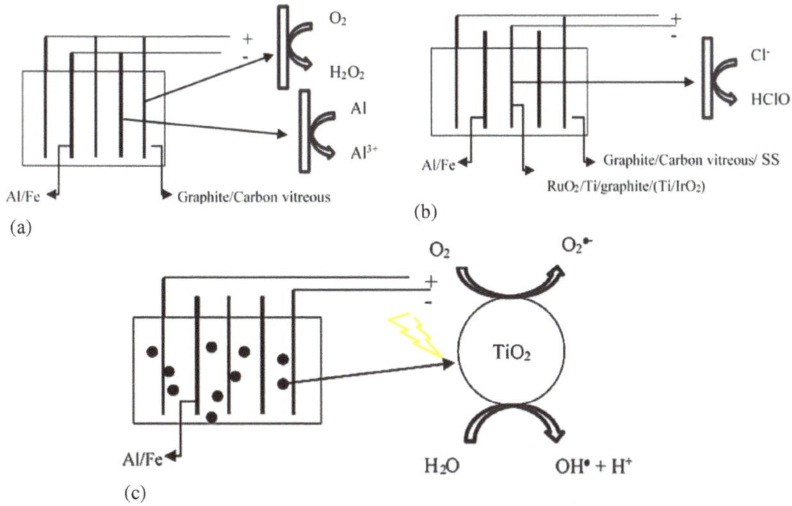

Figure 6-1. Combined electro-coagulation and advanced oxidation process in one pot by in situ generation of oxidant agents and a coagulating agent (Fe^{2+}/Fe^{3+}or Al^{3+}): (a) EC and H_2O_2, (b) EC and $HClO/ClO^-$, (c) EC and $HO^{\bullet}/O_2^{\bullet-}$.

the shortcomings of the single processes, such as partial mineralization by EC and the long treatment time required by AOPs and, thus, leads to a more cost-effective process thanks to a decline in energy consumption (Thiam et al. 2014).

The combined water treatment system can also be applied for human consumption. Sequential electroreduction/EC to eliminate heavy metals, suspended solids, color, and turbidity, followed by ozonation/UV treatment, was performed effectively to remove organic contaminants and ammonia from groundwater. This type of process does not require external chemical addition, can abate a broad range of contaminants, has low sensitivity with regard to water compositions, and produces high-quality effluents (Orescanin et al. 2011b).

This chapter covers both fundamental and recent developments of integrated EC processes, including various treatment techniques such as EC-photo assisted, Sono-EC (SEC), EC-Fenton, EC–EO, EC-peroxidation, and ozone-assisted EC processes. The EC process is able to remove organic, inorganic, and microbial pollutants of water and wastewater by in situ generation of coagulants through electrodes. EO is a method that uses active catalytic electrodes and produces oxidizing agents, and through these, eliminates infections, refractory organic pollutants, and dissolved pollutants. TiO_2-photocatalytic technology can abate hydroxyl radical production for oxidizing and mineralizing contaminants by combining semiconductors and UV radiation. Applying an ultrasonic field on EC can improve efficiency by cleaning the electrodes and by rapidly decomposing organic pollutants. Hydrogen peroxide assists EC to produce a more hyperactive oxidizing agent (hydroxyl radical). Ozonation, along with EC, can produce oxidizing agents by indirect radical-type chain reactions. The electro-Fenton (EF) process leads to hydroxyl radical production by the reaction taking place between metallic ions (produced by scarifying the electrode) and hydrogen peroxide to degrade organic compounds. Biological treatment of wastewater has proved to be effective technology. However, because of some disadvantages such as long treatment time, low efficiency for removal of organic substances (bioresistant pollutants), commissioning, and maintaining operating conditions, this method should be combined with EC to overcome these drawbacks. The mentioned methods can be combined (in one step or sequential) and applied with EC effectively to treat water and wastewater considering the optimum operating conditions and mainly choosing suitable electrodes, current density, pH, and energy consumption.

6.2 ADVANTAGES AND DISADVANTAGES OF ELECTRO-COAGULATION VERSUS ADVANCED OXIDATION PROCESS

The electrochemical processes (EO, EC, EF, photocatalysis-TiO_2, and peroxidation) have the common advantage of eliminating both organic pollution and the color of WW. Thanks to the combination of direct and indirect reactions, EO is able to degrade both organic and inorganic compounds such as NH_4. However, indirect oxidation requires high concentrations of chlorides to form reactive intermediates. In this process, the risk of the formation of toxic organochlorine compounds linked to high concentrations of chloride ions also exists. EF is particularly well known for its high performance in terms of COD removal and increased biodegradability of the effluent. However, EF requires an acidic pH and addition of hydrogen peroxide or ferric ions in certain quantities. Its inefficiency in removing ammonia nitrogen means that, in combination with another process, it is capable of treating NH_4. In regard to EC, the combination of electrochemical and physicochemical processes results in the effective elimination of turbidity and high-molecular-weight organic compounds. EC also produces fewer metallic residues compared with the chemical coagulation (CC) process. However, like EF, the management of sludge loaded with metallic wastes is a problem to be taken into account. Ammonia nitrogen is poorly eliminated during EC. Some authors argue that the removal of NH_4 observed in EC is related to the phenomenon that transforms NH_4 into NH_3 gas (Dia et al. 2016). Peroxidation is relatively costly, whereas the photo-assisted process comparatively has a lower operating cost. Peroxidation leads to an enhancement of flocculation by the oxidation of metals, and the photo-assisted TiO_2 method can mineralize a broad range of pollutants. However, the introduction of foreign species to a pure TiO_2 structure is required to enhance efficiency.

6.3 ELECTRO-COAGULATION AND TiO_2 PHOTO-ASSISTED PROCESS

6.3.1 Introduction to the Photocatalysis Process and Hybrid Technique with Electrocoagulation

The photo-electrocatalytic oxidation (PECO) technique, integrating both electrolytic and photocatalytic processes, has been recognized as a reliable and cost-effective water treatment process. This process represents the recombination of the electron–hole pairs (e^-_{CB} / h^+_{VB}) photogenerated on the catalyst surface, and accordingly, a higher efficiency in terms of the degradation of pollutants can be achieved (Daghrir et al. 2013b). The photocatalytic process is a method that applies a combination of radiation and a semiconductor, such as TiO_2 and ZnO, to produce highly oxidant radicals that are able to decompose organic matter. The sequential EC/photocatalytic coupled process can be used for different wastewater treatments (Suárez-Escobar et al. 2016).

Titanium dioxide (TiO_2) is a catalyst with special characteristics such as high photochemical stability, low toxicity, and reduced costs. The absorption of photons with an energy higher than 3.2 eV leads to the initiation of excitation and charge separation. Consequently, an electron at the conductive band (e^-_{CB}) and a positive hole at the valence band (h^+_{VB}) emerge. This h^+_{VB} will react with adsorbed water molecules and OH^- ions at the catalyst surface and generate $^{\bullet}OH$ radicals as a powerful oxidizing agent. Moreover, e^-_{CB} can produce an anion radical superoxide with the reaction of oxygen molecule, which results in hydrogen peroxide formation and subsequently the hydroxyl radical creation. The dissolved oxygen is crucial for the heterogeneous photocatalysis process because of the prevention of the recombination process on TiO_2 (h^+_{VB} / e^-_{CB}) and keeps the electroneutrality of the catalyst. It is worth mentioning that the COD must be lower than 800 mg L^{-1} for this kind of process to achieve the best results (Boroski et al. 2009).

Because pure TiO_2 requires the use of UV light and the recombination rates of high photogenerated electrons and holes, which is seen as a drawback, a number of studies have been performed on the development of novel TiO_2 particles to absorb visible light by introducing cadmium sulfide (CdS, with a band gap of 2.4 eV), SnO_2, WO_3, FeO_3, and Bi_2S_3 into the titanium dioxide matrix as TiO_2/M_xO_y or TiO_2/M_xS_y. For instance, the couple TiO_2/CdS can be applied to increase the degradation of organic matter such as acid orange II, organic dye drimaren red, phenazopyridine, methyl orange, and so on. by photocatalytic and photo-electrocatalytic processes (Daghrir et al. 2013c). The following equation shows the effect of the addition of CdS to TiO_2:

$$\mathrm{CdS(e^- + h^+) + TiO_2 \rightarrow CdS(h^+) + TiO_2} \tag{6-1}$$

Hydroxyl radicals ($OH^{\bullet}$), superoxide radical anions ($O_2^{-\bullet}$), hydrodioxyl radicals ($HO_2^{\bullet}$), and (h^+_{VB}) are the oxidizing agents in the UV/TiO_2 system. H_2O_2 is also produced. Fe^{2+} and Fe^{3+} are generated in the EC process. Thus, the combination of UV/TiO_2 and EC can cause a Fenton or Fenton-like reaction. The reaction mechanism of the TiO_2/UV system can be represented as follows (Wu et al. 2008):

$$\mathrm{TiO_2} + hv \rightarrow \mathrm{TiO_2(e_{CB^-} + h^+_{VB})} \tag{6-2}$$

$$\mathrm{TiO_2(h^+_{VB}) + H_2O \rightarrow TiO_2 + H^+ + OH^{\bullet}} \tag{6-3}$$

$$\mathrm{TiO_2(h^+_{VB}) + OH^- \rightarrow TiO_2 + OH^{\bullet}} \tag{6-4}$$

$$\mathrm{TiO_2(e_{CB^-}) + O_2 \rightarrow TiO_2 + O_2^{-\bullet}} \tag{6-5}$$

$$O_2^{-\bullet} + H^+ \rightarrow HO_2^{\bullet} \quad (6\text{-}6)$$

$$\text{Contaminants (such as dye)} + OH^{\bullet} \rightarrow \text{Degradation products} \quad (6\text{-}7)$$

$$\text{Contaminants (such as dye)} + h_{VB^+} \rightarrow \text{Oxidation products} \quad (6\text{-}8)$$

$$\text{Contaminants (such as dye)} + e_{CB^-} \rightarrow \text{Reduction products} \quad (6\text{-}9)$$

Moreover, the other mechanism was suggested as follows:

$$e_{CB^-} + H^+ + O_2^{-\bullet} \rightarrow HO_2^- \quad (6\text{-}10)$$

$$HO_2^- + H^+ \rightarrow H_2O_2 \quad (6\text{-}11)$$

$$H_2O_2 + e_{CB^-} \rightarrow OH^- + OH^{\bullet} \quad (6\text{-}12)$$

$$2HO_2^{\bullet} \rightarrow H_2O_2 + O_2 \quad (6\text{-}13)$$

$$H_2O_2 + O_2^{-\bullet} \rightarrow OH^- + OH^{\bullet} + O_2 \quad (6\text{-}14)$$

The EC method can also be followed by UV irradiation. Sulfate radical ($SO_4^{\bullet-}$) is a powerful oxidant with redox potential ($E_0 = 2.5$ to 3.1 V), which is even stronger than hydroxyl radical ($E_0 = 1.8$ to 2.7 V). The residual transitional metals can be harmful to human health; however, the UV irradiation method is an environmentally friendly technique for the radical production of sulfate (Jaafarzadeh et al. 2016).

$$S_2O_8^{2-} + UV \rightarrow SO_4^{\bullet -} + SO_4^{\bullet -} \quad (6\text{-}15)$$

$$HSO_5^- + UV \rightarrow HO^{\bullet} + SO_4^{\bullet -} \quad (6\text{-}16)$$

Wastewater must be pretreated (e.g., by EC) before applying this process, because of the absorption of UV radiation by color and turbidity and consequently preventing the activation of oxidants. Table 6-1 shows recent advances in photocatalysis and EC.

6.3.2 Kinetic Model

The pseudo-second-order model has been proposed in terms of total organic carbon (TOC) as follows (Suárez-Escobar et al. 2016):

$$\frac{1}{[TOC]_t} = \frac{1}{[TOC]_o} + kt \quad (6\text{-}17)$$

where

$[TOC]_t$ = TOC value at the time t (in minutes),
$[TOC]_o$ = TOC value at the initial time, and
k = Constant for the model.

Table 6-1. Recent Studies on Combined Electro-Coagulation Processes.

Process type	*Wastewater type*	*Operating conditions*	*Removal efficiency%*	*Advantages and feasibility*	*Estimated cost \$US m^{-3} and/or energy consumption*	*Reference*
Sequential integrated photo-Fenton-EC process	Tannery effluent	Photo-Fenton: pH: 3, $Fe^{2+} = 0.4$ g L^{-1}, $H_2O_2 = 15$ g L^{-1}), time: 540 min EC: pH: 8.3, electrode: Al, current: 68 mA cm^{-2}, time: 15 min	—	A significant improvement in COD removal compared with the conventional method but not significant in terms ofTSS, TFS, and TVS, turbidity, and chromium concentration EC-photo-Fenton is less expensive than the conventional method	Photo-Fenton: 64.13 EC: 2.09 Combined: 66.22 Conventional: 69.06 By considering both the operational and sludge disposal	Módenes et al. (2012)
Coupled EC-EF	Domestic wastewater	pH: 7.0 Electrode: iron (anode) and carbon vitreous (cathode) Current: 0.34 A dm^{-2} Time: 60 min	COD: 50.07 TSS: 90.40 Turbidity: 70.80	A favorable secondary treatment to simultaneously remove organic, inorganic, and microbial pollutants	0.21 (including electrode cost, energy consumption, and sludge disposal)	Daghrir and Drogui (2013a)
EC-EF	Landfill leachate	EC: pH: 6.54, electrode: Ir, current: 30 mA cm^{-2}, time: 180 min EC-EF: Time: 60 min, H_2O_2: 5,000 ppm	EC: COD: 65.85 EC-EF: COD: 74.21	The EF method is superior to the conventional Fenton method not only from the conversion point of view but also from the economic point of view	0.418 kg COD/kg Al 11.092 kW·h/kg COD	Orkun and Kuleyin (2012)
Sequential EC and EF	Car wash wastewater	EC: pH: 7.3, electrode: iron, current: 4.2 mA cm^{-2}, time: 20.3 min EF: pH: 3, electrode: current: 2 A Time: 10 min, H_2O_2: 500 mg L^{-1}	COD: 80.8, Phosphate: 4.9 Turbidity: 85.5	—	—	Mirshahghassemi et al. (2017)

(*Continued*)

Table 6-1. Recent Studies on Combined Electro-Coagulation Processes. (Continued)

Process type	*Wastewater type*	*Operating conditions*	*Removal efficiency%*	*Advantages and feasibility*	*Estimated cost \$US m^-3 and/or energy consumption*	*Reference*
EC and EO	Restaurant wastewaters	pH: 7.0 Electrode: Fe/Al–Gr Current: 0.4 A Time: 90 min	O&G: 98 COD: 90 BOD: 86 Turbidity: 98 SS: 98 Soluble phosphate: 88	The EC–EO process is a practical method for the treatment of RWW by a lower total cost and less sludge production	1.56 includes energy and electrode consumption, chemicals, and sludge disposal	Daghrir et al. (2012)
Sequential EC–EO process	Soluble coffee effluent	pH: 7.98 Electrode: Al/Gr Current EC: 149.2 A m^{-2} Current EO: 500 A m^{-2} Time EC: 62 min Time EO: 53 min	Total decolorization COD: 74 TOC: 63.5 highly oxidized biocompatible (BOD5/COD = 0.6)	Savings of 25% operational cost and 34% of savings in total energy consumption compared with EO	Total operation cost (TOpC): 8.9 Energy: 45.28 (kW·h/m^3) Including consumption of energy, anode material, and the amount of produced sludge	Ibarra-Taquez et al. (2017)
EC–EO–EF (ECEO–EF)	Effluent of an anaerobic reactor	pH: 7, electrode: Al, stainless steel, and RuO_2/Ti plates, current 3 A, time 60 min	Ammonia: 98 Phosphate: 98 COD: 72	This technology is capable of sufficiently removing both phosphate and ammonia	—	Mahvi et al. (2011)
Integrated ozone-EC process	Industrial effluent	pH: 7, electrode: iron, current 3 A/dm^3, time 5 h	COD: 82.5	Ozone-assisted EC process yielded higher pollutant removal than ozone and EC processes alone	Power: 6.35 kW h/m^3	Asaithambi et al. (2016)
Ozone-assisted EC	Synthetic wastewater containing Acid Brown 214	pH: 9.5, electrode: iron, current 15 mA cm^{-2}, time 30 min	Complete decolorization	The hybrid process at a low operation cost with the decolorization yield of 100%	Energy consumption of 7.4 kW · h/kg	Behin et al. (2015)
Combined EC-electro-peroxidation Process	Domestic wastewaters	pH: 7.2, electrode: mild steel (anode) and vitreous carbon (cathode), current 40 mA cm^{-2}, time 60 min	(COD): 67 ± 9, (SS): 98 ± 2, Turbidity: 55 ± 10, color: 61 ± 9, P_{tot}: 97 ± 0, pathogens: 99 ± 1	Capable of simultaneously removing inorganic, organic, and microbial pollutants from many wastewaters	0.211 ± 0.02 includes energy and electrode consumption and metallic sludge disposal	Senghor et al. (2015)

Peroxi-EC/ electro-oxidation electroflotation process	Olive mill wastewater	pH: 4; electrodes: aluminum (Al), stainless steel, and RuO_2/Ti plates; current: 40 mA cm^{-2}; time: 30 min; H_2O_2 1,000 mg L^{-1}; NaCl: 1 g L^{-1}	COD: 96 BOD: 93.6 Polyphenol: 94.4 Color: 91.4 Turbidity: 88.7 SS: 97 O & G: 97.1	Effective and feasible process for pretreating olive mill wastewater, making possible a post-treatment of the effluent in a biological system	Specific energy consumption (SEC) for COD removal: 6.82 kW·h/kg SEC for polyphenol: 92.33 kW·h/kg	Esfandyari et al. (2015)
Peroxi -EC process	Synthetic wastewater containing azithromycin (pharmaceutical)	pH: 3, electrodes: iron Current: 20 mA cm^{-2}, Time: 60 min, H_2O_2: 2 mg L^{-1}	COD: 95.6	The process is technologically feasible	Mean energy consumption = 2. 5kW·h	Yazdanbakhsh et al. (2015)
Sequential EC-photocatalytic process	Lithographic wastewater	EC: pH: 8.23, electrode: iron time: 20 min, Current: 125 A m^{-2}, pH: 8, catalyst: TiO_2 0.09%, time: 45 min	TOC: 74.43 Photo: 26.78 EC: 58.42	Removal of color, turbidity, and TOC	—	Suárez-Escobar et al. (2016)
Sequential EC and TiO_2 photo-assisted	Wastewater effluents from pharmaceutical and cosmetic industries	EC: pH = 6, electrode: iron Time: 90 min, current 763 A m^{-2} Photoassisted: pH = 3, time 4 h Catalyst: TiO_2 0.25 g L^{-1} H_2O_2: 10 mmol L^{-1}	EC: COD: 91 Turbidity: 86 Combined: COD: 97	Refractory compounds were degraded by the combined process compared with EC. Potential application on an industrial scale.	—	Boroski et al. (2009)
EC and ultrasonic processes	Synthetic wastewater containing humic acid (HA)	pH: 7, electrode: Pt/Al Time: 15 min, Voltage: 10 V 24 kHz frequency	HA removal: 96.5 Without Ultrasound	Simultaneous use of both methods, decreased removal efficiency	—	Asgharian et al. (2017)
sono-EC	Synthetic wastewater containing RB19	pH: 5, electrode: Al Time: 60 min, Current: 18 mA cm^{-2}, Ultrasound power: 150 W	RB 19 removal: 97	The total increase in removal efficiency was 14%. Increase in both energy efficiency and removal efficiency was noticed	\$US 0.373/m^3 including energy consumption and electrode costs for 5 min operation	He et al. (2016)

In this model, we assume that the adsorption rate on the photocatalyst is proportional to the square of the vacant adsorption sites. On the contrary, the Langmuir–Hinshelwood (L–H) kinetic model has been frequently used for the photo-electrocatalytic degradation of organic compounds. The following equation represents the L–H kinetic model (Daghrir et al. 2013b):

$$r = -\frac{dC}{dt} = k_r\theta = \frac{k_r KC}{1+KC} \tag{6-18}$$

where

k_r = Reaction rate constant,
K = Reactant adsorption constant,
θ = Fraction of the surface of the catalyst covered by a pollutant, and
C = Concentration of a pollutant at time t.

At low pollutant concentration, the L–H model can be reduced to the pseudo first-order kinetic model, whereas at a higher concentration of the contaminant, C can be substituted by C_0 (the initial concentration of the pollutant).

The pseudo first-order kinetics was also proposed for the decolonization of wastewater (Wu et al. 2008). The reaction rate constants were compared for different combinations of the UV/TiO_2/EC system, and the results showed the following order: UV/TiO_2/EC (0.89 h^{-1}) > UV/TiO_2 (0.35 h^{-1}) > UV/EC (0.26 h^{-1}) ≥ EC (0.22 h^{-1}). In another study, first-order law was applied successfully. The rate constants decreased as the wavelengths reduced (from 350 to 254 nm), and the highest oxidation was observed at the highest electronic density. The rate constant was also increased by the addition of hydrogen peroxide because of the hindrance of the recombination of the e^-_{CB} / h^+_{VB} pair by electron acceptance and reaction with superoxide radicals (Boroski et al. 2009).

6.3.3 Effective Parameters

pH and wavelength impact. $Fe(OH)^{2+}$ is the predominant species at the pH range between 2.5 and 5.5 and can absorb the wavelength of light between 300 and 400 nm and form hydroxyl radicals as follows:

$$Fe(OH)^{2+} + h\nu \rightarrow Fe^{2+} + OH^{\bullet} \tag{6-19}$$

Other species, like $Fe(OH)_2^+$, which mostly present at a pH greater than 5.5, are less photoreactive. Besides, Fe^{2+} ions cannot absorb light beyond 300 nm and do not participate in the photolysis process when the light wavelength is more than 300 nm (Wu et al. 2008).

pH has effects on the adsorption capacity and dissociation of pollutants, the charge distribution on the photocatalyst surface, and the oxidation potential of the valence band. The point of zero charge (PZC) of the photocatalyst TiO_2 has a pH value of around 6 (pH_{PZC}). The maximum photo-electrocatalytic efficiency of chlortetracycline (CTC) (97.3%) has been observed without any adjustment of pH (5.51). If the pH values are less than 6.0, the surface of the catalyst becomes positively charged ($TiOH^{2+}$), which leads to the promotion of electrostatic repulsion toward cationic species; when the pH values are more than the PZC of TiO_2, the surface becomes negatively charged (TiO^-), which results in electrostatic repulsion toward anionic species (Daghrir et al. 2013b).

Electrode material and concentration. The characteristics of TiO_2 such as crystallization, grain size, specific surface area, morphology, and porosity have effects on the photocatalytic activity of titanium. One of the main disadvantages of the photocatalytic process is the fast recombination of the photogenerated e^-/h^+. This phenomenon can be solved by the addition of an electron acceptor like hydrogen peroxide, which acts by catching the electron in the conduction band and increasing the oxidation rate of contaminants. Cathode materials such as graphite, carbon felt, activated carbon fiber, and vitreous carbon (VC) are the alternatives for the electrochemical production of H_2O_2 (Daghrir et al. 2013b). As mentioned, a drawback of pure TiO_2 is that it requires the use of the UV spectrum and the recombination rates of high photogenerated electrons and holes. Therefore, the development of a novel TiO_2 photocatalyst by introducing the narrow bad gap semiconductor (M_xO_y/TiO_2 or M_xS_y/TiO_2) or doping TiO_2 by the use of either anions or cations are suggested (Daghrir et al. 2013c). It is expected that by the increase in catalyst concentration, the photodegradation will intensify because of the enhancement of the available photoactive sites. Nevertheless, a very high amount of solid catalyst particles can make light penetration via a heterogeneous solution difficult, leading to a light scattering occurrence, which weakens photoexcitation. For instance, in a study of pharmaceutical and cosmetic wastewater treatment, 0.25 g L^{-1} was observed as the optimum catalyst concentration (Boroski et al. 2009).

6.3.4 Application in Wastewater Treatment

Photocatalytic processes are usually applied to the treatment of wastewater containing dyes because of their ability to mineralize almost all colors. Furthermore, photocatalytic treatment can be used in municipal wastewater treatment to remove organic compounds and pathogen microorganisms. The system of integrated EC, followed by the photocatalytic process, has been used for the treatment of pharmaceutical and cosmetic wastewater, paper industry wastewater, and lithographic wastewater (Boroski et al. 2009, Suárez-Escobar et al. 2016). The application of the sequential technique on pharmaceuticals and cosmetic effluents demonstrated that EC could remove mainly colloid organics and suspended materials, whereas the photocatalytic process removed the remaining refractory compounds (Boroski et al. 2009). In another study, a combined EC-TiO_2/UV system was applied to treat methyl orange wastewater. High color removal efficiency (98%) was obtained by optimizing a current density of 125 A m^{-2}, an electrode distance of 2.5 cm, an electrolyte KCl concentration of 0.5 g L^{-1}, a pH of 6.86, and a TiO_2 concentration of 100 mg L^{-1} (in a homogenous phase). The combined system showed higher efficiency compared with the individual EC and TiO_2/UV techniques (Zhang et al. 2013).

6.4 SONO-ELECTRO-COAGULATION PROCESS

6.4.1 Ultrasound Process and the Hybrid Technique with Electro-Coagulation

By applying US in aqueous media, three phenomena have been recognized. In the inside of cavitation, bubbles collapse at high temperature and pressure. Consequently, volatile compounds are pyrolyzed, and thermal decomposition of H_2O leads to the generation of hydroxyl radicals. Thermal decomposition of organic pollutants occurs at the interface of the gas bubbles and liquids. Finally, in the bulk solution, organic substances are degraded by $^{\bullet}OH$ (Asgharian et al. 2017).

Passivation of the electrode after a long operation period is one of the drawbacks of EC. This phenomenon can cause an increase in the resistance, lead to a decrease in coagulation ions in the solution, and an increase in the cell voltage and operation time, and finally reduce the current efficiency. An ultrasonic process can be integrated into the EC process to solve this issue. Moreover, the US leads to a decrease in the size of the metal hydroxide, which can adsorb more contamination (He et al. 2016). Treatment of wastewater in an electrolytic cell by US can enhance the kinetics and the efficacy of the process because of the destruction of the layer formed at the electrodes, reduction in the diffusion thickness of the electrical double layer on the surface of the electrode, activation of ions directly at the electrodes, activation of the electrode surface by affecting the crystal structure of the electrode, and an increase in the local temperature at the electrode surface. On the contrary, ultrasonic waves may have some drawbacks such as causing the devastation of a part of metal hydroxide, which leads to a shrinking of solid particles and reduction in the effectiveness of colloidal formation, destruction of a part of the absorbed layer on colloidal particles, and disruption of the movement processes in the medium (Kovatcheva and Parlapanski 1999).

It should be noted that an integrated sono-EC is not effective for the removal of humic acid (HA) from wastewater. The study dealing with a combined EC and US process showed that this method led to a decrease in the removal efficiency of HA because of the fact that the US wave destroyed the clusters formed by EC, and consequently, coagulated HA dissolved and returned to the solution. These results demonstrated that Pt/Al electrodes are effective and show the best performance at a neutral pH in the EC process for the removal of HA (Asgharian et al. 2017).

6.4.2 Kinetics of the Sono-Electro-Coagulation Process

The kinetic model of the sono-EC process can be evaluated for various current densities and pH values by pseudo-first-order and variable-order kinetics. The following equations describe the simplified pseudo-first-order kinetic model (He et al. 2016):

$$\frac{dq}{dt} = k_1(q_e - q_t) \tag{6-20}$$

$$\ln\left(\frac{q_e - q_t}{q_e}\right) = -k_1 t \tag{6-21}$$

where k_1 is the pseudo-first-order rate constant; q_t and q_e are the adsorbed values at time t and equilibrium condition, respectively. The variable-order kinetic model for different coagulated particles can be derived from the Langmuir equation,

$$\frac{-dS}{dt} = \varepsilon_{\mathrm{Al}}\varepsilon_c \frac{nI}{ZFV}\frac{\Gamma_{\max} kS}{1+kS} \tag{6-22}$$

where

$S =$ Pollutant concentration,
$\varepsilon_{\mathrm{Al}} =$ Efficiency of the formation of hydro-pollutant-metal coagulated matter,
$\varepsilon_c =$ Current efficiency,
$n =$ Number of cells,
$I =$ Current intensity,
$Z=$ Valence of the metal of the electrode,
$F =$ Faraday constant, 96,487 C mol^{-1},
$V =$ Volume of the solution;
$C_{\max} =$ Maximum value (in mole) of the pollutant removed per mole of metal ions at equilibrium condition, and
$k =$ Langmuir constant.

6.4.3 Effect of Operating Parameters

The study conducted for the removal of Rhodamine 6G dye showed that sonolysis cannot remove the pollutant because coagulation was not started. Although both EC and combined sono-EC were effective, at an acidic solution, no differences were observed between EC and sono-EC using aluminum anode. On the contrary, in alkaline media, the efficiency of treatment using sono-EC was much higher. In fact, by improving the agitation and subsequently by increasing the frequency and the intensity of the collisions of coagulant and the pollutants, the efficiency is enhanced. By using an iron electrode, efficiency enhancement was observed at both acidic and basic conditions. Indeed, by using Fe electrodes, Fe^{2+} ions, and Al electrodes, Al^{3+} ions are generated; therefore, iron (II) will be easily oxidized to iron (III) by dissolved oxygen, and consequently, oxyhydroxide coagulants will be produced via pH-dependent equilibria (Raschitor et al. 2014).

pH effect. In a study, the Reactive Blue 19 dye (RB19) removal was investigated, and the results showed that the abatement rate was higher at a pH of 5 than at a pH of 7 or 9. The reason for this may be a slower dissolution of the scarifying electrode at a higher pH and the consequent reduction in the formation of hydroxide flocs. However, at a pH of 5, aluminum ions can more efficiently attract RB19 (negatively charged) (He et al. 2016).

Effect of current density. In sono-EC, the current density is the most crucial factor. By enhancing the current density, the formation of metal ions and metal hydroxide is promoted, and consequently, pollutant removal efficacy increases (He et al. 2016).

Effect of ultrasonic power. By increasing the US intensity, the passive film of the electrode surface becomes thinner and results in a decrease in resistance and energy consumption (He et al. 2016). The low frequency of 20 to 100 kHz is typically applied for cleaning electrodes, thanks to cavitation impact. Nevertheless, by increasing the US frequency above 1 MHz, cavitation loses its effects severely because of other parameters such as vibration acceleration and acoustic streaming. At low frequency, the bubbles' residence time is higher than that of the produced radicals (remaining in bubbles), whereas at high frequency, because of a lower residence time of the bubbles, oxidative radicals are released into wastewater. Therefore, at a lower frequency, less degradability of pollutants is observed, although a highly effective electrode cleaning can be realized (Asgharian et al. 2017). Table 6-1 shows recent studies on this area and some of the effective parameters.

6.5 ELECTRO-COAGULATION-FENTON PROCESS

6.5.1 Electro-Fenton Process and the Hybrid Method with Electro-Coagulation

The Fenton technique has been used to degrade various organic compounds in wastewater. In the traditional Fenton process, hydrogen peroxide H_2O_2 reacts with Fe^{2+} ferrous ion to form the hydroxyl radical $OH^\bullet$, which is a strong nonselective oxidant ($E \approx 2.73$ V) capable of degrading most of the organic compounds up to the final stage of mineralization (Dia et al. 2016).

$$H_2O_2 + Fe^{2+} \rightarrow Fe^{3+} + OH^\bullet + OH^-, \quad k = 70\ M^{-1}\ s^{-1} \tag{6-23}$$

Apart from the main Fenton reaction, a series of secondary reactions will take place in the reaction medium (Umar et al. 2010).

$$Fe^{3+} + H_2O_2 \rightarrow Fe^{2+} + HO_2^\bullet + H^+, \quad k = 10^{-3} - 10^{-2}\ M^{-1}\ s^{-1} \tag{6-24}$$

$$OH^\bullet + H_2O_2 \rightarrow HO_2^\bullet + H_2O, \quad k = 3.3 \times 10^7\ M^{-1}\ s^{-1} \tag{6-25}$$

$$OH^\bullet + Fe^{2+} \rightarrow Fe^{2+} + OH^-, \quad k = 3.2 \times 10^8\ M^{-1}\ s^{-1} \tag{6-26}$$

$$Fe^{3+} + HO_2^\bullet \rightarrow Fe^{2+} + O_2H^+ \tag{6-27}$$

$$Fe^{2+} + HO_2^\bullet \rightarrow Fe^{3+} + H_2O_2 \tag{6-28}$$

$$2HO_2^\bullet \rightarrow H_2O_2 + O_2 \tag{6-29}$$

Even if the rate of formation of the hydroxyl radical in the first reaction is fairly rapid ($k = 70\ M^{-1}\ s^{-1}$), the ferrous ion and hydrogen peroxide react very rapidly with $OH^\bullet$, $k = 3.3 \times 10^7\ M^{-1}\ s^{-1}$ for H_2O_2 and $k = 3.2 \times 10^8\ M^{-1}\ s^{-1}$ for Fe^{2+}.

Moreover, the reaction of the regeneration of Fe(II) from Fe(III) by hydrogen peroxide is carried out with slow kinetics ($k = 10^{-3}$ to 10^{-2} M^{-1} s^{-1}). These phenomena can limit the availability of the hydroxyl radical in the solution and consequently reduce the efficiency of the degradation of the pollutants. Moreover, the large quantities of Fe(II) and peroxide added to maintain the productivity of the hydroxyl radical may lead to a surplus production of ferric hydroxide sludge following the coagulation of colloids by Fe(III) (Dia et al. 2016).

To increase the efficiency of the conventional Fenton's process, a new technique is developed, that is, EF, which combines the EO process and the Fenton reaction. The main advantages of this method are the in situ generation of Fe(II), the electrochemical regeneration of Fe(II) from Fe(III), and the in situ formation of hydrogen peroxide from the reduction of dissolved oxygen.

$$O_2 + 2H^+ + 2e^- \rightarrow H_2O_2 \tag{6-30}$$

EF operation can be carried out in different possible scenarios (Qiang et al. 2003): Addition of Fe(II) and peroxide and regeneration of Fe(II) by cathodic reduction of Fe(III); addition of peroxide and anode formation of Fe(II), followed by regeneration of Fe(II) by cathodic reduction of Fe(III); addition of Fe(II), cathodic formation of peroxide from dissolved oxygen and regeneration of Fe(II) by cathodic reduction of Fe(III); peroxide cathodic formation from dissolved oxygen and anodic formation of Fe(II), followed by regeneration of Fe(II) by cathodic reduction of Fe(III).

The electrodes used to produce EF differ according to the requirements of the process. Thus, iron and its alloys are widely used as sacrificial anodes for the electrochemical generation of ferrous ions. Different types of cathodes such as stainless steel, titanium, graphite, carbon felt, and carbon PTFE (Polytetrafluoroethylene) are used to reduce Fe(III) to Fe(II). To minimize the oxidation of Fe(II) to Fe(III) at the anode, an anode to cathode area ratio of 3/8 has been recommended. In regard to the cathodic generation of hydrogen peroxide, the electrodes most commonly encountered are cross-linked carbonaceous carbon and polytetrafluoroethylene carbon (Dia et al. 2016).

Although EF and PEF are useful and effective technologies, their application is limited because of the comparatively high operation costs incurred from a long treatment time (Thiam et al. 2014). Table 6-1 demonstrates some recent research on combined treatment systems.

Hydrogen peroxide (H_2O_2 as an oxidant) can be produced at the cathode. Numerous cathode materials such as graphite, VC, and carbon felt can be utilized for this purpose. The capacity and ability of the cathode in an electrolytic cell for generating H_2O_2 is a function of current intensities and electrolysis time proportionally (Daghrir and Drogui 2013a).

6.5.2 Effective Parameters

The main factors influencing the effectiveness of treatment are pH, current intensity, reagent doses used, electrolyte type, electrode material, H_2O_2/Fe^{2+} molar ratio, and interelectrode distance.

A higher current leads to a smaller EC reactor; however, a too high current may reduce efficiency and increase the operating cost. If a salt such as NaCl is added as the electrolyte, it results in a decrease in power consumption (because of conductivity improvement) and in situ chlorine/hypochlorite (as an oxidizing agent) production. It can lead to an enhancement in the color removal efficiency because of mass transport promotion by transmission in the diffusion layer of the anode. pH adjustment leads to the generation of metal hydroxide and the amount of hypochlorous acid generated. A higher pH may unfavorably affect EC and cause a decrease in the formation of chlorine/hypochlorite and metal hydroxide flocs (Kabdaşli et al. 2007).

pH effect. pH is one of the most influential factors impacting the efficiency of the Fenton process. Low pH values are favorable for the formation of hydrogen peroxide because the reduction of oxygen to peroxide takes place in an acid medium. The optimum pH range is limited to only between 2 and 4.5. pH values above or below the optimum range will affect the quality of treatment. At a high acidic condition (pH <2), hydroxyl radical trapping by H^+ protons is accelerated, and ferrous ions create stable complexes with hydrogen peroxide, leading to limiting the Fenton reaction. If the pH is higher than the optimum pH (pH >4.5), the regeneration of Fe(II) from Fe(III) is limited by the coagulation of colloids by Fe(II) and Fe(III) ions. Other detrimental effects associated with the increase in pH are autodecomposition of hydrogen peroxide into H_2O and O_2, trapping of hydroxyl radicals by carbonate and bicarbonate ions $\left(CO_3^{2-} \text{ and } HCO_3^-\right)$, and the reduction of the oxidation potential of the hydroxyl radical (Dia et al. 2016).

The influence of pH (between 6 and 8) on COD removal efficiency by EC treatment was studied for car wash wastewater. The optimum pH removal was 7.3 for a maximum COD removal of 80.8%, representing the maximum flocculation of $Fe(OH)_2$ and $Fe(OH)_3$ (Mirshahghassemi et al. 2017). Furthermore, an experiment of an EF process with a pH ranging from 2 to 5 was conducted, and a maximum COD removal of 90.4% was achieved at pH $= 3$. However, as the pH increased, the efficiency decreased because of the formation of $Fe(OH)_3$, which is less active.

Effect of current intensity. In general, the increase in the electric current has a positive effect on the efficiency of the EF process. It makes it possible to form more hydrogen peroxide from the reduction of dissolved O_2. The Fe(II) generation rates at the sacrificial anode are proportional to the current. Nevertheless, a high current value may not have a significant effect on process performance. COD abatement can be limited for a current intensity value greater than 2 A with an iron anode. Indeed, beyond this intensity, parasitic reactions such as the oxidation of hydrogen peroxide to oxygen and the reduction of H^+ to H_2 resulting in a reduction in the formation of hydroxyl radicals lead to a decrease in the acidity of the environment. An optimum value of current intensity must be carefully determined to have maximum removal efficiency, while minimizing power consumption and the production of metallic residues after the neutralization phase. For instance, the optimal current densities were obtained in the treatment of leachate by EF are 100 and 42 mA cm^{-2}, respectively (Dia et al. 2016).

Combined EC and EF processes were performed for domestic wastewater treatment using iron anode and VC cathode. The applied current density led to a simultaneous generation of Fe^{2+} and H_2O_2. Current densities ranging between 0.0 and 0.34 A dm^{-2} were applied in an electrolysis time of 60 min. By increasing the current density, an enhanced removal of COD, turbidity, and TSS was observed, which showed the promotion of the generation of reactive oxygen species. In fact, the applied current density in the electrolytic cell prompted the dissolution of the iron electrode into the solution and a cathodic reduction of O_2. The dissolved oxygen was electrochemically reduced to H_2O_2. Hence, the Fenton's reagent, including Fe^{2+} and H_2O_2, resulted in the formation of hydroxyl radicals. In addition, the electrogenerated Fe^{2+} ions reacted with hydroxides (OH^-) ions and produced $Fe(OH)_2$ in situ, which led to a modification and oxidation of the structure of organic particles (Daghrir and Drogui 2013a).

Effect of H_2O_2 and Fe ions. The amount of reactants and the molar ratio of hydrogen peroxide and ferrous ions have a major influence on the efficiency of the Fenton reaction. If one of the two reagents is in excess or insufficient, it may have a negative impact on treatment. Low concentrations of peroxide and Fe(II) limit the production of the hydroxyl radicals responsible for depollution. In contrast, the high concentrations of ferrous ions favor an overproduction of sludge. An excess of hydrogen peroxide results in the flotation of the sludge via the oxygen bubbles caused by the decomposition of the peroxide. The amount of reagent used is directly related to the COD load of the effluent. For instance, by fixing the molar ratio of H_2O_2/Fe(II) as 12, a linear increase is observed in COD removal efficiency with an increased addition of peroxide. However, the COD removal rate stagnates above the concentration of 0.34 mol L^{-1} peroxide. The same experiment shows an improvement in the quality of treatment with an increase in the quantity of Fe(II) up to a limit value. These results prove the existence of an optimal H_2O_2/Fe(II) ratio for EF treatment; however, this value is changed for each process (Dia et al. 2016).

In a study using the EF process, the effect of hydrogen peroxide concentration on COD removal efficiency showed that increasing the H_2O_2 concentration led to an improvement in efficiency. However, at a concentration of $H_2O_2 > 500$ mg L^{-1}, the efficiency reduced. This happened because of the hydroxyl radical scavenging impact of hydrogen peroxide and hydroxyl radical recombination (Mirshahghassemi et al. 2017). For the treatment of landfill leachate, the efficiency of COD abatement augmented with an increase in hydrogen peroxide dosage, although the removal efficiencies remained almost constant with the use of 5,000 and 10,000 ppm of hydrogen peroxide dosage (Orkun and Kuleyin 2012).

Effect of interelectrode distance. An optimum interelectrode distance ranging between 1 and 2.5 cm is favored. In some configurations, ferrous ions are regenerated by the cathodic reduction of ferric ions; thus a very weak interelectrode distance would favor the anodic oxidation of Fe(II). The interelectrode distance does not have a significant influence on the efficiency of treatment, but a large distance between the electrodes can lead to additional energy consumption (Dia et al. 2016).

6.5.3 Photo-Fenton-Electro-Coagulation Process

The photo-Fenton process ($H_2O_2/UV/Fe^{2+}$) involves hydroxyl radical ($HO^•$) production by the photolysis of hydrogen peroxide (H_2O_2/UV) and Fenton reaction (H_2O_2/Fe^{2+}). The quantities of $HO^•$ formed in this process are higher at acidic pH. However, the H_2O_2/UV process does not follow the function of pH values. This process can effectually eliminate toxic and refractory organics, nitrogen-containing organics, color, and organophosphorus pesticides at the pH values of 3, 3, 3.5, and 7, respectively (Kang et al. 2000).

$$H_2O_2 + Fe^{2+} \rightarrow HO^• + Fe^{3+} + OH^- \quad (6\text{-}31)$$

$$H_2O_2 + UV \rightarrow 2HO^• \quad (6\text{-}32)$$

$$Fe^{3+} + H_2O + UV \rightarrow HO^• + Fe^{2+} + H^+ \quad (6\text{-}33)$$

The photo-Fenton process has been used for the treatment of tannery, food industry, wood processing wastewaters, landfill leachate, pharmaceutical wastewater, and dyes. An integration of EC and the ionizing radiation method has been suggested to treat highly colored and polluted industrial wastewater. A study on the integration of both EC and photo-Fenton processes for minimizing the final produced sludge was performed by using tannery effluent. By applying the photo-Fenton process (for mineralization of organic matter) and then the EC process (removal of suspended and dissolved substances), a low amount of sludge was achieved compared with the first EC and second EF process. The optimal condition for the photo-Fenton process was achieved by solar irradiation in acidic media by the addition of Fe^{2+} (0.5 g L^{-1}) and H_2O_2 (30 g L^{-1}) at an initial pH of 3 (Módenes et al. 2012).

6.5.4 Comparative Studies

A comparison between EC (steel (SS 104) or aluminum electrodes), CC-flocculation [using ferric chloride ($FeCl_3 \cdot 6H_2O$), ferrous sulfate ($FeSO_4 \cdot 7H_2O$), alum ($Al_2(SO_4)_3 \cdot 18H_2O$)], and Fenton oxidation ($FeSO_4 \cdot 7H_2O$ and H_2O_2) processes for color removal from dyebath effluent was performed. In terms of COD removal efficiency, both Fenton's oxidation and EC showed the highest efficiencies, whereas EC with the Al electrode revealed the best performance for biodegradability, but the lowest yield for color removal. The highest sludge generation was observed using the SS electrode, although 100% removal of color was achieved (Kabdaşli et al. 2007).

6.6 ELECTRO-COAGULATION-ELECTRO-OXIDATION PROCESS

6.6.1 Electro-Oxidation Processes and the Combined Technique with Electro-Coagulation

A hybrid system of EC–EO is able to remove contaminants such as phosphate and ammonia from wastewater. By imposing the electric current through the electrodes, (RuO_2/Ti anode and the SS cathode), the sacrificial electrode (aluminum) placed between the two mentioned electrodes is supposed to have negative and positive sides. The anode and cathode generate hydrogen and oxygen gases. Sometimes, NaCl should be added to maintain a certain concentration of Cl^- in the solution for hypochlorous acid (HClO) generation during treatment (Mahvi et al. 2011). In some cases, sequential treatment can be applied to obtain the best results. For instance, Cr-EDTA removal using the coagulation method is not effective based on ligand exchange between Cr(III) and Fe(II) or Fe(III). However, a complete removal of the recalcitrant Cr fraction can be achieved using EO and ozonation treatment (using BDD electrode), followed by EC (iron electrode), to decompose Cr-EDTA (Durante et al. 2011).

The anodic oxidation process has been used for a few decades for the treatment of color and certain organic pollutants such as phenol, cyanides, and aniline. It is often used to reduce the load of COD and ammonia nitrogen or as a tertiary treatment for the degradation of recalcitrant organic pollutants. Two types of reactions are recognized in this process: the direct oxidation reaction and the indirect oxidation reaction. In a direct oxidation reaction, oxidation is carried out either by electrochemical conversion or by electrochemical combustion. During electrochemical conversion, nonbiodegradable organic compounds are partially oxidized to more biodegradable compounds, whereas during electrochemical combustion, organic pollutants are completely degraded into CO_2 and H_2O. The direct reaction is carried out in two steps: In the first step, the anodic oxidation of the molecule of water leads to the formation of the reactive species $OH^\bullet$ which is adsorbed on an active site of $M(OH^\bullet)$; in the second step, the hydroxyl radical oxidizes the organic pollutant R into RO. The reaction between the oxidized organic compounds "RO" and the hydroxyl radicals can lead to complete oxidation (Dia et al. 2016, Drogui et al. 2007).

$$H_2O + M \rightarrow M(OH^\bullet) + H^+ + e^- \tag{6.34}$$

$$R + M(OH^\bullet) \rightarrow M + RO + e^- \tag{6-35}$$

$$R + M(OH^\bullet) \rightarrow M + mCO_2 + nH_2O + H^+ + e^- \tag{6-36}$$

Competition can occur during oxidation and, thus, reduce the effectiveness of pollutant degradation (Drogui et al. 2007). The formation of oxygen is an example of parasitic reactions.

$$H_2O + M(OH^\bullet) \rightarrow M + O_2 + 3H^+ + 3e^- \tag{6-37}$$

During indirect oxidation, organic substances are destroyed by reactive intermediates such as hydrogen peroxide H_2O_2, ozone O_3, peroxydisulfuric acid $H_2S_2O_8$, hypochlorous acid HClO, hypobromous acid HBrO, and other oxidants formed by EO of the inorganic compounds present in the solution.

$$2SO_4^{2-} + 2H^+ \rightarrow H_2S_2O_8 + 2e^- \tag{6-38}$$

$$Cl^- + 2H_2O \rightarrow HClO + H_3O^+ + 2e^- \tag{6-39}$$

$$O_2 + 2H^+ + 2e^- \rightarrow H_2O_2 \tag{6-40}$$

Because the goal of electrochemical treatment is to degrade organic pollutants, it is preferable to use electrodes that have a high oxygen overvoltage. Some of the most commonly used anodes are ruthenium coated with titanium and titanium oxides (DSA), Sn–Pb–Ru (SPR) ternary electrodes coated with titanium, BDD, graphite, and so on. The use of these electrodes with high oxygen overvoltage implies the generation of reactive oxygen species, such as the hydroxyl radical, responsible for the degradation of organic matter. These materials also exhibit chemical stability with respect to acidic and alkaline media. The choice of the electrode can significantly influence the performance of treatment. For instance, by using different types of electrodes at the anode, decreasing rates of COD abatement are obtained with the order SPR > DSA > PbO_2/Ti > Graphite. A 92% reduction in COD and an almost complete elimination of ammonia nitrogen are obtained with the anode in SPR and a current density of 150 mA cm^{-2}. The satisfactory abatement rates obtained with the SPR electrode are related to the production of intermediate oxidants such as HClO formed during the indirect oxidation. According to the authors, the SPR has more capacity to generate intermediate oxidants than the other studied electrodes. To combine both EO and EC effects, iron and aluminum are used as soluble anodes with a COD removal percentage ranging from 30% to 50%. The BDD electrode is used for the treatment of ammonia nitrogen, with the performance exceeding 95% removal efficiency. The EC–EO is a process that is able to simultaneously produce oxidant and coagulant agents by use of either iron or aluminum electrodes in a bipolar or graphite electrode in a monopolar mode in an electrolytic cell. A combination of these two processes (EC–EO) leads to the removal of both insoluble and soluble pollutants such as colloid particles, color, O&G, COD, BOD, suspended organic pollutants, and phosphate from wastewater (Dia et al. 2016). Various operating parameters such as electrode materials (iron or aluminum), current intensity (0.2 to 1.5 A), treatment time (20 to 90 min), initial pH (6 to 10), and concentration of electrolytes NaCl and Na_2SO_4 (20 to 100 mg L^{-1}), affect treatment efficiency and have to be optimized to reduce cost and increase the effectiveness for wastewater treatment (Daghrir et al. 2012).

Sequential EC–EO is also an effective technique for wastewater treatment, simply by switching the electrode polarities (Al/Gr) and changing the current density (a higher current density is required for EO). An analysis of the results showed that pollutants with molecular weights in the range of 10 to 30 kDa can be removed effectively by EC, whereas EO is able to mineralize the contaminants with molecular weights of more than 30 kDa. From an economic point of view, this combined process consumes 25% less energy compared with the EO process (Ibarra-Taquez et al. 2017). Table 6-1 shows the latest studies utilizing hybrid systems.

6.6.2 Effective Factors

In a study, the EC–EO process was capable of appropriately removing phosphate and ammonia. The optimum performance of the system was observed at a neutral pH. Increasing the current density and treatment time led to the enhancement of overall performance, and high performance was achieved even at a high concentration of pollutants (Mahvi et al. 2011). An investigation of domestic wastewater treatment showed that the most important factors were current density and treatment time in a system consisting of Fe or Al/Gr. High chlorine and metallic ions were produced at the optimum conditions (0.7 A and 39 min). Using response surface methodology (RMS), it was revealed that the effect of these two main parameters was about 90%, whereas the effect of the material of scarifying electrodes and the other factors was around 10% (Daghrir et al. 2013d).

The efficacy of the Ti/RuO_2, Ti/Pt, and Ti/Pt/Ir electrodes for textile wastewater treatment was investigated by researchers. The results showed that COD could be removed up to 85%. In a sequential EC followed by EO treatment of textile wastewater, graphite electrode showed a higher efficiency for the anodic degradation of organics compared with RuO_2/IrO_2/TaO_2-coated titanium. In such a process, Cl^- ion reduction during EO led to the production of free chlorine and an increase in COD removal efficiency. The total nitrogen was removed effectively using graphite, whereas it was eliminated by 40% by the RuO_2/IrO_2/TaO_2-coated titanium (Raju et al. 2008).

Oxidant concentrations. The concentration of chloride ions (Cl^-) and the applied current density also play important roles in the efficiency of treatment. The anodic oxidation of chloride ions to chlorine gas is followed by its disproportionation in solution and leads to the formation of hypochlorous acid (HClO). However, chlorine can react with organic compounds and form organochlorine compounds, some of which are potentially carcinogenic. In a study, it was found that the concentration of chloride ions has a positive effect on COD abatement up to a concentration of 5,000 mg L^{-1}. Beyond this value, the excess addition of chloride does not improve the performance of the process. In the presence of sulfate ion, persulfuric acid ($H_2S_2O_8$) can be generated on a BDD anode. Persulfuric acid is a potent oxidant that can be used for the oxidation of organic matter. In sum, organic compounds can be directly oxidized at the anode and indirectly oxidized in the solution by hypochlorous acid or by persulfuric acid (Dia et al. 2016).

Current density. Like all electrochemical processes, the applied intensity is a main factor in the operation. Increasing the current density results in an improved decontamination performance in terms of COD, ammonia, and color. According to a study, by increasing the current density from 50 to 150 mA cm^{-2}, the ammonia nitrogen removal rate doubled from 40% to 80%, whereas COD reduction increased by 25% to 35%. However, the increase in the current density may cause the solution to dye on corrosion of the anode. These adverse effects associated with the corrosion of electrodes point to the importance of choosing electrodes that exhibit both a high oxygen overvoltage and good chemical stability. It should be noted that working with low current densities can be detrimental to

EO operation (Dia et al. 2016). An investigation of the impact of current intensity on the treatment of restaurant wastewater by using EC–EO demonstrated that there is an optimum current (0.4 A) for the removal of COD, TSS, and turbidity. It is worth mentioning that for choosing the optimum current, the economics of energy consumption must be considered as well. In fact, by increasing the current over 0.4 A, the solid particles could not be separated successfully in the sedimentation section (Daghrir et al. 2012).

Treatment time. Treatment time also affects the effectiveness of the EO process. In general, its increase favors the degradation of pollutants. By increasing the treatment time from 0.5 to 4 h, a 5% to 30% increase in COD removal efficiency and about 20% increase in ammonia nitrogen removal efficiency were observed (Dia et al. 2016). The treatment time can also influence the production rate of active chlorine and Al^{3+} ions in the EC–EO process. In general, COD decreases, whereas metallic sludge production increases as time passes. The treatment time should allow the scarifying electrode to produce sufficient amounts of coagulant ions (Daghrir et al. 2012).

pH effect. The influence of pH on treatment remains a controversial subject. According to a study, only a slight increase in COD degradation was obtained after varying the pH of the leachate from 8.3 to 3.0. However, some researchers suggested acidic pH for optimizing power consumption and COD abatement. The modest increase in COD removal efficiency in acidic media is attributed to the low concentration of carbonate and bicarbonate ions found at low pH values. These alkaline species can rapidly react with produced hydroxyl radicals and, thus, reduce the effectiveness of degradation of organic compounds (Dia et al. 2016). Initial pH influences the stability of hydroxide species, amends the particles' surface charge, and impacts the final efficiency of the process. The maximum COD removal efficiency is reported at a neutral pH. Moreover, the pH increases during the treatment because of an increase in hydroxide ions by the reduction of water at the cathode (Daghrir et al. 2012).

Conductivity effect. The conductivity of wastewater affects the current intensity, applied voltage, and consequently energy consumption. Some electrolytes such as Na_2SO_4 or NaCl can be added to the solution to improve conductivity. However, the costs of sludge disposal will increase by such addition. If the conductivity of wastewater is relatively high, it is not necessary to adjust initial conductivity (Daghrir et al. 2012).

Electrode materials. It has been reported that for restaurant wastewater treatment, an aluminum electrode is more efficient than an iron electrode in terms of COD removal. At the same applied current (1.0 A), the amount of sludge produced during the treatment of wastewater by iron and aluminum electrodes was 2.45 and 4.35 kg m^{-3}, respectively. A higher residual sludge was generated with aluminum electrodes because of the superior electrical conductivity of the aluminum electrode to that of the iron electrode. Aluminum electrodes may be more appropriate for anodic dissolution than iron electrodes because of their higher conductivity, and because, in the EC process, when the conductivity of the electrolyte or electrode declines, ohmic resistance and energy consumption intensify, which leads to a higher amount of metallic sludge production (Daghrir et al. 2012).

6.6.3 Kinetic Model

The first-order kinetic model has been proposed in terms of COD removal efficiency (Linares-Hernández et al. 2010)

$$\frac{dC_{\mathrm{COD}}}{dt} = -k[C_{\mathrm{COD}}] \tag{6-41}$$

where

C_{COD} = Concentration of COD (mg L^{-1}),
t = Time (h), and
k = Rate constant (h^{-1}).

Furthermore, the kinetic power law expressions have been reported for the sequential EC–CO process for the removal of COD and TOC of soluble coffee effluents as follows:

$$r_{\mathrm{COD}}^{\mathrm{EC}} = k_1(C_{\mathrm{COD}}^{\mathrm{EC}})^{5.2} \tag{6-42}$$

$$r_{\mathrm{COD}}^{\mathrm{EO}} = k_2(C_{\mathrm{COD}}^{\mathrm{EO}})^{2.0} \tag{6-43}$$

$$r_{\mathrm{TOC}}^{\mathrm{EC}} = k_3(C_{\mathrm{TOC}}^{\mathrm{EC}})^{5.9} \tag{6-44}$$

$$r_{\mathrm{TOC}}^{\mathrm{EO}} = k_4(C_{\mathrm{TOC}}^{\mathrm{EO}})^{2.0} \tag{6-45}$$

The reaction order of the EC process was higher than that of the EO process for both COD and TOC removal efficiencies, showing a swift drop in COD and TOC removal at the first 20 min of treatment in the EC process, whereas the order of reaction of the EO process is the second order for both TOC and COD ($n = 2.0$). This fact can be attributed to the more complex mechanism and behavior of EC (aggregation, precipitation, and adsorption) compared with the EO process (direct and indirect oxidation) (Ibarra-Taquez et al. 2017).

6.6.4 Performance and Efficiency in Terms of Coagulant and Oxidant Agents

The process performance of chlorine production is assessed by Faraday efficiency (η). Faraday efficiency values for the chlorine evolution reaction can be computed by using the following equation (Daghrir et al. 2012):

$$\eta = \frac{n_e F C_i V}{I t M_i} \tag{6-46}$$

where

V = Volume of the solution (L),
F = Faraday's constant (96,485 C mol^{-1}),
I = Applied current (A),
C_i = Concentration of active chlorine or the concentration of iron (or aluminum) (g L^{-1}),
n_e = Number of electrons transferred during the reaction at the electrode,
t = Electrolysis time (s), and
M_i = Molecular weight (g mol^{-1}).

Moreover, the efficacy of the EC–EO process depends on the concurrent production of an oxidant agent (e.g., $HClO/ClO^-$) and a coagulating agent (Fe^{2+}/Fe^{3+} or Al^{3+}). Active chlorine concentration can be considered as the sum of the following three species: Cl_2, HClO, and ClO^-. In the pH range of 6 to 9, active chlorine can be created by hypochlorous acid (HClO) and hypochlorite ions (ClO^-).

6.6.5 Application in Wastewater Treatment

Restaurant wastewater treatment by the EC-CO method demonstrated that aluminum electrodes were more effective than iron electrodes for the removal of contaminants COD_T (90%), BOD (86.0%), and O&G (98.1%) at optimum conditions. Moreover, the treatment cost, including electrode costs, chemicals, energy, and sludge disposal, could be estimated as \$US 1.56 ± 0.01 m^{-3} (Daghrir et al. 2012). Textile wastewater was treated using an EC–CO system. The aluminum electrode proved more effective than the iron electrode, and by applying 0.6 A current at a 90-min treatment time, a high removal of COD, TSS, BOD, O&G, soluble phosphate, and turbidity was achieved. An energy analysis revealed that this process cost \$US 2.03 m^{-3} (Naje and Abbas 2013). For treatment of dairy wastewater, EC was effective for the removal of colloidal and suspended particles; on the contrary, it was not effective enough to reduce COD (only half of the COD was eliminated). Using EO, around 40% of COD was removed. However, the coupled process was able to remove 60% of COD, colloidal and suspended particles, color, turbidity, phosphorus, K^+, and NTK (Chakchouk et al. 2017). In another study, the treatment of laundry wastewater (LWW) by ultrafiltration (UF) and subsequently the treatment of UF concentrate with electro-coagulation/electro-oxidation (EC/EO) resulted in treated water for discharge into the sewage network by the following characterization: COD <500 mg L^{-1}, TSS <200 mg L^{-1}, and nonylphenol ethoxylate ($NPEO_{3-17}$) <200 μg L^{-1}. The optimum operating conditions of the EC–EO cell were provided as a bipolar configuration of Fe (anode) and Gr (cathode) electrode at a current density of 40 mA cm^{-2} and at a treatment time of 45 min (Khosravanipour Mostafazadeh et al. 2019).

6.7 ELECTRO-COAGULATION-PEROXIDATION PROCESS

6.7.1 Peroxidation Process and the Combined Technique with Electro-Coagulation

Electroperoxidation (EP) is a process in which hydrogen peroxide H_2O_2 is electrochemically produced at the cathode electrode to oxidize the contaminants in wastewater (Drogui et al. 2001). In the EC-peroxidation (EC-P) process, H_2O_2 reacts with the Fe^{2+} ion, and consequently, H_2O_2 is decomposed to $OH^{\bullet}$ radicals. Hydrogen peroxide can be generated in the electro-coagulation-electroperoxidation (EC-EP) cell using VC as a cathode by oxygen reduction. The produced Fe^{3+} forms $Fe(OH)_3$ as precipitates in the solution (Senghor et al. 2015).

$$O_2 + 2H^+ + 2e^- \rightarrow H_2O_2, \quad E^{\bullet} = 0.695\ \text{V/SHE} \tag{6-47}$$

$$Fe^{2+} + H_2O_2 + H^+ \rightarrow Fe^{3+} + OH^{\bullet} + H_2O, \quad E^{\bullet} = 0.77\ \text{V/SHE} \tag{6-48}$$

H_2O_2 can also be added to the EC process. The hydrogen peroxide forms the hydroxyl radical at the cathode. In the presence of the scarifying aluminum electrode, the Al^{3+} ions produced in the anode can be reduced and reacted with the hydroxyl radical and an organic compound R as follows (Esfandyari et al. 2015):

$$H_2O_2 + e^- + H^+ \leftrightarrow H_2O + OH^{\bullet} \tag{6-49}$$

$$3H_2O_2 + Al^0 + 3H^+ \leftrightarrow 3H_2O + 3OH^{\bullet} + Al^{3+} \tag{6-50}$$

$$RH + OH^{\bullet} \leftrightarrow R^{\bullet} + H_2O \tag{6-51}$$

$$R^{\bullet} + O_2 \leftrightarrow ROO^{\bullet} \tag{6-52}$$

$$ROO^{\bullet} + RH \leftrightarrow R^{\bullet} + ROOH \tag{6-53}$$

Several advantages have been identified for this kind of process. It is more efficient for pollutant removal from wastewater by using coagulants without introducing sulfate or chloride ions. In this way, fewer metal residuals and sludge can be generated compared with the use of metal salts. H_2O_2 is externally added into the EC system, which leads to the production of a highly active oxidant [hydroxyl radical ($^{\bullet}OH$)], which degrades the organic species in the presence of the Fenton reaction (Yazdanbakhsh et al. 2015).

EC-EP has some merits compared with the alternate Fenton method, such as high performance to remove biorefractory compounds and less H_2O_2 consumption. This technique can be applied on carcinogenic materials, radioisotopes, pharmaceuticals, BTEX, VOCs, SVOCs, PAHs, TPHs, landfill leachate, and nitrogen compounds (Environmental-expert 2017). Although peroxi-coagulation (PC) in one pot is a simultaneous coagulation and oxidation process of the larger organic pollutant process, it has some disadvantages such as causing a destruction of hydroxyl radicals (because of excessive iron ions) and a reduction in the cathode lifetime and ability to generate H_2O_2 (because of the adsorption and abrasion of coagulated particle hydroxides on the cathode surface) (Thiam et al. 2014). Thus, the sequential combination may be more effective, whereas the aim of the treatment, the type of wastewater, and the operating conditions should be considered to select the combined simultaneous or sequential process. In a study, the COD removal efficiency was about 80% employing EC; however, it increased to 90% by the addition of H_2O_2 into the solution (Roa-Morales et al. 2007).

6.7.2 Effective Factors

Effect of pH. The solubility of aluminum hydroxide changes in terms of the pH of the solution. In acidic conditions, the oxidation of the aluminum electrode reacts with water to produce hydrogen. Also, aluminum ions are present in the solution, and by oxidizing at the anode electrochemically, $Al(OH)_3(s)$ is formed

$$2Al + 6H_2O \rightarrow 2Al^{3+} + 3H_2 + 6OH^- \qquad (6\text{-}54)$$

Various compounds can be produced depending on the pH values under this condition. At low Al concentrations (5×10^{-3} M), Al^{3+}, $Al(OH)^{2+}$, $Al(OH)_3$, $Al_{13}(OH)_4^{5+}$, and $Al_{13}(OH)_{32}^{7+}$ are formed. For instance, for treatment of olive mill wastewater, at pH 4, the main types of species were Al^{3+} and $Al(OH)_3$, whereas at pH 4.5 and above, only $Al(OH)_3$ was observed, which can be in the form of soluble, insoluble, or colloidal particles (Domı́nguez et al. 1998). Moreover, the highest amount of $^{\bullet}OH$ from H_2O_2 under acidic conditions results in the maximum oxidation of organic compounds (Esfandyari et al. 2015). In a study, the best pollutant removal efficiency was obtained in a pH range of 3 to 8, and the optimum pH was 4.0, which shows that the acidic pH led to the best results (Roa-Morales et al. 2007).

Effect of current density. Current density directly influences the coagulant dosage rate, bubble production rate, and the size and growth of flocs, which consequently affect the efficiency of the process (Yazdanbakhsh et al. 2015). The effect of current density for the treatment of domestic wastewater using VC and Fe electrodes on the removal rates of CODs, SS, P_{tot}, color, and turbidity showed that an increase in current density (2 to 9 mA cm^{-2}) led to a slight increase in COD removal efficiency, not a significant increase in SS, and an enhancement in turbidity and color removal, but had no effect on P_{tot} removal. Thus, there is an optimum current density (in this case, 4.0 mA cm^{-2}) for such a process (Senghor et al. 2015).

Treatment time effect. The removal of contaminants is quite rapid when the process commences; however, the treatment efficiencies become progressively stable after a certain treatment time, which shows that there is an optimum treatment time (Senghor et al. 2015). The same trend has also been observed by other researchers (Yazdanbakhsh et al. 2015).

6.7.3 Application in Wastewater Treatment

EC-EP systems with iron electrodes have been effectively applied for the treatment of industrial wastewater containing aniline, herbicides, and biorecalcitrant pollutants, and those with aluminum electrodes have been utilized for the removal of dyes, pollutants in surface water, and clay in potable water. For the treatment of wastewater from pasta and cookie processing, the experiment in a pilot batch scale at pH 4 and 18.2 mA m^{-2} current density resulted in removal efficiencies of 90% for COD, 96% for BOD, 95% for TS, and 99.9% for fecal coliforms (Roa-Morales et al. 2007). The treatment of tertiary domestic wastewater by the EC-EP technique confirmed that this process can be effective for the removal of BOD_5 and SS to meet the allowable limits. However, the residual fecal coliform concentration was higher than the limiting value at optimal operating conditions (current density of 4 mA cm^{-2} over a 60 min treatment time) using the Fe/CV system. The cost of operation, including energy and electrode consumption and sludge disposal, was estimated around \$US 0.211 ± 0.02 m^{-3} (Senghor et al. 2015). Peroxi-EC process (PEP) is efficient and less expensive when it is applied for detoxifying herbicide, pesticide, and insecticide in wastewater, polishing the biological effluents of wastewater from the petrochemical industry, and treatment of pharmaceutical wastewater. In a recent study, azithromycin COD removal of 95.6% was achieved successfully under the optimal conditions [pH = 3.0, current density = 20 mA cm^{-2}, hydrogen peroxide concentration = 2 m ML^{-1}, operation time = 60 min, and energy consumption = 2 kW·h (kg COD)$^{-1}$] (Yazdanbakhsh et al. 2015).

6.7.4 Kinetic Model

A linear pseudo-second-order equation has been proposed for this type of process

$$\frac{t}{q_t} = \frac{1}{k_2 q_e^2} + \frac{t}{q_e} \tag{6-55}$$

where

q_e and q_t = Adsorbed contaminants at equilibrium,
t = Electrolysis time, and
k_2 = Rate constant that depends on conditions like ionic strength.

The higher the values of k_2, the higher the oxidizing capability and regeneration of hydroxyl radicals (Yazdanbakhsh et al. 2015).

6.8 OZONATION-ELECTRO-COAGULATION PROCESS

6.8.1 Theory of the Ozone Treatment Process and the Integrated Technique with Electro-Coagulation

The ozone oxidation process can be performed directly or indirectly (radical mechanism), as can be seen in the following equations:

$$O_3 + 2H^+ + 2e^- \rightarrow O_2 + H_2O, \quad E = 2.07\ \text{V} \tag{6-56}$$

$$O_3 + H_2O \rightarrow O_2 + 2HO^{\bullet}, \quad k = 1.1 \times 10^{-4}\ M^{-1}\ s^{-1} \tag{6-57}$$

$$O_3 + OH^- \rightarrow O_2^{\bullet -} + HO_2^{\bullet}, \quad k = 70\ M^{-1}\ s^{-1} \tag{6-58}$$

$$O_3 + OH^{\bullet} \rightarrow O_2 + HO_2^{\bullet} \leftrightarrow 2O_2^{\bullet -} + H^+ \tag{6-59}$$

$$O_3 + HO_2^{\bullet} \rightarrow 2O_2 + HO^{\bullet}, \quad k = 1.6 \times 10^9\ M^{-1}\ s^{-1} \tag{6-60}$$

$$2HO_2^{\bullet} \rightarrow O_2 + H_2O_2 \tag{6-61}$$

The efficiency of ozonation will be enhanced by the activation of a hydroxyl radical–mediated mechanism resulting from Fe ions in electrocoagulated wastewater. In fact, Fe(II) ions catalyze the decomposition of hydrogen peroxide to hydroxyl radicals, leading to an enhancement in hydroxyl radical generation and, accordingly, to an improvement in the effectiveness of the process (Hernández-Ortega et al. 2010).

As is well known, EC produces a significant amount of sludge. Ozone-assisted EC is an alternate process to reduce sludge generation. Fe^{2+} can react with an oxygen molecule and finally produce $Fe(OH)_3$. Ozone can directly react with Fe^{2+} to form the intermediate $(FeO)^{2+}$ and subsequently produce $^{\bullet}OH$ radicals as an oxidizing agent to remove the color and COD of wastewater. The following reactions represent the coupled ozone-EC system (Asaithambi et al. 2012):

$$O_3 + Fe^{2+} \rightarrow O_2 + FeO^{2+} \tag{6-62}$$

$$FeO^{2+} + H_2O \rightarrow Fe^{3+} + HO^{\bullet} + HO^- \tag{6-63}$$

$$FeO^{2+} + Fe^{2+} + 2H^+ \rightarrow 2Fe^{3+} + H_2O \tag{6-64}$$

This hybrid process has several advantages in comparison with physicochemical techniques, such as no requirement for flocculating agents, no necessity for pH adjustment, lower sludge production, easy separation of flocs by filtration, lower bound water content, a higher surface of flocs, and appropriate for reuse (because it is more clear) (Orescanin et al. 2011a). Although ozonation is an effective method for removing the color and turbidity of wastewater, it is not usable for reducing persistent organic compounds. Thus, an integrated EC-ozonation can compensate this drawback. It was observed that this integrated process could lead to lower sludge production in wastewater treatment (Bernal-Martínez et al. 2010) and a colorless solution with lower phenol and contaminants than that with EC or ozonation separately. A sequential application of ozonation after EC makes it possible to remove all microorganisms (García-García et al. 2014). Table 6-1 presents different studies and effective parameters in combined treatment systems.

6.8.2 Kinetic Model

The pseudo-first-order kinetic model has been proposed for the ozone-assisted EC process based on COD removal efficiency (Asaithambi et al. 2012)

$$\frac{C}{C_0} = e^{-k_{obs}t} \tag{6-65}$$

The rate constant, k_{obs}, is a function of initial COD concentration C_A and current density CD. The rate constant can be presented as follows:

$$k_{obs} = m[C_A]^a[CD]^b \tag{6-66}$$

The constants a and b are estimated through time versus concentration and current density plots.

6.8.3 Crucial Parameters

Current density. In combined electrochemical treatment, current density, and wastewater concentration are significant factors for controlling the reaction rate. COD removal efficiency increases as current density increases because a higher amount of Fe^{2+} is generated at the scarifying electrode, which results in coagulation and floatation by the production of more bubbles at the cathode. However, COD removal efficiency and power consumption decrease with an increase in the effluent COD concentration at a constant current, which is in accordance with Faraday's Law (Asaithambi et al. 2016). An increase in current density leads to efficiency enhancement, whereas it results in an increase in power consumption and the energy consumption of O-EC (more than that of the EC process). The interesting point is that as current density increases, the relative power consumptions of EC-O and EC decrease (Behin et al. 2015). However, an increase in the current density from 20 to 30 and 40 mA cm^{-2} for industrial wastewater treatment showed that at 20 mA cm^{-2}, the maximum COD removal efficiency was observed, whereas usually an increase in current density in EC leads to enhanced efficiency (Bernal-Martínez et al. 2010).

pH and distance between electrodes. The variation of pH in a range of 3 to 11 for the treatment of industrial effluents showed that the COD removal efficiency enhanced at a pH range of 0 to 7; nevertheless, it decreased with the pH values of 7 to 11 because of the change in iron electrodissolution. Furthermore, the COD removal efficiency diminished by increasing the interelectrode distance, which consequently led to an increase in energy consumption. In fact, the electrical resistance is proportional to the distance between the electrodes, which is in accordance with Faraday's law (Asaithambi et al. 2016). Conversely, it has been reported that the optimum pH for color removal using the EC-O process is 9.5. It can be explained by the fact that hydroxyl radicals can be formed from ozone at pH values above 7 to 8 (alkaline condition); however, at an acidic pH, the species that exists most is molecular ozone, which has a lower oxidation potential.

Although the dissolved ozone concentration is higher at a lower pH, the efficiency of treatment at an alkaline pH was higher than that of the operation at a neutral pH (Behin et al. 2015).

For industrial treatment using Ozone-EC, the maximum COD removal efficiency was obtained at pH 7, whereas the maximum color removal efficiency was achieved at pH 9. At a high pH in the presence of an iron electrode, some hydroxide ions are oxidized at the anode, resulting in a decrease in the production of iron ions. Also, $Fe(OH)_6^{3-}$ or $Fe(OH)_4^-$ ions are most probably present at a high pH, which causes a reduced color removal efficiency (Bernal-Martínez et al. 2010).

6.8.4 Comparison and Application of Ozonation, Electro-Coagulation, and Ozone-Electro-Coagulation Processes

Ozone-assisted EC processes have several advantages. They do not require the external addition of flocculent; pH does not change significantly during the treatment, and the effluent is clear, colorless, and odorless. A comparison of decolorization of wastewater containing Acid Brown 214 showed that the efficiency of the combined EC-O process was the highest (100%) and the EC process had the lowest efficiency (45.72), whereas the efficacy of the ozonation process was 82.88 after 30 min (Behin et al. 2015). The results of treatment of distillery industrial wastewater revealed that EC was more efficient than the ozone treatment process, and the combined process has a higher efficiency than both EC and ozonation for the removal of color and COD. Boat pressure washing wastewater (BPWW) was treated by a combined ozonation/EC process. The wastewater contained mainly heavy metals and organic pollutants. The results showed that this treatment could effectively remove heavy metals (V, Cr, Fe, Cu, Zn) by coagulation/flocculation and organic contaminants (COD and TOC) by both ozone and chlorine/hypochlorite oxidation to meet the regulated values (Orescanin et al. 2011a).

6.9 COMBINED ELECTRO-COAGULATION AND BIOLOGICAL TREATMENT (ELECTRO-BIO SYSTEM)

Combined biological and EC techniques have been investigated for the treatment of various types of wastewater such as bleaching effluent, graywater, industrial effluents, municipal wastewater, and synthetic wastewater. The results of these studies revealed the removing ability of organic substances, suspended solids, and phosphorus. To treat landfill leachate, the biofiltration process displayed adequate results in terms of ammonia, BOD_5, turbidity, and phosphorus elimination, whereas the COD removal efficiency was relatively low (around 13%). By applying EC with a magnesium-based anode to biotreated landfill leachate, the residual COD was reduced (about 53%), and likewise, 85% of the true color was removed under a current density of 10 mA cm^{-2} and a treatment time of 30 min (Oumar et al. 2016).

Although electrochemical processes are expensive compared with biological processes, applying electrochemical techniques is proper and essential when an effluent is nonbiodegradable. One of the economical and feasible processes is treating wastewater containing slowly biodegradable contaminants by combined electro-bio processes. The electrochemical process improves the biodegradability of an effluent. Among several electrochemical and chemical processes [EC, EO, CC] used as a pretreatment step of the biological treatment of kraft bagasse pulp bleach wastewater, EC was markedly more effective from the viewpoint of operation cost. The operation cost of the EC–Bio combined process was \$US 0.17 m^{-3}, much less than that of the CC–Bio system (\$US 0.86 m^{-3}) and the EO–Bio system (\$US 0.98 m^{-3}) for an almost equal pollutant removal. The effectiveness of the EC–Bio arrangement was higher than that of other systems for equal removal of contaminants (80% COD elimination) (Antony and Natesan 2012).

In a study, the EC, solar photo-Fenton, and biological oxidation processes were combined to treat textile wastewater. Even if the EC treatment could eliminate organic matter effectively, the effluent of the process was not acceptable for discharge into surface water.

By EC treatment, followed by a sedimentation of organic matter, 36% COD removal efficiency was achieved. Thus, an enhancement in the COD removal efficiency of up to 65% was achieved by applying a photo-Fenton reaction operation. In the next step, the activated sludge (AS) biological process (biological oxidation) was used under aerobic conditions. Finally, a COD value (139 mg L^{-1}) lower than a recommended maximum limit (150 mg L^{-1}) (set by regional environmental legislation) was obtained after 48 h of operation. Hence, with the integrated treatment system, degradation of all organic impurities was performed by using the mentioned sequential processes.

Biodegradation index percentage (D_t) can be determined by using the following equation. It is considered appropriate in the range of $D_t > 70\%$ by using the Zahn–Wellens biodegradability test.

$$D_t = \left(1 - \frac{C_t - C_B}{C_A - C_{BA}}\right) \times 100 \qquad (6\text{-}67)$$

where C_A and C_{BA} are the dissolved organic carbon DOC (mg L^{-1}) in the sample and in the blank, measured 3 h after beginning the experiment; and C_t and C_B are the DOC (mg L^{-1}) in the sample and the blank, respectively, measured at time t (Manenti et al. 2014).

In another research work, the coupling of EC and biological treatment of mature landfill leachate from a nonhazardous waste storage was accomplished. The samples contained a comparatively low COD and a low BOD_5/COD ratio (i.e., low biodegradability) but a high inorganic compound such as NH_4^+. The results revealed that during EC, 33% and 56% of COD removal efficiencies were achieved by applying 23 and 95 A m^{-2} respectively, although the removal of ammonium was not observed. In this case, biological treatment [including or excluding AS] led to the removal of 62% ammonium after 166 h. By using AS, nitrification efficiency enhanced by 10%. Hence, the incorporated treatment (EC as a pretreatment) can be a suitable approach for removing both organic and nonorganic pollutants (Djelal et al. 2015).

6.10 COMPARATIVE STUDIES

In a study, a comparison of EC and biological treatment was presented for LWW treatment. In the biological treatment step, wastewater is treated with a *Bacillus* strain of aerobic bacteria, which is especially suitable for the degradation of fats, lipids, protein, detergents, and hydrocarbons. EC was done by using aluminum metal under a controlled voltage. The efficiency of each methodology was evaluated by measuring the surfactant concentration, COD, and TDS. A noteworthy decrease in these substances was detected after 12 h of biological treatment, whereas a remarkable reduction in surfactants was observed within the first 0.5 h through EC, which was attributed to the rapid generation of aluminum hydroxyl species (Ramcharan and Bissessur 2017a).

The advantages of biological treatment of wastewater consist of a decreased amount of contaminated and detrimental chemicals coupled with easy-to-implement green processes. In fact, waste is recycled into organism-based biomass via biological treatment and can be disposed into the environment naturally and simply. However, the main drawbacks associated with this process can be presented as a large space requirement for storage, longer operation time requirement, and restriction of its application to a broad range of effluent types that sometimes require the addition of special bacterial strains (Ramcharan and Bissessur 2017b).

Municipal wastewater treatment was performed by using three different processes: EC (with two aluminum electrodes and an applied current density of 3 mA cm^{-2}), CC, and the biological method. The results showed that COD and TSS removal efficiencies were relatively similar in the order of 78% to 84% and 90% to 97%, respectively. In both cases, the highest removal percentage was achieved by using EC. The most important point in this comparison is the treatment time, which is significant from the economic point of view. Biological treatment took around 4 and 2.5 h for COD and TSS removal, respectively, whereas it took around 45 min and 1.5 h for CC and EC, respectively. However, CC and EC produced a higher sludge volume than the biological treatment (Hesham Ahmad et al. 2016).

6.11 BIOFILTRATION-ELECTRO-COAGULATION COUPLING

Other treatment pathways involve coupling the electrochemical and biological processes. Yu et al. (2005) studied, in particular, the coupling of a unit of biofiltration and an EC unit for the elimination of phosphorus in the processing channels of small installations. The EC cell was used upstream of the biofiltration unit. The EC cell was composed of iron electrodes. The coagulating agent (Fe^{2+}) was obtained by an anodic dissolution of the iron electrode according to Equation (6-68)

$$Fe_{(s)} \rightarrow Fe^{2+} + 2e^{-} \quad (6\text{-}68)$$

Ferrous ions form at the anode, followed by the oxidation of ferrous ions to ferric iron in the presence of oxygen. The Fe^{3+} and Fe^{2+} ions react with the phosphate ions to form $FePO_4$ and/or $Fe_3\,(PO_4)_2$ compounds.

$$Fe^{2+} + \frac{1}{4}O_2 + \frac{1}{2}H_2O \rightarrow Fe^{3+} + OH^{-} \quad (6\text{-}69)$$

$$Fe^{3+} + PO_4^{3-} \rightarrow FePO_4 \quad (6\text{-}70)$$

$$3Fe^{2+} + 2PO_4^{3-} \rightarrow Fe_3(PO_4)_2 \quad (6\text{-}71)$$

The electrocoagulated effluent was fed to a decanter for solid/liquid separation. The decanted effluent was then fed to the biofiltration unit for the reduction of BOD_5 and COD. In general, these biological processes are effective when effluents contain biodegradable and nontoxic compounds. In contrast, some potentially toxic or refractory compounds (hormones and hormone derivatives, antibiotics, cosmetics, pharmaceuticals, etc.) can be found in domestic wastewater and they disrupt the functioning of these biological units. To limit or prevent the emergence of these various types of organic contaminants in the environment, wastewater treatment systems (domestic or municipal) should adapt to provide treated water that continuously complies with the discharge standards. To do this, electrolytic techniques combining both the direct and indirect oxidation effects of electric current can be put to good use by subjecting these waste effluents to further processing before their rejection in the receiving medium.

A few studies were conducted on the coupling of EC and biological processes. In a study, the EC and EO processes were used to treat leachate with a COD of 11,720 mg L^{-1} after a biological treatment step. Iron and BDD were used as anodes for EC and EO, respectively. Stainless steel was used as a cathode in both treatment modes. The effluent was first electrocoagulated for one hour with a voltage of 4 V without stirring. After an intermediate stage of filtration, the permeate obtained was subjected to 8 h of EO with an electrical density of 30 mA cm^{-2}. The successive treatment of EC and EO resulted in a reduction of 81% in COD and 50% in TOC. In another study, leachate was treated initially by a nitrification/denitrification process, followed by membrane filtration using two UF and nanofiltration modules. The concentrate obtained had a COD of 6,200 mg L^{-1}, containing a large part of the large refractory molecules that were not degraded by the biological chain and not retained by the filters. To treat the concentrate obtained after the biological treatment and filtration step, the researchers used an EC process with four aluminum plates (two anodes and two cathodes) having a total area of

252 cm^2 in a monopolar configuration. With a current density of 15.9 mA cm^{-2} and a processing time of 30 min, 45% of COD and 60% of color was eliminated (Dia et al. 2016).

6.12 ADVANTAGES AND DISADVANTAGES OF BIOLOGICAL AND ELECTRO-COAGULATION PROCESSES

Biological techniques cannot always meet the quality requirements of effluents containing bioresistant species, such as organic acids (propionic acid, valeric acid) and antibiotic (tetracycline or CTC) compounds. To resolve such issues, combined treatment systems can be applied, such as physicochemical treatment, followed by biological treatment. For example, the combined EC method, in conjunction with biofiltration, works for the treatment of liquid swine manure (LSM). In such a system, the effluent of a biofiltration system (e.g., trickling filter) is sent to the EC system to remove the refractory organic compounds. The biofiltration system utilizes indigenous microflora fixed on organic support–encompassed peat and wood shavings to treat raw pig manure. This method is effective for the sorption of biodegradable organic compounds. In one experiment, COD and BOD were reduced by 96% to 98% and 90% to 95%, respectively. However, residual COD caused by refractory organic compounds was not oxidized. The highest elimination of refractory organic compound was achieved with an aluminum bipolar (Al-BP) electrode and iron monopolar electrode (Fe-MP) configuration at current intensities of 0.5 and 2.0 A in 30 min of treatment, respectively, which resulted in 68% COD and 87% BOD removal efficiencies (Laridi et al. 2005).

Integrated biological and EC processes signify an alternate methodology for the efficient treatment of landfill leachate. As a matter of fact, biological methods can efficiently remove nitrogen and BOD, whereas the EC method is able to eliminate metals, phosphorus, suspended solids, and biorefractory organic compounds like humic substances. EC as a physicochemical technique is often applied as a tertiary treatment in biological treatment downstream.

The disadvantage of such a combined system is the formation of ammonium through EC treatment. During the biological process, nitrification occurs at the aerobic condition (ammonia is oxidized to nitrate), but in the EC process, nitrates are electrochemically reduced to ammonia via the following reactions on the cathode:

$$NO_3^- + H_2O + 2e^- \leftrightarrow NO_2^- + 2OH^- \tag{6-72}$$

$$NO_2^- + 5H_2O + 6e^- \leftrightarrow NH_3 + 7OH^- \tag{6-73}$$

To meet the effluent standard of nitrogen (e.g., 10 mg NH_4-N/L), ammonia production from nitrate reduction must be restricted by identifying the appropriate cathode material. The results of the experiments of using different cathode materials (Fe, Al, SS, Cu, and Ti/Pt) on biofiltrated landfill leachate

(BFLL) revealed that the Ti/Pt cathode showed the lowest rate of ammonia formation, followed by Cu, Al, SS, and Fe (Dia et al. 2017a).

The biological method could be applied as a pre- or post-treatment according to the requirement of the treatment. As a pretreatment step, a reduction of mineral and biodegradable organic contaminants is performed to reduce the load of the subsequent nonbiological treatment. Otherwise, as a post-treatment system, it is used after AOPs or physical/chemical treatment. Thus, biological treatment leads to a reduction in operating costs. Such processes can be applied to treat biodegradable organic compounds and other inorganic pollutants such as NH^{4+} and PO_4^{3-}. However, if the biological system is used as a pretreatment process, recalcitrant organic molecules such as humic-like substances remain soluble in the aqueous effluent. These substances can act as precursors of unwanted organochlorine by-products. Residual organic matter needs to be reduced or removed before discharge to prevent this phenomenon. One of the methods to eliminate such pollutants is EC.

In a study, organic compounds such as HA, fulvic acid (FA), and hydrophilic compounds (Hyl) were fractionated to evaluate the efficiency of EC on each fraction of BFLL. The characterization of BFLL showed that humic substances (HA and FA) constituted around 90% of dissolved organic matter (DOM) in terms of TOC and COD, whereas the residual organic matter mostly comprised of Hyl compounds. By using iron and aluminum electrodes, 70% and 65% of the total COD were abated during EC. HA was completely removed, whereas FA and Hyl fractions were moderately removed (Dia et al. 2017b).

6.13 CONCLUSION

Combined EC processes can be effectively applied to various types of effluents such as DWW, RWW, Dairy WW, food industry, industrial effluents, distillery, tannery, textile, pharmaceutical wastewaters, landfill leachate, and so on to remove organic and inorganic contaminants. Understanding the types of contaminants and mechanisms of pollutant abatement for each specific process, toxicity, wastewater volume, and so on are necessary to choose the best technique and apply the optimum conditions for treatment. Crucial factors affecting the process efficiency and economic aspect are the current density, treatment time, type of electrodes, conductivity of the solution, chemical additive concentration (H_2O_2, NaCl, TiO_2 catalyst), pH, and configuration of the reactor (such as the interdistance of electrodes). Electrode types, combined iron/aluminum as scarifying electrodes (anode/cathode) and graphite/VC/TiO_2/Ti/Pt/Ti/IrO_2 (cathode/anode), are among the most important factors to be considered for performing a combined process in different bipolar/monopolar modes of operations. The optimization of certain parameters is essential for achieving high efficiency in electrochemical treatment with acceptable levels of energy consumption and the total cost. Either the pseudo-second-order equation or the pseudo-first-order equation is mostly used to predict the kinetics of the reactions in such a process. Both EC and biological treatment methods have their pros and cons; thus, coupling these two processes could result in an improvement in treatment and help overcome the challenges involved. EC can be applied as a pre- or post-treatment step depending on the wastewater type.

NOMENCLATURE

AOP = Advanced oxidation process;
BDD = Boron-doped diamond;
BFLL = Biofiltrated landfill leachate;
BOD = Biochemical oxygen demand;
BPWW = Boat pressure washing wastewater;
COD = Chemical oxygen demand;
DOM = Dissolved organic matter;
DSA = Ruthenium coated with titanium and titanium oxides;
DWW = Domestic wastewater;
EC = Electro-coagulation;
EF = Electro-Fenton;
EO = Electro-oxidation;
EC-O = Electro-coagulation-ozonation;
EC-P = Electro-coagulation-peroxidation;
EP = Electroperoxidation;
EPO = Electro-photo-oxidation;
FA = Fulvic acid;
HA = Humic acid;
Hyl = Hydrophilic compounds;
LSM = Liquid swine manure;
LWW = Laundry wastewater;
O&G = Oil and grease;
PECO = Photo-electrocatalytic oxidation;
PEF = Photoelectron-Fenton;
PZC = Point of zero charge;
RMS = Response surface methodology;
RWW = Restaurant wastewater;
SEC = Sono-electro-coagulation;
SPR = Sn–Pb–Ru;
SS = Suspended solids;
TDS = Total dissolved solids;
TOC = Total organic carbon;
US = Ultrasound;
VC = Vitreous Carbon;
WW = Wastewater.

References

Ahmad, H., W. K. Lafi, K. Abushgair, and J. M. Assbeihat. 2016. "Comparison of coagulation, electrocoagulation and biological techniques for the municipal wastewater treatment." *Int. J. Appl. Eng. Res.* 11 (22): 11014–11024.

Antony, S. P., and B. Natesan. 2012. "Optimization of integrated electro-bio process for bleaching effluent treatment." *Ind. Eng. Chem. Res.* 51 (24): 8211–8221.

Asaithambi, P., A. R. A. Aziz, and W. M. A. B. W. Daud. 2016. "Integrated ozone—Electrocoagulation process for the removal of pollutant from industrial effluent: Optimization through response surface methodology." *Chem. Eng. Process.* 105: 92–102.

Asaithambi, P., M. Susree, R. Saravanathamizhan, and M. Matheswaran. 2012. "Ozone assisted electrocoagulation for the treatment of distillery effluent." *Desalination* 297: 1–7.

Asgharian, F., M. R. Khosravi-Nikou, B. Anvaripour, and I. Danaee. 2017. "Electrocoagulation and ultrasonic removal of humic acid from wastewater." *Environ. Prog. Sustainable Energy* 36 (3): 822–829.

Aziz, A. R. A., P. Asaithambi, and W. M. A. B. W. Daud. 2016. "Combination of electrocoagulation with advanced oxidation processes for the treatment of distillery industrial effluent." *Process Saf. Environ. Prot.* 99: 227–235.

Behin, J., N. Farhadian, M. Ahmadi, and M. Parvizi. 2015. "Ozone assisted electrocoagulation in a rectangular internal-loop airlift reactor: Application to decolorization of acid dye." *J. Water Process Eng.* 8: 171–178.

Bernal-Martínez, L. A., C. Barrera-Díaz, C. Solís-Morelos, and R. Natividad. 2010. "Synergy of electrochemical and ozonation processes in industrial wastewater treatment." *Chem. Eng. J.* 165 (1): 71–77.

Boroski, M., A. C. Rodrigues, J. C. Garcia, L. C. Sampaio, J. Nozaki, and N. Hioka. 2009. "Combined electrocoagulation and TiO_2 photoassisted treatment applied to wastewater effluents from pharmaceutical and cosmetic industries." *J. Hazard. Mater.* 162 (1): 448–454.

Chakchouk, I., N. Elloumi, C. Belaid, S. Mseddi, L. Chaari, and M. Kallel. 2017. "A combined electrocoagulation–electrooxidation treatment for dairy wastewater." *Braz. J. Chem. Eng.* 34 (1): 109–117.

Daghrir, R., and P. Drogui. 2013a. "Coupled electrocoagulation–electro-Fenton for efficient domestic wastewater treatment." *Environ. Chem. Lett.* 11 (2): 151–156.

Daghrir, R., P. Drogui, J. F. Blais, and G. Mercier. 2012. "Hybrid process combining electrocoagulation and electro-oxidation processes for the treatment of restaurant wastewaters." *J. Environ. Eng.* 138 (11): 1146–1156.

Daghrir, R., P. Drogui, and M. A. El Khakani. 2013b. "Photoelectrocatalytic oxidation of chlortetracycline using Ti/TiO_2 photo-anode with simultaneous H_2O_2 production." *Electrochim. Acta* 87: 18–31.

Daghrir, R., P. Drogui, and D. Robert. 2013c. "Modified TiO_2 For environmental photocatalytic applications: A review." *Ind. Eng. Chem. Res.* 52 (10): 3581–3599.

Daghrir, R., P. Drogui, and F. Zaviska. 2013d. "Effectiveness of a hybrid process combining electro-coagulation and electro-oxidation for the treatment of domestic wastewaters using response surface methodology." *J. Environ. Sci. Health Part A* 48 (3): 308–318.

Dia, O., P. Drogui, G. Buelna, and R. Dubé. 2017a. "Strategical approach to prevent ammonia formation during electrocoagulation of landfill leachate obtained from a biofiltration process." *Sep. Purif. Technol.* 189: 253–259.

Dia, O., P. Drogui, G. Buelna, R. Dubé, and B. S. Ihsen. 2017b. "Electrocoagulation of bio-filtrated landfill leachate: Fractionation of organic matter and influence of anode materials." *Chemosphere* 168: 1136–1141.

Dia, O., P. Drogui, R. Dubé, and G. Buelna. 2016. "Utilisation des procédés électrochimiques et leurs combinaisons avec les procédés biologiques pour le traitement des lixiviats de sites d'enfouissement sanitaires—Revue de littérature." *Rev. Sci. Eau* 29 (1): 63–89.

Djelal, H., Y. Lelievre, and C. Ricordel. 2015. "Combination of electro-coagulation and biological treatment by bioaugmentation for landfill leachate." *Desalin. Water Treat.* 54 (11): 2986–2993.

Domínguez, C., J. García, M. A. Pedraz, A. Torres, and M. A. Galán. 1998. "Photocatalytic oxidation of organic pollutants in water." *Catal. Today* 40 (1): 85–101.

Drogui, P., J.-F. Blais, and G. Mercier. 2007. "Review of electrochemical technologies for environmental applications." *Recent Pat. Eng.* 1 (3): 257–272.

Drogui, P., S. Elmaleh, M. Rumeau, C. Bernard, and A. Rambaud. 2001. "Hydrogen peroxide production by water electrolysis: Application to disinfection." *J. Appl. Electrochem.* 31 (8): 877–882.

Durante, C., M. Cuscov, A. A. Isse, G. Sandonà, and A. Gennaro. 2011. "Advanced oxidation processes coupled with electrocoagulation for the exhaustive abatement of Cr-EDTA." *Water Res.* 45 (5): 2122–2130.

Environmental-expert. 2017. "Electrochemical wastewater treatment: Electrocoagulation, electrooxidation, electroperoxi-coagulation." Accessed June 27, 2017. www.environmental-expert.com.

Esfandyari, Y., Y. Mahdavi, M. Seyedsalehi, M. Hoseini, G. H. Safari, M. G. Ghozikali, et al. 2015. "Degradation and biodegradability improvement of the olive mill wastewater by peroxi-electrocoagulation/electrooxidation-electroflotation process with bipolar aluminum electrodes." *Environ. Sci. Pollut. Res.* 22 (8): 6288–6297.

Farhadi, S., B. Aminzadeh, A. Torabian, V. Khatibikamal, and M. Alizadeh Fard. 2012. "Comparison of COD removal from pharmaceutical wastewater by electrocoagulation, photoelectrocoagulation, peroxi-electrocoagulation and peroxi-photoelectrocoagulation processes." *J. Hazard. Mater.* 219–220: 35–42.

García-García, P., F. N. Arroyo-López, and F. Rodríguez-Gómez. 2014. "Partial purification of iron solutions from ripe table olive processing using ozone and electro-coagulation." *Sep. Purif. Technol.* 133: 227–235.

Ghernaout, D. 2013. "Advanced oxidation phenomena in electrocoagulation process: A myth or a reality?" *Desalin. Water Treat.* 51 (40–42): 7536–7554.

He, C.-C., C.-Y. Hu, and S.-L. Lo. 2016. "Evaluation of sono-electrocoagulation for the removal of Reactive Blue 19 passive film removed by ultrasound." *Sep. Purif. Technol.* 165: 107–113.

Hernández-Ortega, M., T. Ponziak, C. Barrera-Díaz, M. A. Rodrigo, G. Roa-Morales, and B. Bilyeu. 2010. "Use of a combined electrocoagulation–ozone process as a pre-treatment for industrial wastewater." *Desalination* 250 (1): 144–149.

Ibarra-Taquez, H. N., E. GilPavas, E. R. Blatchley III, M-Á Gómez-García, and I. Dobrosz-Gómez. 2017. "Integrated electrocoagulation-electrooxidation process for the treatment of soluble coffee effluent: Optimization of COD degradation and operation time analysis." *J. Environ. Manage.* 200: 530–538.

Jaafarzadeh, N., M. Omidinasab, and F. Ghanbari. 2016. "Combined electrocoagulation and UV-based sulfate radical oxidation processes for treatment of pulp and paper wastewater." *Process Saf. Environ. Prot.* 102: 462–472.

Kabdaşli, I., I. Arslan-Alaton, B. Vardar, and O. Tünay. 2007. "Comparison of electrocoagulation, coagulation and the Fenton process for the treatment of reactive dyebath effluent." *Water Sci. Technol.* 55 (10): 125–134.

Kang, S.-F., C.-H. Liao, and S.-T. Po. 2000. "Decolorization of textile wastewater by photo-Fenton oxidation technology." *Chemosphere* 41 (8): 1287–1294.

Khosravanipour Mostafazadeh, A., A. T. Benguit, A. Carabin, P. Drogui, and E. Brien. 2019. "Development of combined membrane filtration, electrochemical technologies, and adsorption processes for treatment and reuse of laundry wastewater and removal of nonylphenol ethoxylates as surfactants." *J. Water Process Eng.* 28: 277–292.

Kovatcheva, V. K., and M. D. Parlapanski. 1999. "Sono-electrocoagulation of iron hydroxides." *Colloids Surf., A* 149 (1–3): 603–608.

Laridi, R., P. Drogui, H. Benmoussa, J.-F. Blais, and J. C. Auclair. 2005. "Removal of refractory organic compounds in liquid swine manure obtained from a biofiltration process using an electrochemical treatment." *J. Environ. Eng.* 131 (9): 1302–1310.

Linares-Hernández, I., C. Barrera-Díaz, B. Bilyeu, P. Juárez-GarcíaRojas, and E. Campos-Medina. 2010. "A combined electrocoagulation–electrooxidation treatment for industrial wastewater." *J. Hazard. Mater.* 175 (1–3): 688–694.

Mahvi, A. H., S. J. A. d. Ebrahimi, A. Mesdaghinia, H. Gharibi, and M. H. Sowlat. 2011. "Performance evaluation of a continuous bipolar electrocoagulation/electrooxidation–electroflotation (ECEO–EF) reactor designed for simultaneous removal of ammonia and phosphate from wastewater effluent." *J. Hazard. Mater.* 192 (3): 1267–1274.

Manenti, D. R., A. N. Módenes, P. A. Soares, F. R. Espinoza-Quiñones, R. A. R. Boaventura, R. Bergamasco, et al. 2014. "Assessment of a multistage system based on electrocoagulation, solar photo-Fenton and biological oxidation processes for real textile wastewater treatment." *Chem. Eng. J.* 252: 120–130.

Mirshahghassemi, S., B. Aminzadeh, A. Torabian, and K. Afshinnia. 2017. "Optimizing electrocoagulation and electro-Fenton process for treating car wash wastewater." *Environ. Health Eng. Manage.* 4 (1): 37–43.

Módenes, A. N., F. R. Espinoza-Quiñones, F. H. Borba, and D. R. Manenti. 2012. "Performance evaluation of an integrated photo-Fenton–electrocoagulation process applied to pollutant removal from tannery effluent in batch system." *Chem. Eng. J.* 197: 1–9.

Naje, A. S., and S. Abbas. 2013. "Combination of electrocoagulation and electro-oxidation processes of textile wastewaters treatment." *Civ. Environ. Res.* 3 (13): 61–74.

Naje, A. S., S. Chelliapan, Z. Zakaria, A. Ajeel Mohammed, and A. Alaba Peter. 2017. "A review of electrocoagulation technology for the treatment of textile wastewater." *Rev. Chem. Eng.* 33 (3): 263–292.

Orescanin, V., R. Kollar, and K. Nad. 2011a. "The application of the ozonation/electrocoagulation treatment process of the boat pressure washing wastewater." *J. Environ. Sci. Health Part A* 46 (12): 1338–1345.

Orescanin, V., R. Kollar, and K. Nad. 2011b. "The electrocoagulation/advanced oxidation treatment of the groundwater used for human consumption." *J. Environ. Sci. Health Part A* 46 (14): 1611–1618.

Orkun, M. O., and A. Kuleyin. 2012. "Treatment performance evaluation of chemical oxygen demand from landfill leachate by electro-coagulation and electro-Fenton technique." *Environ. Prog. Sustainable Energy* 31 (1): 59–67.

Oumar, D., D. Patrick, B. Gerardo, D. Rino, and B. S. Ihsen. 2016. "Coupling biofiltration process and electrocoagulation using magnesium-based anode for the treatment of landfill leachate." *J. Environ. Manage.* 181: 477–483.

Qiang, Z., J.-H. Chang, and C.-P. Huang. 2003. "Electrochemical regeneration of Fe^{2+} in Fenton oxidation processes." *Water Res.* 37 (6): 1308–1319.

Raju, G. B., M. T. Karuppiah, S. S. Latha, S. Parvathy, and S. Prabhakar. 2008. "Treatment of wastewater from synthetic textile industry by electrocoagulation–electrooxidation." *Chem. Eng. J.* 144 (1): 51–58.

Ramcharan, T., and A. Bissessur. 2017a. "Treatment of laundry wastewater by biological and electrocoagulation methods." *Water Sci. Technol.* 75 (1–2): 84–93.

Ramcharan, T., and A. Bissessur. 2017b. "Electrocoagulative and biological treatment of laundry wastewater." In Chap. in *Biological wastewater treatment and resource recovery.* London: IntechOpen.

Raschitor, A., C. M. Fernandez, I. Cretescu, M. A. Rodrigo, and P. Cañizares. 2014. "Sono-electrocoagulation of wastewater polluted with Rhodamine 6G." *Sep. Purif. Technol.* 135: 110–116.

Roa-Morales, G., E. Campos-Medina, J. Aguilera-Cotero, B. Bilyeu, and C. Barrera-Díaz. 2007. "Aluminum electrocoagulation with peroxide applied to wastewater from pasta and cookie processing." *Sep. Purif. Technol.* 54 (1): 124–129.

Senghor, F., P. Drogui, and B. Seyhi. 2015. "A combined electrocoagulation-electroperoxidation process for the tertiary treatment of domestic wastewaters." *Water Air Soil Pollut.* 226 (11): 373.

Shivayogimath, C. B., and S. R. Vinayak Rao. 2016. "Treatment of dairy effluent by electrocoagulation-fenton oxidation." *Int. J. Innovative Res. Sci. Eng. Technol.* 5 (8): 15047–15053.

Suárez-Escobar, A., A. Pataquiva-Mateus, and A. López-Vasquez. 2016. "Electrocoagulation—Photocatalytic process for the treatment of lithographic wastewater. Optimization using response surface methodology (RSM) and kinetic study." *Catal. Today* 266: 120–125.

Thiam, A., M. Zhou, E. Brillas, and I. Sirés. 2014. "Two-step mineralization of Tartrazine solutions: Study of parameters and by-products during the coupling of electrocoagulation with electrochemical advanced oxidation processes." *Appl. Catal., B* 150–151: 116–125.

Torres-Sánchez, A. L., S. J. L. C. Catalina de la Rosa, M.-V. María, M.-S. María, and J. M. Peralta-Hernández. 2014. "Electrocoagulation process coupled with advance oxidation techniques to treatment of dairy industry wastewater." *Int. J. Electrochem. Sci.* 9: 6088–6102.

Umar, M., H. A. Aziz, and M. S. Yusoff. 2010. "Trends in the use of Fenton, electro-Fenton and photo-Fenton for the treatment of landfill leachate." *Waste Manage.* 30 (11): 2113–2121.

Wu, C.-H., C.-L. Chang, and C.-Y. Kuo. 2008. "Decolorization of Procion Red MX-5B in electrocoagulation (EC), UV/TiO_2 and ozone-related systems." *Dyes Pigm.* 76 (1): 187–194.

Yazdanbakhsh, A. R., M. R. Massoudinegad, S. Eliasi, and A. S. Mohammadi. 2015. "The influence of operational parameters on reduce of azithromyin COD from wastewater using the peroxi-electrocoagulation process." *J. Water Process Eng.* 6: 51–57.

Yu, M. J., J. S. Koo, G. N. Myung, Y. K. Cho, and Y. M. Cho. 2005. "Evaluation of bipolar electrocoagulation applied to biofiltration for phosphorus removal." *Water Sci. Technol.* 51 (10): 231–239.

Zhang, Y., Y.-q. Cong, and Q. Wang. 2013. "Electrocoagulation-TiO_2 photo-assisted combined system applied to methyl orange wastewater removal." *Environ. Eng. Manage. J.* 12 (3): 517–526.

CHAPTER 7

Combined Electro-Oxidation Processes

Sushil Kumar, Bhagyashree Tiwari,
Patrick Drogui, R. D. Tyagi

7.1 INTRODUCTION

The rapidly growing population, industrialization, urbanization, and economic development increase water requirements for various purposes. Approximately 40% of the world's population suffer from water scarcity, and it is estimated to rise to 60% by 2025 (Sato et al. 2013). This increase in water consumption eventually leads to a rise in wastewater generation. For instance, in Canada, approximately 85 km^3 of wastewater is generated per year (Sato et al. 2013).

Wastewater contains many organic and inorganic compounds, which could be toxic. Numerous investigations from the last four decades reported the presence of pollutants in wastewater, which include pharmaceuticals, microfibers, personal care products, endocrine-disrupting chemicals, and others (Jiang et al. 2013). The constituents of wastewater and its concentration vary significantly and depend on the source (such as the residential sector or industry). For instance, the wastewater generated from residential sectors contains organic and inorganic chemicals, detergents, soaps, and pathogenic and nonpathogenic micro-organisms, whereas the wastewater discharged from any industry could be highly diverse and different from that of the residential sectors and depends on the type of processing and product manufactured in the industry. For example, the waste stream from any pharmaceutical-producing industry contains a high concentration of pharmaceuticals, solvents, residual products, metals, halides, nitrates, and many more (Tiwari et al. 2020).

In general, wastewater is treated by conventional technologies, which include chemical coagulation, aerobic treatment, and anaerobic biological treatment. These treatment techniques are not efficient enough for removal and/or neutralization of complex dissolved organic pollutants (e.g., pharmaceuticals, polyaromatic hydrocarbons, endocrine-disrupting chemicals, etc.) (Jiang et al. 2013; Zolfaghari et al. 2016). Thus, other advanced treatment techniques have been investigated for the removal of complex organic pollutants from water and wastewater. These treatment techniques include adsorption using activated carbon, EO, membrane filtration, photocatalysis, and ozonation (Särkkä et al. 2015). Among these technologies, EO has been extensively studied for the treatment of water and wastewater (Särkkä et al. 2015). For instance, Lang et al. (2020) investigated the treatment of marine aquaculture wastewater using the electrochemical oxidation process. Their study reported the effective removal of NH^{4+}–N and NO^{2}–N (Lang et al. 2020). The sequential coagulation and electrochemical oxidation process was applied for the treatment of industrial textile wastewater. The coupled process was reported to remove 93.4% of chemical oxygen demand (COD), 100% color, and 75% of total organic carbon (TOC) (GilPavas et al. 2018).

The EO treatment process involves the oxidation of organic pollutants either by direct oxidation of hydroxyl radicals at the anode surface or by indirect oxidation in which oxidants are produced by chemical compounds (Särkkä et al. 2015). The factors that determine the EO treatment efficiency are electrode material, electrochemical cell design, mass transport regime, electrode potential, current density, current input, and electrolysis medium (Martínez-Huitle et al. 2015). The abatement efficiency of the EO process greatly depends on the electrode material because it provides physicochemical and mechanical stability. The current density and electrode potential regulate the reactivity and rate of the electrochemical reaction and, thus, affect the treatment efficiency. The EO treatment processes are highly efficient, easy to automate, versatile, and economical, and requiring only mild operation conditions. Despite these advantages, low current efficiency, the short lifetime of the electrode, high cost of the certain electrodes, unequal current distribution, weak mass transfer, and difficulty in the treatment of large volumes of water with low pollutant concentration limit the application of EO (Pan et al. 2019). Thus, numerous studies have investigated the coupling of the EO process with other chemical and biological treatment processes such as membrane filtration, Fenton process, electroperoxane processes and so on (Pan et al. 2019, Särkkä et al. 2015, Zolfaghari et al. 2016). However, to the best of our knowledge, there is a lack of amalgamation of these coupled process studies in the literature. Thus, this chapter provides a review and extensive discussion on the coupling of the EO process with other chemical treatment processes. This chapter also presents the factors affecting these processes and case studies.

7.2 ELECTRO-OXIDATION AND TIO_2 PHOTO-ASSISTED PROCESSES

Over a long period of time, semiconductor photocatalysis has received much attention because of its exceptional efficiency for environmental cleansing (Bhatkhande et al. 2002, Hoffmann et al. 1995, Mills and Le Hunte 1997). Until now, more than 190 types of semiconductors have been declared suitable as photo catalysts. TiO_2 is one of the most used semiconductor photo catalysts, but on its own, it is not considered a feasible alternative. For achieving high efficiency during the process, it is necessary to couple the photocatalysis method with UV light. Now, a growing interest is seen for using solar energy as the photon source along with low-cost catalysts, but the efficiency of the solar photocatalysis of pure titanium oxide is lower than that of the ultraviolet source. In the catalysis of titanium oxide, two types of reactions are involved, namely, adsorption–desorption equilibrium and dark reaction photocatalysis under irradiation (Bhatkhande et al. 2002).

Titanium dioxide appears in amorphous form and crystallizes in three different polymorphs in nature according to the order of abundance, rutile, anatase, and brookite. The structures of anatase and rutile are tetrahedral, and brookite has an orthorhombic shape. Photocatalyst Aeroxide P-25 powder, which has a greater surface area (49 $m^2\,g^{-1}$), is used as a standard material during the reaction of titanium oxide. Many studies have hypothesized that the best photocatalytic action for TiO_2 is completed with an optimum ratio of anatase and rutile. Degussa, P-25, comprises anatase and rutile in a ratio of about 3:1. It is made up of 20% to 30% rutile and 70% to 80% anatase TiO_2 with particle sizes in the range of 12 to 20 nm (Hassan et al. 2016). The greater photoactivity of TiO_2-P25 is predominant because of its crystalline composition of rutile and anatase. It has been assumed that the minor band gap of rutile absorbs the photons and produces electron–hole pairs. Then, electron movement takes place from the rutile catalyst band to electron traps in the anatase phase (Hassan et al. 2016), which prevents recombination and permits the hole to transfer to the surface of the particle and react with OH^- or H_2O to produce the $^{\bullet}OH$ radical (Hurum et al. 2003).

As explained in Equations (7-1) to (7-4), when TiO_2 captivates UV light (<380 nm), electrons are endorsed from the valence band to the conduction band to produce electron–hole pairs (e^-/h^+). The oxidation of organic compounds is mainly due to positive hole and/or hydroxyl radical ions, whereas electrons mostly reduce molecular oxygen to superoxide radical anions, which can lead to a number of reactive oxygen species ($^{\bullet}OH$, $O_2^{\bullet -}$, $HO_2^{\bullet}$, etc.) These radicals oxidize a large range of organic contaminants to nontoxic inorganics such as CO_2, H_2O, and other minerals (Hurum et al. 2003). This mechanism can be clearly seen in Figure 7-1.

$$TiO_2 + h\upsilon(E_G \geq 3.2\ eV) \rightarrow TiO_2 + (e_{CB} + h_{VB}) \tag{7-1}$$

$$TiO_2 + h_{VB} + HO^- \rightarrow TiO_2 + HO^{\bullet} \tag{7-2}$$

$$TiO_2 + e_{CB} + O_2 \rightarrow O_2^{\bullet -} \tag{7-3}$$

$$R + HO^{\bullet} \rightarrow R(HO^{\bullet}) \rightarrow R_1 + R \rightarrow \text{Inorganic substances (i.e.,} CO_2 + H_2O) \tag{7-4}$$

The degradation of titanium oxide allows us to develop an economical, nontoxic, high photoactive, eco-friendly, and sustainable technology, which leads to zero-waste production in the water and wastewater treatment method. Because of this advantage, many researchers have applied this method for the treatment of municipal solid waste leachate or pretreatment to increase the biodegradability of refractory compounds (Malpass et al. 2007).

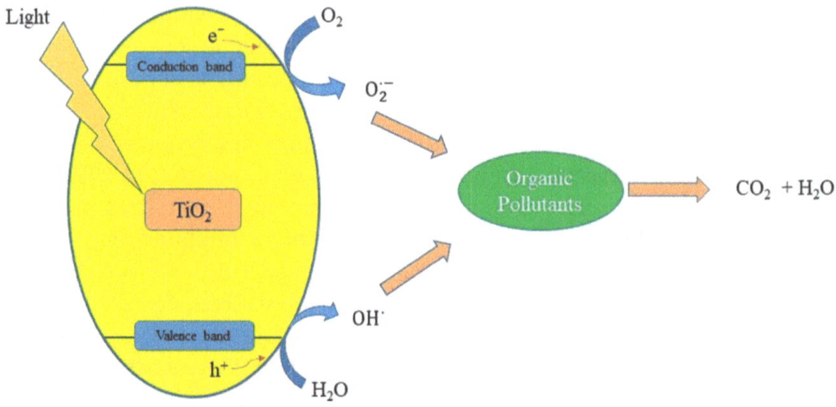

Figure 7-1. TiO_2 photo-assisted process.

In a study, Ratiu et al. (2011) examined the effect of photocatalysis using zeolite-supported TiO_2 on the degradation efficiency of 4-AP in an aqueous solution by electrochemical oxidation by means of a boron-doped diamond (BDD) electrode. The results show the importance of the operating parameters in achieving high electrochemical and photocatalytically assisted electrochemical degradation efficiency. The electrochemical degradation of 4-AP on the BDD electrode was enhanced by the use of the pulsed potential, which allowed both in situ electrochemical cleaning of the electrode and an indirect oxidation of 4-AP by the oxygen evolution reaction. The application of photocatalysis to the 4-AP electrochemical degradation using Z-TiO_2 revealed an improved effect when the anodic potential was set at +1.25 V versus Ag/AgCl in the water constancy region close to the oxygen development potential (Ratiu et al. 2011).

The catalysts used in photocatalytic reactions are invariably semiconductor materials. The overall progress of realization of different photocatalysts can be reported in three generations.

The first generation (1975 to 1985) is focused primarily on the studies that provided an understanding of the semiconductor/solution interface under irradiation from studies of monocrystalline semiconductor surfaces and the exploration of different types of semiconductors. At the time, it was confirmed that polycrystalline materials would be more appropriate for most of the applications, and the first generation of photocatalysis work focused on suspensions of fine particles. The second generation (1986 to 2000) focused on polycrystalline thin films and efforts to modify semiconductors for visible light utilization to explore the scope of photocatalysis. Inthis period industry emerged, and dye-sensitized photocatalysts were extensively evaluated for the destruction of dyes in waste streams. In the third-generation period since 2000 to date, the focus is on nanotechnology in photocatalysis for energy conversion and controlling air/water pollution (Langford 2012). Different UV active semiconductor photocatalysts have been developed and can be classified into four groups based on their electronic configuration: (1) d0 metal (Ti^{4+}, Zr^{4+}, Nb^{5+}, Ta^{5+}, W^{6+}, and Mo^{6+}) oxide photocatalysts; (2) d10 metal (In^{3+}, Ga^{3+}, Ge^{4+}, Sn^{4+}, and Sb^{5+}) oxide photocatalysts; (3) f0 metal (Ce^{4+}) oxide photocatalysts; and (4) a small group of nonoxide photocatalysts (Ismail and Bahnemann 2014). Numerous n-type semiconductor metal sulfides (CdS) and oxides (TiO_2, ZnO, CdS, SnO_2, and WO_3) are regarded as the most promising and efficient materials for the photocatalytic degradation of organic waste and photo-assisted solar hydrogen generation on the basis of their band positions with reference to the normal hydrogen electrode. However, most of the previously mentioned catalysts have some drawbacks such as (a) a wider band gap, making it difficult for them to use the visible radiations of the solar spectrum (b) lower stability in an aqueous medium, leading to an agglomeration of, and, hence, a reduction of, reaction sites, and (c) rapid bulk and surface e^-/h^+ recombination rates (Kuvarega et al. 2015). TiO_2 is one of the most effective semiconductor photocatalysts having high chemical stability, photocatalytic activity, nontoxicity, and low cost. TiO_2 has seven polymorphs. However, only rutile, anatase, and brookite exist in nature (Chu et al. 2004). TiO_2 has generated considerable interest because of its photochemical, photocatalytic, and photovoltaic properties (Souvereyns et al. 2013). Research studies on the development of noble photocatalysts that can be used in the slurry and immobilized system have great importance. However, immobilized photocatalysts eliminate the necessity of agitation and minimize agglomeration. In addition, recycling becomes comparatively easy (Wang et al. 2015b). Numerous research efforts have proved that nano-TiO_2 is more effective as a photocatalyst than its bulk counterpart. Compared with other photocatalysts for H_2 production, TiO_2 has received more attention from researchers. The photocatalytic performance of TiO_2 semiconducting nanoparticles of less than 10 nm has shown significant improvement because of the quantum size effect (Anpo and Takeuchi 2003).

7.3 SONO-ELECTRO-OXIDATION PROCESS

The electrochemical process is one of the promising, effective, economical, and eco-friendly methods for water and wastewater treatment (Martínez-Huitle and Ferro 2006). Recent advancements show that a combination of the electrochemical process with other oxidation processes is now emerging as a promising treatment method having many advantages such as high oxidation ability, high reaction potential, and high efficiency (Zhao et al. 2008). In recent days, various types of advanced oxidation processes are used for the treatment of water and wastewater. Among several different techniques, the ultrasonic technique is widely used for the degradation of contaminants and for increasing the efficiency of other oxidation processes. Considerable research shows that ultrasound, when combined with various catalysts, can be effectively used for the treatment of water and wastewater (Ince and Tezcanlí 2001, Tauber et al. 2005). However, research also indicates that ultrasound alone cannot be effective for treating wastewater containing complex organic and inorganic pollutants (Gogate and Pandit 2004). Recent advancements indicate that many efforts have been made to improve the efficiency of the ultrasonic process by combining it with other oxidation processes (Hai et al. 2007). The sono-electro-oxidation process is the result of hybrid technology, which combines the ultrasonic process with the electro-oxidation process.

The ultrasonic process is employed to overcome the problem of passivation of electrodes, which is one of the major problems in the EO process (Compton et al. 1997). Electrode fouling is controlled by the effect of ultrasound, and this is called *sonoelectrochemistry*. The method of producing ultrasound is a physical technique based on the method of generating, developing, and foundering the cavities of streams and gases, and this process is known as *cavitation*. In the field of chemistry, ultrasound is widely used for the synthesis of organic matter, catalyst preparation, polymerization, sonolysis, and sonoelectrosynthesis (Stoppa 2012).

The combination of the ultrasonic technique with an EO process leads to a powerful method for contaminant degradation (Mason and Lorimer 2002). The electrochemical mechanism improves by the application of powerful ultrasound, which ultimately affects the physical and chemical processes involved in the treatment process (Mason and Lorimer 2002).

7.3.1 Degradation of Contaminants Using Sono-Electro-Oxidation Processes

As discussed previously, when power ultrasound and electrochemical techniques are combined together and are applied for the treatment of water and wastewater, they provide a powerful and effective tool for the degradation of pollutants. This is due to the fact that ultrasound is responsible for the generation of hydroxyl radicals, increase of mass transport, and activation of the electrode surface.

Based on the theory, various research studies have been done so far (Lorimer et al. 2000). Lorimer et al. (2000) used this technique for the degradation of acidic dye, Sandolan yellow. During the experiment, platinum was used as an electrode and sodium chloride as an electrolyte. During the process, chlorine was generated at the anode, and hydroxide ions were generated at the cathode, which resulted in the in situ production of hypochlorite ion [Equations (7-5) to (7-7)], which is considered as a powerful oxidant (Lorimer et al. 2000).

Reaction at the anode:

$$2Cl^- \rightarrow Cl_2 + 2e \tag{7-5}$$

Reaction at the cathode:

$$2e + 2H_2O \rightarrow 2OH^- + H_2 \tag{7-6}$$

The overall reaction:

$$2OH^- + Cl_2 \rightarrow Cl^- + OCl^- + H_2O \tag{7-7}$$

Another study done by Garbellini et al. (2010) showed that the pentachlorophenol can be effectively degraded by using ultrasound at a fixed frequency of 20 kHz with a combination of potentiostatic electrolyses. During the process, the electrode surface was not cleaned purposely. The effect of ultrasound caused surface activation of the electrode, and simultaneously, the generation of hydroxyl radicals helped achieve a higher degradation of pesticides (Garbellini et al. 2010). Siddique et al. (2011) used ultrasound technology with the electrochemical process for the degradation of unhydrolyzed and hydrolyzed forms of reactive blue (RB) 19 dye. During the process, lead oxide was used as an electrode. The whole process was carried out at various frequencies and at different pH values. The results obtained during this research showed that 90% color removal and 56% TOC removal was achieved during the process (Siddique et al. 2011).

The sono-EO process has been recently used with so many other processes for the degradation of pollutants. The technology provides a powerful tool for the degradation of organic compounds (Stoppa 2012). Abdelsalam and Birkin (2002) used electrochemically generated H_2O_2 with power ultrasound for the degradation of organic dye molecules. During the process, it was observed that the degradation of the dye species was enhanced by the presence of ultrasound, Fe^{2+}, and hydrogen peroxide ions. Long-term exposure to the solution resulted in the removal of 60.7% COD from the solution. In another similar study, Oturan et al. (2008) used a sono-electro-Fenton process for the treatment of wastewater containing organic pollutants. During the process, platinum was used as the anode and carbon as the cathode. The process showed a significantly high efficiency for the degradation of organic pollutants (Oturan et al. 2008).

7.3.2 Advantages of the Sono-Electro-Oxidation Process

Sono-EO processes can be effectively used for the treatment of toxic waste at moderate conditions, and these are also capable of treating effluents having low conductivities (Rivera et al. 2009). Electricity is used for these processes and no additional chemicals are needed; thus, these are considered environmentally friendly. Sono-EO processes are also effective in treating wastewater containing dyes (Rivera et al. 2009).

7.4 ELECTROCHEMICAL-PEROXIDATION PROCESS

Electrochemical peroxidation (ECP) is one of the patented processes that have been used to degrade organic pollutants in various aqueous matrixes (groundwater, water from septic tank, slurries, pulp and paper waste) in various studies (Gholikandi and Kazemirad 2018). A by-product of the process, colloidal iron hydroxide, is an insatiable metal scavenger that has been employed to eliminate heavy metals from aqueous solutions (Arienzo et al. 2001) and, thus, may deliver a promising solution for waters copolluted by metals and organic compounds.

The term "*ECP*" is defined as a process in which hydrogen peroxide (H_2O_2) is electrochemically produced at the cathode electrode to oxidize the contaminants present in a solution (Turkay et al. 2017b). In the presence of ferrous ion, H_2O_2 breaks to form $OH^{\bullet}$ radicals during Fenton's reaction (Turkay et al. 2017a). To avoid adding the catalyst (Fe^{2+} ions) to a solution and to reduce the treatment cost, it would be advantageous to generate Fe^{2+} in situ via the EC process.

$$Fe \rightarrow Fe^{2+} + 2e^{-} \tag{7-8}$$

$$4Fe^{2+} + 10H_2O + O_{2(g)} \rightarrow 4Fe(OH)_{3(S)} + 8H^{+} \tag{7-9}$$

$$2H_2O + 2e^{-} \rightarrow 2OH^{-} + H_{2(g)} \tag{7-10}$$

$$H_2O_2 + Fe^{2+} \rightarrow Fe^{3+} + {}^{\bullet}OH + OH^{-} \tag{7-11}$$

$$O_2 + 2H^{+} \rightarrow 2e^{-} + H_2O_2 \tag{7-12}$$

$$Fe^{3+} + e^{-} \rightarrow Fe^{2+} \tag{7-13}$$

$$BDD + H_2O \rightarrow BDD({}^{\bullet}OH) + H^{+} + e^{-} \tag{7-14}$$

ECP provides an improved rate of pollutant oxidation in concentrated wastewater over the Electro-Fenton process (Babuponnusami and Muthukumar 2014, Moussavi and Aghanejad 2014). In addition, it provides a choice for the treatment of wastewater having high concentrations of toxic organic compounds (Moussavi and Aghanejad 2014). ECP is also one of the suitable options for effective treatment of industrial wastewater containing recalcitrant pollutants such as those produced in paperboard recycling plants (Moussavi and Aghanejad 2014).

H_2O_2 can be generated in the bulk from the two-electron reduction of oxygen (directly injected as pure gas or bubbled air) at the cathode surface in acidic/neutral media with $E = 0.68$ V/SHE (Volts/standard hydrogen electrode). H_2O_2 production and stability depend on factors such as cell configuration, cathode properties, and operational conditions. To attain high H_2O_2 electrogeneration efficiency, the contact between cathode, oxygen, and water must be maximized. For this reason, porous cathodes such as gas-diffusion electrodes (GDEs) and three-dimensional electrodes of high specific surface areas are preferred for H_2O_2 electrogeneration. Because carbon is nontoxic and exhibits a high overpotential for H_2 evolution and low catalytic activity for H_2O_2 decomposition, along with relative good stability, conductivity, and chemical resistance, carbonaceous cathodes have been widely employed for H_2O_2 electrogeneration with high efficiency (Pérez et al. 2019).

H_2O_2 itself is only a moderately strong oxidant, exhibiting E (H_2O_2/H_2O) = 1.77 V/SHE in an acidic medium and E (H_2O_2/OH^-) = 0.88 V/SHE in an alkaline medium. It can oxidize only reduced sulfur compounds, cyanides, some organics such as aldehydes, formic acid, and some nitro-organic and sulfo-organic compounds. For this reason, electrochemical processes with H_2O_2 electrogeneration are usually performed in the presence of the Fe^{2+} ion to yield the Fenton's reagent (Brillas et al. 2009).

7.5 ELECTRO-PEROXONE PROCESS

Ozone has been used as a source of chemical reagent for the treatment of water and wastewater for a long time because of its potentially high oxidant and disinfectant properties. Ozone can be directly responsible for the degradation of pollutants or can generate hydroxyl radicals, which offer a further breakdown of target contaminants. However, because of its life cycle, ozone breaks down into oxygen just after its production. This creates some complications to ozonation processes, such as putting a question mark over the economical viability of the processes and creating a requirement for on-site production (Turkay et al. 2017a, Wang et al. 2015a). For the production of a highly concentrated ozone, dry air is passed through a strong electric field. Theoretically, 1 kW·h of electric energy is needed for the production of 1,058 g of ozone, but practically, the value turns out to be 150 g kW·h (Turkay et al. 2017b). This indicates that the whole process requires highly economical resources because of the high cost generated by the low efficiency of ozone production. Also, because of its highly selective nature of oxidants, some pollutants are resistant to ozonation. It has also been seen in the literature that for some pollutants such as pharmaceutical products, synthetic dyes, and phthalates, mineralization efficiency becomes low when only ozone is used for the treatment of water and wastewater. In some cases, toxic by-products may also form when high doses of ozone are used.

The Electro-peroxone (E-peroxone) process is an innovation in the electrochemical advanced oxidation process (EAOP) that is empowered by in situ production of H_2O_2 from cathodic O_2 reduction through conventional ozonation. The electronically produced H_2O_2 can considerably increase ozone (O_3) conversion to hydroxyl radicals ($OH^\bullet$), thus greatly improving contaminant degradation and TOC mineralization by the E-peroxone process than by conventional ozonation. Because of the high efficiency of pollutant degradation, the E-proxone process can also reduce the reaction time and energy utilization for the treatment processes. In addition, bromate formation can be reduced during the E-peroxone process for water containing bromide. During the process, all oxidants (O_3, H_2O_2, and $OH^\bullet$) are generated on site at a controlled rate by the use of only clean oxygen and electric power. In addition, the E-peroxone process does not produce any secondary pollutants and neither does it require any chemical or catalyst. With a minimum cost of upgradation, the conventional ozonation process can be retrofitted to the E-peroxone process by simply adding a low-cost carbon-based cathode. Therefore, the E-peroxone process can provide a convenient and low-cost approach to significantly improve the performances of the present ozonation systems in numerous ways and, thus, turn out to be one of the promising EAOPs for the treatment of water and wastewater.

7.5.1 Mechanism of the Process

The effective generation of hydrogen peroxides from cathodic oxygen reduction plays an important role in the E-peroxone process. However, numerous reactions may occur in competition with oxygen reduction to hydrogen peroxide at the

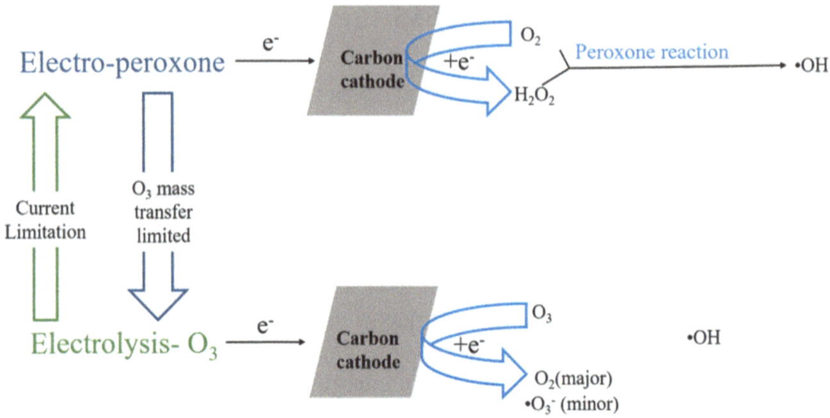

Figure 7-2. Electro-peroxone process.

cathode throughout the E-peroxone process. For example, H_2 evolution leads to a further reduction of electrogenerated hydrogen peroxide and ozone reduction. Considering the high potential for hydrogen evolution and significantly low catalytic activity for hydrogen peroxide decomposition, carbon-based electrodes are normally introduced to generate H_2O_2 from cathodic oxygen reduction. The reaction mechanism can be seen in Figure 7-2 and the equations that follow.

During the process, the reaction continues through the anion of hydrogen peroxide and is tentatively recommended as the first step for electron transfer [Equations (7-15) to (7-17)] (Merényi et al. 2010).

$$H_2O_2 \rightleftharpoons H_2O^- + H^+ \tag{7-15}$$

$$H_2O^- + O_3 \rightarrow HO_2^{\bullet} + O_3^- \tag{7-16}$$

$HO_2^{\bullet}$ is in equilibrium with $O_2^{\bullet-}$ [$pK_a(HO_2^{\bullet}) = 4.8$, equilibrium (7-17)], and $O_3^{\bullet-}$ has been recommended for protonation [Equation (7-18)] and successful decomposition into O_2 plus $OH^{\bullet}$ [Equation (7-19)], whereas $O_2^{\bullet-}$ undergoes electron transfer to O_3 [Equation (7-20)], thereby producing another $OH^{\bullet}$ (via $O_3^{\bullet-}$) (Merényi et al. 2010).

$$HO_2^{\bullet} \rightleftharpoons O_2^{\bullet-} + H^+ \tag{7-17}$$

$$O_3^{\bullet-} + H^+ \rightleftharpoons HO_3^{\bullet} \tag{7-18}$$

$$HO_3^{\bullet} \rightarrow O_2 + OH^{\bullet} \tag{7-19}$$

$$O_2^{\bullet-} + O_2 \rightarrow O_2 + O_3^{\bullet-} \tag{7-20}$$

The last equation shows that two moles of $OH^{\bullet}$ should be produced per one mole of hydrogen peroxides and two moles of ozone [overall reaction Equation (7-21)].

$$H_2O_2 + 2O_3 \rightarrow 2OH^{\bullet} + 3O_2 \tag{7-21}$$

7.5.2 Advantages of the E-Peroxone Process

One of the main advantages of the E-peroxone process is that it can be easily applied in conventional ozonation systems by simply introducing a pair of electrodes in the ozone contactor, which requires very less capital cost and modification (Li et al. 2013). The leftover oxygen from the ozonation process can be utilized as an ideal O_2^- source for cathodic hydrogen peroxide generation during the E-peroxone process. The process can greatly enhance the degradation of O_3^- resistant contaminants and the mineralization of solution TOC (Wang et al. 2015a). In addition, the electrogenerated hydrogen peroxide can significantly reduce the amount of bromate (BrO_3^-) generated during the treatment of bromide (Br^-)-containing water (Yao et al. 2017).

7.6 ELECTRO-FENTON PROCESS

Among most of the AOP, the Fenton process is highly promising because of its ease of use, low cost, high efficiency, and the nontoxicity of the Fenton's reagents (ferrous ion and hydrogen peroxide) (Nidheesh and Gandhimathi 2012, Wang and Jiang 2007). During the process, Fe^{2+} present in the system catalyzes H_2O_2 into $OH^\bullet$, which is a highly reactive and potential oxidizing agent that can react with most of the organic compounds containing C–H and C–C bonds at almost diffusion-controlled rates ($>10^{-9}$ M^{-1} S^{-1}) (Büyüksönmez et al. 1999, Pignatello et al. 2006). Fenton's reagents eliminate organic pollutants under normal temperatures and pressure conditions and permit high depuration with comparatively low cost and abundant materials (Duesterberg and Waite 2006, Zhang et al. 2019). In addition, both Fe^{2+} and Fe^{3+} ions behave like coagulants, and coagulation phenomena occurring during the Fenton processes can further eliminate organic pollutants (Badawy and Ali 2006, Collins and Bolton 2016). With the potential oxidation performance and a basic understanding of the chemical phenomena, the Fenton process has been deliberated as one of the most promising AOPs and has also been implemented in water and wastewater treatment processes (Babuponnusami and Muthukumar 2014, Feng et al. 2010).

However, still a few disadvantages exist in conventional Fenton-based processes, which include the storage and transportation of highly concentrated hydrogen peroxide, the fast consumption of catalysts, and the production of iron sludge with the additional work of its disposal (Collins and Bolton 2016). Therefore, to overcome these drawbacks, new technology has been developed, while still making beneficial use of the high oxidation potential of the Fenton process. Electro-Fenton has generated great interest for its ability to eliminate organic pollutants from water and wastewater (Borole et al. 2018, Feng et al. 2010). In an Electro-Fenton process, pollutants are removed by Fenton's reagents along with anodic oxidation on the anode surface. Because of the development of refractory carboxylic acids, anodic oxidation alone is not sufficient to mineralize most aromatic contaminants. But with the production of $OH^\bullet$, the Electro-Fenton process can yield good mineralization efficiency in removing organic pollutants. Unlike the conventional processes, the Electro-Fenton process avoids the storage and transportation of external hydrogen peroxides with in situ-produced H_2O_2 (Collins and Bolton 2016, He and Zhou 2017) and, hence, is an environmentally friendly technology as the requirement of chemicals has been reduced. Although Electro-Fenton provides significant advantages over the conventional Fenton process, it has some limitations too. The slow production of hydrogen peroxide caused by the low solubility of oxygen, low current efficiency at a higher pH, and higher operational costs are some of the limitations of this process (Babuponnusami and Muthukumar 2014).

7.6.1 Process Mechanism

The mechanism of a Fenton process is based on the oxidation of $OH^•$, as seen in Figure 7-3, which is one of the prevailing oxidizing agents. In a Fenton process, $OH^•$ is produced through a Fenton reaction [Equation (7-22)] and functions as the starting reaction of the chain Fenton reactions [Equations (7-22) to (7-26)] (Brillas et al. 2009)

$$H_2O_2 + Fe^{2+} \rightarrow Fe^{3+} + OH^• + OH^- \quad (7\text{-}22)$$

$$OH^• + H_2O_2 \rightarrow HO_2^• + H_2O \quad (7\text{-}23)$$

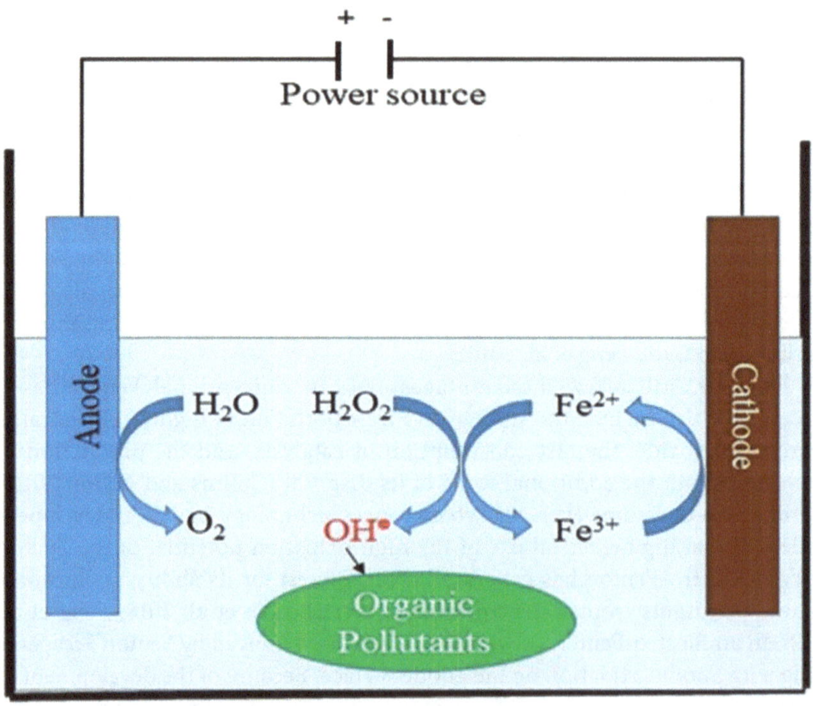

Figure 7-3. Electro-Fenton process.

The Fenton reaction offers a simple arrangement to produce $OH^{\bullet}$. However, the process requires high quantities of reagents to generate enough $OH^{\bullet}$. The simple arrangement indicates the economic viability of the process, but on the flip side, the creation of ferric sludge requires supplementary processing. More important, high reagent concentrations increase the rate of degenerative reactions [Equations (7-22) and (7-23)], resulting in low efficiency

$$Fe^{3+} + HO_2^{\bullet} \rightarrow Fe^{2+} + O_2 + H^+ \quad (7\text{-}24)$$

$$Fe^{2+} + HO_2^{\bullet} \rightarrow Fe^{3+} + HO_2^- \quad (7\text{-}25)$$

$$Fe^{2+} + OH^{\bullet} \rightarrow Fe^{3+} + OH^- \quad (7\text{-}26)$$

Equations (7-22) to (7-24) form a cycle where oxygen is released, and Equations (7-25) and (7-26) function as end reactions. The active intermediate species, $OH^{\bullet}$, is known as a strong oxidizing agent that can effectively degrade most organic compounds into innocuous compounds through dehydrogenation or hydroxylation (Collins and Bolton 2016, Özcan et al. 2009, Zhang et al. 2019). These reactions are twice as destructive as they destroy $OH^{\bullet}$ and expend reagents. Therefore, the decrease in mineralization efficiency leads to the development of oxidation reaction intermediates that are highly toxic (He and Zhou 2017). To prevent the occurrence of such reactions, it is required to control Fenton's reagent.

The aim of developing the Electro-Fenton process is to overcome some of the disadvantages of the process through a controlled synthesis of hydrogen peroxides in water or wastewater meant for treatment and a catalysis of the Fenton reaction by the regeneration of Fe^{2+}. Hydrogen peroxides are produced by a two-electron reduction of oxygen molecules from compressed air [Equation (7-27)] in an electrochemical cell in an acidic medium of around pH 3 (He and Zhou 2017, Oturan and Oturan 2018).

Reaction at the cathode:

$$O_2 + 2H^+ \rightarrow 2e^- + H_2O_2 \quad (7\text{-}27)$$

The Fenton reaction then produces Fe^{2+} that can be reduced at the cathode [Equation (7-28)] at the equal potential of aiding hydrogen peroxide generation to regenerate ferric ions (Oturan and Oturan 2018):

Reaction at the sacrificial anode:

$$Fe^{3+} + e^- \rightarrow Fe^{2+} \quad (7\text{-}28)$$

Fenton's reagent generation rate [Equations (7-27) and (7-28)] can be controlled mainly by applying the appropriate amount of current between electrodes of a two-electrode cell or by controlling the cathode potential in the case of three-electrode cells (Collins and Bolton 2016, Oturan and Oturan 2018, Zhang et al. 2019).

The in situ production of hydrogen peroxides is an elemental phenomenon in the Electro-Fenton process. Therefore, this process requires suitable cathode materials. At the initial stage of developing cathode material used for the in situ generation of H_2O_2, a mercury pool was selected (Oturan and Oturan 2018). Because of its toxic nature, this cathode was later changed to different carbonaceous materials such as the carbon sponge, reticulated vitreous carbon, carbon felt, and carbon–PTFE–O_2 gas-diffusion cathode (Collins and Bolton 2016, Oturan and Oturan 2018). Later, other carbon materials such as graphite, nanotubes, and carbon fibers were used for the cathode material.

The selection of anode material is also very important during the Electro-Fenton process because heterogeneous hydroxyl radicals ($OH^{\bullet}$) can also be produced at the anode surface from water oxidation [Equation (7-29)]:

$$H_2O \rightarrow OH^{\bullet} + H^+ + e^- \tag{7-29}$$

The amount of $OH^{\bullet}$ and its involvement in the EO of organic matter potentially depend upon the type of anode material used (Oturan and Oturan 2018).

7.6.2 Advantages and Disadvantages of the Electro-Fenton Process

The main advantages of the Electro-Fenton process include the following: (1) on-site production of H_2O_2, (2) minimal risks associated with the handling the system, (3) no requirement for the transportation and storage of hydrogen peroxides, (4) continuous redevelopment of Fe^{2+} on the cathode, and (5) low iron sludge production (Nidheesh and Gandhimathi 2012; Özcan et al. 2008, 2009; Zhang et al. 2019).

The disadvantages are the following: (1) low yield of hydrogen peroxide; (2) low unit cell body throughput; (3) low current density; and (4) low conductivity (Özcan et al. 2009, Zhang et al. 2019).

7.7 ELECTRO-OXIDATION FILTRATION PROCESS

The anodic oxidation process or EO process consists of direct electron transfer from the pollutant to the anode via the production of hydroxyl radical ($OH^{\bullet}$). The $OH^{\bullet}$ produced at the surface of the anode is a highly reactive species and a strong oxidant, which allows the degradation of organic compounds. The lifetime of the $OH^{\bullet}$ radical is short, and, thus, $OH^{\bullet}$ radicals are present in a thin layer ($<1\ \mu m$) close to the anode surface (Trellu et al. 2018). Therefore, the oxidation reaction is not homogenous and occurs at or near the electrode surface. The side reaction (for instance, the oxygen evolution reaction) reduces the efficiency of current for oxidation. Moreover, the mass transport in the EO process is limited by diffusion through a thick stagnant boundary layer (approximately 10 to 100 μm).

Research investigations suggested that the use of porous electrodes in a flow-through configuration might increase the electroactive surface area and reduce the mass transfer/diffusion limitation (Trellu et al. 2018). Reactive electrochemical membranes (REMs) are a novel technology because they simultaneously filter and electrochemically oxidize organic pollutants. The REMs utilize low direct current and provide a highly enhanced mass transport of pollutants and a high electrochemically active surface area (Trellu et al. 2018).

The electrochemical activity of the REM process is greatly impacted by the morphology and physicochemical properties of the membrane material. Currently, the REM process involves the use of a carbon-based membrane and a ceramic membrane for EO. The conductive carbon material used for the synthesis of REM membranes includes graphene, graphite, and multiwall carbon nanotubes. The ease of ceramic processing aids in the pilot-scale fabrication of REMs with the added advantage of regulating the pore size (Trellu et al. 2018). The small pore diameters of the REM membranes allow a rapid mass transport of radicals, providing a high specific surface area for the electrochemical reaction. However, the corrosion of these membranes at a high electrode voltage potentially affects their treatment efficiency. Trellu et al. (2018) suggested that an ideal REM must be environmentally friendly, cost-effective, possess high reactivity, and have a long lifetime, as well as having a suitable porous structure to mitigate the limitation of diffusion. A study investigated the possibility of using TiO_2 anodes as REMs for water treatment (Zaky and Chaplin 2013). Ti oxides have an oxygen evolution potential and $^{\bullet}OH$ production rate similar to the BDD electrode. Ti oxide electrodes (Ti_4O_7) produce a physisorbed hydroxyl radical. Zaky and Chaplin (2013) reported that the electroactive surface area in an REM was 619 times the nominal geometric surface area. The electroactive surface area in the REM is the area available for reaction, and it is estimated by using electrochemical impedance spectroscopy and cyclic voltammetry (Trellu et al. 2018). Fouling and corrosion are the two factors that challenge REM feasibility at the pilot scale. Both ceramic- and carbon-based membranes experience fouling; however, research investigation has highlighted that ceramic membranes (such as TiO_2) are less susceptible to fouling because of their better hydrophilic properties compared with carbon-based membranes (Trellu et al. 2018). Although carbon-based REMs possess high permeability, high conductivity, and small pore size, corrosion at a high anode potential is the main challenge faced by these REMs. The high oxygen potential and monolithic structure of ceramic REMs help reduce corrosion and allow the electrogeneration of hydroxyl radicals. Although REMs offer increased mass transfer rates, an improvement in the reactivity of REMs is required.

7.8 MEMBRANE TECHNOLOGY COUPLED WITH THE ELECTROCHEMICAL PROCESS

Membrane technology is a promising technique for water and wastewater treatment as it offers many advantages like continuous operation, generation of high-quality permeate, low cost, and low energy requirements, easy scale-up, and no chemical requirement. In the last few decades, various membrane treatment processes have been developed for the decontamination of water,

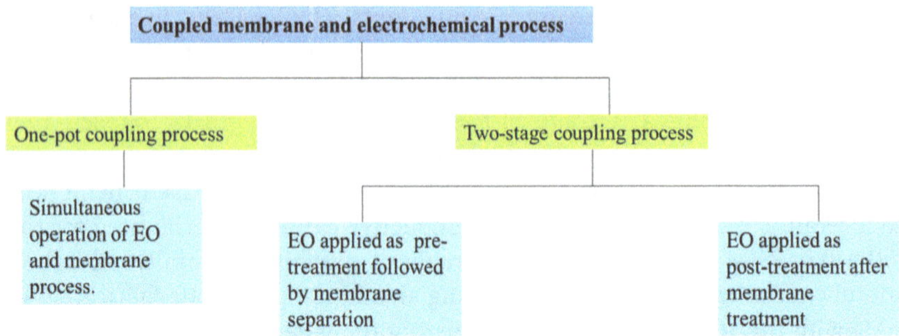

Figure 7-4. Modes of coupled electrochemical oxidation and membrane technology.

such as membrane bioreactors, membrane distillation, reverse osmosis, particle filtration, nanofiltration, ultrafiltration, microfiltration, and forward osmosis. As discussed in the previous section, the major limitation of the membrane process is the fouling of the membrane module because of cake formation, pore blockage, and adsorption (Pan et al. 2019). At the same time, the weak mass transfer and high energy consumption limit the application of EO.

The combination of membrane technology and EO have gained attention in the last few decades as it offers effective removal of toxic pollutants with low sludge production (Pan et al. 2019). The coupling of membrane technology with EO can be carried out in two modes, that is, through the one-pot coupling process and the two-stage coupling process. Figure 7-4 provides an overview of the coupling modes, and the details of each mode are provided subsequently.

7.8.1 One-Pot Coupling Process

The one-pot coupling process consists of the simultaneous treatment of water by membrane separation and EO in a single unit. This technique provides enhanced membrane filtration because of the presence of an electric field and helps in achieving high treatment efficiency and permeation flux. The in-situ EO and micro-flow disturbance reduce fouling and concentration polarization and increase the life span of the membrane. Moreover, the footprint of the one-pot technique is less because of its compact size as electrodes and membranes are placed in one single reactor (Pan et al. 2019). In this approach, nonconductive membranes are used for separation, such as polyamide membrane, polyvinylidene membrane, Al_2O_3 ultrafiltration membrane, and so on.

7.8.2 Two-Stage Coupling Process

The two-stage coupling process consists of two stand units for membrane treatment and EO. In the two-stage coupling process, the two integrations are studied. In the first integration, EO treatment is carried out before membrane separation. This pretreatment strategy is used for water containing high concentrations of organic pollutants (and, thus, can cause rapid membrane fouling). In the second approach, EO acts as a post-treatment strategy after membrane separation

(Pan et al. 2019). For the two-stage coupling, in general, conductive membranes such as the conductive polymer-based membrane, metal/metal oxide membrane, and carbon-based membranes are used. When EO is used as a pretreatment process in the two-stage coupling process, it effectively reduces the permeation resistance and membrane fouling. Moreover, EO is reported to be an effective post-treatment strategy for the treatment of concentrated stream generated after membrane separation.

7.8.3 Coupled Biological and Electro-Oxidation Process—Case Studies

The wastewater generated from the textile industry is composed of complex organic pollutants, detergents, dyes, surfactants, and inorganic salts, all of which constitute a risk to the environment (Moreira et al. 2017). A study reported that because of the high concentration of dyes in the waste stream of the textile industry (approximately 300 mg L^{-}) (Senthilkumar et al. 2012), wastewater affects the growth of aquatic biota and inhibits photosynthesis in the aquatic environment. The conventional wastewater treatment processes of the textile industry waste stream result in high sludge production or release of toxic chemicals in the environment as most of the dyes are believed to be carcinogenic (Moreira et al. 2017).

The biological treatment process has been proven to treat textile industry wastewater effectively; however, the discharge of a large quantity of sludge and the requirement of a longer treatment duration are the main drawbacks of the biological system. Thus, combined EO and biological processes for the treatment of textile industry wastewater were investigated by many studies (Moreira et al. 2017, Senthilkumar et al. 2012). Senthilkumar et al. (2012) studied the one-pot and two-stage coupling of EO and aerobic biodegradation of dye wastewater. The study utilized the aerobic oxidation process using selected dye degrading bacteria for the biological process and ruthenium coated on a titanium metal (RuO_x–TiO_x) anode for EO. The study reported that the two-stage coupled approach resulted in a color removal efficiency of 98.5% at an optimum current density of 6.84 A dm^{-2} and was more efficient than the one-pot technique, which yielded similar color removal efficiency but required an optimum current density of 10.2 A dm^{-2}. The study further showed the coupled process reduced energy consumption significantly compared with individual treatment and resulted in improved treatment efficiency (Senthilkumar et al. 2012).

Landfill leachate is the complex wastewater generated by the deposition of municipal and solid waste. The characteristics of landfill leachate vary greatly in terms of composition, moisture of solid waste, climate conditions, and age of the landfill. Landfill leachate is reported to contain a high concentration of organic matter [e.g., COD = 3,385 to 5,000 mg L^{-1}], hydrocarbons, plasticizers, phenols, humic and fulvic acid, and so on (Moreira et al. 2017). Because of high ammonia concentration, low phosphorus concentration, and low biological oxygen demand (BOD), the conventional biological treatment processes are inefficient for the treatment of landfill leachates. Membrane bioreactor (MBR) processes are extensively investigated for the treatment of landfill leachate because of their ability to remove the high concentration of ammonia by the biological nitrification process; however, the generation of an excess amount of sludge has led researchers to investigate the coupled process for landfill leachate treatment. In this regard, Zolfaghari et al. (2016) investigated the coupled sequential MBR process and EO for landfill leachate treatment. The study utilized the two-stage coupled approach and investigated EO both as pretreatment and post-treatment options before and after MBR treatment, respectively. The study showed that the application of EO as a post-treatment strategy reduces its effectiveness, and as a result, the toxicity of treated effluents increases because of the formation of organochlorinated compounds. The study suggested the application of EO before MBR treatment as the oxidation of high-molecular-weight organic compounds in EO increases TOC removal in the MBR. Further, the by-products of EO (organochlorinated compounds) are also efficiently removed in MBR treatment via adsorption and air stripping (Zolfaghari et al. 2016).

7.9 ADVANTAGES AND DISADVANTAGES OF ELECTRO-OXIDATION VERSUS ADVANCED OXIDATION PROCESSES

In many water and wastewater treatment processes, a number of contaminants exist that are difficult to remove by physical, chemical, or biological processes alone. In recent years, there has been an increasing concern over the presence of pharmaceutical medicines in drinking water and aquatic environments (Klavarioti et al. 2009). Pesticides get trapped in runoff from agriculture farms into freshwater bodies. Personal care products are typically washed away into drains and finally into the system to which they are connected. Landfill leachate contains toxic compounds that can contaminate groundwater sources. Such pollutants fall into the class of micropollutants, because their concentrations are very low (Schrab et al. 1993). Their size is one of the reasons why they are so hard to eliminate from water and wastewater by certain processes (Schrab et al. 1993). Removal of these micropollutants requires more powerful oxidation processes such as advanced oxidation processes (AOPs) (Sievers 2011). Advanced electrochemical oxidation processes involve the use of the hydroxyl radical to attack the pollutants present in wastewater. This radical can be produced at the electrode surface such as in the direct electro-oxidation processes or in bulk solution such as in the indirect electrolytic processes. The aim in either case is to try to mineralize pollutants. This technology should be used for the following reasons: (1) to ensure good quality of treated wastewater for recycling into the original production process, (2) to eliminating further water polishing steps, and (3) to avoid sludge generation and the need for using final sludge disposal methods and the resulting environmental impact. In real applications, the AOP has been used to mineralize pollutants that are difficult to oxidize by means of other processes, such as those existing in textile, tannery, petrochemical, and pharmaceutical wastewaters among others.

This section outlines the advantages and disadvantages of EO and advanced EO processes, and Table 7-1 summarizes the related information.

7.9.1 Advantages of the AOP

Rapid reaction rate. OH radicals have some of the fastest rates of reaction among the oxidants used for the treatment of water and wastewater because of their high oxidation potential and their nonselective nature. Fast reactions in such processes lead to much lower retention times than those in other conventional water treatment methods (Collins and Bolton 2016).

Small land area requirement. Because of the oxidation potential of the OH molecule, advanced oxidation process components do not need much land area for the development of the desired flow rate for the system (Crini and Lichtfouse 2019).

Mineralization of organic compounds. AOPs can transform the organic matter present within the water into stable inorganic compounds such as water, carbon dioxide, and salts. The extremely reactive nature of OH radicals signifies that these radicals will attack almost any organic compound or material without discriminating and, therefore, can eliminate many different pollutants in one reactor vessel (Ray 1998).

Can act as a disinfectant. When an AOP is used with UV disinfection, the oxidation potential of AOP systems makes them efficient for acting as a disinfection system for any pathogenic microorganism that may be present in water or wastewater (Mahy et al. 2019).

No sludge production. During the water and wastewater treatment process, the AOP does not transform contaminants into another phase. As a result, no sludge production occurs, unlike in the other treatment techniques, which generate solids like sludge that need to be filtered out and disposed of (Suty et al. 2004).

Does not generate concentrates. Treatment methods such as membranes cause increased concentrations of waste pollutants, because they simply separate clean water from contaminated compounds. The AOP directly reacts with these contaminants and converts them to harmless compounds, leading to reduced concentrations in the effluent.

No production of hazardous substance. One of the main concerns with chlorine disinfection is the formation of highly toxic by-products during the treatment process (Collins and Bolton 2016). To prevent the formation of these by-products, an extra dechlorination step is frequently required before something can be accomplished with the treated water. The biggest matter of concern is bromate formation and the presence of excess peroxide, but these can be effectively controlled with a well-designed AOP system (Crini and Lichtfouse 2019) as OH radicals can combine to produce water molecules.

7.9.2 Disadvantages of the AOP

High capital cost. Perhaps the biggest disadvantage of the AOPs is their cost. The most significant cost is the operating and maintenance cost that arises out of the cost of power supply and chemicals needed to operate the system (Collins and Bolton 2016).

Table 7-1. Advantages and Disadvantages of Electro-Oxidation and Advanced Electro-Oxidation Processes.

Advantages	*Disadvantages*
EO	
• Environmentally compatible process • Provides palatable, odorless water • No need for additional chemicals • Simple equipment, easy operation, and short retention time • Robustness, as the reaction can be stopped and started by just switching off the power supply. • Works at less temperature and pressure • Suitable for varieties of pollutants and possible to treat small- and large-quantity pollutants • Sludge formed is stable, easily settable, and easy to dewater. • Less sludge production • No secondary pollution • Can also be operated in remote areas with solar energy	• High operation and cost • High energy consumption • Sacrificial anode needs to be replaced regularly • Electricity cost may be high at some places • To apply this process, effluents need to be conductive in nature • Electrode fouling because of the deposition of pollutants on electrodes • Gelatinous hydroxide may lead to solubilization in some cases
Advanced EO	
• Rapid reaction rate • Less land area required • Can act as a disinfectant • Does not produce concentrates • Mineralization of organic compound occurs • No production of hazardous substances • Capable of degrading soluble and insoluble dyes from industrial wastewater • No formation of bromate by-products • Recycling of catalysts is possible • No sludge generation in the case of the H_2O_2/O_3 process	• High capital cost • Electric power is compulsory to operate the system • Complex chemistry behind the process • Removal of residual hydrogen peroxide is necessary • During the Fenton process, a low pH (>2.5) is needed. Lowering the pH can be a costly proposition • Ozone may form toxic by-products • During the Fenton process, iron sludge is generated

Source: Crini and Lichtfouse (2019), Liu et al. (2010), Tan Tien (2017), Williams (1971), Woisetschläger et al. (2013)

Büyüksönmez, F., R. Rynk, T. F. Hess, and E. Bechinski. 1999. "Occurrence, degradation and fate of pesticides during composting: Part I: Composting, pesticides, and pesticide degradation." *Compost Sci. Util.* 109 (12): 6570–6631.

Chu, R., J. Yan, S. Lian, Y. Wang, F. Yan, and D. Chen. 2004. "Shape-controlled synthesis of nanocrystalline titania at low temperature." *Solid State Commun.* 130 (12): 789–792.

Collins, J., and J. R. Bolton. 2016. *Advanced oxidation handbook*. Denver: American Water Works Association.

Compton, R. G., J. C. Eklund, and F. Marken. 1997. "Sonoelectrochemical processes: A review." *Electroanalysis* 9 (7): 509–522.

Crini, G., and E. Lichtfouse. 2019. "Advantages and disadvantages of techniques used for wastewater treatment." *Environ. Chem. Lett.* 17 (1): 145–155.

Duesterberg, C. K., and T. D. Waite. 2006. "Process optimization of Fenton oxidation using kinetic modeling." *Environ. Sci. Technol.* 40 (13): 4189–4195.

Feng, C.-H., F.-B. Li, H.-J. Mai, and X.-Z. Li. 2010. "Bio-electro-Fenton process driven by microbial fuel cell for wastewater treatment." *Environ. Sci. Technol.* 44 (5): 1875–1880.

Garbellini, G. S., G. R. Salazar-Banda, and L. A. Avaca. 2010. "Effects of ultrasound on the degradation of pentachlorophenol by boron-doped diamond electrodes." *Port. Electrochim. Acta* 28 (6): 405–415.

Gholikandi, G. B., and K. Kazemirad. 2018. "Application of electrochemical peroxidation (ECP) process for waste-activated sludge stabilization and system optimization using response surface methodology (RSM)." *Water Sci. Technol.* 77 (6): 1765–1776.

GilPavas, E., I. Dobrosz-Gómez, and M. Á. Gómez-García. 2018. "Optimization of sequential chemical coagulation—Electro-oxidation process for the treatment of an industrial textile wastewater." *J. Water Process Eng.* 22: 73–79.

Gogate, P. R., and A. B. Pandit. 2004. "A review of imperative technologies for wastewater treatment II: Hybrid methods." *Adv. Environ. Res.* 8 (3–4): 553–597.

Hai, F. I., K. Yamamoto, and K. Fukushi. 2007. "Hybrid treatment systems for dye wastewater." *Crit. Rev. Env. Sci. Technol.* 37 (4): 315–377.

Hassan, M., Y. Zhao, and B. Xie. 2016. "Employing TiO_2 photocatalysis to deal with landfill leachate: Current status and development." *Chem. Eng. J.* 285: 264–275.

He, H., and Z. Zhou. 2017. "Electro-Fenton process for water and wastewater treatment." *Crit. Rev. Env. Sci. Technol.* 47 (21): 2100–2131.

Hoffmann, M. R., S. T. Martin, W. Choi, and D. W. Bahnemann. 1995. "Environmental applications of semiconductor photocatalysis." *Chem. Rev.* 95 (1): 69–96.

Hurum, D. C., A. G. Agrios, K. A. Gray, T. Rajh, and M. C. Thurnauer. 2003. "Explaining the enhanced photocatalytic activity of Degussa P25 mixed-phase TiO_2 using EPR." *J. Phys. Chem. B* 107 (19): 4545–4549.

Ince, N. H., and G. Tezcanlí. 2001. "Reactive dyestuff degradation by combined sonolysis and ozonation." *Dyes Pigm.* 49 (3): 145–153.

Ismail, A. A., and D. W. Bahnemann. 2014. "Photochemical splitting of water for hydrogen production by photocatalysis: A review." *Sol. Energy Mater. Sol. Cells* 128: 85–101.

Jiang, J.-Q., Z. Zhou, and V. K. Sharma. 2013. "Occurrence, transportation, monitoring and treatment of emerging micro-pollutants in waste water—A review from global views." *Microchem. J.* 110: 292–300.

Klavarioti, M., D. Mantzavinos, and D. Kassinos. 2009. "Removal of residual pharmaceuticals from aqueous systems by advanced oxidation processes." *Environ. Int.* 35 (2): 402–417.

Kuvarega, A. T., R. W. M. Krause, and B. B. Mamba. 2015. "Evaluation of the simulated solar light photocatalytic activity of N, Ir co-doped TiO_2 for organic dye removal from water." *Appl. Surf. Sci.* 329: 127–136.

Lang, Z., M. Zhou, Q. Zhang, X. Yin, and Y. Li. 2020. "Comprehensive treatment of marine aquaculture wastewater by a cost-effective flow-through electro-oxidation process." *Sci. Total Environ.* 722: 137812.

Langford, C. H. 2012. "Photocatalysis—A special issue on a unique hybrid area of catalysis." *Catalysts* 2 (3): 327–329.

Li, Z., S. Yuan, C. Qiu, Y. Wang, X. Pan, J. Wang, et al. 2013. "Effective degradation of refractory organic pollutants in landfill leachate by electro-peroxone treatment." *Electrochim. Acta* 102: 174–182.

Liu, H., X. Zhao, and J. Qu. 2010. "Electrocoagulation in water treatment." In *Electrochemistry for the environment*, edited by C. Comninellis and G. Chen. 245–262. New York: Springer.

Lorimer, J. P., T. J. Mason, M. Plattes, and S. S. Phull. 2000. "Dye effluent decolourisation using ultrasonically assisted electro-oxidation." *Ultrason. Sonochem.* 7 (4): 237–242.

Mahy, J. G., C. Wolfs, A. Mertes, C. Vreuls, S. Drot, S. Smeets, et al. 2019. "Advanced photocatalytic oxidation processes for micropollutant elimination from municipal and industrial water." *J. Environ. Manage.* 250: 109561.

Malpass, G. R. P., D. W. Miwa, A. C. P. Miwa, S. A. S. Machado, and A. J. Motheo. 2007. "Photo-assisted electrochemical oxidation of atrazine on a commercial Ti/$Ru_{0.3}Ti_{0.7}O_2$DSA electrode." *Environ. Sci. Technol.* 41 (20): 7120–7125.

Martínez-Huitle, C. A., and S. Ferro. 2006. "Electrochemical oxidation of organic pollutants for the wastewater treatment: Direct and indirect processes." *Chem. Soc. Rev.* 35 (12): 1324–1340.

Martínez-Huitle, C. A., M. A. Rodrigo, I. Sirés, and O. Scialdone. 2015. "Single and coupled electrochemical processes and reactors for the abatement of organic water pollutants: A critical review." *Chem. Rev.* 115 (24): 13362–13407.

Mason, T., and J. P. Lorimer. 2002. *Applied sonochemistry: The uses of power ultrasound in chemistry and processing.* Practical Sonochemistry. Hoboken, NJ: Wiley.

Merényi, G., J. Lind, S. Naumov, and C. V. Sonntag. 2010. "Reaction of ozone with hydrogen peroxide (peroxone process): A revision of current mechanistic concepts based on thermokinetic and quantum-chemical considerations." *Environ. Sci. Technol.* 44 (9): 3505–3507.

Mills, A., and S. Le Hunte. 1997. "An overview of semiconductor photocatalysis." *J. Photochem. Photobiol., A* 108 (1): 1–35.

Moreira, F. C., R. A. R. Boaventura, E. Brillas, and V. J. P. Vilar. 2017. "Electrochemical advanced oxidation processes: A review on their application to synthetic and real wastewaters." *Appl. Catal., B* 202: 217–261.

Moussavi, G., and M. Aghanejad. 2014. "The performance of electrochemical peroxidation process for COD reduction and biodegradability improvement of the wastewater from a paper recycling plant." *Sep. Purif. Technol.* 132: 182–186.

Nidheesh, P. V., and R. Gandhimathi. 2012. "Trends in electro-Fenton process for water and wastewater treatment: An overview." *Desalination* 299: 1–15.

Oturan, M. A., I. Sirés, N. Oturan, S. Pérocheau, J.-L. Laborde, and S. Trévin. 2008. "Sonoelectro-Fenton process: A novel hybrid technique for the destruction of organic pollutants in water." *J. Electroanal. Chem.* 624 (1–2): 329–332.

Oturan, N., and M. A. Oturan. 2018. "Electro-Fenton process: Background, new developments, and applications." In *Electrochemical water and wastewater treatment*, edited by C. A. Martínez-Huitle, M. A. Rodrigo, and O. Scialdone, 193–221. Oxford, UK: Butterworth-Heinemann.

Özcan, A., Y. Şahin, A. S. Koparal, and M. A. Oturan. 2009. "A comparative study on the efficiency of electro-Fenton process in the removal of propham from water." *Appl. Catal. B* 89 (3–4): 620–626.

Özcan, A., Y. Şahin, and M. A. Oturan. 2008. "Removal of propham from water by using electro-Fenton technology: Kinetics and mechanism." *Chemosphere* 73 (5): 737–744.

Pan, Z., C. Song, L. Li, H. Wang, Y. Pan, C. Wang, et al. 2019. "Membrane technology coupled with electrochemical advanced oxidation processes for organic wastewater treatment: Recent advances and future prospects." *Chem. Eng. J.* 376: 120909.

Pérez, J. F., J. Llanos, C. Sáez, C. López, P. Cañizares, and M. A. Rodrigo. 2019. "Towards the scale up of a pressurized-jet microfluidic flow-through reactor for cost-effective electro-generation of H_2O_2." *J. Cleaner Prod.* 211: 1259–1267.

Pignatello, J. J., E. Oliveros, and A. MacKay. 2006. "Advanced oxidation processes for organic contaminant destruction based on the Fenton reaction and related chemistry." *Crit. Rev. Env. Sci. Technol.* 36 (1): 1–84.

Ratiu, C., F. Manea, C. Lazau, C. Orha, G. Burtica, I. Grozescu, et al. 2011. "Photocatalytically-assisted electrochemical degradation of *p*-aminophenol in aqueous solutions using zeolite-supported TiO_2 catalyst." *Chem. Pap.* 65 (3): 289–298.

Ray, A. K. 1998. "A new photocatalytic reactor for destruction of toxic water pollutants by advanced oxidation process." *Catal. Today* 44 (1–4): 357–368.

Rivera, M., M. Pazos, and M. A. Sanromán. 2009. "Improvement of dye electrochemical treatment by combination with ultrasound technique." *J. Chem. Technol. Biotechnol.* 84 (8): 1118–1124.

Särkkä, H., A. Bhatnagar, and M. Sillanpää. 2015. "Recent developments of electro-oxidation in water treatment—A review." *J. Electroanal. Chem.* 754: 46–56.

Sato, T., M. Qadir, S. Yamamoto, T. Endo, and A. Zahoor. 2013. "Global, regional, and country level need for data on wastewater generation, treatment, and use." *Agric. Water Manage.* 130: 1–13.

Schrab, G. E., K. W. Brown, and K. C. Donnelly. 1993. "Acute and genetic toxicity of municipal landfill leachate." *Water Air Soil Pollut.* 69 (1–2): 99–112.

Senthilkumar, S., C. A. Basha, M. Perumalsamy, and H. J. Prabhu. 2012. "Electrochemical oxidation and aerobic biodegradation with isolated bacterial strains for dye wastewater: Combined and integrated approach." *Electrochim. Acta* 77: 171–178.

Siddique, M., R. Farooq, Z. M. Khan, Z. Khan, and S. F. Shaukat. 2011. "Enhanced decomposition of reactive blue 19 dye in ultrasound assisted electrochemical reactor." *Ultrason. Sonochem.* 18 (1): 190–196.

Sievers, M. 2011. "Advanced oxidation processes." In *Treatise on water science*, edited by P. Wilderer, 377–408. Amsterdam, Netherlands: Elsevier.

Souvereyns, B., et al. 2013. "Hydrothermal synthesis of a concentrated and stable dispersion of TiO_2 nanoparticles." *Chem. Eng. J.* 223: 135–144.

Stoppa, G. 2012. "Ultrasound in electrochemical degradation of pollutants." In *Electrolysis*, edited by J. Kleperis and V. Linkov, 205–226. London: IntechOpen.

Suty, H., C. De Traversay, and M. Cost. 2004. "Applications of advanced oxidation processes: Present and future." *Water Sci. Technol.* 49 (4): 227–233.

Tan Tien, T. 2017. "Electrochemical water treatment technology in Viet Nam: Achievement & future development." *Sci. J. Chem.* 5 (6): 87–94.

Tauber, M. M., G. M. Guebitz, and A. Rehorek. 2005. "Degradation of azo dyes by laccase and ultrasound treatment." *Appl. Environ. Microbiol.* 71 (5): 2600–2607.

Tiwari, B., P. Drogui, and R. D. Tyagi. 2020. "Removal of emerging micro-pollutants from pharmaceutical industry wastewater." In *Current developments in biotechnology and bioengineering*, edited by S. Varjani, A. Pandey, R. D. Tyagi, H. H. Ngo, and C. Larroche, 457–480. Amsterdam, Netherlands: Elsevier.

Trellu, C., B. P. Chaplin, C. Coetsier, R. Esmilaire, S. Cerneaux, C. Causserand, et al. 2018. "Electro-oxidation of organic pollutants by reactive electrochemical membranes." *Chemosphere* 208: 159–175.

Turkay, O., S. Barişçi, and M. Sillanpää. 2017a. "E-peroxone process for the treatment of laundry wastewater: A case study." *J. Environ. Chem. Eng.* 5 (5): 4282–4290.

Turkay, O., Z. G. Ersoy, and S. Barışçı. 2017b. "Review—The application of an electro-peroxone process in water and wastewater treatment." *J. Electrochem. Soc.* 164 (6): E94–E102.

Wang, H., S. Yuan, J. Zhan, Y. Wang, G. Yu, S. Deng, et al. 2015a. "Mechanisms of enhanced total organic carbon elimination from oxalic acid solutions by electro-peroxone process." *Water Res.* 80: 20–29.

Wang, J., B. He, and X. Z. Kong. 2015b. "A study on the preparation of floating photocatalyst supported by hollow TiO_2 and its performance." *Appl. Surf. Sci.* 327: 406–412.

Wang, J., and J. H. Jiang. 2007. "Experimental study on laundry wastewater treatment by ultrafiltration–ultraviolet disinfection process for recycling and reuse." *Chem. Eng.* 35 (1): 56–59.

Williams, T. R. 1971. "Fundamentals of analytical chemistry (Skoog, Douglas A.)." *J. Chem. Educ.* 48 (1): A48.

Woisetschläger, D., B. Humpl, M. Koncar, and M. Siebenhofer. 2013. "Electrochemical oxidation of wastewater—Opportunities and drawbacks." *Water Sci. Technol.* 68 (5): 1173–1179.

Yao, W., Q. Qu, U. von Gunten, C. Chen, G. Yu, and Y. Wang. 2017. "Comparison of methylisoborneol and geosmin abatement in surface water by conventional ozonation and an electro-peroxone process." *Water Res.* 108: 373–382.

Zaky, A. M., and B. P. Chaplin. 2013. "Porous substoichiometric TiO_2 anodes as reactive electrochemical membranes for water treatment." *Environ. Sci. Technol.* 47 (12): 6554–6563.

Zhang, M.-h., H. Dong, L. Zhao, D.-x. Wang, and D. Meng. 2019. "A review on Fenton process for organic wastewater treatment based on optimization perspective." *Sci. Total Environ.* 670: 110–121.

Zhao, G., S. Shen, M. Li, M. Wu, T. Cao, and D. Li. 2008. "The mechanism and kinetics of ultrasound-enhanced electrochemical oxidation of phenol on boron-doped diamond and Pt electrodes." *Chemosphere* 73 (9): 1407–1413.

Zolfaghari, M., K. Jardak, P. Drogui, S. K. Brar, G. Buelna, and R. Dubé. 2016. "Landfill leachate treatment by sequential membrane bioreactor and electro-oxidation processes." *J. Environ. Manage.* 184: 318–326.

CHAPTER 8

Environmental Applications of Electro-Coagulation Processes

Bhagyashree Tiwari, Anita Talan,
Patrick Drogui, R. D. Tyagi

8.1 INTRODUCTION

One of the major challenges that mankind is facing today is to provide clean water to the growing population around the world. The need for clean water is critical as water bodies are being constantly polluted because of the indiscriminate discharge of industrial effluents as well as other anthropogenic activities and other natural processes. Some unknown geochemical processes have contaminated groundwater with arsenic in many countries. This problem is present not only in developing countries but also in developed countries such as the United States. The reuse of wastewater has become an absolute necessity because clean water is not found everywhere, leading to an urgent need to develop innovative and inexpensive techniques for the treatment of water and wastewater.

To date, a wide range of wastewater treatment techniques are known, including biological processes for nitrification, denitrification, and phosphorus removal as well as physiochemical processes (e.g., filtration, air stripping, ion exchange, chemical precipitation, chemical oxidation, carbon adsorption, ultrafiltration, reverse osmosis, electrodialysis, volatilization, and gas stripping), which often necessitates the use of chemical reagents. Apart from these, very promising techniques based on electrochemical technology are being developed and are improved in such a way that they do not require the addition of chemicals. These techniques include electro-coagulation (EC), electroflotation, and electrodecantation (Emamjomeh and Sivakumar 2009). Of these, EC has seen profitable commercialization but received less scientific attention. EC has the potential to eliminate the disadvantages of classical treatment techniques. The advantages of EC include the following:

- It requires simple equipment and is easy to operate.
- After EC treatment, the water is clear, colorless, odorless, and palatable.

- It is a low-sludge-producing technique.
- Sludge formed by EC is easily settable and is de-watered as it is composed of metallic oxides and hydroxides.
- Flocs formed are similar to chemical flocs, except that electrocoagulate flocs are larger and acid-resistant, contain less bound water, show more stability and, therefore, can be easily separated by filtration.
- It produces less total dissolved solids (Meunier et al. 2006) compared with other chemical treatments, leading to low water recovery cost.
- This process has the advantage of removing the smallest colloidal particles as the applied electric field sets them in faster motion facilitating coagulation.
- This process avoids the use of chemicals, and, therefore, the problem of neutralizing them does not arise.
- Pollutants are carried on top of the solution (because of the produced gas bubbles), from where pollutants can be easily collected, concentrated, and removed.
- Less maintenance is required as an electrolytic cell is controlled electrically with no moving parts.
- Solar energy can also be used for running the process if the process is used in rural areas where electricity is scarce.

Numerous studies have investigated the potential of EC for the removal of organic and inorganic contaminants (heavy metals, ammonia, pharmaceuticals, hydrocarbons, etc.) from water and wastewater (Al-Shannag et al. 2015, Banerji and Chaudhari 2016, Hakizimana et al. 2017, Nidheesh and Singh 2017, Song et al. 2017). For instance, Akbal et al. (2011) investigated the removal of copper, chromium, and nickel from metal plating wastewater using EC. This chapter provides a brief discussion of the treatment of wastewater originating from various sources using EC. Sections 8.2 and 8.3 discuss the treatment of metal ion and anionic species such as fluoride and phosphate from wastewater. Subsequently, the procedure of EC for landfill leachate treatment and agroindustrial wastewater is provided in Sections 8.4 and 8.5. Following agroindustrial wastewater treatment, the treatment of wastewater originating from the cheese whey industry, slaughterhouses, restaurants, textile industry, and laundry is discussed in Sections 8.6 to 8.10, respectively. Section 8.11 provides the details of the implementation of the EC process for drinking water treatment. Finally, a summary of the chapter is provided in Section 8.12.

8.2 REMOVAL OF HEAVY METALS FROM WASTEWATER

The growth and development of various industrial sectors such as paper, fertilizer, metal plating, batteries, tanneries, and so on has led to an increase in the use of heavy metals for various activities. Industrial activities such as mining, mineral processing, and extractive metallurgical operations result in a release of heavy metals into the environment via the discharge of wastewater. Arsenic, zinc, chromium, nickel, cadmium, and so on are frequently found in wastewater. These compounds are resistant to biodegradation because of their high atomic mass, complex chemical structure, and high solubility in water. Moreover, these persistent toxic compounds have a bioaccumulative tendency, which raises concern over their detrimental effects on living organisms. These heavy metals are carcinogenic and cause various health problems. For instance, lead and chromium affect the central nervous system and human physiology, cause skin irritation and kidney and lung carcinoma, and damage the liver and reproductive system (Fu and Wang 2011).

To remove these heavy metals from wastewater, numerous treatment processes have been investigated (Al-Shannag et al. 2015). These treatment processes include chemical precipitation, adsorption, biosorption, membrane filtration, reverse osmosis, ion exchange, and electrochemical treatment. Among various treatment processes, the precipitation of heavy metals by adjusting the pH of wastewater using chemical coagulants to form insoluble hydroxide is a commonly used effective treatment process. In this precipitation process, aluminum and iron salts are used to remove heavy metals as colloidal matter. However, the formation of by-products as secondary pollutants and the high cost associated with chemicals limit the chemical coagulation process for sustainable implementation. In contrast, EC is a more reliable and effective treatment process as it does not require the addition of chemical coagulants for precipitation and generates high-quality effluents with low sludge production. In EC, an electric field is applied between the cathode and the anode to produce hydrogen ions (H^+) and O_2; metals are oxidized to their cationic states at the anodic side in water, and, at the same time, water is reduced at the cathodic side. This oxidation and reduction process generates flocculating agents in water, which adsorb and coagulate pollutants and then remove them from water.

Many studies demonstrate the successful application of EC for the removal of heavy metals from wastewater emanating from various sources such as municipalities, the pulp and paper industry, the textile industry, and so on. This section discusses the removal of the most abundant and toxic heavy metals such as arsenic and zinc using EC.

8.2.1 Arsenic

Arsenic is the 53rd most abundant compound in the earth's crust. Arsenic exists in four oxidation states, namely, metallic, arsenide, arsenite, and arsenate. Arsenic and its derivatives are used for various purposes such as (1) for the production of glass, paper, semiconductors, antifouling paints, and leather preservatives; (2) as wood preservatives, pesticides, and insecticides; and (3) for the processing of dyes and pigments and in the tanning process (Nidheesh and Singh 2017). World Health Organization (WHO) identifies arsenic as a Group 1 human carcinogen. Because of its carcinogenic and toxic nature, the regulatory bodies of various countries and WHO recommend a permissible limit of up to 10 μg L^{-1} for arsenic concentration in drinking water (Nidheesh and Singh 2017). The removal of arsenic contaminants from water is governed by separation techniques because the degradation of arsenic is impossible. Moreover, the removal of arsenite (trivalent arsenic) compound can be done only after its oxidation into arsenate (pentavalent) compounds. Thus, numerous oxidation treatment processes (e.g., chemical oxidation, air oxidation) coupled with separation techniques (coagulation, sorption, membrane filtration) have been used around the globe for arsenic elimination from water. EC has been proven to be an efficient treatment process for the elimination of arsenic from water bodies.

In water, arsenic manifests in various forms; it gets protonated at acidic conditions, whereas it exists in deprotonated form in a basic medium. The series of reactions for the deprotonation of arsenite and arsenate is as follows:

$$As(OH)_3 \longrightarrow H^+ + AsO(OH)_2^- \tag{8-1}$$

$$AsO(OH)_2^- \longrightarrow H^+ + AsO_2(OH)^{2-} \tag{8-2}$$

$$AsO_2(OH)^{2-} \longrightarrow H^+ + AsO_3^{3-} \tag{8-3}$$

$$AsO(OH)_3 \longrightarrow H^+ + AsO_2(OH)_2^- \tag{8-4}$$

$$AsO_2(OH)_2^- \longrightarrow H^+ + AsO_3(OH)^{2-} \tag{8-5}$$

$$AsO_3(OH)^{2-} \longrightarrow H^+ + AsO_4^{3-} \tag{8-6}$$

Iron and aluminum electrodes are commonly used electrodes for arsenic removal in EC because of their low cost, easy availability, and higher efficiency for arsenic decontamination. These two electrodes (Al and Fe) can remove 99.9% of arsenate from water. Apart from these two electrodes, titanium, magnesium, zinc, copper, and brass electrodes have also been demonstrated to eliminate arsenate from water; however, their removal efficiency is less than that of an iron electrode. To perform complete oxidation of arsenite to arsenate, either the addition of chemical oxidants or the anodic oxidation (AO) process has been studied.

Chemical oxidants such as hypochlorite are added before or during the EC process, which significantly govern conversion. However, the amount of oxidants required is very high, whereas the AO process generates hydroxyl radicals with a high oxidation potential process, which converts arsenite to arsenate easily without any chemical compound. EC is more efficient than chemical coagulation because most arsenate species are insoluble at a higher pH and the addition of chemical coagulants decreases the pH of the water, which does not favor the generation of aluminum hydroxide precipitate or insoluble ferric hydroxide. However, in EC, the pH of the solution increases with electrolysis time, in turn, forming an insoluble arsenate complex and speeding up arsenic removal.

It can be interesting to combine EC with microfiltration techniques to remove arsenic from drinking water in the continue mode operation. Molgora et al. (2013) used this approach to study the effectiveness of such a combination to efficiently remove arsenic (As) from contaminated water (with an initial concentration of 100 μg As L^{-1}) and to reach the levels (10 and 25 μg As L^{-1}) recommended by WHO and the Health Ministry of Mexico for drinking water, respectively.

8.2.2 Zinc and Copper

Zinc is the 25th most abundant element in the earth's crust and an essential element needed by the human body in a small quantity as a cofactor for certain metabolic reactions. Salts of zinc sulfide and oxide are used for making ceramics, white paints, rubber, wood preservers, and dying fabrics. Water streams flowing from zinc chemical industries are the primary source of zinc contamination in water bodies. Zinc concentrations are increasing unnaturally. The accepted concentration of Zn in drinking water is 0.5 mg L^{-1} (Mansoorian et al. 2014). Most zinc enters the environment during industrial activities such as mining, coal and waste combustion, and steel processing. Some soils are completely contaminated with zinc because either zinc-contaminated sewage sludge from the industrial zone may be used as a fertilizer or zinc has to be mined or refined in these soils.

In general, zinc settles down in the bottom of water bodies, which causes bioaccumulation of zinc in aquatic organisms. The toxicity of zinc in water is influenced by water hardness and pH. In zinc of lower toxicity, water has a higher hardness and a lower pH, and in zinc of higher toxicity, water has a lower hardness and a higher pH. In general, zinc and its salts have high acute and chronic toxicity to the aquatic system in polluted waters. Numerous physical methods such as reverse osmosis, ion exchange coagulation, solvent extraction, and sequestration have their own limits. To remove heavy metals, the EC treatment process involves various physical and chemical processes such as cathode reduction, electrophoresis migration, anode oxidation, coagulation, and surface absorption. EC helps achieve an efficiency of approximately 98.3% removal of zinc at an acidic pH of 4 with a current density of 6 mA cm^{-2} using iron rods, whereas only 95.8% efficiency of zinc removal is reported with a stainless electrode at a current density of 8 mA cm^{-2} (Mansoorian et al. 2014). The iron electrode is more effective than stainless steel because of the fact that ferrohydroxide compounds generated from the anode are nonsoluble in a wide pH range.

Meunier et al. (2004) used an EC system to simultaneously remove Cu and Zn in leachate from sewage sludge. Sludge leachate was an effluent generated by a chemical leaching process for metal extraction from sewage sludge. Sludge was previously acidified to a pH of 2.0 using H_2SO_4. Subsequently, $FeCl_3$ and H_2O_2 were successively mixed with the sludge in appropriate concentrations required for Cu and Zn solubilization. After 3 to 4 h of leaching, the leached sludge was conditioned by adding a cationic polymer. Sludge was then dehydrated by using a deep bed filter, followed by centrifugation. The acid leachate initially contained 29.2 ± 2.9 mg Cu L^{-1} and 7.28 ± 0.59 mg Zn L^{-1}. The sludge leachate was then subjected to treatment using an EC cell comprising mild steel electrodes operated at a current intensity of 0.8 A for 30 min. The removal of Cu and Zn from the leachate varied from 92.4% to 98.9% for Cu and from 69.8% to 76.6% for Zn. An amount of 55 kg TDS^{-1} (total dissolved solid) of metallic sludge was produced. The EC system involved a total cost of $CAN 21.24 and 13.09 per ton of dry sludge treated, including only energy consumption and metallic sludge disposal.

8.3 REMOVAL OF ANIONIC SPECIES FROM WASTEWATER

8.3.1 Fluoride

Fluoride is an essential trace element required by the human body for normal bone function and for the formation of dental enamel. The presence of fluoride in groundwater is due to the slow and steady dissolution of fluorine-containing rocks. Fluoride is widely used in coal-fired power stations, semiconductor manufacturing, ceramic production, and aluminum smelters. Therefore, the wastewater emanating from these industries may cause environment pollution and risk to human health. High fluoride concentration in the human body causes brain disorder, infertility, arthritis, thyroid disorder, and dental fluorosis. The WHO-recommended permissible amount of fluoride in drinking water is 1.5 mg L^{-1} (Behbahani et al. 2011). The fluoride treatment process is divided into two: precipitation and sorption. Lime precipitation is a widely used technique for fluoride decontamination from water. However, lime precipitation can reduce fluoride concentration only up to 1 mg L^{-1}. The second process, that is, removal via sorption, is an expensive one because of the cost associated with column regeneration (Emamjomeh et al. 2011).

Researchers have explored the potential of EC for fluoride decontamination as EC produces destabilizing flocculants that aid in pollutant removal by charge neutralization. In general, an aluminum electrode is used for fluoride removal. If a high concentration of fluoride is present in water, Al^{3+} forms the AlF_6^{3-} ion, which is further converted into insoluble Na_3AlF_6 because of the presence of sodium ion in electrolyte solution (Vasudevan et al. 2011). The reaction for fluoride removal in EC is as follows:

Cathode:

$$2H_2O + 2e^- \longrightarrow H_2(g) + 2OH^- \quad (8\text{-}7)$$

Anode:

$$Al \longrightarrow Al^{3+} + 3e^- \quad (8\text{-}8)$$

$$Al^{3+} + 6F \longrightarrow AlF_6^3 \quad (8\text{-}9)$$

$$AlF_6^3 + 3Na^+ \longrightarrow Na_3AlF_6 \quad (8\text{-}10)$$

In electrolytes:

$$Al^{3+}(aq) + 3H_2O \longrightarrow Al(OH)_3 + 3H^+ \tag{8-11}$$

$$nAl(OH)_3 \longrightarrow Al_n(OH)_{3n} \tag{8-12}$$

8.3.2 Phosphate

Eutrophication is the leading cause of impairment of many water bodies, and it occurs because of the presence of the high concentration of phosphates in water. The use of fertilizers in the agriculture and animal husbandry sectors is the main factor contributing to the accumulation of phosphates in the environment. Phosphorus exists in three different forms in the environment, namely orthophosphate, polyphosphate, and organic phosphate. Among these three forms, orthophosphate is the most abundant form of phosphorus found in wastewater. The range of phosphorus concentration in wastewater emanating from domestic areas and industrial plants is around 3 to 15 mg L^{-1}. However, the permissible limit for phosphorus concentration in the discharge of treated effluents is 1 mg L^{-1} or much lower (e.g., <0.1 mg L^{-1} in the areas of the Great Lakes).

Conventionally, chemical treatment using aluminum sulfate and ferric chloride is a widely used technique for phosphorus precipitation. The disadvantages associated with chemical precipitation include the requirement of neutralization of effluents and sludge handling/disposal because of the lack of other efficient treatment processes. Although biological treatment is less expensive than chemical precipitation; biological treatment requires a transfer of the orthophosphate ion from liquid to sludge, which results in additional costs for sludge treatment.

During the removal of phosphorus by EC, the ions of aluminium or iron generated from their respective electrodes interact with the anionic species of phosphate present in water, such as PO_4^{3-}, HPO_4^{2-}, and $H_2PO_4^-$, forming insoluble complexes. EC efficiently removes total phosphorus P_{tot} of around 96% (Tran et al. 2012). The chemical reactions that occur between iron and phosphorus ions in EC are as follows:

$$3Fe^{2+} + 2PO_4^{3-} \longrightarrow Fe_3(PO_4)_2 \tag{8-13}$$

$$Fe^{3+} + PO_4^{3-} \longrightarrow FePO_4 \tag{8-14}$$

8.4 TREATMENT OF LANDFILL LEACHATE

Landfilling, composting, and incineration are widely used treatment strategies for the management of solid waste generated from various sources. Every year, approximately 335 million tons of solid waste are disposed by landfill (Mandal et al. 2017), which generates landfill leachate (LL) with highly complex compositions (e.g., heavy metals, polyaromatic hydrocarbons, organic and inorganic fractions, xenobiotic compounds). The crucial challenges associated with LL treatment are the varied composition and toxicity exhibited by ammonia, a diverse concentration of pollutants, and the presence of humic and fulvic acids. In general, the biological treatment of LL for the removal of ammonia and membrane filtration are globally used; however, the presence of inhibitory compounds such as polychlorinated biphenyls and polyaromatic hydrocarbons inhibits microbial activity and, thus, ultimately affects treatment efficiency. An efficient LL treatment requires the treatment techniques to be able to withstand the varied concentrations, composition, and toxicity of LL and to generate high-quality permeate.

It is quite challenging to treat LL by using only one treatment process. Therefore, a combined approach is required to efficiently treat LL and to comply with the discharge standards. Recently, EC and biofiltration (BF) processes were sequentially used to treat LL (Dia et al. 2017). EC was able to remove $37 \pm 2\%$ of the initial total COD. A fractionation of organic compounds showed that EC was particularly efficient in removing insoluble COD and humic acids. In addition, other pollutants such as turbidity, true color, Zn, and phosphorus were significantly reduced by EC with $82 \pm 2.7\%$, $60 \pm 13\%$, $95 \pm 2.6\%$, and $82 \pm 5.5\%$ of removal, respectively. The subsequent treatment by the BF process led to a complete removal of ammonia (>99% removal) and a partial removal of dissolved organic compounds ($42 \pm 7\%$ of COD removal). Therefore, the EC/BF hybrid process can form the basis of a process capable of removing organic and inorganic pollutants from many refractory wastewaters (mature landfill leachates and industrial and municipal wastewaters).

Likewise, EC was used as a tertiary process for LL treatment (Dia et al. 2017). EC was employed to treat residual organic matter from an LL pretreated by an aerated biofilter system. Organic matter [humic acids (HA), fulvic acids (FA), and hydrophilic compounds (Hyl)] were fractionated to estimate the efficiency of EC on each fraction. Initial characterization of the biofiltrated LL showed that humic substances (HA + FA) represented nearly 90% of TOC. The effects of current densities, type of anode (aluminum versus iron), and treatment time on the efficiency of COD removal were investigated. The best COD removal efficiencies were recorded at a current density ranging between 8.0 and 10 mA cm^{-2} during a 20 min treatment. Under these conditions, 70% and 65% of COD were removed using aluminum and iron electrodes, respectively. Fractionation of organic matter after EC treatment revealed that HA was completely removed using either an aluminum or an iron anode. However, FA and Hyl fractions were partially removed, with the percentages varying from 57% to 60% and 37% to 46%, respectively. FA and Hyl removal were quite similar using either an aluminum or an iron anode. Likewise, a significant decrease in 254 nm absorbance was recorded (UV_{254} removal of 79% to 80%) using either type of anode. These results proved that EC is a suitable and efficient approach for treating the residual refractory organic matter from an LL previously treated by a biological system.

During EC, the formed gases accumulate at the electrode surface, which causes electrical resistance, resulting in decreased ionic transfer. To prevent electrical resistance caused by accumulation and an uneven distribution of gas bubbles, Niza et al. (2020) tested the efficiency of EC h by using vibration-induced electrode plates for LL treatment. The study showed that the vibration-induced electrode plates increased the treatment efficiency of EC for COD and color removal from LL. EC resulted in 90% of color and up to 35% of COD removal at a vibration intensity of 2.8 V and an applied current of 2.5 A in 35 min (Niza et al. 2020).

8.5 TREATMENT OF WASTEWATER FROM AGROFOOD INDUSTRIES

One of the major contributors to worldwide pollution is waste generated by agroindustries. The effluents generated are a hazard to the environment and require appropriate management approaches. An increase in negative consequences on the environment has made the environmental regulatory authorities across the globe set strict criteria for waste disposal from industries.

About 65% to 70% of the organic pollutants released in water bodies are from food and agroindustries such as sugar, dairy, fruit canning, distillery, meat processing, and pulp and paper. Wine production is one of the leading agroindustries in the Mediterranean countries and many other parts of the world exercising an increasing influence on the economy of these countries (Teh et al. 2016). Similarly, the pulp and paper industry is the most significant industry in many countries of the world, and its products are an important constituent of industrial trade in terms of the value of production. All agroindustries generate huge volumes of wastewater originating from several washing operations such as crushing, pressing, and cleaning of fermenters, containers, other equipment and surfaces. Olive oil production and its increasing use across the globe have gained fundamental economic importance. The surplus production of olive oil generates huge volumes of polluted wastewater, mainly known as palm oil mill effluent. Similarly, coffee-producing countries such as Brazil, Vietnam, and Colombia are producing final effluents that are highly toxic, with the potential to explode into a huge environmental problem, thus creating a need for their treatment with cost-effective technologies. The composition and concentration of different agrofood industries vary from low- to high-strength substrates. The main parameters of agrofood industrial wastewater, such as total solids (Morozesk et al. 2016), total nitrogen, biochemical oxygen demand (BOD), chemical oxygen demand (COD), and total phosphorus also vary based on the type of industry.

From the public health point of view, rules and regulations on treating and utilizing the components of wastewater quickly and efficiently are getting stricter nowadays. With the rapid development of technology, substantial research is being done to utilize agroindustry waste and optimize its disposal without affecting the environment. Anaerobic digestion is being considered as an environmentally friendly process of green biotechnology to treat agroindustrial waste. Carbon emission and carbon footprints of water utilities are also important issues for consideration. In this context, the prospective threat of the release of harmful gases and chemicals from the wastewater treatment plants of small and large agroindustries is real, and, therefore, it is highly essential to take steps to counter this threat(Shak and Wu 2014). For this purpose, the anaerobic treatment process is preferred over aerobic treatment because no aeration is required in an anaerobic process, and the methane gas generated is used as an energy source. This process yields balanced energy production, and the generated methane can replace fossil fuel to some extent. Therefore, the anaerobic process has been considered as a sustainable wastewater treatment technology that is cost-effective and environmentally friendly.

Drogui et al. (2008) investigated the treatment of wastewater originating from meat processing, cereal products, food beverages, and abattoir industries using an EC process with Fe and Al electrodes in monopolar and bipolar configuration. In the monopolar configuration, eight pieces of an iron or aluminum electrode were individually connected to power supply, whereas in the bipolar configuration, only two outmost electrodes were physically connected to power supply among the array of eight electrodes. In the initial study, maximum organic removal efficiencies of 85.1% and 86.3% were observed for abattoir wastewater using the iron and aluminum electrodes, respectively. A further study was performed using abattoir wastewater, and it was found that the aluminum electrode in the bipolar configuration was the best process configuration. Moreover, the main advantages of EC is that it does not require acidification and conductivity enhancement.

8.6 CHEESE WHEY WASTEWATER

Cheese whey wastewater (CWW) is a strong organic and saline effluent whose treatment is highly important. Approximately 145×10^6 tons of wastewater per year is produced from the dairy industries around the world. Whey wastewater originates from the dairy industry and mainly contains lactose (44 to 52 g L^{-1}), proteins (6 to 8 g L^{-1}), and mineral salts (4 to 9 g L^{-1}) (Guven et al. 2008). However, its composition varies because of the percentage of raw milk used, the amount of nonvalorized cheese used, and the volume of cleaning water (Carvalho et al. 2013). According to industrial data, CWW is four times the volume of processed milk. The COD and BOD of CWW range from 4,000 to 5,000 mg L^{-1} and from 1,000 to 3,000 mg L^{-1}, respectively (Kushwaha et al. 2011). Because of the high pollution load of CWW, various biological and physiochemical methods for CWW treatment were investigated, and it was found that anaerobic digestion was the most effective and widely used treatment process. However, high concentrations of fats and proteins in CWW inhibit the treatment process and ultimately result in low removal efficiency. As an answer to this, aerobic treatment of CWW in sequence batch reactors helped achieve high efficiencies in the removal of COD, nitrogen, and phosphorus of approximately 80%, 96%, and 98%, respectively (Kushwaha et al. 2011). However, aerobic treatment has disadvantages such as long treatment time and large volume of sludge production and these factors limit its application. Moreover, the aerobic treatment process requires nutrients as a supplement for the growth of microorganisms because most CWW is deproteinated.

EC offers many advantages over biological processes for the treatment of CWW, such as easy operation with a low sludge production rate and the use of simple equipment. Moreover, the application of response surface methodology (RSM) for an experimental design of the EC process helps in the optimization of treatment conditions. Guven et al. (2008) optimized electrochemical treatment using the RSM approach. Their study investigated the treatment of CWW using an iron electrode at 25 °C with an applied voltage of 11.29 V. A maximum COD removal efficiency of 53.32% was achieved in 8 h of reaction time in a continuous mode.

8.7 SLAUGHTERHOUSE WASTEWATER

A slaughterhouse generates large amounts of wastewater from the slaughtering process and cleaning of the process. Therefore, it is important to reuse this water and recover value-added products from meat processing wastewater from the agribusiness point of view. Treatment methods used for slaughterhouse wastewater are comparable with those used in municipal wastewater treatment, including primary, secondary and tertiary treatment. The main purpose of the primary treatment methods is to separate solids and liquids and to remove BOD by approximately 30%. Mainly, screeners, sieves, and strainers are used for primary treatment, but treatment methods such as homogenization, equalization, and flotation are also used for primary treatment (Kobya et al. 2006). After primary treatment, the effluents are treated using other methods that are basically divided into four categories: physicochemical treatment, advanced oxidation processes (AOPs), biological treatment, and combined processes (Sahu et al. 2014).

Physicochemical treatment methods. The most commonly used physicochemical treatment methods are (1) coagulation-flocculation and sedimentation, (2) EC, (3) dissolved air flotation (DAF), and (4) membrane processes, and these are described following. In a coagulation process, the colloidal particles of slaughterhouse wastewater are grouped into larger particles known as flocs. The colloidal particles are negatively charged, which makes them stable and resistant to aggregation. Therefore, positively charged coagulants are used to destabilize the colloidal particles to form flocs, which are then removed by sedimentation. Inorganic metal-based coagulants are mostly used because of their high removal efficiencies; some examples are aluminum sulfate, aluminum chlorohydrate, ferric chloride, ferric sulfate, and polyaluminum chloride (Bustillo-Lecompte and Mehrvar 2015).

EC is employed on a large scale as it is a cost-effective technology for removing organics, heavy metals, and pathogens by inducing chemical current without chemical addition. Fe^{3+} and Al^{3+} are used as different electrode materials, and other electrode types, including P and TiO_2, can interact with H^+ or OH^- ions in an acidic or alkaline medium, respectively. EC has high removal efficiencies up to 96%. Slaughterhouse wastewater contains varied and high amounts of organic matter (proteins, blood, fat, etc.). The EC process was tested for the treatment of slaughterhouse wastewater using either mild steel or aluminum electrodes arranged in a bipolar (BP) or monopolar configuration system (Asselin et al. 2008). The results showed that the best performance was obtained using the mild steel BP electrode system operated at a current intensity of 0.3 A, through 60 or 90 min of treatment. Under these conditions, removal efficiencies of 86 $\pm$ 1% and 99 $\pm$ 1% were recorded for BOD and oil and grease, respectively, whereas 50 $\pm$ 4% and 82 $\pm$ 2% removal efficiencies were recorded for soluble COD and total COD, respectively. EC is also efficient for decolorization (red color) and clarification of effluents. Removal efficiencies of 89 $\pm$ 4% and 90 $\pm$ 4% were recorded for total suspended solids and turbidity, respectively. EC operating under optimal conditions involves a total cost of \$US 0.71 m^{-3} of treated PS effluents. This cost includes energy and electrode consumption, chemicals, and sludge disposal.

DAF is a method of separation of solids and liquids by introducing air. Fats and grease particles are carried to the surface, creating a sludge cover that can be continuously removed using scum scraping. Flocculants and blood coagulants are added to enhance the effectiveness of the DAF treatment for COD and BOD removal.

Nowadays, membrane processes are in great demand for meat processing effluents and are becoming an alternate treatment method. Different membrane processes such as microfiltration, nanofiltration, reverse osmosis, and ultrafiltration are in current use. The rate of efficiency of using membranes is 90% and above for removing particulates, colloids, macromolecules, organic matter, and pathogens. However, for nutrient removal, membrane technology is not effective, and, therefore, it is coupled with conventional processes (Bustillo-Lecompte and Mehrvar 2015). Another drawback of using membrane technology is the fouling of membranes while treating high-strength wastewater because of the formation of biofouling layers on the membranes, which leads to limiting the permeation rate.

Biological processes. Primary treatment and physicochemical processes do not treat slaughterhouse wastewater completely to a standard set by regulations. Thus, secondary treatment is required for the removal of leftover waste after primary treatment. Biological processes like anaerobic treatment and aerobic treatment and those with trickling filters, activated sludge bioreactors, and wetlands are used for organic and nutrient removal with efficiencies up to 90%. Here, we discuss the following biological treatment methods: (1) anaerobic treatment; (2) aerobic treatment; and (3) constructed wetlands.

Anaerobic treatment is highly desirable because of its effectiveness in treating highly concentrated industrial effluents. In an anaerobic process, organic compounds are degraded by anaerobic bacteria in the absence of oxygen to produce CO_2 and CH_4. Anaerobic systems have the advantages of realizing low sludge production, minimum energy requirements, potential resource recovery, and high COD removal. The anaerobic devices used in the process for the treatment of meat processing effluents include an anaerobic baffled reactor (Goris and Diekert 2016), anaerobic digester, anaerobic filter, anaerobic lagoon, septic tanks (ST), and an upflow anaerobic sludge blanket. However, in anaerobic treatment, a complete stabilization of organic compounds is difficult due to the high organic strength of slaughterhouse wastewater. Therefore, an additional treatment step is required to eliminate the organics, nutrients, and pathogens after treatment (Bustillo-Lecompte and Mehrvar 2015). The process needs a large space and a higher residence time to achieve complete treatment efficiency, and these factors affect the economic feasibility of anaerobic treatment. Thus, a combined anaerobic and aerobic process is essential to achieve maximum efficiency for the treatment.

Aerobic treatment is basically employed to remove nutrients after primary treatment. The oxygen requirement and treatment time are dependent on the strength of the slaughterhouse waste after anaerobic treatment. Aerobic treatment is an advantageous treatment process because of low odor production, fast biological growth rate, fast adaptation to temperature fluctuations, and loading rate. The operating costs of aerobic systems are higher than those of anaerobic systems because of the high maintenance and energy requirements for artificial oxygenation. For aerobic treatment, different devices for unit operations are needed for slaughterhouse wastewater treatment, such as rotating biological contractors and sequencing batch reactors.

Constructed wetlands follow the same degradation mechanisms of natural wetlands for water contamination, integrating biological and physicochemical processes from the interaction of flora, fauna, soil, microorganisms, and atmosphere for the adsorption, biodegradation, filtration, photo-oxidation, and sedimentation of organics and nutrients. Organic and nutrient removal efficiencies are achieved in the range of 78% to 99%. Constructed wetlands are simple wastewater treatment systems with low process and maintenance costs with few negative impacts on the environment.

Advanced oxidation processes. Advanced oxidation processes (AOPs) are recommended for primary and secondary treatment for maximum removal of organics and pathogens. These processes are diverse and include gamma radiation, ozonation, ultrasound technology, UV/H_2O_2, ultraviolet radiation (UV), ozone, and photocatalysis for the oxidation and degradation of organic matter present in slaughterhouse wastewater. AOPs are used to inactivate pathogens without adding extra chemicals and prevent the formation of hazardous by-products similarly to other disinfection methods such as chlorination. These processes have high reaction rates and very short treatment times. Photo-oxidation using UV/H_2O_2 and photocatalysis using photo-Fenton-based-processes are mostly used for slaughterhouse wastewater treatment. However, these processes are highly expensive if implemented alone (Bustillo-Lecompte et al. 2016). Thus, these processes are combined with biological processes for maximum efficiency.

Combined processes. To achieve maximum efficiency for pathogen and organic removal, it is necessary to implement combined processes that are economically and operationally beneficial for slaughterhouse wastewater treatment. These combined processes couple the advantages of various technologies to treat high-strength industrial wastewaters. A study investigated the coupling of EC and electroflotation (EF) for the treatment of poultry slaughterhouse wastewater (Paulista et al. 2018). The EC to EF ratio during the treatment was maintained by current reversal between aluminium and graphite electrodes. The study reported that the EC–EF ratio of 3/5 (51.8 mg L^{-1} Al/64 N mL L^{-1} additional EF gas) resulted in a maximum removal of COD between 76% and 85%, color (93% to 99%) and turbidity (95% to 99%). The study further showed that the coupled process reduces electrode consumption and the cost associated with sludge disposal (Paulista et al. 2018). Emerick et al. (2020) used the ultrasound-assisted EC process for the treatment of swine slaughterhouse wastewater. The study compared the electrocoagulation flotation (ECF) process with the ultrasound-assisted EC process. Although both processes yielded similar results in terms of turbidity (97.4%), color (97%), BOD (93%), and oil and grease removal (90%), the treatment time of EC reduced from 60 min (in ECF) to 25 min with the use of the coupled treatment strategy.

8.8 RESTAURANT WASTEWATER

Restaurants generate a large volume of wastewater containing high concentrations of organic matter including proteins, carbohydrates, detergents, and total suspended solids containing fat, oil, and greases. A number of technologies that are used for the treatment of restaurant wastewater are available. Among them, biological processes are widely used because of their cost-effectiveness on a large scale. However, the long hydraulic retention times and large working volumes make these processes less attractive than physicochemical treatments in which retention times are shorter. Electrochemical treatment is the most favored over other physicochemical treatments because of the use of simple equipment, brief retention times, and ease of operation, which reduce the operating costs and, hence, makes it an economical process for large-scale applications. EC processes have also attracted great attention for treating restaurant wastewater. In this process, a coagulant agent is required for the treatment and is generated in situ via an anodic dissolution of aluminum or iron oxides. The metal ions form flocculates of the contaminants and the hydrogen bubbles that are produced at the cathode transport light solids and hydrophobic materials such as fats, oils, and greases to the surface of the liquid where they are skimmed. A combination of processes such as electroflotation with EC is also used in treating restaurant wastewater by increasing the flotation of flocs formed in the EC unit. Electro-oxidation is also an electrochemical technique that is used to remove dissolved pollutants from restaurant wastewater by oxidation of pollutants directly or indirectly. The aforementioned combined process is effective due to its capability to produce oxidants and coagulant agents by using either iron or aluminum electrodes arranged in a bipolar configuration or graphite electrodes arranged in monopolar configuration in the same electrolytic cell simultaneously.

High oil and grease contents are detected in restaurant wastewater. EC is more effective in treating small and light suspended particles because of the electroflotation effect. It is widely accepted to treat various types of wastewaters, especially restaurant wastewater, because of its versatility and compatibility. This method is advantageous when compared with conventional methods in terms of ease of operation, simple equipment, less retention time, no addition of chemicals, fast sedimentation of electrocoagulated flocs, and reduced sludge production (Saththasivam et al. 2016). Therefore, EC is used to remove contaminants effectively. Mainly, aluminum and iron metals are used as electrodes and their generated cations are used to make flocs [Equations (8-15) to (8-17)]. In this process, the coagulant Al^{3+} is produced by the aluminum electrode because of electrochemical reactions. The solubility of metal hydroxide complexes depends on pH and ionic strength (Rincón and La Motta 2014). At a pH range of 6 to 7, insoluble flocs are formed. The gases produced at the cathode during the electrolysis of water and metal dissolution make the flocs float. The chemical reactions that occur at the aluminum electrode in the EC process are as follows:

Anode:

$$Al \longrightarrow Al^{3+} + 3e^{-} \tag{8-15}$$

Cathode:

$$3H_2O + 3e^{-} \longrightarrow 3/2H_2 + 3OH^{-} \tag{8-16}$$

Solution:

$$Al^{3+} + 3H_2O \longrightarrow Al(OH)_3 + 3H^{+} \tag{8-17}$$

A study investigated the effect of the electrode type, initial pH value, current density, and polymer concentration on the performance efficiency of the EC process applied for the treatment of tertiary oil recovery wastewater (Liu et al. 2019). The study reported that using EC at a current density of 4 mA m^{-2} with an aluminum electrode, 98.99% of oil and 99.74% of turbidity removal efficiencies were achieved. A pilot-scale study on the treatment of restaurant wastewater using the combined EC–EF process utilized the RSM for the optimization of treatment conditions (Qin et al. 2013). The study revealed that the combined EC–EF process was cost-effective and could be implemented for full-scale treatment of restaurant wastewater.

8.9 TREATMENT OF TEXTILE WASTEWATER

Textile wastewater has high BOD/COD, high color, and high TDS. The wastewater generated from the cotton dyeing industry is extremely polluted because of the presence of reactive dyes, which are not treated by biological processes. The wastewater from textiles affects the growth and development of aquatic organisms, leading to an imbalance in the environment. To reduce the toxicity of textile wastewater, many processes such as physical, chemical, biochemical, and hybrid are required to treat this wastewater in an economic and efficient way (Khandegar and Saroha 2013). The various treatment processes for textile wastewater include (1) physical methods (e.g., adsorption, EC/flocculation, filtration), (2) chemical or advanced oxidation methods, and (3) biological methods. The EC/flocculation and enzyme-based biological methods is elaborated here.

To overcome the drawbacks of conventional wastewater treatment technologies, EC has been introduced to treat textile wastewaters alone or in combination with conventional processes. Being a simple, reliable, and cost-effective method without any further addition of chemicals, EC has been applied to achieve higher removal efficiency. It also reduces the amount of sludge to be disposed of. EC utilizes the direct current source between metal electrodes immersed in wastewater. The electrical current introduced to the EC system causes the dissolution of metal plates into wastewater, and these metal ions at various pH conditions can form a wide range of flocs and metal hydroxides that are easily precipitated and adsorb the soluble contaminants and aggregates of the suspended particles present in wastewater (Brillas and Martínez-Huitle 2015, Liang et al. 2014). Aluminum and iron are the mostly used metals in EC, and the associated reactions are as follows:

Anode:

$$Al_{(s)} \longrightarrow Al^{3+} + 3e^- \tag{8-18}$$

Cathode:

$$3H_2O + 3e^- \longrightarrow 3/2H_{2(g)} + 3OH^- \tag{8-19}$$

$$Al^{3+} + 3H_2O \longrightarrow Al(OH)_3 + 3H^+ \tag{8-20}$$

The freshly formed amorphous aluminum hydroxide flocs have large surface areas that are responsible for fast adsorption of soluble organic compounds and trapping of colloidal particles. These flocs are removed by sedimentation or flotation from an aqueous medium.

The combined treatment strategy of the EC process with the electrochemical advanced oxidation process has been the subject of research in many studies (Khandegar and Saroha 2013, Zazou et al. 2019). Zazou et al. (2019) investigated the coupling of EC with AO, electro-Fenton (EF), and peroxi-coagulation. Their study showed that the sequential EC–EF process is effective for the treatment of textile industry wastewater. The coupled EC-EF process resulted in the removal of 100% of turbidity, 97% of total organic carbon, and 100% of color. The study further showed that the combined process decreased the energy consumption and reaction time and resulted in a complete decolorization of wastewater at 0.45 to 1.5 kW·h kg^{-1} energy.

Table 8-1. Fungi Responsible for the Degradation of Dyes.

Fungi	*Dye degraded*
Aspergillus niger	Remazol Brilliant Blue R (RBBR) and Acid Red 299 (NY1)
Lichen *Permelia perlata*	Disperse dye Solvent Red 24
Coriolopsis sp. (1c3)	Triphenylmethane dyes [crystal violet (CV), methyl violet (MV), cotton blue (CB), and malachite green (MG)]
Pleurotus eryngii	Naphthalene
Penicillium simplicissimum	Triphenylmethane dyes [crystal violet (CV), methyl violet (MV), cotton blue (CB), and malachite green (MG)]

Enzyme-based biological methods include the use of (1) fungal, (2) algal, and (3) pure culture and mixed culture for dye degradation. In changing environmental conditions, fungal culture has an ability to acclimate its metabolism for its survival. Intracellular and extracellular enzymes help in the metabolic activity of fungi. Enzymes such as lignin peroxidase, manganese peroxidase, and laccase have been used for dye degradation in textile wastewater (Holkar et al. 2016). The list of fungi responsible for the degradation of dyes is given in Table 8-1.

Algae are now seeing a growing importance in the area of degradation of textile wastewater, and several species (Table 8-1) are recognized for dye degradation through three different mechanisms: (1) consumption of dyes for their growth; (2) transformation of dyes to noncolored intermediates; and (3) chromophore adsorption on algae. Biodegradation and biosorption are applied in treating textile wastewater using algae. In biodegradation, the enzymes break the bonds of the chemical structure of the dyes to transform the dyes into other chemical compounds. In biosorption, dye moves from the water phase to the solid phase.

Aerobic, anaerobic, and facultative microbes, either in pure or in mixed form, are used for the degradation of dyes during textile wastewater treatment. Azo dyes are degraded by bacterial processes because of the reductive breakage of azo bonds (–N=N–) by azo-reductase enzymes under anaerobic conditions. The breakages of azo bonds are responsible for the formation of other colorless toxic intermediates, which can be further treated by anaerobic or aerobic processes (Holkar et al. 2016).

Microbial fuel cell (MFC) systems have been used for the treatment of textile wastewater. In an MFC system, the electrochemically active microbes oxidize various organic compounds of textile wastewater in the anode chamber, and the generated protons and electrons are transported to the cathode chamber where oxygen is reduced to water. The anode and cathode chambers in an MFC are separated using membranes. The electricity generated can be easily harvested using the external resistor placed between anode and cathode. The major disadvantage of using an MFC on a large scale is the cost of MFC membranes and low production of power. Recent studies (Patade et al. 2016) show various configurations that are used for electricity generation employing MFC systems. The commercialization of MFCs is mainly based on membranes with low resistance, high selectivity, low cost, and stability. The MFC membranes are used to transport ions from one chamber to the other.

8.10 TREATMENT OF LAUNDRY WASTEWATER

Laundry wastewater accounts for 10% of pollutant load in municipal wastewater treatment plants. The composition of laundry wastewater depends on the cleaning material used for washing and laundered materials. In general, the COD, BOD, and TS of laundry wastewater are in the range of 800 to 1,000, 300 to 500, and 50 to 250 mg L^{-1}, respectively (Bering et al. 2018). The wastewater originating from laundry industries contains a high concentration of cationic and anionic surfactants. The conventional coagulation, adsorption, chemical oxidation, and flotation processes are not able to remove surfactants and suspended solids efficiently.

To overcome the drawbacks of the conventional methods, laundry wastewater is treated using a new bipolar EC and electroflotation process. In this new bipolar method, both processes are carried out simultaneously in a single reactor by optimizing the reactor parameters such as pH, hydraulic retention time (HRT) (Sillanpää et al. 2018), and current density. The reactors for this process are specially designed for the effective removal of turbidity, COD, phosphates, and surfactants at a short HRT in a wide range of pH conditions. In the bipolar configuration and the EC-electroflotation system, three aluminum plates are used between two titanium electrodes having opposite charges and undergo anodic reactions on the cathode and cathodic reactions on the anode (Šostar-Turk et al. 2005). The main anodic reactions are as follows:

Titanium anode:

$$2H_2O \longrightarrow O_2 + 4H^+ + 4e^- \tag{8-21}$$

Aluminum cathode:

$$Al \longrightarrow Al^{3+} + 3e^- \tag{8-22}$$

Aluminum is hydrolyzed and polymerized under specific pH conditions, which gives rise to the formation of certain species such as $Al(OH)^{2+}$, $Al_2(OH)_2^{4+}$, and various charged hydroxyl cationic complexes that can effectively remove pollutants and contaminants by adsorption and neutralizing the charge. Contaminants are removed after precipitation. In this process, titanium is used as it is cost-effective at a large scale. Several other reactions also take place in this bipolar system and, thereby, increase the ohmic resistance in the electrochemical cell. Therefore, alternating current is provided to precipitate the carbonate salt on the electrode surface. The implementation of the EC process for the treatment of laundry wastewater is cost-effective because the high salt concentration in laundry wastewater provides better conductivity to the reaction and, thus, decreases the energy consumption. The combined EC and EF process was studied for the treatment of laundry wastewater for the recycling of water for cleaning purposes (Dimoglo et al. 2019). The batch study consisting of an aluminum electrode with varying pH, current density, and treatment time reveals that the coupled process required a current density of 5.26 mA cm^{-2} and a processing time of 5 min for 90% removal of surfactants, turbidity, and color. The study further showed that the combined processes reduced electricity consumption and utilized 1.25 kW·h m^{-3} for effective removal.

8.11 DRINKING WATER TREATMENT

The treatment of drinking water is important for domestic and industrial usage as hard water causes many problems. The high demand for drinking water necessitates the use of modern and cost-effective technologies for water softening. EC is an advanced technology that combines the functions of the conventional methods of coagulation, flotation, and electrochemical with three steps: (1) coagulant formation through a dissolution of metal ions of the anode electrode, (2) destabilization of pollutants and suspended particles, and (3) de-emulsification and aggregation of insoluble phases and floc formation.

The destabilization of pollutants, suspended particles, and the de-emulsification mechanism use double-layer compression and ion neutralization in water and wastewater for floc formation. Iron electrodes are basically used in EC for water treatment (Sillanpää et al. 2018). The reactions occurring at the iron electrode are as follows:

Anode:

$$Fe_{(s)} \longrightarrow Fe^{2+} + 2e^{-} \quad (8\text{-}23)$$

$$Fe^{2+} + 2OH^{-} \longrightarrow Fe(OH)^{-} \quad (8\text{-}24)$$

Cathode:

$$2H_2O + 2e^{-} \longrightarrow H_2 + 2OH^{-} \quad (8\text{-}25)$$

Overall:

$$Fe + 2H_2O \longrightarrow Fe(OH)_2 + H_2 \quad (8\text{-}26)$$

The dissolution of iron electrodes leads to the release of ferro ions into the solution through the electrolytic oxidation of the anode electrode and produces metal hydroxides on reaction with hydroxide monomers and polymers. This dissolution is completely based on the pH of the solution, and flocs are created through this dissolution. Gases are produced through anode and cathode electrodes during electrolysis that causes flotation and better removal of contaminants and pollutants. This process is combined with other techniques for easy operation. The process has the advantages of short resistance time, no chemical requirement, low sludge production, sludge stability, proper sedimentation of flocs, is easy for sludge dewatering, and environmental compatibility (Ghosh et al. 2008). This process reduces the contaminants and suspended particles caused by floc formation and removes flocs through sedimentation. This is an efficient method to remove water hardness based on pH and optimal current density.

8.12 SUMMARY

The need for clean water is critical as water bodies are being constantly polluted because of the indiscriminate discharge of industrial effluents, anthropogenic activities, as well as natural processes. The term *wastewater treatment* is defined as the process or a combination of different processes that is used for the removal of pollutants and contaminants, which can be discharged to the environment after treatment. Historically, wastewater treatment started in the mid-nineteenth century; however, the two world wars delayed its progress. The outbreak of diseases such as cholera in the mid-nineteenth century led to the development of wastewater treatment plants. The first centralized septic tanks and trickling sand filters were developed in the 1860s in the United States (Yadav et al. 2020). Presently, numerous techniques and devices are being used for the treatment of wastewater, such as activated sludge process, membrane bioreactors, sequence batch reactors, chemical oxidation process, AOPs, electrocoagulation (EC), electro-oxidation (EO), and so on. Many studies demonstrate the successful implementation of the EC process for the removal of metals, organic and inorganic compounds originating from various sources such as the municipal, pulp and paper, and textile industries. In EC, the electric field is applied between the cathode and the anode to produce hydrogen ions (H^+) and O_2 and metals are oxidized to their cationic state at the anodic side in water. At the same time, water is reduced at the cathodic side, and this oxidation and reduction process generates flocculating agents in water. The formation and migration of flocculating agents lead to the adsorption and coagulation of pollutants on the flocculating agents and then removed from water. EC has proven to be a more reliable and effective treatment process than other conventional processes, because it does not require the addition of chemical coagulants for precipitation and generates high-quality effluents with low sludge production. However, the high treatment cost caused by a long reaction time, electrical resistance, and high energy consumption limit the implementation of the EC process on a large scale. The coupling of the EC process with other electrochemical treatment processes such as electroflotation, AO, and electro-Fenton could be done to overcome the limitations. The potential of EC treatment alone and as a coupled treatment strategy has been investigated for the decontamination of various compounds such as arsenic, zinc, copper, fluoride, phosphate, and many more. The treatment of wastewater originating from meat processing, cereal products, food beverages, and abattoir industries using EC provides a greater insight into the whole process.

ABBREVIATIONS

ABR = Anaerobic baffled reactor
AOPs = Advanced oxidation processes
BOD = Biochemical oxygen demand
COD = Chemical oxygen demand
CWW = Cheese whey wastewater
DAF = Dissolved air flotation
HRT = Hydraulic retention time
LL = Landfill leachate
MFC = Microbial fuel cell
RSM = Response surface methodology
ST = Septic tanks
TDS = Total dissolved solids
UV = Ultraviolet radiations

References

Al-Shannag, M., Z. Al-Qodah, K. Bani-Melhem, M. R. Qtaishat, and M. Alkasrawi. 2015. "Heavy metal ions removal from metal plating wastewater using electrocoagulation: Kinetic study and process performance." *Chem. Eng. J.* 260: 749–756.

Akbal, F., and S. Camcı. 2011. "Copper, chromium and nickel removal from metal plating wastewater by electrocoagulation." *Desalination* 269 (1–3): 214–222.

Asselin, M., P. Drogui, H. Benmoussa, and J.-F. Blais. 2008. "Effectiveness of electrocoagulation process in removing organic compounds from slaughterhouse wastewater using monopolar and bipolar electrolytic cells." *Chemosphere* 72 (11): 1727–1733.

Banerji, T., and S. Chaudhari. 2016. "Arsenic removal from drinking water by electrocoagulation using iron electrodes—An understanding of the process parameters." *J. Environ. Chem. Eng.* 4 (4): 3990–4000.

Behbahani, M., M. R. A. Moghaddam, and M. Arami. 2011. "Techno-economical evaluation of fluoride removal by electrocoagulation process: Optimization through response surface methodology." *Desalination* 271 (1–3): 209–218.

Bering, S., J. Mazur, K. Tarnowski, M. Janus, S. Mozia, and A. W. Morawski. 2018. "The application of moving bed bio-reactor (MBBR) in commercial laundry wastewater treatment." *Sci. Total Environ.* 627: 1638–1643.

Brillas, E., and C. A. Martínez-Huitle. 2015. "Decontamination of wastewaters containing synthetic organic dyes by electrochemical methods. An updated review." *Appl. Catal., B* 166–167: 603–643.

Bustillo-Lecompte, C. F., S. Ghafoori, and M. Mehrvar. 2016. "Photochemical degradation of an actual slaughterhouse wastewater by continuous UV/H_2O_2 photoreactor with recycle." *J. Environ. Chem. Eng.* 4 (1): 719–732.

Bustillo-Lecompte, C. F., and M. Mehrvar. 2015. "Slaughterhouse wastewater characteristics, treatment, and management in the meat processing industry: A review on trends and advances." *J. Environ. Manage.* 161: 287–302.

Carvalho, F., A. R. Prazeres, and J. Rivas. 2013. "Cheese whey wastewater: Characterization and treatment." *Sci. Total Environ.* 445–446: 385–396.

Dia, O., P. Drogui, G. Buelna, R. Dube, and B. S. Ihsen. 2017. "Electrocoagulation of bio-filtrated landfill leachate: Fractionation of organic matter and influence of anode materials." *Chemosphere* 168: 1136–1141.

Dimoglo, A., P. Sevim-Elibol, Ö. Dinç, K. Gökmen, and H. Erdoğan. 2019. "Electrocoagulation/electroflotation as a combined process for the laundry wastewater purification and reuse." *J. Water Process Eng.* 31: 100877.

Drogui, P., M. Asselin, S. K. Brar, H. Benmoussa, and J. F. Blais. 2008. "Electrochemical removal of pollutants from agro-industry wastewaters." *Sep. Purif. Technol.* 61 (3): 301–310.

Emamjomeh, M. M., and M. Sivakumar. 2009. "Review of pollutants removed by electrocoagulation and electrocoagulation/flotation processes." *J. Environ. Manage.* 90 (5): 1663–1679.

Emamjomeh, M. M., M. Sivakumar, and A. S. Varyani. 2011. "Analysis and the understanding of fluoride removal mechanisms by an electrocoagulation/flotation (ECF) process." *Desalination* 275 (1–3): 102–106.

Emerick, T., J. L. Vieira, M. H. L. Silveira, and J. J. João. 2020. "Ultrasound-assisted electrocoagulation process applied to the treatment and reuse of swine slaughterhouse wastewater." *J. Environ. Chem. Eng.* 8 (6): 104308.

Fu, F., and Q. Wang. 2011. "Removal of heavy metal ions from wastewaters: A review." *J. Environ. Manage.* 92 (3): 407–418.

Ghosh, D., C. R. Medhi, and M. K. Purkait. 2008. "Treatment of fluoride containing drinking water by electrocoagulation using monopolar and bipolar electrode connections." *Chemosphere* 73 (9): 1393–1400.

Goris, T., and G. Diekert. 2016. "The genus Sulfurospirillum." In *Organohalide-Respiring Bacteria*, edited by L. Adrian and F. Löffler, 209–234. Berlin: Springer.

Guven, G., A. Perendeci, and A. Tanyolaç. 2008. "Electrochemical treatment of deproteinated whey wastewater and optimization of treatment conditions with response surface methodology." *J. Hazard. Mater.* 157 (1): 69–78.

Hakizimana, J. N., B. Gourich, M. Chafi, Y. Stiriba, C. Vial, P. Drogui, et al. 2017. "Electrocoagulation process in water treatment: A review of electrocoagulation modeling approaches." *Desalination* 404: 1–21.

Holkar, C. R., A. J. Jadhav, D. V. Pinjari, N. M. Mahamuni, and A. B. Pandit. 2016. "A critical review on textile wastewater treatments: Possible approaches." *J. Environ. Manage.* 182: 351–366.

Khandegar, V., and A. K. Saroha. 2013. "Electrocoagulation for the treatment of textile industry effluent—A review." *J. Environ. Manage.* 128: 949–963.

Kobya, M., E. Senturk, and M. Bayramoglu. 2006. "Treatment of poultry slaughterhouse wastewaters by electrocoagulation." *J. Hazard. Mater.* 133 (1–3): 172–176.

Kushwaha, J. P., V. C. Srivastava, and I. D. Mall. 2011. "An overview of various technologies for the treatment of dairy wastewaters." *Crit. Rev. Food Sci. Nutr.* 51 (5): 442–452.

Liang, C.-Z., S.-P. Sun, F.-Y. Li, Y.-K. Ong, and T.-S. Chung. 2014. "Treatment of highly concentrated wastewater containing multiple synthetic dyes by a combined process of coagulation/flocculation and nanofiltration." *J. Membr. Sci.* 469: 306–315.

Liu, F., Z. Zhang, Z. Wang, X. Li, X. Dai, L. Wang, et al. 2019. "Experimental study on treatment of tertiary oil recovery wastewater by electrocoagulation." *Chem. Eng. Process. Process Intensif.* 144: 107640.

Mandal, P., B. K. Dubey, and A. K. Gupta. 2017. "Review on landfill leachate treatment by electrochemical oxidation: Drawbacks, challenges and future scope." *Waste Manage.* 69: 250–273.

Mansoorian, H. J., A. H. Mahvi, and A. J. Jafari. 2014. "Removal of lead and zinc from battery industry wastewater using electrocoagulation process: Influence of direct and alternating current by using iron and stainless steel rod electrodes." *Sep. Purif. Technol.* 135: 165–175.

Meunier, N., P. Drogui, C. Gourvenec, G. Mercier, R. Hausler, and J. F. Blais. 2004. "Removal of metals in leachate from sewage sludge using electrochemical technology." *Environ. Technol.* 25 (2): 235–245.

Meunier, N., P. Drogui, C. Montané, R. Hausler, G. Mercier, and J.-F. Blais. 2006. "Comparison between electrocoagulation and chemical precipitation for metals removal from acidic soil leachate." *J. Hazard. Mater.* 137 (1): 581–590.

Mólgora, C. C., A. M. Domínguez, E. M. Avila, P. Drogui, and G. Buelna. 2013. "Removal of arsenic from drinking water: A comparative study between electrocoagulation-microfiltration and chemical coagulation-microfiltration processes." *Sep. Purif. Technol.* 118: 645–651.

Morozesk, M., M. M. Bonomo, L. D. Rocha, I. D. Duarte, E. R. Zanezi, H. C. Jesus, et al. 2016. "Landfill leachate sludge use as soil additive prior and after electrocoagulation treatment: A cytological assessment using CHO-k1 cells." *Chemosphere* 158: 66–71.

Nidheesh, P. V., and T. S. A. Singh. 2017. "Arsenic removal by electrocoagulation process: Recent trends and removal mechanism." *Chemosphere* 181: 418–432.

Niza, N. M., M. S. Yusoff, M. A. A. M. Zainuri, M. I. Emmanuel, A. M. H. Shadi, and M. A. Kamaruddin. 2020. "Performance of batch electrocoagulation with vibration-induced electrode plates for landfill leachate treatment." *J. Water Process Eng.* 36: 101282.

Patade, S., K. Silveira, A. Babu, F, D'Costa, and Y. Mhatre. 2016. "Bioremediation of dye effluent waste through an optimised microbial fuel cell." *Int. J. Adv. Res. Biol. Sci* 3 (5): 214–226.

Paulista, L. O., P. H. Presumido, J. D. P. Theodoro, and A. L. N. Pinheiro. 2018. "Efficiency analysis of the electrocoagulation and electroflotation treatment of poultry slaughterhouse wastewater using aluminum and graphite anodes." *Environ. Sci. Pollut. Res.* 25 (20): 19790–19800.

Qin, X., B. Yang, F. Gao, and G. Chen. 2013. "Treatment of restaurant wastewater by pilot-scale electrocoagulation–electroflotation: Optimization of operating conditions." *J. Environ. Eng.* 139 (7): 1004–1016.

Rincón, G. J., and E. J. La Motta. 2014. "Simultaneous removal of oil and grease, and heavy metals from artificial bilge water using electro-coagulation/flotation." *J. Environ. Manage.* 144: 42–50.

Sahu, O., B. Mazumdar, and P. Chaudhari. 2014. "Treatment of wastewater by electrocoagulation: A review." *Environ. Sci. Pollut. Res.* 21 (4): 2397–2413.

Saththasivam, J., K. Loganathan, and S. Sarp. 2016. "An overview of oil–water separation using gas flotation systems." *Chemosphere* 144: 671–680.

Shak, K. P. Y., and T. Y. Wu. 2014. "Coagulation–flocculation treatment of high-strength agro-industrial wastewater using natural *Cassia obtusifolia* seed gum: Treatment efficiencies and flocs characterization." *Chem. Eng. J.* 256: 293–305.

Sillanpää, M., M. C. Ncibi, A. Matilainen, and M. Vepsäläinen. 2018. "Removal of natural organic matter in drinking water treatment by coagulation: A comprehensive review." *Chemosphere* 190: 54–71.

Song, P., Z. Yang, G. Zeng, X. Yang, H. Xu, L. Wang, et al. 2017. "Electrocoagulation treatment of arsenic in wastewaters: A comprehensive review." *Chem. Eng. J.* 317: 707–725.

Šostar-Turk, S., I. Petrinić, and M. Simonič. 2005. "Laundry wastewater treatment using coagulation and membrane filtration." *Resour. Conserv. Recycl.* 44 (2): 185–196.

Teh, C. Y., P. M. Budiman, K. P. Y. Shak, and T. Y. Wu. 2016. "Recent advancement of coagulation–flocculation and its application in wastewater treatment." *Ind. Eng. Chem. Res.* 55 (16): 4363–4389.

Tran, N., P. Drogui, J.-F. Blais, and G. Mercier. 2012. "Phosphorus removal from spiked municipal wastewater using either electrochemical coagulation or chemical coagulation as tertiary treatment." *Sep. Purif. Technol.* 95: 16–25.

Vasudevan, S., B. S. Kannan, J. Lakshmi, S. Mohanraj, and G. Sozhan. 2011. "Effects of alternating and direct current in electrocoagulation process on the removal of fluoride from water." *J. Chem. Technol. Biotechnol.* 86 (3): 428–436.

Yadav, B., A. K. Pandey, L. R. Kumar, R. Kaur, S. K. Yellapu, B. Sellamuthu, R. D. Tyagi, and P. Drogui. 2020. "Introduction to wastewater microbiology: Special emphasis on hospital wastewater." In *Current Developments in Biotechnology and Bioengineering*, 1–41. Elsevier.

Zazou, H., H. Afanga, S. Akhouairi, H. Ouchtak, A. A. Addi, R. A. Akbour, et al. 2019. "Treatment of textile industry wastewater by electrocoagulation coupled with electrochemical advanced oxidation process." *J. Water Process Eng.* 28: 214–221.

CHAPTER 9

Environmental Applications of Electro-Oxidation Processes

Mitra Ebrahimi Gardeshi, Mahdieh Khajvand,
Ali Khosravanipour Mostafazadeh,
Patrick Drogui, R. D. Tyagi

9.1 INTRODUCTION

Electrochemical (EC) oxidation is an effective method for treating different wastewaters (WWs) and disinfecting water. The intrinsic advantage of this technology is its compatibility with the environment. Because the electron is used as a clean reagent, process-integrated environmental protection is ensured by developing new processes or products with less harmful impact. Another benefit of this method is that it operates at ambient temperatures and atmospheric pressure. The capability for using it in different compositions and flow rates of influents is a positive aspect of this technology. Its adaptation to various applications and integration with other technologies is another feature of this technology. Electro-oxidation (EO) is known as *green technology* because, in general, auxiliary chemicals are not required for it and it does not produce waste. EO is used as a technique for WW treatment, mainly for industrial effluents. The common layout consists of two electrodes, operating as cathode and anode, which are connected to a power source. This chapter aims to provide an overview of the EO processes employed to remove environmental pollutants from various WWs.

9.2 REMOVAL OF PERSISTENT ORGANIC POLLUTANTS FROM WASTEWATERS

Persistent organic pollutants (POPs) are environmentally refractory and bioaccumulative chemicals. They have become a global environmental concern because of long-distance transportation, persistence against biological degradation, an inclination for bioaccumulation in fatty tissues, and high toxicity, even at sorely low concentrations (Pariatamby and Kee 2016). These chemical substances cause cancer and are responsible for the disruption of the immune system and the central nervous system. According to the Stockholm Convention, an international environmental treaty, the production and use of some POPs such as Aldrin, Chlordane, dichlorodiphenyltrichloroethane, and Mirex is either banned or restricted because of their adverse effects on humans and ecosystems (Badmus et al. 2018). POPs enter WW treatment facilities from rain, agricultural, domestic, and industrial WW (Katsoyiannis and Samara 2004). In general, WW treatment is conducted based on the nature of the pollutants. Because of the biorefractory character of these pollutants, they cannot be decomposed by conventional water treatments. Therefore, using physicochemical processes causes the contaminants to transfer from water to another phase and create new waste (Oturan 2014). Recently, there has been an increased interest in employing the EO process for the removal of persistent organic contaminants from WW. Concisely, EO is carried out despite strong oxidants, mainly hydroxyl radicals. The in situ EO enhances the efficiency of the abatement of organic substances (Karpińska and Kotowska 2019). Electrochemical technologies provide an effective solution for the removal of the organic matter of WW because electrons offer a clean, versatile, and cost-effective reagent (Martínez-Huitle and Panizza 2018). EO techniques are implemented for the mineralization of diverse organic pollutants, including toxic and recalcitrant ones, to water and carbon dioxide, and inorganics or the production of innocuous substances. The development of EO offers new-generation technologies for the removal of organic compounds from WW. Various industries such as textile, olive oil, pulp and paper mill, and tannery effluents have used this technique for the treatment process. In this process, active $OH^{\bullet}$ radical species are produced, which can oxidize resistant organic compounds (Fallahzadeh and Omidi 2019).

The EC of POPs in WW can be performed through two different approaches: (1) Direct oxidation: Pollutants are destroyed on the anode surface without the participation of other substances. Direct oxidation is attained in two ways, electrochemical conversion and combustion; and (2) Indirect oxidation: Organic pollutants are oxidized through a mediator that is electrochemically generated to fulfill the oxidation. The agents include chlorine and hypochlorite, hydrogen peroxide, ozone, and metal mediators such as Ag^{2+} (Deng and Englehardt 2007, Martínez-Huitle and Panizza 2018).

This process depends on various factors, including the generation of hydroxyl radicals, the nature of the anode electrode, and the process competition with the oxygen evolution reaction (Martinez-Huitle and Ferro 2006). The development of novel electrode materials has a significant effect on improving treatment efficiency. The most extensively studied anodes for the removal of POPs are graphite, Ti-Pt/β-PbO_2, boron-doped diamond (BDD), Ti/SnO_2-Sb/PbO_2, Ti/$TiRuO_2$, Ti/IrO_2, Ti/Pt-Ir, Ti/PbO_2, Ti/PdO-Co_3O_4, Ti/RhO_x-TiO_2, lead, Ta/Ir, Ru/Ir, Pt/Ir, IrPt, IrO_2, carbon, and so on (Särkkä et al. 2015).

A main disadvantage of EO is that it is associated with a slow mass transfer that limits its large-scale application (Trellu et al. 2018). Besides, other drawbacks are related to the costs of electrical energy; adding electrolytes when WW with low conductivity is treated and fouling of organic material on the surface of the electrode can reduce the activity of, and shorten the lifetime of, the electrode. However, EO offers numerous advantages that outweigh the drawbacks, including the following:

- Increase in biodegradability of WW after treatment: This property is employed to perform experiments by combining EO and biological processes such as the bio-electro-Fenton process (Oturan et al. 2018).
- Relatively simple process: Oxidants are generated in situ during the treatment either directly at the surface of the anode or indirectly from some electroactive species in the treated water (Särkkä et al. 2015). Because of the use of simple equipment, process control is easy.
- Possibility of avoiding the addition of chemical reagents: By direct oxidation, there is no need to add chemicals to treat solutions and producing secondary pollution (Trellu et al. 2018).
- Environmental compatibility: This process is considered as a green treatment because of the nonproduction of any sludge. Besides, the electron is a clean and effective reagent, whose reactivity may be tuned by choosing a proper electrocatalyst. Thus, EO process does not generate pollutants.
- Versatility: Electrochemical methods can eliminate various contaminants and treat different volumes of WW from the lab scale to industrial applications. Besides, several techniques can be utilized, such as direct and indirect oxidations (Martinez-Huitle and Ferro 2006).
- Relatively safe process: As the process operates under room temperature and pressure, it is relatively safe and high-energy efficiency is achieved.

Brillas et al. (2005) employed Pt and BDD anodes to remove paracetamol by EO in a stirred tank reactor by applying various pH, current, and temperature ranges. They reported 98% and 19% total organic carbon (TOC) removal by BDD and Pt, respectively, at 300 mA for a 360 min treatment time (Brillas et al. 2005). Murugananthan et al. (2010) investigated the effect of electrolyte, pH of aqueous medium, and applied current density on ketoprofen's (KP) mineralization behavior. They achieved 100% TOC removal in a stirred EO tank reactor with a BDD anode and applying 49.5 mA for 1,020 min on a KP solution. The solution used contained 250 mL of 5 μM drug in 0.1 M Na_2SO_4 and pH 6.0. They reported that a complete mineralization of KP was obtained using Na_2SO_4 as a supporting electrolyte (Murugananthan et al. 2010). Farhat et al. (2015) employed EO by BDD and added inert nitrate anolytes and sulfate solutions as background electrolytes for the removal of seven different persistent organic contaminants, namely, carbamazepine; *N*,*N*-diethyl-*meta*-toluamide; diatrizoate; iopromide; tribromophenol; triclopyr; and triclosan. They reported that utilizing sulfate increased EO rates 10 to 15 times higher for the target contaminants in comparison with adding nitrate anolyte (Farhat et al. 2015). Ren et al. (2019) employed the combined reaction system ultrasound and electrochemical oxidation (US–EC) using stainless-steel electrodes to improve the degradation of chlorpyrifos (pesticide). They achieved 93.3% removal using the US–EC process; however, the EC and US systems alone yielded 72.8% and 50.98% removal (Ren et al. 2019). A summary of some EO studies for the removal of POPs from various WWs is presented in Table 9-1.

Table 9-1. EO Treatment for the Removal of POPs from Wastewater.

Name of pollutant	*Technology*	*EO conditions*	*Removal efficiency of the system*	*Reference*
Paracetamol	Electro-oxidation of synthetic wastewater	Anode = Pt and BDD Current = 300 mA Time = 360 min	TOC removal with the BDD anode: 98% TOC removal with the Pt anode: 19%	Brillas et al. (2005)
Ketoprofen	Electro-oxidation of synthetic wastewater	Anode = Pt and BDD Current density = 8.8 mA cm^{-2} Na_2SO_4 = 0.1 M; pH = 6.0 Ketoprofen = 5 μM Time= 780 min	TOC removal with the BDD anode: 100%	Muruganathan et al. (2010)
Chlorpyrifos	Integrated system ultrasound and electrochemical oxidation of synthetic wastewater	Anode: stainless-steel $V = 20$ V; $T = 25°C$ Ultrasonic power = 200 W Na_2SO_4 = 2 mg L^{-1}	Chlorpyrifos removal: 93.3%	Ren et al. (2019)
NPEO3-17	Combined system of ultrafiltration and electro-oxidation of real laundry wastewater	Anode = BDD Current = 12 A Treatment time = 45 min	NPEO3-17 removal: 97%	Mostafazadeh et al. (2019)
Polychlorinated dibenzo-*p*-dioxins and dibenzofurans (PCDD/Fs)	Electrochemical oxidation of landfill leachate	Anode = BDD Time = 180 min	1,2,3,4,6,7,8-HpCDD removal: 73% OCDD removal: 71%	Vallejo et al. (2013)
Pentachlorophenol (PCP)	Indirect electrochemical oxidation of synthetic wastewater	Anode = Ti/RuO_2 Current density = 10.0 mA cm^{-2} Cl^- = 10 mmol L^{-1}, PCP = 50 mg L^{-1} Time = 60 min pH = 8.0, $T = 30°C$	PCP conversion:100%	Ma et al. (2014)
Thiodan 35 CE (endosulfan)	Coupled system of electro-oxidation and ozonation	Anode = BDD Thiodan = 20 mg L^{-1} $T = 30$ min	After 30 min, the sample did not show any toxicity	Landeros et al. (2017)

Electrochemical methods offer an efficient means of treating WW. Recently, EO has received attention in the area of WW treatment, and especially in the removal of POPs, it is being seen as a promising option.

9.3 REMOVAL OF PATHOGENS FROM WASTEWATER

Pathogens include viruses, bacteria, fungi, protozoans, helminths (worms), allergens, and toxins. These pathogens cause disease in living organisms. The presence of pathogens in WW causes serious health problems for the personnel of the treatment plants and community members living in the neighborhood of the plants. Pathogens reach surface and groundwater from several sources. Domestic and hospital WW contain large numbers of pathogens. Pathogenic agents are found in feces and urine from humans and animals. A wide range of pathogens are present in WW; however, some of them are more common. The viruses commonly found in WW include hepatitis, human immunodeficiency virus, influenza, and adenovirus, which can cause diseases such as hepatitis, acquired immunodeficiency syndrome, flu, and upper and lower respiratory tract distress. The prevalent bacteria in sewage are Leptospira, Salmonella, Bacillus, and Vibrio. These can lead to diseases such as leptospirosis (Weil disease), paratyphoid fever, typhoid fever, anthrax, and cholera. Besides, pathogenic fungi, parasitic helminths, and protozoans have been found in relatively large numbers in WW (Gerardi and Zimmerman 2004).

To remove pathogens, tertiary treatment methods such as ultrafiltration, ozonation, adsorption, chlorination, and UV light disinfection are employed. Chlorination is the usual method used for disinfection. However, it suffers many serious drawbacks. First, the reaction of chlorine with organic compounds that exist in water causes the production of hazardous by-products such as organochlorinated compounds. Moreover, the mentioned compounds change the taste and odor of water. Recently, EO has been developed for its application, along with other techniques, to enhance purification. The mechanism of electrochemical disinfection is direct oxidation on the anode surface and indirect oxidation in the solution. This technology can be used for the disinfection of WW without the presence of chloride to avoid the formation of toxic chlorine by-products (Ganizadeh et al. 2017).

For disinfection in drinking water treatment, the EO method is quite practical. It can produce hydroxyl radicals at the anode and generate chlorine/hypochlorite during indirect oxidation. The main mechanism for using EO to treat drinking water is inactivation of bacteria by hydroxyl radicals generated by water discharge and also using direct oxidation at the electrode surface (Jeong et al. 2007). Obviously, various electrodes have different abilities to produce radicals. For example, BDD electrodes constitute hydroxyl radicals and dimensionally stable anodes activate chlorine, depending on the type of electrolyte solution used (Jeong et al. 2009). Ma et al. (2011) investigated the use of the hemin/graphite electrode. This electrode is preferable because when low potential is applied without any addition of chloride, sizable compounds such as H_2O_2 and OH at the cathode surface are produced.

Disinfection has synergistic effects on the oxide anode, the electric field, and the radicals formed during treatment. The amount of nitrite, ammonia, and mono-chloramine residues formed in drinking water disinfection depends on the treatment conditions and original water quality parameters (Bergmann and Rollin 2007). Chlorine dioxide is a main by-product that can be produced under several conditions in the EO process:

- First zone: This is the contact plane and vicinity of the anode with acidic pH because of the main reaction of oxygen formation [Equation (9-1)]

$$H_2O \rightarrow O_2 + 4H^+ + 4e^- (1.229\ V) \tag{9-1}$$

- Second zone: This is the contact plane and vicinity of the cathode [Equation (9-2)]

$$2H_2O + 2e^- \rightarrow H_2 + 2OH^- (-0.828\ V) \tag{9-2}$$

- Third zone: The bulk of electrolyte can be considered. The acceptable value for pH is between 6.5 and 9.5 for drinking water regulations.

The performance of EO crucially depends on the characteristics of the initial effluents, the cell configuration, and the applied current density. In general, the disinfection rate increases by raising the current density. However, in a higher current density, the said relation becomes less effective, and it leads to the dissolution and corrosion of anode electrode in water (Cano et al. 2012). The anodes used for the disinfection of microbial water by EO include Pt, BDD, Ti/RiO_2, $Ti/Pt\text{-}IrO_2$, graphite felt, $Ti/SnO_2/IrO_2/Sb_2O_5$, Ti/RuO_2, and Ti_nO_{2n-1} for the removal of *Escherichia coli*, Bacteriophage MS2, *Pseudomonas aeruginosa*, *Legionella*, *Bacillus* sp., Coliform bacteria, *Deinococcus geothermalis E5005* and so on (Särkkä et al. 2015).

Frontistis et al. (2011) reported the elimination of *E. coli* from secondary treated effluents by EC. They utilized BDD as an anode and zirconium as a cathode. The applied current density was 2.1 mA cm^{-2}, and a complete removal of *E. coli* was achieved in just 1.5 min (Frontistis et al. 2011). Cano et al. (2012) investigated the disinfection process for reuse of treated municipal WWs using EC by the BDD anode. They applied current density in the range of 1.3 to 13.0 A m^{-2}, and without adding any external chemical reagent, they achieved a complete removal of *E. coli*. They claimed that to assure the complete disinfection of a particular treated WW, power consumption of around 0.2 $kW \cdot h\ m^{-3}$ is sufficient (Cano et al. 2012). Bruguera-Casamada et al. (2017) investigated the disinfection of 100 mL of 10^6 CFU mL^{-1} *Pseudomonas aeruginosa* suspensions by EC. They applied BDD, IrO_2-based, and RuO_2-based anodes and a stainless-steel cathode. They obtained total inactivation in 60 min using a BDD anode and in 30 min using IrO_2- or RuO_2-based anodes. They also reported faster disinfection in the presence of NaCl and HClO. They took less than 30 and 5 min for complete inactivation with BDD and IrO_2- or RuO_2-based anodes (Bruguera-Casamada et al. 2017). A summary of some EO studies for the removal of pathogens from various WWs is presented in Table 9-2.

Table 9-2. EO Treatment for the Removal of Pathogens from Wastewater.

Name of the pollutant	*Technology*	*EO conditions*	*Removal efficiency of the system*	*Reference*
Escherichia coli O157:H7	Electrochemical oxidation of iceberg lettuce wash water	Anode = BDD *E. coli* = 10^5 CFU mL^{-1} Time = 0.9, 25, 60, and 90 min for water with COD = 60, 300, 600, and 800 mg L^{-1}	*Escherichia coli* reduction: 5 log units	López-Gálvez et al. (2012)
Escherichia coli	Electro-oxidation of urban wastewaters	Anode = BDD Current density = 1.3–13.0 A m^{-2} Energy consumption = 0.2 kW h m^{-3}	*Escherichia coli* deactivation: 100%	Cano et al. (2012)
Escherichia coli	Electro-oxidation of synthetic water	Anode = Pt Current density = 100 mA cm^{-2} Time = 180 min *E. coli* = 10^3 CFU mL^{-1} KH_2PO_4 = 0.2 M	*Escherichia coli* deactivation: 90%	Jeong et al. (2007)
Deinococcus geothermalis E50051, Pseudoxanthomonas taiwanensis JN1109, Meiothermus silvanus B-R2A5-50.4	Electro-oxidation of synthetic paper mill wastewater	Anode = mixed metal oxide Current density = 50 mA cm^{-2} Time = 3 min	Inactivated effectively >2 log	Särkkä et al. (2008)

(*Continued*)

Table 9-2. EO Treatment for the Removal of Pathogens from Wastewater. (Continued)

Name of the pollutant	*Technology*	*EO conditions*	*Removal efficiency of the system*	*Reference*
Pseudomonas aeruginosa	Electrochemical oxidation of synthetic wastewater	Anode = BDD, IrO_2, RuO_2 *Pseudomonas aeruginosa* = 10^6 CFU mL^{-1}	Total inactivation using a BDD = 60 min Total inactivation using an IrO_2 or RuO_2 = 30 min	Bruguera-Casamada et al. (2017)
Coliform bacteria	Electrochemical oxidation of biologically treated domestic wastewater	Anode = BDD Current density = 120 mA cm^{-2} Time = 10 min	Coliform bacteria reduction: 4 log units	Schmalz et al. (2009)
Escherichia coli	Electrochemical oxidation of synthetic urine wastewaters	Anode = BDD, IrO_2, RuO_2, and Pt Electric charge = 1.34 Ah dm^{-3}	*Escherichia coli* deactivation: 100%	Dbira et al. (2019)
Escherichia coli	Electro-oxidation of synthetic urine wastewaters	Anode = mixed metal oxide Current density = 30.33 mA cm^{-2} Time = 45 min	*Escherichia coli* deactivation: 100%	Singla et al. (2020)

EO has shown qualified results in WW disinfection because of its ease of use, efficiency, and limited operational space. However, examples to show that EC treatment has been employed successfully in full-scale applications are few. The major obstacles to the commercialization of electrochemical processes are low current efficiencies and limited space–time yields (i.e., the amount of removed pollutants per unit volume and unit time), which in turn makes the process energy-intensive (Radjenovic and Sedlak 2015).

9.4 TREATMENT OF LANDFILL LEACHATE

In most countries, the oldest and most common method of waste disposal is landfill. Because of the physicochemical and biological changes in solid waste, the organic fraction of the waste is degraded in landfills, generating a highly contaminated liquid called *leachate*. The production of leachate is also as a result of the percolation of rainwater into landfills. A landfill produces a large volume of polluted leachate, which causes potential hazards for public health and ecosystems (Li et al. 2010).

Landfill leachate contains large amounts of organic contaminants and ammonia nitrogen (as the two principal chemicals) along with inorganic salts, chlorinated organic compounds, and heavy metals. Organic pollutants in leachate are identified by chemical oxygen demand (COD), 5 day biological oxygen demand (BOD_5), and TOC. In general, the characteristics of leachate are a function of age, precipitation, weather variation, waste type, and composition. In particular, the quality of landfill leachate changes over time (Bashir et al. 2013). Based on the age of a landfill, leachate is classified as young, intermediate, and old. Table 9-3 presents the characteristics of the various leachate types.

The young landfill leachate contains a high concentration of COD and BOD_5 and moderately high content of ammonium nitrogen. Besides, the ratio of BOD_5-to-COD is high. In contrast, the old landfill leachate is characterized by low COD and BOD_5-to-COD ratio. The decomposition of protein in waste releases ammonia, and, therefore, caution must be exercised. The concentration of ammonia does not show apparent changes over time. Besides, old leachate has a broad range of molecular weights with a large high-molecular-weight fraction (77% $>$ 10,000 Da) compared with young leachate (70% $<$ 500 Da) (Deng and Englehardt 2007, Li et al. 2010).

Landfill leachate is regarded as *too polluted WW* because of the high amount of organic substances and NH_3–N in it. Therefore, the treatment of landfill leachate is essential to meet the required standards before discharge to receiving water. EO is an admitted technology for the treatment of landfill leachate and can significantly reduce concentrations of organic pollutants, ammonia, and color in leachate. Various methods have been investigated for the treatment of leachate since the 1970s. Biological purifications are applicable for a high BOD_5-to-COD ratio. However, these methods fail because of the presence of low BOD_5-to-COD or

Table 9-3. Landfill Leachate Classification versus Age.

Type of leachate	*Young*	*Intermediate*	*Old*	*References*
Age (year)	<5	5–10	>10	Amokrane et al. (1997), Baig et al. (1999), Li et al. (2010)
pH	<6.5	6.5–7.5	>7.5	Abbas et al. (2009), Bhalla et al. (2013), Pasalari et al. (2019)
COD (mg L^{-1})	>10,000	4,000–10,000	<4,000	Adhikari and Khanal (2015), Bhalla et al. (2013), Li et al. (2010), Renou et al. (2008)
BOD_5/COD	>0.3	0.1–0.3	<0.1	Adhikari and Khanal (2015), Baig et al. (1999), Bhalla et al. (2013), Renou et al. (2008)
TOC/COD	<0.3	0.3–0.5	>0.5	Abbas et al. (2009), Li et al. (2010), Lim et al. (2016), Pasalari et al. (2019)
Ammonia nitrogen (mg L^{-1})	<400	400	>400	Abbas et al. (2009), Li et al. (2010), Lim et al. (2016)
Organic compounds	80% VFA	5%–30% VFA + HA + FA	HA + FA	Abbas et al. (2009), Adhikari and Khanal (2015), Li et al. (2010); Renou et al. (2008)
Biodegradability	High	Medium	Low	Adhikari and Khanal (2015), Bhalla et al. (2013), Pasalari et al. (2019), Renou et al. (2008)

Note: VFA = volatile fat acids; HA = humic acid; FA = fulvic acid.

a high amount of toxic metals. Physicochemical processes are employed to degrade a wide variety of refractory compounds in leachate, and the removal efficiency of organic substances in the EO of leachate is superior to that in other techniques (Deng and Englehardt 2007). This can be attributed to the high reactivities of oxidant agents such as hydroxyl radicals. EC of contaminants of landfill leachate is also performed through two approaches as mentioned previously: direct and indirect oxidation (Li et al. 2010).

The performance of EO depends on the anode material used, current density, chloride concentration, and electrolyte concentration. Various anode materials have been investigated for the EC of leachate, and the result shows that anodes with a high anodic oxygen evaluation potential yield high COD and NH_3–N removal efficiencies. However, the effect of cathode materials has not been studied extensively. Besides, an increase in current density also improves COD and NH_3–N removal rates. The presence of Cl^- in solution enhances indirect oxidation and, consequently, facilitates the removal of contaminants. As mentioned, the addition of Cl^- releases dangerous by-products. It should be noted that Cl^- is present in landfill leachate, but the concentration is not sufficient for significant indirect oxidation. The addition of electrolytes can lead to some improved efficiencies in pollutant removal (e.g., by adding SO_2^{-4}, the elimination of COD increases). Moreover, by adding H_2O_2, both COD and NH_3–N removal rates improve. The pH is not of an apparent effect on pollutant removal. Finally, pretreatment can lead to improved removal efficiency and decrease consumption energy. Carbon adsorption mostly removes low-molecular-weight organics and is a preferred option to be combined with EO, which is suitable for the destruction of high-molecular-weight organics (Deng and Englehardt 2007).

Anode materials for landfill leachate treatment are usually Ti–TiO_2–RuO_2–IrO_2, graphite carbon, aluminum, Ti/PbO_2, BDD, iron, PbO_2/Ti, binary Ru–Ti-oxide-coated titanium (DSA type), and ternary Sn–Pd–Ru-oxide-coated titanium. Cathodes are stainless steel, graphite carbon, aluminum, and iron (Bashir et al. 2013).

Turro et al. (2011) employed EO for the treatment of stabilized landfill leachate with a COD of 2,960 mg L^{-1}. They used Ti/IrO_2–RuO_2 as an anode and investigated the effect of electrolysis time, temperature, current density, pH of the initial effluents, $HClO_4$ concentration, and the addition of NaCl or Na_2SO_4 on performance. They concluded that the essential parameters that influenced the process were the pH of the effluents and the addition of salts. They obtained 90% COD, 65% total carbon, and complete color and total phenols removal at a 240 min treatment time, with 32 mA cm^{-2} current density, and a temperature of 80 °C. The electricity consumption indicated the removal of 35 kW · h/kg of COD (Turro et al. 2011). Fernandes et al. (2012) carried out EO as a post-treatment of biological treatment of a leachate from an intermunicipal sanitary landfill. They used BDD to eliminate the remaining biorefractory organic matter from the biological treatment effluent. They investigated the effect of the dilution of the leachate sample and the applied current density on the performance of EO. They found either an increase in the dilution or a dramatic increase in energy consumption

Table 9-4. EO Treatment of Landfill Leachate.

Name of pollutant	*Technology*	*EO conditions*	*Removal efficiency of the system*	*Reference*
Ammonium	Electro-oxidation raw landfill leachate	Anode = BDD Current densities = 90 mA cm^{-2} $N{-}NH_4^+$ = 1,900 mg L^{-1} Time = 360 min	Ammonium removal: 100%	Cabeza et al. (2007)
COD, total carbon (TC), total phenols (TPh), and color	Electro-oxidation of stabilized landfill leachate	Anode = $Ti/IrO_2{-}RuO_2$ Current density = 32 mA cm^{-2} Time = 240 min, T = 80°C	COD removal: 90% TC removal: 65% Color removal: 100% TPh removal: 100%	Turro et al. (2011)
COD and ammonium	Combined system of activated sludge, Fenton oxidation, and electro-oxidation	Anode = BDD COD = 4,430 mg L^{-1} Ammonium = 1,225 mg L^{-1}	COD <160 mg L^{-1} Ammonium < 15 mg L^{-1}	Urtiaga et al. (2009)
COD and ammonium	Combined system of biological treatment, ultrafiltration, disc tube reverse osmosis (RO), and electro-oxidation	Anode = BDD Current density = 50 mA cm^{-2} Flow velocity = 6 L h^{-1} pH = 5.16 Time = 6 h Energy consumption = 223.2 kWh m^{-3}	COD removal: 87.5% Ammonium removal: 74.06%	Zhou et al. (2016)
COD, ammonium, color, and turbidity	Combined system of coagulation-flocculation (alum) and electrochemical oxidation	Anode = BDD Current density of 50 mA cm^{-2} for 90 min, followed by that of 25 mA cm^{-2} for 90 min	COD removal > 90% Complete elimination of color, turbidity, and ammonium	de Oliveira et al. (2019)
COD and TOC	Combined system of ultrafiltration and electro-oxidation	Anode = BDD Current density of 360 A m^{-2} Time = 4 h	COD removal: 92% TOC removal: 86.8%	Ukundimana et al. (2018)

because of the current intensity. On the contrary, the mineralization of organic matter improved after the dilution of leachate (Fernandes et al. 2012). Ding et al. (2018) used an integrated electro-oxidation/electro-coagulation/electroreduction (EO/EC/ER) process for tertiary landfill leachate treatment. They assessed the influence of operation conditions such as anode materials, interelectrode gap, and WW flow velocity on the performance of the process. They achieved 50% to 60% removal of organic matter and 100% ammonia removal with the EO/EC/ER process in a scaled-up reactor with an energy consumption of 15 $kW \cdot h\ m^{-3}$ (Ding et al. 2018). Table 9-4 presents the treatment of landfill leachate by EO.

As previously mentioned, the main advantage of EO is to oxidize organic pollutants into water and carbon dioxide by avoiding the transfer of contaminants from one phase to another. Also, the process is carried out at room temperature and atmospheric pressure, which prevents the volatilization and discharge of unreacted waste (Deng and Englehardt 2007). Although high energy consumption and potential for the formation of chlorinated organics may limit the application of EO techniques, EO is a powerful technology, especially for a low BOD_5-to-COD ratio or a highly toxic landfill leachate for which treatment by biological processes proves to be a failure or is ineffective.

9.5 TREATMENT OF AGRICULTURAL AND AQUACULTURE WASTEWATERS

The main sources of pollution in agricultural WW are toxic substances, organic and inorganic compounds such as heavy metals, pesticides or plant protection products, phenols, and derivatives that are used in farming for improving crop yield. An increase in food production will lead to a consequent increase in the volume of sewage and the cost of disposal. The majority of pesticides used are organochlorine molecules. This compound has been prohibited because of its high toxicity. Therefore, organochlorine compounds have been substituted by organophosphorus molecules, which have a lower persistence and are more selective in nature (Floesser-Mueller and Schwack 2001).

The type and nature of agroindustrial waste affect treatment processes. The conventional physical, biological, and chemical methods can be used for treating agricultural WW. The self-purification capacity of streams, the permissible levels of pollutants in water bodies, and the economic interests of industries impact the selection of treatment processes. Advanced technologies such as membrane separation processes [e.g., reverse osmosis (RO) and ultrafiltration] and advanced oxidation processes are being increasingly surveyed for WW treatment (Pouran et al. 2015).

The common methods for agricultural WW treatment are the use of physicochemical and biological processes. The organic load or toxicity of WW (and suspended solids) from different industries can be reduced by these chemical treatments. However, the yields of the two methods decline in the presence of biorefractory compounds such as pesticides and, too often, secondary waste is produced (Ikehata and Gamal El-Din 2005). For solving this problem, several processes have been developed. Electrochemical advanced oxidation processes (EAOPs) can help overcome this problem. Easy implementation, no chemical addition, and high efficiency are advantages of this process (Brillas et al. 2009, Zhu et al. 2009). Furthermore, because of the presence of high salinity and variation in WW flow, this treatment is not prone to failure (Vijayaraghavan et al. 2008).

During advanced oxidation process processes, hydroxyl-free radicals are generated, which have a high electrochemical oxidant potential for a complete mineralization of pollutants (Durán et al. 2015).

Treatment of WW is one of the challenges facing the aquaculture industry. Most of the aquaculture farms operate on batch exchanges and conventional biological treatments cannot be used in these farms, because first, there are wide fluctuations in the flow rate. Second, aquaculture WW is saline in nature. This salt compound prevents the application of biological treatment methods unless special bacterial halophilic strains. Therefore, electrochemical methods of treating aquaculture WW are favored because either they are not prone to failure on account of the presence of high salinity or there is variation in WW flow or strength. High salinity ensures excellent electric conductivity that can reduce energy consumption, and the high chloride concentration improves indirect oxidation through the electrogeneration of strong oxidants such as hypochlorous acid (HOCl); these are the advantages of electrochemical treatment. Also, this method prevents the production of toxic chlorine gas that is used in the breakpoint chlorination process, because the oxidants are electrolytically generated in situ, and this generation can be handled by the operating conditions of electrolysis (Alfafara et al. 2004). The in situ generation of HOCl is advantageous, because the aqueous solution of sodium hypochlorite is safer to use than chlorine gas. The present-day commercial-scale electrolytic systems are more rugged because of the availability of dimensionally stable electrode material (e.g., Pt and Ti) and they are not prone to experience loss in the electrode area. Moreover, online adjustment of electrolyte pH remarkably increases the efficiency of the process. A constraint in the EC process arises if the WW that is to be treated is low in salinity. During the electrolytic oxidation process, sodium chloride (NaCl) is added to improve conductivity and to generate chlorine. Hence, the added NaCl should be removed before discharge into inland water; otherwise, it may lead to pseudo hardness (Vijayaraghavan et al. 2008).

The removal of ammonia and nitrite in the treatment of fish culture water has been surveyed by several researchers based on the following methods: selective ion exchange (Miladinovic and Weatherley 2008), flocculation (Ebeling et al. 2005), the use of sand beds (Palacios and Timmons 2001), and the use of membrane bioreactors (Sharrer et al. 2007).

Physicochemical and biological are two main categories of treatment of aquaculture WW. The conventional methods have the ability to remove nitrate and organic carbon, but during this process, sludge is produced. So, recently, EO treatment was used for removing the mentioned compounds. The reduction of nitrate ions to nitrate and finally to nitrogen gas on the surface of the cathode can be realized by using EO. The efficiency of this process depends on the electrode material, current input, pH, conductivity, the concentration of NaCl, and the initial concentration of nitrate ions. Nitrite ions (NO_2^-) is highly soluble in water and can react with water to produced nitrogen gas, ammonia, and hydroxylamine (NH_2OH). The desired result is the production of nitrogen gas from nitrate, but usually ammonia is formed. Sodium bicarbonate ($NaHCO_3$) is necessary to adjust the pH during electrochemical treatment because the electrolyte gradually becomes alkaline. The two main end products are ammonia and nitrite, but they negatively affect the efficiency of the whole process, thus posing a major limitation. To negate this impact, chloride salt is widely added to control the reaction. In this process, at the anode, chlorine is oxidized and it reacts with water, and as a result, HOCl is formed. The hypochlorite ions then react with nitrite and ammonia to produce nitrate and nitrogen. Table 9-5 summarizes different treatment methods for agricultural and aquaculture WW.

9.6 TREATMENT OF PETROLEUM WASTEWATERS

Various organic and inorganic components in oilfield WW or produced water (PW) are present that can mix with groundwater and soil and pollute the surfaces (Fakhru'l-Razi et al. 2009). PW components can have some effects on the environment as follows: an increase in salinity; dispersed and soluble oil contributions in marine ecosystems; the inclusion of other compounds from treating chemicals; a higher concentration of heavy metals than in seawater; and the presence of radionuclides.

The operational conditions and chemicals used in process facilities for treatment of PW is different by changing the nature of the producing/storage formation which are withdrawn. The composition of PW can be considered as oil and/or gas production (Ekins et al. 2007). PW includes the following compounds in most streams:

1. Dissolved and dispersed oil compounds [a mixture of benzene, toluene, ethylbenzene, and xylenes, naphthalene, phenanthrene, dibenzothiophene, polyaromatic hydrocarbons (PAHs), and phenols];
2. Dissolved formation minerals;
3. Production chemical compounds (including some chemicals that are added to treat or prevent operational problems), treatment chemicals (production treating, gas processing, and stimulation) and production-treating chemicals (scale and corrosion inhibitors, biocides, emulsion breakers, antifoam);
4. Production solids (including solid formation, corrosion, and scale products, bacteria, waxes, and asphaltenes); and
5. Dissolved gases. Water cannot dissolve all hydrocarbons; thus, most of the oil is dispersed in water; and the dissolved and suspended oil present in PW (prior to treatment) depend on the presence of several factors (Fakhru'l-Razi et al. 2009).

Table 9-5. Different Treatment Methods for Agricultural and Aquaculture Wastewater.

Type of treatment		*Method*	*Advantages*	*Disadvantages*	*Reference*
Conventional method	Physicochemical treatments	Ion exchange (IE) Reverse osmosis (RO) Electrodialysis (ED) Activated carbon adsorption	High permeability efficiency of selective ions Low production costs Very little feed pretreatment is required	The cost of operation is higher because of high energy expenditure Fouling is also an issue as it affects the membrane performance and increases the complexity in the membrane operations The generation of toxic by-products	Ekins et al. (2007)
	Biological treatment	Trickling filters Fluidized-bed reactor Rotating biological contactor Biofloc technology wetlands			
Electrochemical technology			High efficiency Ambient operating conditions Small equipment sizes Minimal sludge generation and rapid start-up	Production of ammonia and nitrite that limit its efficiency. The addition of chloride salts partially solves the problem, but the nitrate removal efficiency remains poor.	Ekins et al. (2007)

Various physical, chemical, and biological methods can be used to reduce the oil content and salinity of PW from both offshore and onshore facilities (Fakhru'l-Razi et al. 2009). The basic process involved in all these methods is the transfer of contaminants from WW to other media. Among these different techniques, the electrochemical method, because of its selectivity, cost-effectiveness, and versatility, attracts more attention than the other methods for treating WW in the output stream of refineries (Fu et al. 2014).

In offshore facilities, compact chemical and physical methods are preferred because of space constraints. However, both the capital cost of these physical methods and the price of the chemicals for removing hazardous sludge in the chemical method are high. Thus, the use of these methods is limited. These methods are also not applicable in cases where there are suspended oil or organic and inorganic components. Therefore, an effective method for removing suspended compounds is biological treatment.

For the treatment of petroleum refinery effluents by electrochemical methods, the electrode material is the main factor that can affect process efficiency because it can alter both the oxidation mechanism and anodic reaction. Ruthenium has good properties with excellent conductivity (Terezo and Pereira 2002). In a laboratory pilot-scale plant that included double anodes with active metal and graphite and iron as cathode and a noble metal content catalyst with a large surface, the COD and BOD of oilfield-PW could be reduced by over 90% in 6 min. In these processes, the produced Mn^{2+} ions oxidized and coagulated organic pollutants including bacteria (Ma and Wang 2006). The BBD electrode can be effectively used in water treatment because of its good electroconductivity and chemical inertness (Hupert et al. 2003). Table 9-6 summarizes different treatment methods for petroleum WW.

9.7 TREATMENT OF TEXTILE WASTEWATER

The textile industry consumes a large amount of water and generates large volumes of recalcitrant dyes and other textile processing chemicals in textile WW. This effluent containing high dissolved solids and inappropriate pH, COD, and color. (Bisschops and Spanjers 2003). Therefore, it is necessary to treat textile WW before its discharge to the environment because it can damage the environment (Bisschops and Spanjers 2003, Kim et al. 2003, Ölmez et al. 2007).

Textile WW contains different dyes and other chemicals such as salts, fabric softeners, and antiwrinkle agents; pesticides; coupling agents; polishing and coating agents; surfactants; and so on. Thus, the treatment efficiency is limited by the specific properties of these chemicals, and also, the electrochemical reactions will be highly complex. Different techniques are available for the treatment of textile WW, such as biological processes, adsorption, membrane process, and chemical coagulation, which are all complex in nature. For removing COD, color, and other contaminants from textile WW, traditional biological and

Table 9-6. Different Treatment Methods for Petroleum Wastewater.

Type of treatment	*Method*	*Advantages*	*Disadvantages*	*Reference*
Conventional method	Gravity separation and skimming, air flotation, coagulation, de-emulsification, and flocculation	Simplicity and adaptability	• Low efficiency • High operation cost • Corrosion and generates a considerable quantity of secondary pollutants (chloride and sulfate in coagulation–precipitation) • Large volumes of sludge or waste that pose serious environmental problems • Because of the presence of sulfide that affects the oxygen transfer of the biosystems, only a partial inhibition of biodegradation takes place, leading to inefficiency	Ekins et al. (2007)
	Biological treatment			
Electrochemical technology		Efficient in removing soluble inorganic solids, suspended solids, and oil and grease Relatively low amount of sludge	To achieve high efficiency, it is necessary to combine this method with other techniques	Ekins et al. (2007)

physicochemical methods have been used. Synthetic dyes are aromatic in nature, contain antimicrobial agents, and, thus, are resistant to biological treatment methods (O'Neill et al. 1999).

Also, the aforementioned techniques suffer from several technical problems. The price of chemical coagulants and adsorbents, as well as the cost to handle membrane fouling, is highly expensive, and these are the critical drawbacks of chemical coagulation, adsorption, and membrane processes. In addition, production of secondary pollutants is another drawback of these techniques (Cañizares et al. 2006, Kaur et al. 2015). For example, aluminum sulfate as an inorganic coagulant can remain in treated water as a trace, and its entry into the human body can lead to disorders. Although the biological method is environmentally friendly, many synthetic dyes are nonbiodegradable in nature and, therefore, need additional treatment for removing effluents. Furthermore, a long reaction time and high energy are required for aerobic and anaerobic biological treatments (Kushwaha et al. 2010). The drawbacks of these techniques encouraged researchers to search for other techniques.

By advanced oxidation methods, the limitations of the traditional methods can be overcome. In the EO method, electrons are main reagents that oxidize pollutants into carbon dioxide and water or other oxides (Chen 2004) without generating secondary pollutants. EO has maximum efficiency in terms of COD and color removal (Vlyssides et al. 1999).

The research on a continuous EO reactor shows that the main process parameters such as pH, current (i), and electrolysis time (t) can affect removal of COD and color, energy consumed, and efficiency of EO treatment for textile WW.

The nature of electrodes and electrolytes are the main factors that impact the performance of EO in nonbiodegradable WW such as textile WW. The EO of textile WW in batch reactors has been investigated using different types of electrocatalysts (Doğan and Türkdemir 2005, Kaur et al. 2015, 2017, Kim et al. 2003, Körbahti and Tanyolaç 2009). Although coated electrodes have low performance in WW with a high chloride content, some anodes are highly efficient toward the oxidation of pollutants such as Ti/PbO_2, Ti/IrO_2, BDD, Ti/SnO_2, graphite, and Ti/RuO_2. PbO_2 and SnO_2 anodes have a high oxygen overpotential (1.9 V) (Anglada et al. 2009, Chen 2004), whereas BDD anodes possess a greater oxygen overpotential (2.7 V), and coupled with high performance, BDD electrodes are expensive (Britto-Costa and Ruotolo 2012). Because of their low oxygen overpotential, Ti/IrO_2 and graphite electrodes show poor performance (Pulgarin et al. 1994, Rodgers et al. 1999). Ti/RuO_2 anodes have high mechanical and chemical strength with an oxygen overpotential of approximately 2.0 V and are dimensionally stable. In addition, they produce strong oxidants (Cl_2, HOCl, and ClO^-) with a high concentration of chloride (Kaur et al. 2017).

Hydroxyl radicals ($^{\bullet}OH$) are produced on the anode as a result of electrolysis, which are adsorbed on the surface of the anode (Ti/RuO_2) and oxidize pollutants by the direct oxidation mechanism (Chen 2004, Santos et al. 2010). Electrochemically generated oxidants such as Cl_2, HOCl, and ClO^-, $^{\bullet}OH$ radicals, H_2O_2 and $^{\bullet}OCl$

radicals and so on may be involved in the degradation of pollutants via both direct and indirect oxidation methods. It has been reported that chlorine indirectly degrades pollutants but has a drawback of possibly generating carcinogenic/toxic chlorinated organic compounds.

Most researchers have used batch EO reactors to treat simulated textile WW. Batch EO treatment is not preferable because WW produced from textile industries has high flow rates and a high pollution load (Bisschops and Spanjers 2003, Kim et al. 2003). A few studies (Körbahti and Tanyolaç 2009; Mukimin et al. 2015, 2017; Sakalis et al. 2005) are available for continuous EO textile WW treatment. Sakalis et al. (2005) and Körbahti and Tanyolaç (2009) conducted continuous EO experiments with simulated textile WW.

Mukimin et al. (2015) treated textile WW in a continuous mode using an electrocatalytic tube reactor with a Ti/PbO_2 cylinder electrode to remove BOD, COD, TSS, NH_3, phenol, and H_2S. PbO_2-coated anodes have been reported to be electrochemically instable because of electrochemical corrosion and a lower performance in WW with a high chloride content (Cossu et al. 1998), which may further contaminate the treated WW with Pb. Mukimin et al. (2017) used Ti/$RuIrO_2$ mesh cylinder electrodes for the treatment of printing WW to remove COD and color. Moreover, interesting results on degraded/transformed products were presented on the basis of indirect oxidation, and various chlorinated organic compounds were identified.

Table 9-7. Different Treatment Methods for Textile Wastewater.

Type of treatment	*Method*	*Limitation*	*Reference*
Conventional method	Integration of biological, chemical, and physical methods	• Because of the large variability of the composition of textile wastewaters, these methods are inadequate and insufficient • High cost for the treatment of the effluents that are produced • Requiring high land cost • Complexity of handling chemicals	Ekins et al. (2007)
Electrochemical technology	Electrolysis and electrocoagulation	High initial capital costs	Ekins et al. (2007)

For treating WW containing high NaCl salt, as typical textile WW, Ti/RuO_2 anodes are more appropriate. Carcinogenic and toxic chlorinated organic compounds may be produced during the reaction because there is a high amount of NaCl. Table 9-7 summarizes different treatment methods for textile WW.

9.8 TREATMENT OF MUNICIPAL WASTEWATER

Municipal WW is defined as WW that is produced because of human activities in households and industries. Organic and inorganic material, pathogenic organisms, nutrients, and many toxic elements such as heavy metals may be present in municipal WW (Dixit et al. 2011). Some methods are available for treating this type of WW, but some of them will produce sludge that needs disposal. The conventional method for treatment of municipal WW is the biological process. A cost-effective method for the treatment of municipal WW is algae-based treatment, which needs low energy and decreases sludge production. Coliform bacteria, COD, BOD, nitrogen, and phosphorus can be removed by this way (Santhanam 2009). However, some drawbacks in this technology exist, such as operating cost, operating difficulties, low efficiency, requirement of continuous air supply, sensitivity to shock toxic loads, and sludge volume. Also, this method is not appropriate for a large range of organic compounds in WW.

The soluble nitrogen in domestic WW exists as ammonia and urea. These compounds cannot be removed by clarification or adsorption. Air stripping of water above a pH of 10.5, breakpoint chlorination, and selective ion exchange are three physical–chemical methods for removing ammonia (Mercer et al. 1970, O'Farrell et al. 1972, Pressley et al. 1972). Because of the low volume of sludge from electrochemical treatment in comparison with the conventional methods, this method is favorable. Although electrocoagulation can impact different contaminants compared with the EO process, this method is preferable over others. The reason for this is that the amount of sludge that is produced is negligible and the final products are CO_2 and H_2O. Electrolyte type, voltage, and reaction time can impact efficiency. EC can be used to omit irons, silicates, humus, and dissolved oxygen, reduce copper, and decolorization (Chen 2004). In EC, in situ coagulation is produced by the electrolytic oxidation of a suitable anode material. A soluble electrode such as Fe or Al must be selected as the anode material. A rectangular electrolytic cell with two circular anodes and cathodes is used. Different anode materials such as Ti/BDD, Ti/IrO_2, and Ti/IrO_2–RuO_2 are used in this method. The most effective electrode for domestic WW treatment is the Ti/BDD electrode. Direct anodic EO by means of OH^- and indirect EO with mediators such as hydrogen peroxide generated by the cathodic reduction can oxidize domestic WW. In both EO processes, refractory organic compounds can be oxidized because of the formation of powerful oxidizing agents (Daghrir et al. 2014). Table 9-8 summarizes different treatment methods for municipal WW.

Table 9-8. Different Treatment Methods for Municipal Wastewater.

Type of treatment	*Method*	*Limitation*	*Reference*
Conventional method	Aerated biological methods	Requiring continuous air supply High operating costs (skilled labor, energy, etc.) Sensitivity against shock toxic loads Longer treatment time Necessary sludge disposal	Ekins et al. (2007)
Electrochemical technology		Large capital investment Expensive electricity supply	Ekins et al. (2007)

9.9 CHALLENGES AND FUTURE PERSPECTIVE

Reducing the operating costs and the development of efficient and stable electrode materials are the two main challenges in WW treatment. Using flow-through reactors with three-dimensional electrodes is an effective and sound approach. By this way, an increase in efficiency is seen compared with that in the design of a conventional two-dimensional reactor. Reducing the price of electrode material and increasing the stability and performance of electrolytes are other subjects for research. For achieving better and more cost-effective electrodes, it is necessary to improve the electrocatalytic activity and electrochemical stability of the electrode materials. By this way, the operational and capital costs can be reduced.

The formation of toxic by-products by halide-containing WW is another challenge in EO. When oxidation happens, measurements of toxicity, chlorate, and perchlorate are necessary. Electrochemical systems are appropriate for treating hazardous organic waste. For instance, BDD and Ti/SnO_2 can be applied as anode material for treating aqueous film-forming foams that consist of high concentrations of fluorinated surfactants.

In the case of a high concentration of halides, EO may be integrated with other technologies. However, because of increased toxicity, electrochemical treatment is not a viable technical and economical choice for treatment of WW.

The lack of need for storage, a compact and modular reactor design, and the ability to adjust to different organic quantities in WW are technical aspects for research in an electrolyte cell.

9.10 CONCLUSION

The problem of recalcitrant and toxic pollutants in conventional methods of WW treatment can be overcome by using the present-day EC technique, which is sound and effective. This highly practical method has been found useful in different industries such as soil leachate, hospital WW, agricultural and aquaculture WWs, petroleum WW, textile WW, municipal, and domestic. As the electron is a clean reagent, EAOPs do not emit pollution, and this factor is an advantage in this method. The advantages of this technology are high energy efficiency and amenability to automation. Because of the simple equipment that is used, this method is easy to handle and is safe because they operate under ambient temperature and pressure. Because this method can be applied to effluents with COD in the range of 0.1 to 100 g L^{-1}, it is versatile. The costs related to electrical supply, the low conductance of many WWs, requiring the addition of electrolytes, and the loss of activity and shortening of the electrode lifetime by fouling because of the deposition of organic material on the surface are the main problems encountered in this technology. Furthermore, cell configuration, electrode material, electrolyte composition, micro-organism or pollutant type and other experimental parameters such as current density or the temperature of the treated water can impact the efficiency of this method.

NOMENCLATURE

AOP = Advanced oxidation process
BDD = Boron-doped diamond
BOD = Biological oxygen demand
COD = Chemical oxygen demand
EC = Electrocoagulation
EO = Eectro-oxidation
ER = Electroreduction
KP = Ketoprofen
PW = Produced water
PhAC = Pharmaceutically active compound
POP = Persistent organic pollutant
TOC = Total organic carbon
WW = Wastewater

References

Abbas, A. A., G. Jingsong, L. Z. Ping, P. Y. Ya, and W. S. Al-Rekabi. 2009. "Review on landfill leachate treatments." *J. Appl. Sci. Res.* 5 (5): 534–545.

Adhikari, B., and S. N. Khanal. 2015. "Qualitative study of landfill leachate from different ages of landfill sites of various countries including Nepal." *J. Environ. Sci. Toxicol. Food Technol.* 9 (1): 2319–2399.

Alfafara, C. G., T. Kawamori, N. Nomura, M. Kiuchi, and M. Matsumura. 2004. "Electrolytic removal of ammonia from brine wastewater: Scale-up, operation and pilot-scale evaluation." *J. Chem. Technol. Biotechnol.* 79 (3): 291–298.

Amokrane, A., C. Comel, and J. Veron. 1997. "Landfill leachates pretreatment by coagulation-flocculation." *Water Res.* 31 (11): 2775–2782.

Anglada, A., A. Urtiaga, and I. Ortiz. 2009. "Contributions of electrochemical oxidation to waste-water treatment: Fundamentals and review of applications." *J. Chem. Technol. Biotechnol.* 84 (12): 1747–1755.

Badmus, K. O., J. O. Tijani, E. Massima, and L. Petrik. 2018. "Treatment of persistent organic pollutants in wastewater using hydrodynamic cavitation in synergy with advanced oxidation process." *Environ. Sci. Pollut. Res.* 25 (8): 7299–7314.

Baig, S., I. Coulomb, P. Courant, and P. Liechti. 1999. "Treatment of landfill leachates: Lapeyrouse and Satrod case studies." *Ozone: Sci. Eng.* 21 (1): 1–22.

Bashir, M. J. K., H. A. Aziz, S. Q. Aziz, and S. S. Abu Amr. 2013. "An overview of electro-oxidation processes performance in stabilized landfill leachate treatment." *Desalin. Water Treat.* 51 (10–12): 2170–2184.

Bergmann, M. E. H., and J. Rollin. 2007. "Product and by-product formation in laboratory studies on disinfection electrolysis of water using boron-doped diamond anodes." *Catal. Today* 124 (3–4): 198–203.

Bhalla, B., M. Saini, and M. Jha. 2013. "Effect of age and seasonal variations on leachate characteristics of municipal solid waste landfill." *Int. J. Res. Eng. Technol.* 2 (8): 223–232.

Bisschops, I., and H. Spanjers. 2003. "Literature review on textile wastewater characterisation." *Environ. Technol.* 24 (11): 1399–1411.

Brillas, E., I. Sirés, C. Arias, P. L. Cabot, F. Centellas, R. M. Rodríguez, et al. 2005. "Mineralization of paracetamol in aqueous medium by anodic oxidation with a boron-doped diamond electrode." *Chemosphere* 58 (4): 399–406.

Brillas, E., I. Sirés, and M. A. Oturan. 2009. "Electro-Fenton process and related electrochemical technologies based on Fenton's reaction chemistry." *Chem. Rev.* 109 (12): 6570–6631.

Britto-Costa, P. H., and L. A. M. Ruotolo. 2012. "Phenol removal from wastewaters by electrochemical oxidation using boron doped diamond (BDD) and Ti/Ti0.7Ru0.3O2 DSA® electrodes." *Braz. J. Chem. Eng.* 29 (4): 763–773.

Bruguera-Casamada, C., I. Sirés, E. Brillas, and R. M. Araujo. 2017. "Effect of electrogenerated hydroxyl radicals, active chlorine and organic matter on the electrochemical inactivation of *Pseudomonas aeruginosa* using BDD and dimensionally stable anodes." *Sep. Purif. Technol.* 178: 224–231.

Cabeza, A., A. Urtiaga, M.-J. Rivero, and I. Ortiz. 2007. "Ammonium removal from landfill leachate by anodic oxidation." *J. Hazard. Mater.* 144 (3): 715–719.

Cañizares, P., F. Martínez, C. Jiménez, J. Lobato, and M. A. Rodrigo. 2006. "Coagulation and electrocoagulation of wastes polluted with dyes." *Environ. Sci. Technol.* 40 (20): 6418–6424.

Cano, A., P. Cañizares, C. Barrera-Díaz, C. Sáez, and M. A. Rodrigo. 2012. "Use of conductive-diamond electrochemical-oxidation for the disinfection of several actual treated wastewaters." *Chem. Eng. J.* 211–212: 463–469.

Chen, G. 2004. "Electrochemical technologies in wastewater treatment." *Sep. Purif. Technol.* 38 (1): 11–41.

Cossu, R., A. M. Polcaro, M. C. Lavagnolo, M. Mascia, S. Palmas, and F. Renoldi. 1998. "Electrochemical treatment of landfill leachate: Oxidation at Ti/PbO_2 and Ti/SnO_2 anodes." *Environ. Sci. Technol.* 32 (22): 3570–3573.

Daghrir, R., P. Drogui, and J. Tshibangu. 2014. "Efficient treatment of domestic wastewater by electrochemical oxidation process using bored doped diamond anode." *Sep. Purif. Technol.* 131: 79–83.

Dbira, S., N. Bensalah, M. I. Ahmad, and A. Bedoui. 2019. "Electrochemical oxidation/disinfection of urine wastewaters with different anode materials." *Materials* 12 (8): 1254.

Deng, Y., and J. D. Englehardt. 2007. "Electrochemical oxidation for landfill leachate treatment." *Waste Manage.* 27 (3): 380–388.

de Oliveira, M. S., L. F. da Silva, A. D. Barbosa, L. L. Romualdo, G. Sadoyama, and L. S. Andrade. 2019. "Landfill leachate treatment by combining coagulation and advanced electrochemical oxidation techniques." *ChemElectroChem* 6 (5): 1427–1433.

Ding, J., L. Wei, H. Huang, Q. Zhao, W. Hou, F. T. Kabutey, et al. 2018. "Tertiary treatment of landfill leachate by an integrated electro-oxidation/electro-coagulation/electro-reduction process: Performance and mechanism." *J. Hazard. Mater.* 351: 90–97.

Dixit, A., S. Dixit, and C. Goswami. 2011. "Process and plants for wastewater remediation: A review." *Sci. Rev. Chem. Commun.* 11: 71–77.

Doğan, D., and H. Türkdemir. 2005. "Electrochemical oxidation of textile dye indigo." *J. Chem. Technol. Biotechnol.* 80 (8): 916–923.

Durán, A., J. M. Monteagudo, J. Gil, A. J. Expósito, and I. San Martín. 2015. "Solar-photo-Fenton treatment of wastewater from the beverage industry: Intensification with ferrioxalate." *Chem. Eng. J.* 270: 612–620.

Ebeling, J. M., K. L. Rishel, and P. L. Sibrell. 2005. "Screening and evaluation of polymers as flocculation aids for the treatment of aquacultural effluents." *Aquacult. Eng.* 33 (4): 235–249.

Ekins, P., R. Vanner, and J. Firebrace. 2007. "Zero emissions of oil in water from offshore oil and gas installations: Economic and environmental implications." *J. Cleaner Prod.* 15 (13–14): 1302–1315.

Fakhru'l-Razi, A., A. Pendashteh, L. C. Abdullah, D. R. A. Biak, S. S. Madaeni, and Z. Z. Abidin. 2009. "Review of technologies for oil and gas produced water treatment." *J. Hazard. Mater.* 170 (2–3): 530–551.

Fallahzadeh, R. A., and F. Omidi. 2019. "Electro-oxidation as an effective process for removing antibiotics and persistent organic compounds resistant to biodegradation." *J. Environ. Health Sustainable Dev.* 4 (4): 862–865.

Farhat, A., J. Keller, S. Tait, and J. Radjenovic. 2015. "Removal of persistent organic contaminants by electrochemically activated sulfate." *Environ. Sci. Technol.* 49 (24): 14326–14333.

Fernandes, A., M. J. Pacheco, L. Ciríaco, and A. Lopes. 2012. "Anodic oxidation of a biologically treated leachate on a boron-doped diamond anode." *J. Hazard. Mater.* 199–200: 82–87.

Floesser-Mueller, H., and W. Schwack. 2001. "Photochemistry of organophosphorus insecticides." In *Reviews of environmental contamination and toxicology*, edited by G. W. Ware, 129–228. New York: Springer.

Frontistis, Z., C. Brebou, D. Venieri, D. Mantzavinos, and A. Katsaounis. 2011. "BDD anodic oxidation as tertiary wastewater treatment for the removal of emerging micro-pollutants, pathogens and organic matter." *J. Chem. Technol. Biotechnol.* 86 (10): 1233–1236.

Fu, L., G. Lai, P. J. Mahon, J. Wang, D. Zhu, B. Jia, et al. 2014. "Carbon nanotube and graphene oxide directed electrochemical synthesis of silver dendrites." *RSC Adv.* 4 (75): 39645–39650.

Ganizadeh, F., A. Rezaee, and H. Godini. 2017. "Elimination of pathogenic bacteria using electrochemical process containing steel mesh electrode." *J. Adv. Environ. Health Res.* 5 (1): 23–28.

Gerardi, M. H., and M. C. Zimmerman. 2004. *Wastewater pathogens*. Hoboken, NJ: Wiley.

Hupert, M., A. Muck, J. Wang, J. Stotter, Z. Cvackova, S. Haymond, et al. 2003. "Conductive diamond thin-films in electrochemistry." *Diamond Relat. Mater.* 12 (10–11): 1940–1949.

Ikehata, K., and M. Gamal El-Din. 2005. "Aqueous pesticide degradation by ozonation and ozone-based advanced oxidation processes: A review (Part I)." *Ozone Sci. Eng.* 27 (2): 83–114.

Jeong, J., C. Kim, and J. Yoon. 2009. "The effect of electrode material on the generation of oxidants and microbial inactivation in the electrochemical disinfection processes." *Water Res.* 43 (4): 895–901.

Jeong, J., J. Y. Kim, M. Cho, W. Choi, and J. Yoon. 2007. "Inactivation of *Escherichia coli* in the electrochemical disinfection process using a Pt anode." *Chemosphere* 67 (4): 652–659.

Karpińska, J., and U. Kotowska. 2019. *Removal of organic pollution in the water environment.* Basel, Switzerland: Multidisciplinary Digital Publishing Institute.

Katsoyiannis, A., and C. Samara. 2004. "Persistent organic pollutants (POPs) in the sewage treatment plant of Thessaloniki, northern Greece: Occurrence and removal." *Water Res.* 38 (11): 2685–2698.

Kaur, P., J. P. Kushwaha, and V. K. Sangal. 2017. "Evaluation and disposability study of actual textile wastewater treatment by electro-oxidation method using Ti/RuO_2 anode." *Process Saf. Environ. Prot.* 111: 13–22.

Kaur, P., V. K. Sangal, and J. P. Kushwaha. 2015. "Modeling and evaluation of electro-oxidation of dye wastewater using artificial neural networks." *RSC Adv.* 5 (44): 34663–34671.

Kim, S., C. Park, T.-H. Kim, J. Lee, and S.-W. Kim. 2003. "COD reduction and decolorization of textile effluent using a combined process." *J. Biosci. Bioeng.* 95 (1): 102–105.

Körbahti, B. K., and A. Tanyolaç. 2009. "Continuous electrochemical treatment of simulated industrial textile wastewater from industrial components in a tubular reactor." *J. Hazard. Mater.* 170 (2–3): 771–778.

Kushwaha, J. P., V. C. Srivastava, and I. D. Mall. 2010. "Organics removal from dairy wastewater by electrochemical treatment and residue disposal." *Sep. Purif. Technol.* 76 (2): 198–205.

Landeros, C. R., C. E. B. Díaz, A. A. Cháves, and G. R. Morales. 2017. "Evaluation of a coupled system of electro-oxidation and ozonation to remove the pesticide Thiodan® 35 CE (endosulfan) in aqueous solution." *Fuel* 198: 91–98.

Li, W., Q. Zhou, and T. Hua. 2010. "Removal of organic matter from landfill leachate by advanced oxidation processes: A review." *Int. J. Chem. Eng.* 2010: 270532.

Lim, C. K., T. W. Seow, C. H. Neoh, M. H. M. Nor, Z. Ibrahim, I. Ware, et al. 2016. "Treatment of landfill leachate using ASBR combined with zeolite adsorption technology." *3 Biotech* 6 (2): 195.

López-Gálvez, F., G. D. Posada-Izquierdo, M. V. Selma, F. Pérez-Rodríguez, J. Gobet, M. I. Gil, et al. 2012. "Electrochemical disinfection: An efficient treatment to inactivate *Escherichia coli* O157:H7 in process wash water containing organic matter." *Food Microbiol.* 30 (1): 146–156.

Ma, H., and B. Wang. 2006. "Electrochemical pilot-scale plant for oil field produced wastewater by M/C/Fe electrodes for injection." *J. Hazard. Mater.* 132 (2–3): 237–243.

Ma, L., B. Yu, Y. Yu, J. Li, J. Ren, H. Wei, et al. 2014. "Indirect electrochemical oxidation of pentachlorophenol in the presence of different halides: Behavior and mechanism." *Desalin. Water Treat.* 52 (7–9): 1462–1471.

Ma, Q., T. Liu, T. Tang, H. Yin, and S. Ai. 2011. "Drinking water disinfection by hemin-modified graphite felt and electrogenerated reactive oxygen species." *Electrochim. Acta* 56 (24): 8278–8284.

Martinez-Huitle, C. A., and S. Ferro. 2006. "Electrochemical oxidation of organic pollutants for the wastewater treatment: Direct and indirect processes." *Chem. Soc. Rev.* 35 (12): 1324–1340.

Martínez-Huitle, C. A., and M. Panizza. 2018. "Electrochemical oxidation of organic pollutants for wastewater treatment." *Curr. Opin. Electrochem.* 11: 62–71.

Mercer, B., L. Ames, C. Touhill, W. Van Slyke, and R. Dean. 1970. "Ammonia removal from secondary effluents by selective ion exchange." *J. Water Pollut. Control Fed.* 42 (2): R95–R107.

Miladinovic, N., and L. Weatherley. 2008. "Intensification of ammonia removal in a combined ion-exchange and nitrification column." *Chem. Eng. J.* 135 (1–2): 15–24.

Mostafazadeh, A. K., A. T. Benguit, A. Carabin, P. Drogui, and E. Brien. 2019. "Development of combined membrane filtration, electrochemical technologies, and adsorption processes for treatment and reuse of laundry wastewater and removal of nonylphenol ethoxylates as surfactants." *J. Water Process Eng.* 28: 277–292.

Mukimin, A., H. Vistanty, and N. Zen. 2015. "Oxidation of textile wastewater using cylinder Ti/β-PbO_2 electrode in electrocatalytic tube reactor." *Chem. Eng. J.* 259: 430–437.

Mukimin, A., N. Zen, A. Purwanto, K. A. Wicaksono, H. Vistanty, and A. S. Alfauzi. 2017. "Application of a full-scale electrocatalytic reactor as real batik printing wastewater treatment by indirect oxidation process." *J. Environ. Chem. Eng.* 5 (5): 5222–5232.

Murugananthan, M., S. Latha, G. B. Raju, and S. Yoshihara. 2010. "Anodic oxidation of ketoprofen—An anti-inflammatory drug using boron doped diamond and platinum electrodes." *J. Hazard. Mater.* 180 (1–3): 753–758.

O'Farrell, T. P., F. P. Frauson, A. F. Cassel, and D. F. Bishop. 1972. "Nitrogen removal by ammonia stripping." *J. Water Pollut. Control Fed.* 44 (8): 1527–1535.

Ölmez, T., I. Kabdaşlı, and O. Tünay. 2007. "The effect of the textile industry dye bath additive EDTMPA on colour removal characteristics by ozone oxidation." *Water Sci. Technol.* 55 (10): 145–153.

O'Neill, C., F. R. Hawkes, D. L. Hawkes, N. D. Lourenço, H. M. Pinheiro, and W. Delée. 1999. "Colour in textile effluents—Sources, measurement, discharge consents and simulation: A review." *J. Chem. Technol. Biotechnol.* 74 (11): 1009–1018.

Oturan, M. A. 2014. *Electrochemical advanced oxidation technologies for removal of organic pollutants from water.* Berlin: Springer.

Oturan, M. A., P. Nidheesh, and M. Zhou. 2018. "Electrochemical advanced oxidation processes for the abatement of persistent organic pollutants." *Chemosphere* 209: 17–19.

Palacios, G. L., and M. B. Timmons. 2001. "Determining design parameters for recovery of aquaculture wastewater using sand beds." *Aquacult. Eng.* 24 (4): 289–299.

Pariatamby, A., and Y. L. Kee. 2016. "Persistent organic pollutants management and remediation." *Procedia Environ. Sci.* 31: 842–848.

Pasalari, H., M. Farzadkia, M. Gholami, and M. M. Emamjomeh. 2019. "Management of landfill leachate in Iran: Valorization, characteristics, and environmental approaches." *Environ. Chem. Lett.* 17 (1): 335–348.

Pouran, S. R., A. R. A. Aziz, and W. M. A. W. Daud. 2015. "Review on the main advances in photo-Fenton oxidation system for recalcitrant wastewaters." *J. Ind. Eng. Chem.* 21: 53–69.

Pressley, T. A., D. F. Bishop, and S. G. Roan. 1972. "Ammonia–nitrogen removal by breakpoint chlorination." *Environ. Sci. Technol.* 6 (7): 622–628.

Pulgarin, C., N. Adler, P. Peringer, and C. Comninellis. 1994. "Electrochemical detoxification of a 1,4-benzoquinone solution in wastewater treatment." *Water Res.* 28 (4): 887–893.

Radjenovic, J., and D. L. Sedlak. 2015. "Challenges and opportunities for electrochemical processes as next-generation technologies for the treatment of contaminated water." *Environ. Sci. Technol.* 49 (19): 11292–11302.

Ren, Q., C. Yin, Z. Chen, M. Cheng, Y. Ren, X. Xie, et al. 2019. "Efficient sonoelectrochemical decomposition of chlorpyrifos in aqueous solution." *Microchem. J.* 145: 146–153.

Renou, S., J. G. Givaudan, S. Poulain, F. Dirassouyan, and P. Moulin. 2008. "Landfill leachate treatment: Review and opportunity." *J. Hazard. Mater.* 150 (3): 468–493.

Rodgers, J. D., W. Jedral, and N. J. Bunce. 1999. "Electrochemical oxidation of chlorinated phenols." *Environ. Sci. Technol.* 33 (9): 1453–1457.

Sakalis, A., K. Mpoulmpasakos, U. Nickel, K. Fytianos, and A. Voulgaropoulos. 2005. "Evaluation of a novel electrochemical pilot plant process for azodyes removal from textile wastewater." *Chem. Eng. J.* 111 (1): 63–70.

Santhanam, N. 2009. *Oilgae guide to algae-based wastewater treatment.* Chennai, India: Home of Algal Energy.

Santos, I. D., J. C. Afonso, and A. J. B. Dutra. 2010. "Behavior of a Ti/RuO_2 anode in concentrated chloride medium for phenol and their chlorinated intermediates electrooxidation." *Sep. Purif. Technol.* 76 (2): 151–157.

Särkkä, H., A. Bhatnagar, and M. Sillanpää. 2015. "Recent developments of electro-oxidation in water treatment—A review." *J. Electroanal. Chem.* 754: 46–56.

Särkkä, H., M. Vepsäläinen, M. Pulliainen, and M. Sillanpää. 2008. "Electrochemical inactivation of paper mill bacteria with mixed metal oxide electrode." *J. Hazard. Mater.* 156 (1–3): 208–213.

Schmalz, V., T. Dittmar, D. Haaken, and E. Worch. 2009. "Electrochemical disinfection of biologically treated wastewater from small treatment systems by using boron-doped diamond (BDD) electrodes—Contribution for direct reuse of domestic wastewater." *Water Res.* 43 (20): 5260–5266.

Sharrer, M. J., Y. Tal, D. Ferrier, J. A. Hankins, and S. T. Summerfelt. 2007. "Membrane biological reactor treatment of a saline backwash flow from a recirculating aquaculture system." *Aquacult. Eng.* 36 (2): 159–176.

Singla, J., V. K. Sangal, A. Singh, and A. Verma. 2020. "Application of mixed metal oxide anode for the electro-oxidation/disinfection of synthetic urine: Potential of harnessing molecular hydrogen generation." *J. Environ. Manage.* 255: 109847.

Terezo, A. J., and E. C. Pereira. 2002. "Preparation and characterisation of Ti/RuO_2 anodes obtained by sol–gel and conventional routes." *Mater. Lett.* 53 (4–5): 339–345.

Trellu, C., B. P. Chaplin, C. Coetsier, R. Esmilaire, S. Cerneaux, C. Causserand, et al. 2018. "Electro-oxidation of organic pollutants by reactive electrochemical membranes." *Chemosphere* 208: 159–175.

Turro, E., A. Giannis, R. Cossu, E. Gidarakos, D. Mantzavinos, and A. Katsaounis. 2011. "Electrochemical oxidation of stabilized landfill leachate on DSA electrodes." *J. Hazard. Mater.* 190 (1–3): 460–465.

Ukundimana, Z., P. I. Omwene, E. Gengec, O. T. Can, and M. Kobya. 2018. "Electrooxidation as post treatment of ultrafiltration effluent in a landfill leachate MBR treatment plant: Effects of BDD, Pt and DSA anode types." *Electrochim. Acta* 286: 252–263.

Urtiaga, A., A. Rueda, Á Anglada, and I. Ortiz. 2009. "Integrated treatment of landfill leachates including electrooxidation at pilot plant scale." *J. Hazard. Mater.* 166 (2–3): 1530–1534.

Vallejo, M., M. F. San Román, A. Irabien, and I. Ortiz. 2013. "Comparative study of the destruction of polychlorinated dibenzo-*p*-dioxins and dibenzofurans during Fenton and electrochemical oxidation of landfill leachates." *Chemosphere* 90 (1): 132–138.

Vijayaraghavan, K., D. Ahmad, and T. S. Bin Fadzin. 2008. "In situ hypochlorous acid generation for the treatment of brackish shrimp aquaculture wastewater." *Aquacult. Res.* 39 (5): 449–456.

Vlyssides, A. G., M. Loizidou, P. K. Karlis, A. A. Zorpas, and D. Papaioannou. 1999. "Electrochemical oxidation of a textile dye wastewater using a Pt/Ti electrode." *J. Hazard. Mater.* 70 (1–2): 41–52.

Zhou, B., Z. Yu, Q. Wei, H. Long, Y. Xie, and Y. Wang. 2016. "Electrochemical oxidation of biological pretreated and membrane separated landfill leachate concentrates on boron doped diamond anode." *Appl. Surf. Sci.* 377: 406–415.

Zhu, X., J. Ni, and P. Lai. 2009. "Advanced treatment of biologically pretreated coking wastewater by electrochemical oxidation using boron-doped diamond electrodes." *Water Res.* 43 (17): 4347–4355.

CHAPTER 10

Comparative Studies among Electro-Coagulation, Chemical Precipitation, and Adsorption

S. K. Ram, H. Panidepu, C. Vasavi, P. Drogui, R. D. Tyagi

10.1 INTRODUCTION

Water is one of the most important elements of life. The modern world, in spite of achieving tremendous technological advancement, still struggles to efficiently use, reuse, maintain, and sustain all available water resources. According to various sources and an endless number of reports on the problem of water crisis, 663 million people in the world still live without clean water (Hakizimana et al. 2017, Holt et al. 2005). Therefore, to provide safe and clean drinking water to a majority of the world's population, efficient methods of use and reuse of water need to be developed.

Within the ambit of treatment of used water (greywater), many treatment technologies exist, which can make water fit for reuse or at least good enough for human use. Different available treatment technologies can be fundamentally classified as physical, chemical, and biological. In a typical wastewater treatment plant, all three classes of unit operations/processes can be observed. For example, the primary treatment of wastewater is primarily a physical process. Furthermore, all forms of solid–liquid separation via settling [be it after electro-coagulation (EC) or chemical flocculation] or screening are physical treatment processes. Chemical processes are processes that require additives (or chemicals) to remove the dissolved form of pollutants or suspended solids from water. The requirement of the addition of additives makes these processes less attractive. Biological processes fall within the process chain of wastewater treatment in which biological agents (like microorganisms) are used to carry out the process of removal of pollutants out of water.

Unlike the aforementioned conventional treatment technologies, EC is a broad-spectrum technique that falls within the ambit of *advanced treatment technologies* and has been used for the removal of suspended particulate matter from water.

Although the use of electricity in the wastewater treatment system dates back to the nineteenth century, EC has experienced a halt in its progress until recently, presumably because of its high capital cost of implementation (Chen 2004). As an emerging technology for water and wastewater treatment, EC unites the concept of coagulation–flocculation (a physicochemical process) with electrochemistry. As opposed to its chemical counterparts [chemical coagulation (CC) and adsorption], EC is efficient and requires no external additives. The existing literature on the comparison among the classical CC, adsorption, and EC techniques has juxtaposed these technologies for their pollutant removal efficiencies; however, detailed information on a common consensus on which technology is better in terms of energy efficiency, economics, or pollutant removal is not available. In this chapter, we have critically reviewed the existing literature and technically compared the three processes in relation to their various common aspects.

10.2 CHEMICAL PRECIPITATION AND ADSORPTION

10.2.1 Principles of Chemical Precipitation

Chemical precipitation is a process in which dissolved entities of a solution can be extracted in an insoluble form by the addition of certain chemical reagents. The benefit that it renders is the possibility of segregating the insoluble form from the rest with ease. In a solution, for instance, $(SO_4)^{2-}$ and $(PO_4)^{3-}$ are negatively charged soluble ions; they could be made to associate with reagents like barium chloride, which initiates the formation of barium sulfate, leading to the formation of insoluble compounds that can be removed with ease from the suspension. The removal of insoluble compounds from the solution could be facilitated by means of flocculation, filtration, and sedimentation per se. Therefore, chemical precipitation is a phenomenon in which the physical state of impurities is altered through chemical reagents, that is, from soluble to insoluble, and then aiding the separation through sedimentation. Another term, CC, which is also used quite interchangeably with the term *chemical precipitation*, is a phenomenon of clumping together or agglomeration of particles. This process is a combination of two stages, namely, coagulation and flocculation. The coagulating reagents added during coagulation neutralizes the negatively charged colloidal suspension, resulting in the formation of agglomerates. Flocculation is a gentle mixing process that aids in bringing together (through collision) the neutrally charged submicroscopic flocs, thereby forming visible flocs called *pin flocs*. Based on the property of pin flocs, whether they are scum or precipitate, for example, the subsequent physical separation step is followed. A close examination reveals that coagulation appears to be a subset phenomenon and an intrinsic part of precipitation.

Typically, wastewaters contain suspended particles, colloids, and dissolved solids. Suspended particles comprise sand, rock, or vegetable refuge per se. The particle distribution size is, therefore, quite large, ranging from 10 μm to anywhere up to 0.1 mm. Colloids, on the contrary, have a charge associated with them on their surfaces. The particle size is in the range of 10 nm to 10 μm. Owing to their inherent small size, an extremely large volume/surface area is associated with them. In such microscopic scales, surface forces overwhelm gravitational forces emanating from their weights. The charge density distributed over these particles dominates, and the particles tend to repel the other ones while they are present in one another's vicinity. From a process point of view, four stages of coagulation with different principles by its inherent chemistry are available (Binnie et al. 2002).

10.2.1.1 Adding an Extra Layer of Charge

In this stage, the small surface charge on the colloid particles gets subsumed by the addition of larger and stronger electrolytes whose electrostatic forces dominate and can bring about agglomeration of even similarly charged ions. Molecularly, a double-layer compression phenomenon occurs, in which the layer of surface charge on the particles is coated with another stronger charge. This methodology is independent of the concentration of colloidal ions present. Experiments suggest that compounds forming trivalent cations are the most effective coagulants.

10.2.1.2 Neutralization of Charge

In the neutralization stage, the ionic charge that is imparted to the colloidal solution is predominantly opposite to the charge present on the particles. Eventually, a charge neutralization occurs between the particles and the coagulant. This will diminish the repulsive forces among the particles and allow them to come closer together and aggregate. The amount of charge added to the suspension is directly proportional to the ionic charge present in the colloidal solution. In other words, adding opposite charges that are more than necessary could lead to overall charge reversal, cause more repulsion among the particles, and defeat the very purpose of coagulation and precipitation altogether.

10.2.1.3 Precipitate Entrapment

With mixing or flocculation, the floc tends to encompass more colloidal particles from its vicinity and entraps them in the form of carbonates and hydroxides typically. Prima facie, the colloidal particle enables the floc aggregation by acting as a nucleus, which tends to grow in size. An inverse relationship between the dose of the coagulant and the concentration of the colloidal suspension exists. For instance, a solution with less colloidal concentration will require a higher dose of coagulant aid to even begin with the process of floc formation.

10.2.1.4 Large Organic Polymers

In this process, large organic polymers having no unique but multiple charges are added to water such that they develop bridges among particles. Interestingly in this process, large anionic polymers can bring closer both anionic and cationic colloids together.

10.2.2 Critical Parameters Affecting Chemical Coagulation

10.2.2.1 Mixing

A thin line of difference exists between coagulation and flocculation, whereby in the latter process, mixing is an essential process parameter. Mixing with varying the frequency has the potential to cause varying chemical outcomes. In other words, if flocculation is to happen in a reactor with an impeller, the agitation rate of the impeller becomes deterministic. Excessive mixing sometimes can cause a breakage of the flocs, whereas a low rate of mixing diminishes the principal difference between the coagulation and flocculation processes. In flocculation, a velocity gradient develops in the media, which is directly proportional to the unit rate of energy input. The relationship between the power dissipated and the velocity gradient can be given by Equation (10-1).

$$\frac{P}{V} = \mu G^2 \tag{10-1}$$

where

P = Power dissipated (W),
G = Velocity gradient (s^{-1}),
V = Volume of fluid (m^3), and
μ = Dynamic viscosity of water (Pa·s).

The time until which the mixing of the colloidal solution with the coagulant is operated is a critical parameter. The volume of a reactor is divided into mixing zones, and, therefore, apart from the distribution of the velocity gradients, its implicit dependence on time does matter. When it comes to flocculation, the product of the velocity gradient and the retention time does count as Gt.

Mixing is also brought about using *weirs*, whereby the fluid is allowed to pass under pressure through channels. The power needed to enable such mixing is dependent on the pressure head, the flow rate, and the density of fluid [Equation (10-2)]

$$P = \rho g h Q \tag{10-2}$$

where

P = Power consumption (W),
ρ = Density of liquid (kg/m^3),
Q = Flow rate (kg/m^3), and
g = Acceleration caused by gravity (m/s^2).

Sludge-blanket flocculators, sinuous or baffled channels, and paddle flocculators are commonly used flocculators (Hudson and Wolfner 1967).

10.2.2.2 pH of the System

Aluminum and iron salts are typically used to coagulate pollutants with the previously described coagulation principles. It is difficult to clearly draw lines among the processes by which coagulation occurs, especially when the same salts are being used in different coagulation principles. However, from an engineering perspective, in any coagulation process, pH is a decisive parameter. When ferrous or ferric salts like ferrous sulfate and ferric chloride or aluminum salts like aluminum sulfate are added, insoluble hydroxides and carbonates are formed. However, it is important to note that these compounds exist in an insoluble state at a certain range of pH. At a particular pH, it is easier to segregate the flocs; however, to coagulate at other pH values an interplay of other factors will be required, such as mixing time, mixing intensity, and most important, coagulant dose. pH is an essential parameter to save energy input per unit and the overall economic costs. For instance, aluminum salts, when used in the coagulation process, will perform best at a pH of approximately 7 in which the solubility of aluminum hydroxide is minimum. However, if charge neutralization as a principle needs to be harnessed, aluminum salts perform better at a pH of approximately 5 in which the positive charges coat and nullify negative surface charges just as much required to overcome repulsion. Such working knowledge will, thus, guide the engineer to decide on the dosage of alum salts used for coagulation. Similarly, ferrous salts form insoluble derivatives at a pH of around 7 to 10. However, the increase in solubility outside this range is comparatively less than that in aluminum salts, suggesting the broad range of operability of ferrous salts vis-à-vis aluminum salts.

In light of pH fluctuations and manipulations, the buffering capacity of water assumes pivotal precedence so much so that it reflects the resilience of the water body to pH changes nudged by the addition of coagulants. Water rich in dissolved salts and high alkalinity is highly stable to the addition of coagulants and only marginally changes its pH value. However, this also indicates that a heavy dosage of chemicals is required to cause significant changes in the pH value. Contrarily, soft water is highly prone to pH changes with a marginal addition of coagulants.

10.2.2.3 Coagulant and Pollutant Concentration

Different types of coagulants are used in treatment system. In general, chemical coagulants are classified into two groups: (1) aluminum-based salts, and (2) iron-based salts. Iron-based coagulants include ferric sulfate, ferrous sulfate, and ferric chloride. On the contrary, aluminum-based salts such as aluminum sulfate, aluminum chloride, and polyaluminum chloride (PACl) are used. For any given system with a given temperature and pH value, the dosage of the coagulant significantly impacts the removal efficiency. Therefore, it is highly important to optimize the level of dosage for the given type of water composition containing a certain level of pollutants. Further, the type of pollutants and the coagulants themselves will vary the optimized dosage of the coagulants required for the maximum removal of pollutants. However, an underdosage of coagulants will result in poor destabilization of the colloidal suspension, whereas an overdosage will cause excessive sludge production and corrosion of equipment, making the process uneconomical.

Different types of water contain different levels of pollutants and accompanying salts. Soft water and hard water differ in their alkalinity and Ca/Mg concentration. The concentration of pollutants can interact with coagulants, and depending on the mechanism of action, the dosage value of the coagulants is affected.

10.2.2.4 Temperature of the Media

Temperature is one of the important parameters that can impact pollutant removal efficiency. The temperature is more significant when the turbidity of the water is low (Nouri et al. 2015). While using alum, when the temperature is low, aluminum hydroxides strongly hydrate with water and form a highly stable sol. If the water is less than 5°C, the alum of ferric salts does not work properly (Mamchenko et al. 2011). For lower temperature ranges, PACl should be considered as an alternative.

10.2.3 Commonly Used Coagulants

10.2.3.1 Aluminum-Based Coagulants

Alum or aluminum sulfate is the most commonly used coagulant. Although academic literature primarily represents alum as $Al_2SO_4 \cdot 18H_2O$, quite a number of variants to what is supplied in the market are available. The strengths of alum solutions and alum dosages are quite varied, although it is typical to express the strength as a percentage of aluminum oxide (Al_2O_3). Solid alum has a composition of 15% to 17% and is commercially available as $Al_2SO_4 \cdot 14H_2O$. To sum up, due consideration has to be accorded to the alumina content and specific gravity of the solution because they critically determine coagulation behavior.

10.2.3.2 Iron-Based Coagulants

Ferric sulfate and/or chloride are other major coagulants used instead of alum. Alum was found to have some linkage to Alzheimer's disease, and as the aluminum limits in potable water, it started becoming stringent. Ferric sulfate was sought as an alternative. It is advantageous by its inherent nature of being static over a broader pH range.

10.2.3.3 Other Coagulants

Lime as a coagulant is typically used to adjust the pH of the solution. In general, it encompasses both quicklime (CaO) and slaked lime [$Ca(OH)_2$]. The dosage of lime is an important parameter as it has a tendency to induce turbidity in the solution to be treated. Lime as a coagulant aid has been reported to yield inconsistent results because it is also dependent on factors such as temperature rise accruing from a reaction between quicklime and slaked lime and also the slaking time or the time of reaction of quicklime and water. Current research is directed toward producing lime at a macroscopic level such that there is a balance between its intended chemical effect and turbidity.

CC is a well-established technique in the area of wastewater treatment. Successful application of CC has been demonstrated by many researchers for the removal of heavy metals in wastewater (Table 10-1).

10.2.4 Principles of Adsorption

Adsorption is a phenomenon in which ions, gas, molecules, or a phase classified as adsorbate binds to another phase classified as an adsorbent. Adsorption, in

Table 10-1. Application of Coagulants for the Removal of Heavy Metals from Wastewater.

Pollutant	*Initial metal concentration*	*Precipitant*	*Optimum pH*	*Removal efficiency (%)*	*References*
Zn^{2+}	32 mg/L	CaO	9–10	99–99.3	Ghosh et al. (2011)
	450 mg/L	$Ca(OH)_2$	11	99.77	Charerntanyarak (1999)
Cd^{2+}	150 mg/L	$Ca(OH)_2$	11	99.67	
	37 mg/L	$Fe(OH)_3$	11	96.0	Tünay and Kabdaşli (1994).
Mn^{2+}	1,085 mg/L	$Ca(OH)_2$	11	99.30	Charerntanyarak (1999)
Cu^{2+}	16 mg/L	$Mg(OH)_2$	9.5	80	Tünay and Kabdaşli (1994)
Ni^{2+}	51.6 mg/L	NA	7.5	71	Papadopoulos et al. (2004)
	51.6 mg/L	NA	10.5	85	
Cu^{2+}, Zn^{2+}, Cr^{3+}, Pb^{2+}	100 mg/L	CaO	7–11	99.37–99.6	Chen et al. (2009)
Cu^{2+}, Zn^{2+}, Pb^{2+}	0.018, 1.34, 2.3 mM	H_2S	3.0	$100 > 94 > 92$	Alvarez et al. (2007)
Cr^{3+}	5,363 mg/L	CaO and MgO	8.0	>99	Guo et al. (2006)
Hg^{2+}	65.6, 188 mg/L	1,3-Benzenediamidoe thanethiolate	4.7 and 6.4	>99.9	Blue et al. (2008)
Cu EDTA	25, 50, 100 mg/L	1,3,5-Hexahydrotriazin edithiocarbamate	3.0	99.0, 99.3, 99.6	Fu et al. (2007)

conjunction with other processes, plays a pivotal role in the context of water treatment, typically to treat organic pollutants present in water bodies. Large surface area per unit volume of adsorbent forms the crux of the phenomena because it provides numerous sites for adsorbate binding. Adsorption is a technique that is resorted to in later stages of processing after the completion of wastewater treatments with coagulation, flocculation, ozonation, and other physicochemical methods, but with the lingering presence of organic pollutants that are resistant to biological degradation. In other words, adsorption is an attempt to capture pollutants that persist even after the aforementioned processes. Typically, activated charcoal is used to capture pollutants that are organic in nature. The pollutants get trapped in the high surface area that is provided by the adsorbent.

The adsorbent could be either in granular form or in powdered form. Commonly, activated charcoal, clay, polymers and resins, silica gel, alumina, and zeolites are used as adsorbents. Silica gel is a nontoxic and inert adsorbent that could be made in various pore sizes and used to adsorb heavy hydrocarbons and oxygen (Rahman and Padavettan 2012). Clay is used to adsorb edible oils and remove organic pigments. Polymers and resins are used in applications like the recovery of steroids and amino acids. Zeolites are polar and crystalline aluminosilicates and employed primarily in catalysis and in crucial gas or vapor applications such as the removal of carbon dioxide and carbon monoxide from natural gas, drying of refrigerants, and organic liquids (Kladnig 1975). They have a repeated pore network and have a property to release water at higher temperatures. Charcoal is activated by a chemical process of pyrolysis and controlled oxidation and is notably used for adsorption. The physical properties of charcoal such as density, particle size, sieve size, hardness, moisture, ash content, and pore-size distribution determine the nature and efficiency of adsorption (Sundstrom and Klei 1979). Further, adsorption with different polymeric resins can be used to separate specific pollutants.

While using the activated charcoal in granular form, they are packed up in beds sealed from either side, allowing the passage of wastewater through it as inlet and outlet (Figure 10-1). As the adsorbate or organic pollutant traverses through the column, it binds within the pores, thus removing the contaminants in the process. As the process proceeds, saturation of the bed with pollutants occurs such that there is no more space within the column for adsorbate binding. At such a time, *breakthrough* occurs, which is the point at which pollutants in the outlet begin to rise beyond a certain tolerance limit. Breakthrough points are very important in this process, because they indicate the time of substitution of the exhausted bed for a new bed. The saturated bed can be either regenerated or sent to landfills. One method to regenerate a charcoal bed adsorbent is by exposing it to a high temperature with steam. In this manner, the durability of a bed can be increased beyond a few runs. In situ regeneration of the bed can also be implemented by a *backflow* through the column altering the direction of flow using a pump. However, regeneration can typically be done on a finite number of times. Because regeneration cannot be done in real time, adsorption typically is done with more than one column present in backup while one is put under operation. When one column is saturated with pollutants, the flow will be allowed

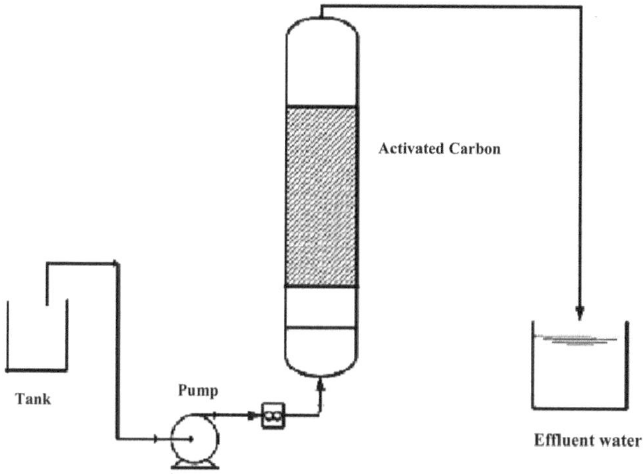

Figure 10-1. Activated carbon columns for the adsorption process.

to pass through a new bed while the former is regenerated; such switching can take place with the goal of optimizing the downstream process (Baker et al. 2003).

10.2.4.1 Adsorption Theory

At the molecular level, adsorption is guided by physical forces such as van der Waals, London forces, and chemical forces. Adsorption can be termed *chemisorption* if the nature of attraction is chemical in nature and *physisorption* if otherwise (Nijkamp et al. 2001). The mathematical description for attraction or repulsion between adsorbate and adsorbent is given by (Figure 10-2) Lennard–Jones "6–12" potential (Darkrim and Levesque 1998). The theory is based on interactions between electron densities of adsorbate and the adsorbent. On a time-averaged basis, symmetry exists in the electron densities of the solute/pollutant/adsorbate. However, at any instant of time, there is asymmetry of electron densities. This asymmetry causes attraction among the molecules as the adsorbate induces a transient dipole moment on the adsorbent molecules. The gist of the Lennard–Jones potential is that attractive forces fall off by the $(1/r)^{12}$ factor of distance separating the molecules, but repulsive forces increase by the $(1/r)^6$ factor of distance separating the molecules. Thus, there can be repulsion or attraction among the molecules depending on the distance among them (London 1937). The Lennard–Jones potential can be quantified as [Equation (10-3)]:

$$\varnothing = 4\in\left[\left(\frac{\sigma}{r}\right)^{12} - \left(\frac{\sigma}{r}\right)^{6}\right] \tag{10-3}$$

where

σ = Collision distance among molecules when attraction forces are zero,
ϵ = Depth of the potential well, and
r = Distance separating two atoms or molecules.

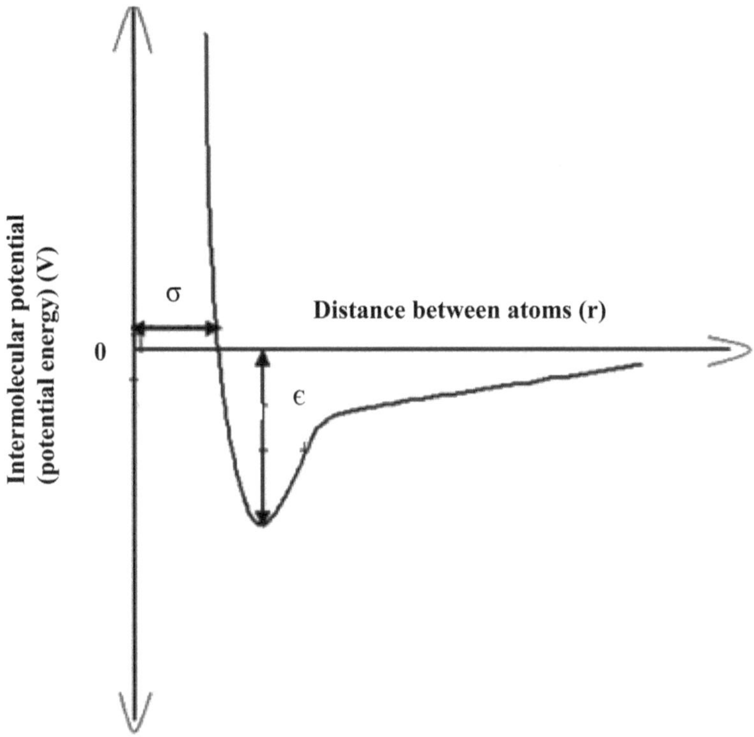

Figure 10-2. Lennard–Jones potential variation with collision diameter. The deeper the well depth (ϵ), the stronger the interaction between the two particles. When the bonding potential energy is equal to zero, the distance of separation, r, will be equal to σ.

10.2.4.2 Adsorption Equilibria

To quantify adsorption and to provide an insight into the adsorption phenomena, many researchers have published adsorption isotherms that show the relationship between the concentration of adsorbent in the solution and the amount of adsorbate adsorbed (mg g^{-1}) at steady state. Steady state is the cornerstone in understanding these isotherms, just as the term isotherm suggests phenomena occurring at constant temperatures. These isotherms are given sufficient time for the equilibria to be reached between the concentrations of adsorbed solute to that in the solution. The commonly read isotherms are linear isotherms, Langmuir isotherms, and Freundlich isotherms (Figure 10-3). A linear isotherm suggests that with an increase in solute concentration in the solution, the quantity of solute adsorbed shall increase. A Langmuir isotherm indicates a monolayer adsorption of solute or pollutant on the surface of the adsorbent. The behavior is that of an S-shaped curve, suggesting the percentage adsorption to increase rapidly initially followed by reaching a finite asymptotic maximum with a further increase in solute concentration. Freundlich isotherms and less-studied [Brunauer, Emmett

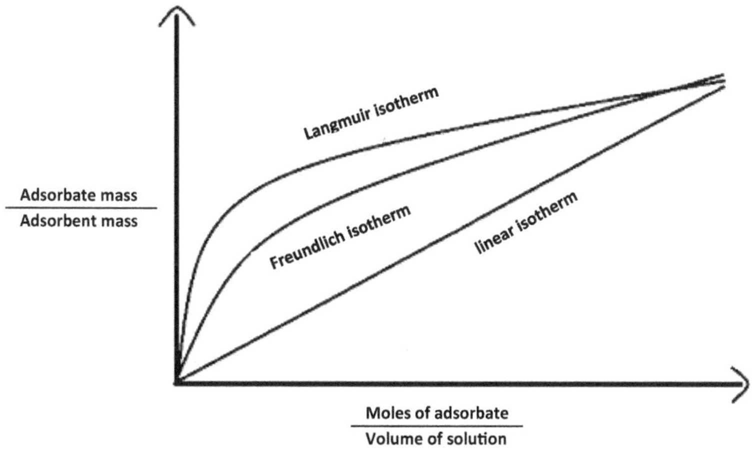

Figure 10-3. Typical types of adsorption isotherm models.

and Teller (BET)] isotherms depict a more general adsorption phenomenon encompassing even multilayer adsorption.

10.2.5 Factors Affecting Adsorption

10.2.5.1 Residence Time and Temperature

The residence time of the pollutant is of paramount importance in adsorption. A higher residence time implies a higher amount of contaminant removal. The viscosity of the flow through the column and the pressure drop in the column are implicitly related to the residence time. *Hydraulic loading* or the flow rate through a column denoted by gpm in.$^{-2}$ (gallon per minute per square inch) is another significant parameter. The pressure drop in the column is also accorded due consideration, and it should not be higher than 1 psi ft^{-2} (Chahbani and Tondeur 2001). A high pressure drop suggests clogging of the column, and, therefore, the flowthrough cannot be maintained by gravitational head alone. The adsorbate concentration is typically measured in mg/L or ppm.

Typically, an increase in temperature favors desorption, whereas a decrease in temperature favors adsorption. Binding between adsorbate and adsorbent is exothermic in nature and resembles condensation in many ways. There is heat of adsorption just as there is latent heat of condensation. An increase in vibrational energies with an increase in temperature is the cause of desorption.

10.2.5.2 Pore Size and Surface Area

The pore size of the adsorbent is, as could be conceivable, deterministic for adsorption. The higher the pore size, the larger the size of adsorbent molecules that can be adsorbed on the surface. Pore sizes can be micro, transitional, and macro pores increasing in that order. The pore-size distribution of an adsorbent is dependent on the source material as well as the method of activation of the material. It is possible to estimate the pore volume of a material by methods such as mercury porosimetry, in which mercury is pushed into the bed as a function of pressure (Giesche 2006). The cumulative pore volume occupied gives an estimate of how the pore sizes are distributed. Depending on the characteristics of adsorbate, adsorbent with a certain pore volume and pore-size distribution is preferred.

Conceivably, the higher the surface area of adsorbent, the higher the amount of adsorbate adsorbed. It is interesting to note that the surface area is not directly estimated by the summation of the internal surface area of the pores, but it is done with respect to the total amount of nitrogen that can be adsorbed at −196°C, which is the boiling point of nitrogen. The choice of nitrogen is premised on its property of low size, inert nature, and low boiling point, thus proving to be an alternative base reference for quantifying adsorption with respect to other adsorbates (Gregg et al. 1967).

10.2.5.3 Solute and Solvent Properties

The size of the solute or pollutant decides the adsorbent to be employed. An increase in solute size probably increases the percentage of adsorption considering other physical properties are constant. However, with dissimilar chemical properties but similar sizes, constitutive properties shall play a pivotal role. Molecules with nitro groups NO_3^- will have a higher tendency to adsorb on charcoal than sulfate and amino or sulfonic groups (Müller et al. 1980). Similarly, aromatic compounds, branched compounds, and the position of substituent groups at the *ortho*, *meta*, or *para* positions are important and deterministic (Snoeyink et al. 1969).

Polarity and nonpolarity of solute is another factor. When solvent carrying adsorbate has a higher affinity for it, adsorption probably decreases because of competing interests. A polar solvent has a stronger tendency to hold the polar pollutant, thereby decreasing the chances of pollutant adsorption on a nonpolar or organic adsorbent. This chemistry can be observed in adsorption of organic solutes on activated charcoal (Müller et al. 1980).

10.2.5.4 pH

It is observed that cognizable solutes are less absorbable than their noncognizable counterparts. These solutes have a charge on them creating electrostatic repulsions among one another while being adsorbed. The net result is lesser adsorption vis-à-vis nonionized and neutral adsorbates. It is important to understand the difference between ionization and dissociation in the context of solvent chemistry. Dissociation can be thought of as a separation of common salt or sodium chloride (NaCl) into NA^+ and Cl^- ions. Ionization happens when ions get protonated or deprotonated. An example of protonation is when carboxyl or COO^- ions get converted to COOH or NH_2 into NH_3^+. Similarly, deprotonation can be understood by a vice versa example. An ion can, thus, have acidic properties or basic properties or both, in which case it will be called *amphoteric*. Acidic compounds adsorb better in a low pH environment and vice versa (Martin and Iwuco 1982). In other words, adsorption of compounds on a bed increases when they have no charge on them, that is, at the isoelectric point. The isoelectric point is the pH at which the net charge on a molecule is zero because the positive and negative charges emanating because of the presence of ions negate each other. Thus, the pH of the solution either protonates or deprotonates ions of the solute. In this context, it is apt to understand that the pK_a value of ions is the pH value at which ions are half ionized and half nonionized and, therefore, indirectly the pK_a value of ions of the adsorbate also implicitly determines adsorption efficiency (Cooney and Wijaya 1987). One can find numerous studies in the literature studying the optimum pH at which adsorption is efficient for a given adsorbate and adsorbent.

10.2.5.5 Competing Solutes

The presence of other solutes apart from adsorbates also compete with them to get adsorbed on the bed. However, it should be noted that not all solutes compete for the same adsorption sites, and similarly not all adsorbents have an identical affinity for the same solutes. It is noted that the presence of inorganic solutes increases adsorption on a bed. For instance, the addition of NaCl, whereas the adsorption of organic solutes on a charcoal bed has been found to increase the packing density of adsorbate on the adsorbent (Randtke and Jepsen 1982). Chemistry dictates that the dissociation of sodium chloride into small Na^+ and Cl^- ions occupy the space between organic solutes on the bed, which thereby reduces the repulsive forces between the adsorbed molecules and, thus, increases the packing density.

Researchers have evaluated a plethora of different adsorbents for wastewater treatment applications. Heavy metal removal is one of the most prominent objectives in many industrial wastewater treatment plants because of the nature of various chemicals used in industrial processes. Table 10-2 lists different studies in which different cheap adsorbents have been enumerated for heavy metal removal.

10.3 ELECTRO-COAGULATION

10.3.1 Electrochemistry of the Electro-Coagulation Process

A basic EC is composed of an EC cell, which is nothing but a rectangular container in which EC will be conducted (Figure 10-4). Depending on the purpose of the process, the geometry and design of this container may vary. The electrolytic cell is provided with two metal electrodes, an anode and a cathode. These metal electrodes are connected to an electricity source. The type of source of electricity varies depending on the configuration and application of the EC process. Iron and aluminum are used extensively as the electrode material because they are relatively cheap, reliable, nontoxic, and abundant and they exist in polyvalent forms forming poly-hydroxy coordinated compounds. The two electrodes are

Table 10-2. Different Types of Adsorbents for the Removal of Heavy Metal Application.

	Adsorption capacity (mg/g)						
Adsorbent	*Pb^{2+}*	*Cd^{2+}*	*Zn^{2+}*	*Cu^{2+}*	*Cr^{6+}*	*Ni^{2+}*	*References*
			Modified natural materials				
Zeolite, clinoptilolite	1.6	2.4	0.5	1.64	—	0.4	Babel and Kurniawan (2003)
Modified zeolite, MMZ	123	—	—	—	—	8	Nah et al. (2006)
HCl-treated clay	—	—	63.2	83.3	—	—	Vengris et al. (2001)
Clay/poly(methoxyethyl)acrylamide	81.02	—	20.6	29.8		80.9	Sölenera et al. (2008)
Calcined phosphate	155.0	—	—	—	—	—	Moufliha et al. (2005)
Activated phosphate	4	—	—	—	—	—	Pan et al. (2007)
			Agricultural and biological wastes				
Maize cope and husk	456	493.7	495.9	—	—	—	Igwe et al. (2005)
Orange peel	—	—	—	—	—	158	Ajmal et al. (2000)
Coconut shell charcoal	—	—	—	—	3.65	—	Babel and Kurniawan (2003)
Pecan shells–activated carbon	—	—	13.9	31.7	—	—	Bansode et al. (2003)
Rice husk	—	2.0	—	—	0.79	—	Bishnoi et al. (2004)
Modified rice hull	—	—	—	—	23.4	—	Tang et al. (2003)
Spirogyra (green alga)	—	—	—	133	—	—	Gupta et al. (2006)
Ecklonia maxima—marine alga	235	—	—	90	—	—	Feng and Aldrich (2004)
Ulva lactuca	—	—	—	—	112.3	—	El-Sikaily et al. (2007)
Oedogonium species	145	—	—	—	—	—	Gupta and Rastogi (2008)
Nostoc species	93.5	—	—	—	—	—	Gupta and Rastogi (2008)
S. bacillus—bacterial biomass	467	85.3	418	381	39.9	—	Ahluwalia and Goyal (2007)
			Modified biopolymers				
Crosslinked chitosan	—	150		164		230	Crini (2005)
Crosslinked starch gel	433	—	—	135	—	—	
Alumina/chitosan composite	—	—	—	200	—	—	

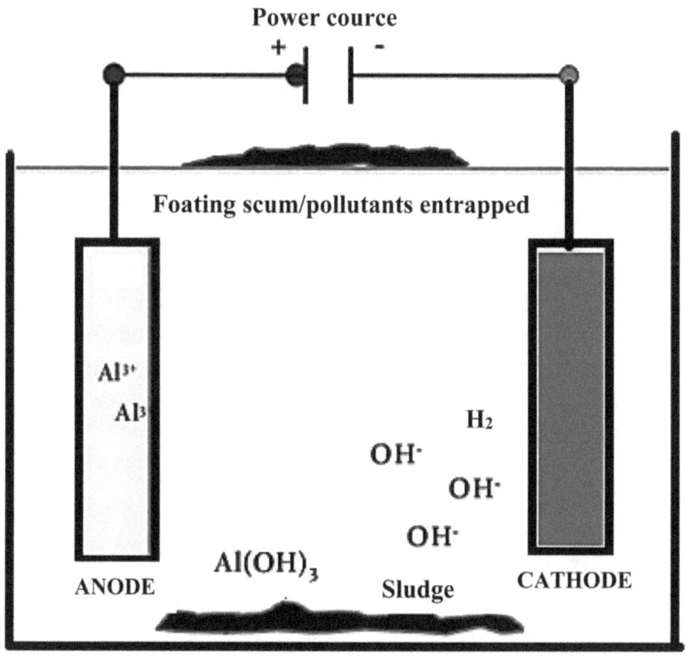

Figure 10-4. Schematic diagram of a typical EC cell.

dipped in the electrolytic cell containing the water or wastewater that needs to be treated. When current is passed through the electrodes using an electricity source, different reactions occur at different electrodes. The anode furnishes the coagulants (M^+ ions) in the EC cell.

The process follows Faraday's laws of electrochemistry, which is given by Equation (10-4).

$$m = \frac{ItM_w}{zF} \tag{10-4}$$

where

I = Current (A),
t = Time (S),
M_w = Molecular weight (g mol^{-1}),
F = Faraday's constant,
z = Number of electrons involved in the electrochemical reaction, and
m = Mass (g) of the anode dissolved in the EC cell.

At the anode, the following reactions occur during the process [Equations (10-5) to (10-10)].

$$Fe(s) \rightarrow Fe^{n+}(aq) + ne^{-1} \tag{10-5}$$

$$4Fe^{2+}(aq) + H_2O + O_2(aq) \rightarrow 4Fe(OH)_3(s) + 8H^+(aq) \tag{10-6}$$

$$Fe^{2+}(aq) + 2OH^- \rightarrow Fe(OH)_2(s) \quad (10\text{-}7)$$

$$4Fe^{2+}(aq) + H_2O + O_2(aq) \rightarrow 4Fe(OH)_3(s) + 8H^+(aq) \quad (10\text{-}8)$$

$$Al(s) \rightarrow Al^{3+}(aq) + 3e^{-1} \quad (10\text{-}9)$$

$$Al^{3+}(aq) + nH_2O \rightarrow Al(OH)_n^{3-n}(s) + nH^+ \quad (10\text{-}10)$$

The existence of iron as ferrous or ferric ion state depends on the pH of the solution and the potential applied in the EC cell. The ionic state of the metal (Fe/Al) can be determined using E-pH charts. Along with the main electrolytic reactions, side reactions occur within the electrolytic cell. These reactions involve the evolution of hydrogen at the cathode with the release of hydroxyl ions [Equation (10-11)], which increases the pH of the solution:

$$2e^- + H_2O \rightarrow H_2 + 2OH^- \quad (10\text{-}11)$$

Quite often the mass of the electrolyte does not match with that estimated by Faraday's law [Equation (10-4)]; this inconsistency is because of the other electrochemical reactions that occur at the anode. Researchers account for this inconsistency with the evolution of oxygen at the anode (Mollah et al. 2004). At a high pH and a sufficiently high anode potential, the water splits into molecular oxygen and protons [Equation (10-12)]:

$$2H_2O \rightarrow O_2 + 4H^+ + 4e^- \quad (10\text{-}12)$$

10.3.2 Destabilization of Colloids

Pollutants in water and wastewater exist in the form of stable suspensions or colloids. In general, these particles are negatively charged. Because of the presence of superficial negative charge, the particles repel one another and remain stable in water. These colloid particles range within the size range of 1 nm to 2 μm. To destabilize the suspension, these electrostatic repulsions among these particles need to be neutralized by applying positively charged ions (often metal ions like aluminum or iron). The application of positively charged metal ions leads to the formation of a double electrostatic layer around the colloidal particles. The inner layer (often called *stern layer*) contains the positively charged ions directly interacting with the colloids' negative charge. The second layer contains positively charged ions that are free to move by diffusion. The maximum potential is at the colloids' surface (called Nernst Potential), and it decreases across the stern layer because of the presence of the positive charge present in the stern layer. The potential at the interface of the stern layer and the outer diffusion layer is called the *zeta potential.*

According to DLVO theory (Missana and Adell 2000), the net interaction energy (V_T) between two colloids is the sum of van der Waals energy (V_A) and electrostatic repulsion (V_R). As the two particles approach each other, the

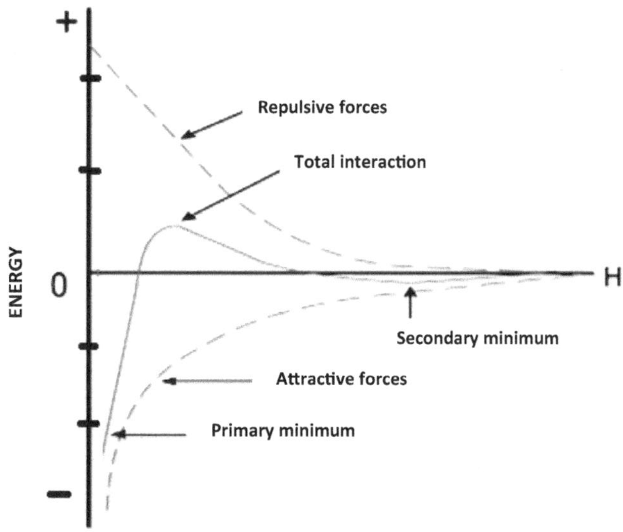

Figure 10-5. The DLVO theory for interaction energies among approaching particles.

attractive force increases more rapidly than the increase in repulsive forces. The total interaction energy is depicted in Figure 10-5. The net interaction energy approaches maximum as the particles come close enough to each other. When coagulants (M^+) are released in EC, they reduce the zeta potential (thereby repulsive forces) and increase the interaction energy among any two particles, allowing them to come closer and interact easily. The coagulants can be furnished by either electrochemical reactions (as in the EC process) or by the addition of metal salts (in CC). Different metal ions have different destabilization effects and potentials. According to the Schultze–Hardly rule, the higher the charge of the metal ion, the higher is its ability to destabilize the suspension (Mollah et al. 2004). After charge neutralization, the particles agglomerate together to form bigger flocs, and a sludge of metal hydroxides containing entrapped pollutants precipitate out of the water.

10.3.3 Critical Parameters of Electro-Coagulation

10.3.3.1 Metal Electrode Type

The type of material for the metal electrode has a significant impact on the overall efficiency of the electrolysis and the amount of coagulants released for the same amount of current passed through an electrolytic circuit. This can be understood by the fact that the difference in the amount of electrolytic production of coagulants (metal ions) will vary by virtue of the metal atom and their ionization energies. If we compare the trivalent state of the two metals (iron and aluminum), we can see that to convert the divalent form of the metal (M^{2+}) to trivalent form (M^{3+}), aluminum requires 2,744.8 kJ/mol, whereas iron requires 2,957 kJ/mol. Thus, in terms of thermodynamics, for the same given amount of electrical energy to an electrode, aluminum will release more coagulants than iron, although it has to be understood that aluminum exists only in its trivalent form, whereas iron can exist in a divalent or trivalent form depending on the relative potential and pH conditions prevailing in the reactor. The second ionization energy of iron is 1,561.9 kJ/mol; therefore, iron seems to have an upper edge when it comes to contributing coagulants for a given amount of electrical energy because of its polyvalent existence. Many studies have compared the efficiencies of the two electrodes and have reported that iron is a better electrode than aluminum in terms of pollutant removal efficiencies (Akbal and Camcı 2010, Bazrafshan et al. 2008, Chafi et al. 2011, Kobya et al. 2003). However, many instances of literature can be found where aluminum has a higher removal efficiency (on COD basis) than iron (Ilhan et al. 2008, İrdemez et al. 2006). A mixed observation can be made throughout the literature, and this can be accounted by the fact that in coagulation, it is not just the coagulant that plays the key role, but that the efficiency depends on the nature of treatment or the kind of interaction that the pollutant has with the coagulant. Katal and Pahlavanzadeh (2011) studied the treatment of pulp and paper wastewater using different combinations of Al/Fe anode–cathode pairs and found that the Al–Al pair had a high color removal efficiency, whereas the Fe–Fe pair was effective in phenol and COD removal. Furthermore, the relative efficiency of the performance of the electrode depends on the speciation of the metal ions because of the existing pH and potential conditions. The speciation of the metal ions can be determined using eV–pH charts for each metal ion. After coagulation, the metal hydroxides interact among one another to form bigger flocs, and their stability depends on the kind of interactions and size of the flocs formed. Thus, the decision lies more on the experimental evidence for each kind of pollutant in a pure matrix of synthetic wastewater or pristine matrices in raw wastewater because of the presence of the accompanying chemical compounds that may influence the coagulation and flocculation processes.

10.3.3.2 Electrode Arrangement

By virtue of the nature of the reactions that occur in EC, the process requires a large surface area to be effective. More than one electrode can be connected in either series or parallel. In the case of a monopolar-parallel connection, all electrodes are connected to the power source directly. However, in series connections, the electrodes are connected to adjacent ones, and the ones in extremity are connected to a power source. A bipolar configuration is also possible when the inner electrodes (sacrificial) are not connected to one another (called bipolar electrodes), while the extremities are connected to a power source (monopolar electrodes). The choice of the electrode configuration is again determined by experimental evidence on the efficiency of the pollutant removal. Several studies have studied different configurations for wastewater treatment. Primarily, the monopolar-parallel arrangement of electrodes yields better treatment efficiencies than the other configurations (Demirci et al. 2015, Golder et al. 2007a,

Kobya et al. 2016). Changing the electrode configurations also affects the energy and cost of operation of the process. Thus, for different pollutant and wastewater matrixes, different configurations should be verified.

10.3.3.3 Power Supply Type

EC typically uses DC current as the power source. During the process, while anode gets consumed (sacrificial electrode), deposition of metal oxide occurs on the cathode (termed *passivation of the cathode*). The passivation of the cathode reduces the efficient flow of electricity through the EC cell. The passivation is counteracted by the addition of NaCl or using alternating pulsed current as the power source. Several studies have demonstrated the effective use of AC power to enhance electrode life (Mollah et al. 2001). Further studies have compared the effectiveness of AC and DC power supplies on the performance of EC, and reportedly, AC power supply shows a higher removal efficiency and lower energy consumption than DC power supply (Eyvaz et al. 2009, Vasudevan et al. 2011).

10.3.3.4 Current Density

The current density applied in an EC is a very important parameter. As the current density in a reactor is increased, the amount of coagulants released increases proportionally. Researchers (Eyvaz et al. 2009, Ilhan et al. 2008, İrdemez et al. 2006) have studied the effects of current density on the efficiency of EC extensively. An increase in the current density after a certain value does not improve the efficiency of EC, or rather any further increase in the current density will lead to loss of energy and reduce the overall efficiency of the process in terms of performance yield per unit energy consumed.

10.3.3.5 Conductivity of Water or Anion Concentration

The presence of different anions has different effects on EC. Researchers have reported that the presence of sulfate ion negatively affects EC, whereas chloride and nitrate ions positively enhance EC by counteracting against sulfate ions and inhibiting passivation of the cathode (Khandegar and Saroha 2013). The conductivity within the electrolytic cell is another highly important factor that directly impacts power consumption for EC by virtue of the resistance (internal resistance of the liquid bulk) rendered by the wastewater matrix. Conductivity is further improved by addition of NaCl or KCl in the water meant for treatment during EC process (Chen 2004, Khandegar and Saroha 2013).

10.3.3.6 Initial pH

The pH of water at the beginning of the EC process is a highly important factor that determines the overall efficiency of pollutant removal. The pH of the solution has a compounding effect on the process as it can impact the speciation of metal ions, rate of metal dissolution, and conductivity of the solution (by the state of ions) and thereby impact energy consumption, passivation of the cathode, zeta potential of the coagulated flocs and, therefore, the complete efficiency of the pollutant removal. The pH can also profoundly impact how the flocs, pollutants, and metal coagulants interact among one another. Almost all studies performed on EC take account of pH and optimize the pH range for optimum treatment conditions (Attour et al. 2016; Cañizares et al. 2008, 2009; Ilhan et al. 2008; Katal and Pahlavanzadeh 2011; Le et al. 2017; Mollah et al. 2004).

10.3.4 Speciation of Aluminum and Iron with pH

The speciation of the coagulant or the metal ion is one of the most important determining factors that can profoundly affect the performance of coagulation of the colloidal pollutant particles during a chemical or EC process equally. The ways in which metal coagulants ionize/dissociate and exist in the chemical process and EC processes are different by the very nature of the method of their generation. This difference has been extensively studied for aluminum ions (Cañizares et al. 2006).

The dosage is the first step in the coagulation process, and a major difference between CC and EC lies here. In CC, metal salts are dissolved at once (batch mode feed), whereas in EC, metals are dissolved at a continuous rate depending on the potential applied in the system. In CC, the metal is inevitably accompanied by anions, whereas EC furnishes metal ions all alone. It is important to note that in spite of this difference, research shows that speciation of aluminum is independent of the manner in which it is introduced; rather it depends on aluminum concentration and pH of the system. Metal concentration and pH are two interconnected parameters in coagulation chemistry because of the amphoteric nature of aluminum oxides. The metal concentration and pH have a wide variation in the two techniques (CC and EC), and this makes all the difference in terms of their respective efficiencies. In EC, the pH quickly increases to higher values, whereas in chemical dosing, the pH decreases to lower values. In the acidic range, the predominant form of aluminum is monomeric cationic hydroxoaluminum (Figure 10-7). When the pH increases, many monomeric forms coexist with an increase in polymeric cations (polynucleate forms) and precipitates. At near neutral pH (7.0), the aluminum exists as aluminum hydroxide predominantly. A further increase in pH results in anionic hydroxoaluminum. It has been reported that the presence of sulfate anions in the system results in precipitate formation.

Different salts of aluminum are used in CC, such as aluminum sulfate, aluminum chloride, and other prepolymerized aluminum. After the addition of aluminum ions, they undergo hydrolysis:

$$\mathrm{Al(OH)_4^-} + \mathrm{H^+} \leftrightarrow \mathrm{Al(OH)_3} + \mathrm{H_2O} \tag{10-13}$$

$$\mathrm{Al(OH)_3} + \mathrm{H^+} \leftrightarrow \mathrm{Al(OH)_2^+} + \mathrm{H_2O} \tag{10-14}$$

$$\mathrm{Al(OH)_2^+} + \mathrm{H^+} \leftrightarrow \mathrm{Al(OH)^{2+}} + \mathrm{H_2O} \tag{10-15}$$

$$\mathrm{Al(OH)^{2+}} + \mathrm{H^+} \leftrightarrow \mathrm{Al^{3+}} + \mathrm{H_2O} \tag{10-16}$$

$$\mathrm{Al(OH)_3(s)} \leftrightarrow \mathrm{Al^{3+}} + 3\mathrm{OH^-} \tag{10-17}$$

$$x\mathrm{Al^{3+}}\, y\mathrm{H_2O} \leftrightarrow \mathrm{Al}_x\mathrm{(OH)}_y^{3x-y} + y\mathrm{H^+} \tag{10-18}$$

In these proposed equations, aluminum is assumed to be monomeric in nature (dilute system), but in reality, aluminum can exist in multimeric forms with polynucleated hydroxyl groups [Equation (10-18)]. This makes it difficult to model the rate kinetics of hydrolysis. Using advanced NMR technologies, researchers have characterized and studied various existing forms of aluminum.

To compare the two techniques in a more justified and rational way, similar pH and aluminum concentrations need to be maintained, which, however, cannot be achieved because pH variation in both the techniques are different. Nevertheless, as studied by Cañizares et al. (2006), aluminum exists in three forms, namely, monomeric, polymeric, and precipitate forms. In EC, when pH rises, precipitate formation is a significant event [the reverse direction of Equation (10-17)]. In an EC system (discontinuous) toward the end, the system predominantly contains aluminum hydroxide precipitates as the pH increases and stabilizes at the maximum value of pH 8 to 9. However, in a CC system, after the addition of salts, the pH slowly falls to 4.5 to 5.0 and, along with aluminum hydroxide precipitates, monomeric hydroxoaluminum ions are significantly present (Figure 10-6).

Thus, it can be concluded that speciation of the metal (aluminum in current discussion) does not depend on the dosing technique but on the conditions maintained in the system. The pH significantly affects the speciation, and it changes differently for the two techniques. Each species of metal ions has its own method to coagulate pollutants, and their relative abundance will significantly affect pollutant removal efficiency.

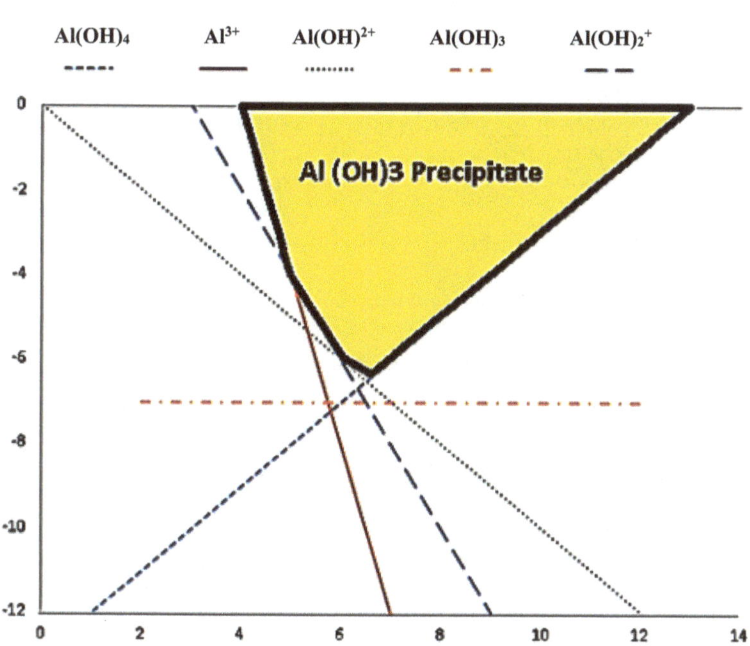

Figure 10-6. Speciation of aluminum ions under different pH conditions.

10.4 COMPARISON BETWEEN ELECTRO-COAGULATION AND CHEMICAL COAGULATION

CC is an age-old technique for removal of suspended particles from water and wastewater; it is well established in the area of water treatment , whereas EC is quite new and is still undergoing renaissance. Many EC-based processes were commissioned and established at its inception, but later, they were forced to be shut down because of economic considerations. Researchers argue that these two techniques are no different, except for the way in which metal ions (or coagulants) are introduced in the wastewater system. This section uncovers the salient differences in the technical aspects of the two processes, and how researchers have attempted to distinguish, characterize, and qualify them separately (Table 10-3).

In CC, metal salts are added at once to water, which inevitably brings along anion counterparts in the salt, which might affect the coagulation process because of the ionic nature of the anions. Unlike CC, EC introduces metal ions slowly and gradually via electrolytic dissolution of the sacrificial anode dipped in water. Golder et al. (2007b) performed a comparative study between CC and EC for Cr(VI) removal and found that the performance of EC was superior to that of CC conducted with the sulfate salt of the metal coagulant.

Table 10-3. Advantages and Disadvantages of EC, CC, and Adsorption Processes.

Treatment method	*Target of removal*	*Advantages*	*Disadvantages*	*References*
Electrocoagulation	Dyes, heavy metals from leachates, municipal wastewater (BOD, COD, and nutrients)	High separation selectivity	High operational cost because of membrane fouling and energy consumption	Mohammadi et al. (2005)
Chemical precipitation	Heavy metals, divalent metals	Low capital cost, simple operation	Sludge generation, extra operational cost for sludge disposal	Kurniawan et al. (2006)
Adsorption	Decolorization, trace impurities like water in oil or oil in water	Low-cost, easy operating conditions, wide pH range, high metal binding capacities	Low selectivity, production of waste products	Aklil et al. (2004), Babel and Kurniawan (2003)

In the case of CC, the coagulants neutralize the surface charges and minimize the repulsive forces to destabilize the suspension. In fact, exactly the same phenomenon occurs in EC as well. However, the difference lies in the latter step when the flocs of coagulated colloidal particles settle under the influence of gravity. In CC, the flocs settle in an unhindered fashion, whereas in EC, electrolytic evolution of gaseous molecules (hydrogen, oxygen) takes place, which causes hindrance to the settling mechanism of the flocs. One might argue that evolution of these gases may remove a fraction of pollutants via floatation mechanism; however, it depends on the interaction between the gas bubbles and the pollutant. There can be no second opinion that hindrance in the settling reduces the settling efficiency and might cause delay or destabilization of the integrity of the flocs.

The dosage of the coagulant is a highly important factor in coagulation. In a CC process, the dosage of the metal coagulant can be manually regulated and controlled by regulating the dosage of the metal salts (or double salts). In the EC process, however, the dosage of metal dissolution depends on other factors that cannot be fine-tuned to perfection. For example, the internal resistance of the bulk liquid will affect the current density and flow, which in turn affect the rate of electrolytic metal dissolution. Controlling the conductivity of bulk liquid in the case of unknown matrices of wastewater seems impractical.

CC and EC behave differently in terms of evolution of pH during the coagulation process (Cañizares et al. 2008). In CC, the pH slowly dips to lower values, whereas in EC, the pH rapidly rises to a stable basic value because of the evolution of OH^- ions [Equation (10-11)]. As discussed previously, the pH of the liquid bulk severely affects the speciation of the metal ions. The turbidity removal patterns in CC as observed during the settling phase follows an exponential decrease right from the point when coagulants are added, but in the case of EC, coagulation takes place in three distinct phases. As EC starts, the concentration of the coagulants slowly builds up, bringing the zeta potential to close to neutral values. This phase is called *lag stage*. After lag stage, turbidity falls rapidly, and this phase is termed *reactive stage*, and finally, turbidity reaches *stable value* and the reduction rate is very slow. This stage is called *stabilizing stage* (Holt et al. 2002). The two different profiles found in typical CC and EC processes are demonstrated in Figure 10-7.

Within the ambit of fluid dynamics and chemical reaction systems, the two processes are completely distinct. A CC system is a completely mixed system, which is vigorously mixed once for a short while and then is left for the solid–liquid separation to occur by virtue of the gravity and size of the flocs. The bulk liquid is preferably left undisturbed, whereas an EC process is essentially mixed by the evolving gaseous byproducts. Therefore, a CC process primarily removes pollutants by *sweep* flocculation caused by the metal hydroxide precipitate, whereas in EC, the pollutant removal occurs by *sweep* flocculation as well as floatation (Yilmaz et al. 2007). Further, the coagulation reaction in CC is like a batch mode reaction that starts with the addition of coagulants and progresses toward a single equilibrium state in which the turbidity of the solution is stabilized and no

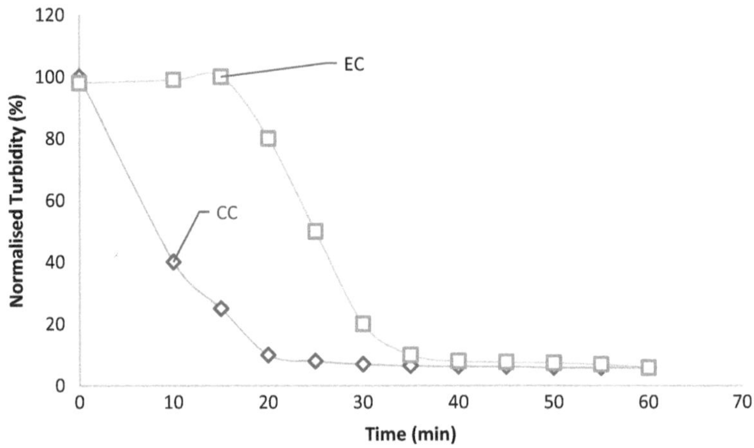

Figure 10-7. Kinetics of turbidity removal in the CC and EC processes.

further coagulation activity occurs, but EC is a dynamically changing process. Because the mode of addition of coagulants is continuous, the equilibrium of EC is continuously shifting and changing until the whole system is exhausted of any reactant. Therefore, EC is quite a dynamic reaction system. Studies (Harif et al. 2012) show that EC and CC are different when it comes to the kinetics of floc formation. EC is reported to produce flocs over a wide range of pH values with faster rates than CC. The floc formation rates and structural pattern in EC points to the DLCA (*diffusion limited cluster aggregation*) mechanism, unlike that in CC, in which the RLCA (*reaction limited cluster aggregation*) mechanism is indicated (Tang et al. 2000). In general, many important factors differentiate EC from CC. Empirical knowledge and experimental evidence are necessary to provide justified evidence while choosing one process over another. In the following section, some existing comparative studies between EC and CC are explored.

10.5 COMPARATIVE STUDIES BETWEEN ELECTRO-COAGULATION AND CHEMICAL COAGULATION

Before making any comparison or understanding the difference between the two processes, it is highly recommended to get behind the inspiration or the motivation behind the development of these new processes. It is very important for the reader to understand why advanced techniques for wastewater treatment are required. Are the classical methods obsolete, insufficient, or unsustainable? To this end, the importance and the motivation of developing EC as a new advanced wastewater technique can be understood by the underlying benefits that this new technique brings along with it. Now, it has been undisputedly accepted that EC is a technique that is better than CC on many counts. EC is considered as among the genre of advanced oxidation processes for water and wastewater treatment. Compared with its chemical counterpart, EC reportedly produces less sludge volume. Coagulants are furnished in situ; therefore, the chances of secondary pollution because of overdosage or the remnant effects of residual chemicals in water are nil. The reactor design can be completely automated, and all the parameters can be easily regulated. Wastewater using EC produces clear, colorless, and odorless effluents. EC can easily target very fine colloidal suspensions, which at times are difficult to remove by CC. The flocs developed by EC process are relatively large in size and more stable than those by CC process. Nevertheless, EC also has some disadvantages such as the requirement to regularly change the sacrificial electrodes and a reduced process efficiency caused by a passivation of the cathode that takes place during the operation. In addition, the process sounds completely impractical and uneconomical for use in places where electricity is expensive.

Although the economy of the process remains a tipping point to decide which way the process will fall, at this dwindling position, one must compare the two processes in terms of their technical feasibility taking account of all the important and decisive factors that can make a process sustainable and thrive or unsustainable and die a natural death. In this context, we require a common ground to compare the two processes to remain unbiased and rational. Researchers have done their best to compare the two processes and put forward an honest opinion regarding their potential and feasibility (Table 10-4).

Larue et al. (2003) studied the efficiency of the treatment steps by comparing characteristics of subsequent solid–liquid separation process. They compared the specific cake resistance of the cake formed during filtration, the height of the sediments after settling down of the flocs, the colloidal concentration in the treated supernatant, and the floc size distribution. The two treatments were compared by making sure that the amount of coagulant (Fe) introduced in both the processes (CC and EC) is equal. According to the settling experiment, the solid–liquid front in EC-treated water was settling rapidly as compared with that in the CC-treated water. The final sediment height in EC-treated water was lower than that in the CC-treated suspension. The flocs refused to settle down until the pH was set at a value of 3.0 , and settling commenced at or above neutral pH, indicating the effect of pH on both EC-treated and CC-treated suspensions. The floc porosity of EC was significantly lower than that of CC. The specific cake resistance was higher for EC-treated suspension than that for its CC counterpart. Microscopic image analysis suggests that the flocs produced by EC ($d_{50\%} = 170$ μm) were larger than those treated by the CC ($d_{50\%} = 100$ μm) suspension.

In another study done by Zhu et al. (2005), for the same given coagulant (Fe) concentration, EC, coupled with microfiltration (MF), removed 4.5 log reduction value as compared with only 2.0 removed by CC-MF. The study confirmed that, in the vicinity of the anode, the concentration of coagulants is higher than the uniform concentration in CC. This gradient of concentration in EC yields superior efficiency on colloidal suspensions as compared with CC in spite of both processes having the same coagulant concentration. It is not always true that EC can provide

Table 10-4. Comparison of Treatment Efficiencies through the Methods of Electro-Coagulation (EC), Chemical Precipitation (CP), Chemical Coagulation (CC), and Adsorption (AD).

Pollutant wastewater (initial concentration)	Initial concentration	pH	Process/material	Removal efficiency	Specific energy consumed (current density)	Operating cost	Treatment time	Mode	Reference
Virus	10^6 pfu/mL	6.3–8.3	Fe-EC (EC-MF)	99.99%	0.25 mA/cm^2	—	—	Batch	Zhu et al. (2005)
		6.3–8.3	CC	49.99%	—	—	—	Batch	
Surface water (EC and CC as membrane pretreatment processes)	DOC 5.0–5.6 mg/L UV_{254}	6.4	Fe-EC	DOC removal—40% UV_{254} removal—50% Soluble Fe^{2+} ions formed. Undesirable for separation	—	—	—	Batch	Bagga et al. (2008)
		6.4	CC ($FeCl_3$)	DOC removal—50% UV_{254} removal—60%	—	—	30 mins	Batch	
Metal industry (oil in water emulsions)	3,000 mg dm^{-3}	6.0	Al-EC	COD removal approximately 90%	10.1 mA/cm^2	—	—	Batch	Cañizares et al. (2008)
		11.5	CC ($AlCl_3$)	COD removal approximately 90%	—	—	—		

Food and beverage industry effluent (industrial fermentation plant)	COD 4,500 mg O_2/L	8.6	Al-EC	COD removal—70%	13.7 mA/cm^2	—	—	Batch	Ryan et al. (2008)
		3.8	CC ($Al_2(SO_4)_3$)	COD removal—50%	—	—	—		
		8.8	Fe-EC	COD removal—63%	13.7 mA/cm^2	—	—		
		2.4	CC ($FeCl_3$)	COD removal—80%	—	—	—		
Heavy metal—Cu	45 mg/L	9.0	Al-EC	100%	5.50 kWh/m^3	0.97 $US /$m^3$	40 min	Batch	Akbal and Camcı (2010)
		9.0	Fe-EC	100%	4.26 kWh/m^3	0.59 $US/$m^3$	40 min		
		11.0	CC	100%	—	1.176 $US/$m^3$	32 min		
		11.0	CC ($FeCl_3$)	100%	—	0.859 $US /$m^3$	32 min		
Heavy Metal—Cr	44.5 mg/L	9.0	Al-EC	100%	5.50 kWh/m^3	0.97 $US /$m^3$	40 min	Batch	
		9.0	Fe-EC	100%	4.26 kWh/m^3	0.59 $US /$m^3$	40 min		
		11.0	CC	100%	—	1.176 $US /$m^3$	32 min		
		11.0	CC ($FeCl_3$)	100%	—	0.859 $US /$m^3$	32 min		
Heavy metal—Ni	394 mg/L	9.0	Al-EC	99.6%	5.50 kWh/m^3	0.97 $US /$m^3$	40 min	Batch	
		9.0	Fe-EC	99.5%	4.26 kWh/m^3	0.59 $US /$m^3$	40 min		
		11.0	CC	98.8%	—	1.176 $US /$m^3$	32 min		
		11.0	CC ($FeCl_3$)	99.0%	—	0.859 $US /$m^3$	32 min		
Heavy metal–As(V)	0.05 mg/L	7.5	Fe-EC	Approximately 70%	28 mA	—	20 min	Continuous	Lakshmanan et al. (2010)
		6.5	CC ($FeCl_3$)	~85–95%	—	—	2 min	Batch	
As (removal through EC and CC in combination with microfiltration, respectively)	100 μg/L	7.0	EC-MF	90.64%	—	0.12 US $/$m^3$	—	Batch	Mólgora et al. (2013)
			CC-MF	96%	—	0.066 $US/$m^3$	—	Batch	

(*Continued*)

Table 10-4. Comparison of Treatment Efficiencies through the Methods of Electro-Coagulation (EC), Chemical Precipitation (CP), Chemical Coagulation (CC), and Adsorption (AD). (Continued)

Pollutant wastewater (initial concentration)	*Initial concentration*	*pH*	*Process/material*	*Removal efficiency*	*Specific energy consumed (current density)*	*Operating cost*	*Treatment time*	*Mode*	*Reference*
Synthetic wastewater (red dye)	100 mg/L	6.1	Al-EC	Color removal 80%–95%	1.5–3.5 kWh/ kg dye (208–310 A/m^2)	0.34–0.52 $US kg dye	14 min	Continuous	Merzouk et al. (2011)
			CC ($Al_2(SO_4)_3 \cdot 18H_2O$)	Color removal 87%	—	0.32 SUS/kg dye	1 h 10 min	Batch	
Synthetic wastewater (acid dye)	50 mg/L	7.0	Al-EC	87.5–93.4%	27.8–99.0 kWh/ kg dye (155–350 A/m^2)	7.04–17.4 US$/ kg dye (0.31–0.8 US$/m^3)	—	Batch	Chafi et al. (2011)
			Fe-EC	90.7–98.1%	26.7–76.6 kWh/ kg dye (155–350 A/v)	4.01–13.8 US$/ kg dye (0.19–0.68 $/m^3)	—	Batch	

Coal mine drainage wastewater (various metals)	500 mL/L	2.43	Fe-EC	28.7–99.96%	1.32–5.6 kWh/m^3 (200–500 A/m^2)	1.09–2.184 \$US/m^3 (0.91–1.93 £/m^3)	—	Batch	Oncel et al. (2013)
			CP (NaOH)	—	—	1.173–7.49 US\$/m^3 (1.0364–6.6177 £/m^3)	50 min	Batch	
Textile wastewater	2031 mg/L COD (under optimum conditions)	6.3	Al-EC (MP-P)	63%	0.72 kWh/m^3 (30 A/m^2)	0.4 US\$/m^3	15 min	Batch	Bayramoglu et al. (2007)
		7.9	Fe-EC (MP-P)	65%	0.68 kWh/m^3 (30 A/m^2)	0.25 US\$/m^3	15 min		
		2.9	CC ($FeCl_3 \cdot 6H_2O$)	71%	—	0.67 US\$/m^3	25 min		
		3.1	CC ($Fe_2(SO_4)_3 \cdot 7H_2O$)	68%	—	0.75 US\$/m^3	25 min		
		4.1	CC ($AlCl_3 \cdot 6H_2O$)	68%	—	0.96 US\$/m^3	25 min		
		4.1	CC ($Al_2(SO_4)_3 \cdot 18H_2O$)	59%	—	0.75 US\$/m^3	25 min		
Removal of Pb^{2+} and Zn^{2+} from acidic soil leachate	—	—	Fe-EC	99.4% (Zn^{2+}) 99.7% (Pb^{2+})	18.5 kWh/tst (68 A/m^2)	35.38 US\$/tst	—	Batch	Drogui et al. (2011)
			CP-$Ca(OH)_2$	98.9% (Zn^{2+}) 99% (Pb^{2+})	88.7 kg $Ca(OH)_2$/tst	38.29 US\$/tst			
			CP (NaOH)	97.2% (Zn^{2+}) 98.3% (Pb^{2+})	65 kg NaOH/tst	50.98 US\$/tst			
			AD (CS)	91.6% (Zn^{2+}) 90% (Pb^{2+})	100 kg CS/tst	39.15 US\$/tst			

Note: tst = ton of soil treated, CS = cocoa shell.

the best solution for positively charged particulate removal. In the case of Golder et al. (2007b), EC was concluded to be a suboptimal solution for Cr(VI) removal as it removed only 53.48% of Cr(VI), although it still outperformed CC, as CC could remove only 11.5% of chromium. The contrast between the efficiencies of these two processes can be best seen during a boron removal study (Yilmaz et al. 2007). Because of the difference in the mode of removal and coagulant addition, EC process could remove 96.6% of boron, while the CC process could remove only 25% of boron. The study concluded that the difference is mainly because of the mechanism of separation of the process. In CC, the main mechanism of removal is sweep flocculation and settling, whereas it is accompanied by flotation in EC. Many researchers have used total coagulating metal concentration as the unbiased basis for comparison. According to Cañizares et al. (2008), the efficiency of such processes is not dependent on the dosing technology but on the total concentration of metal and pH. It should be noted that these results contradict the findings of Zhu et al. (2005), who explained how the method of introduction of metal ions can impact the efficiency of these processes. The study by Cañizares et al. (2008) suggested that for the destabilization of oil–water emulsion, oil concentration must be taken into account. Small differences were observed between CC and EC processes, which can be accounted by the different modes of pH change in the two processes.

Studies by Akbal and Camcı (2010), Oncel et al. (2013), Ryan et al. (2008), and Tran et al. (2012) compared the two processes to treat real matrices of wastewater as well. Ryan et al. (2008) in their study concluded that both EC and CC had comparative efficiencies for the decolorization of molasses wastewater. The decolorization efficiency varied between 88% and 95% for the two processes, but here again, EC was found to be slightly superior to CC. Similar observations were made by (Akbal and Camcı 2010) while treating metal plating wastewater. Both the processes removed almost 99% of metals, but EC produced twice the amount of sludge that CC was producing. Still, EC appears to be an attractive option as it is a quick process providing treatment with only 1.08 kg (Fe)/m^3 of electrodes as compared to 1.5 kg $FeCl_3/m^3$ of wastewater. While comparing the two processes for the treatment of coal mine drainage wastewater (CMDW), Oncel et al. (2013) reported that a huge amount of NaOH was required to adjust the pH of the system and not all metals have the same pH for their optimized removal by chemical precipitation. The removal methods for different metals varied from 20% to 99%. In contrast to CC, almost 100% removal of all metals was done by EC during CMDW treatment.

Chafi et al. (2011) performed a comparative study between EC and CC for the removal of soluble acid dye. The study revealed that unlike CC, dye molecules could either get adsorbed on Al $(OH)_3$ flocs or might get degraded (Azo bond cleavage) by electrochemical actions. EC could remove 98% of the dye, whereas the chemical process could achieve only 53% of removal. Unlike with Azo dye, when the experiments were performed with other dyes (disperse red dyes) by Merzouk et al. (2011), the results were completely different. The CC process can remove 90% of the color from the solution, although EC was still giving superior performance with 95% color removal. This further confirms that the mechanism of destruction of dye molecules or its physical removal depends on the chemical nature of the dye particle. The concentration of pollutant also affects how the comparison between the two processes changes. Tran et al. (2012) studied the mechanism of phosphate removal in municipal wastewater using both EC and CC processes. According to this study, phosphorus removal was higher with EC (98%) as compared to CC (63%) in a low phosphorus contamination range (5 to 10 mg/L), but when the concentration of phosphorus was increased, the performances of both the processes are comparable (95% to 98%). Arsenic is another common pollutant that researchers have used to compare the two processes for their removal efficiencies. A coupled process consisting of microfiltration with either CC or EC was introduced. Studies by Lakshmanan et al. (2010) and Mólgora et al. (2013) largely suggested that EC process has superior performances as compared to CC.

Textile waste consists of dyes and synthetic polymers that add color and turbidity to water. Typically, this water is characterized by high turbidity and COD, which are usually of great chemical complexity.

Lime precipitation has been observed to be an efficient process in treating inorganic effluents with concentrations higher than 1,000 mg/L (Barakat 2011). The adsorption method appears to be more efficient for removing heavy metals from wastewater in lower concentrations (Fu and Wang 2011) and is being widely used. This area has been witnessing good research to bring out low-cost alternatives to replace the commercially available activated carbon. Likewise, Hegazi (2013) proposed the use of economically cheap agricultural products. However, the recovery of heavy metals from the adsorbent and recycling of the adsorbent appears to be expensive through this process. Thus, calling for the need of research into retrieving of heavy metals from adsorbents. Tests have verified that EC appears to be superior to CC in the treatment of organic effluents. In research conducted by Zhu et al. (2005), virus removal was 50% more efficient in EC-MF than in CC-MF.

A closer scrutiny over the literature (Table 10-4) on these processes reveals that researchers have focused more on comparing the classical CC process with the newly established EC process. It is rare to find one single study that has accounted for all the three processes of EC, CC, and adsorption at once, which is understandable, because these processes cannot be compared in an unbiased manner with properly justified common grounds for comparison. The scientific difference in the mechanism of action makes it difficult for researchers to compare these processes altogether, although Drogui et al. (2011) made a comparison between the three processes for removal of lead and zinc ions. Using the same pollutant load, the study concluded that EC and CC were comparable, whereas the efficiency of adsorption was slightly inferior to that of the two. In terms of cost estimation, EC was found to be the most cost-effective method for the heavy metal removal in the studied case.

The three processes of CC, adsorption, and EC share some principal similarities and dissimilarities. These processes alone cannot achieve any true objective and should be followed up by various solid–liquid separation techniques like microfiltration, centrifugation, flocculation, and settling. The three processes differentiate each other by way of the suspended colloid interaction with the causative element of colloidal destabilization and solid–liquid separation. Adsorption works primarily on the physicochemical interaction of colloidal particles with the adsorbent, whereas the other two processes essentially involve neutralization of the surface charge of these colloidal particles. Although mechanistically akin to each other, EC and CC differ in the manner of introduction of the metallic ions, which effectuate the charge neutralization and consequently coagulate the colloidal particles, thereby destabilizing the solution turbidity. Although the coalesced neutralized colloidal particles (in EC and CC) or adsorbent-colloidal particle couples are usually big enough to settle down in the absence of the stabilizing surface charges, the use of polymeric flocculants are recommended to expedite the solid–liquid separation process.

Primarily, all the studies performed to compare EC and CC were done in batch mode, except a few. For a practical industrial application and scale-up of the process, EC needs to be studied in continuous mode. A detailed study should be performed to understand the sensitivity and effectiveness of various parameters in EC (retention time, stabilization period, concentration profiles, concentration development kinetics, concentration gradients, removal kinetics and mechanism, metal consumption rates, hydrodynamic profiles) to affect the pollutant removal efficiency of the process. Such knowledge will help to develop an up-scaled EC process.

10.6 PRACTICAL BASIS OF JUDGMENT: ENERGY AND ECONOMICS COMPARISON

In view of the current global issues that revolve around sustainability and conservation of energy and environment, it is always important to pay attention to the energy efficiency of any developed process. Further, for a process to survive in a profit-driven society, it certainly needs to be economical. To answer the question whether the newly developed technology is feasible and has the potential to replace the existing technology, an economic analysis of the two technologies is inevitable. Many researchers (Table 10-4) performed an economic analysis of EC and CC while performing comparative analysis between them (Akbal and Camcı 2010, Cañizares et al. 2009, Mólgora et al. 2013, Oncel et al. 2013, Ryan et al. 2008).

The total cost of operation is the sum of different costs associated with the EC process operation. The process operation contains various elements that can be expressed in Equation (10-19):

$$\mathrm{TCO} = r_e * E + r_A * A + r_s * S + r_C * C + M + D + L - \mathrm{Am} \qquad (10\text{-}19)$$

where

TCO = Total cost of operation,

r_e = Price of electricity,

E = Amount of electrical energy consumed,

r_A = Price of the anode,

A = Amount of anode material,
r_s = Price of sludge transport and disposal,
S = Amount of sludge produced,
r_c = Rate of the chemicals used,
C = Amount of chemicals used,
M = Maintenance of the unit,
D = Depreciation,
L = Labor cost, and
Am = Amortization cost.

Although the TCO is dependent on many factors as shown in Equation (10-19), the two main factors are the consumption of energy and that of electrodes.

Electrical energy consumption (EEC) and electrode consumption determine the direction of the economic feasibility of the process. EEC can be given as Equation (10-20):

$$\mathrm{EEC} = EI\frac{t_{\mathrm{EC}}}{V_S} \tag{10-20}$$

where
E = Electrical potential (V),
I = Applied current (A),
t_{EC} = Time of electrolysis, and
V_s = Volume of the liquid bulk.

Energy is often reported in the literature in terms of specific electrical energy consumption (SEEC) in which the energy is referred to as the energy per unit of electrode consumed [Equation (10-21)]:

$$\mathrm{SEEC} = nF\frac{E}{3.6\times10^{-3}\ M_w\varphi} \tag{10-21}$$

where
n = Number of electrons ($n=2$ for Fe and $n=3$ for Al),
F = Faraday's constant,
M_w = Molecular weight of the anode material, and
φ = Current efficiency.

The term of efficiency of a system can be calculated using Equation (10-22):

$$\varphi = \frac{\Delta m_{\mathrm{exp}}}{\Delta m_{\mathrm{theo}}} \tag{10-22}$$

where Δm_{exp} is the experimental mass loss of electrode calculated by experimental design, whereas Δm_{theo} is the theoretical mass calculated according to Equation (10-23):

$$\Delta m_{\mathrm{theo}} = \frac{M_w I t_{\mathrm{EC}}}{nF} \tag{10-23}$$

It has been often observed by researchers that the value of electrical efficiency is sometimes greater than 1; this has been accounted by the fact that during electrolysis, the dissolution of coagulants is not only by electrical energy but also by the corrosive environment rendered by the ongoing chemical reactions in the electrochemical cell. Sometimes researchers have expressed specific energy consumption in terms of pollutants removed (SEECP), as determined by Equation (10-24):

$$\text{SEECP} = \frac{EIt_{\text{EC}}}{V_S \sum(YM_{\text{pol}}[\text{Pollutant}]_o)} \tag{10-24}$$

where

Y = Pollutant removal percent,
M_{pol} = Molecular weight of the pollutant,
$[\text{Pollutant}]_o$ = Initial concentration of the pollutant,
V_s = Volume of the liquid, and
m_{pol} = Total mass of pollutant removed.

Electrode consumption in EC is determined by the amount of current used, but this dissolution varies according to the composition of the matrix. The pH conditions and the presence of anions can result in passivation and reduce the efficiency of electrolytic dissolution.

In another study (Oncel et al. 2013), the operating cost of the two processes (EC and CC) were evaluated while treating CMDW. The energy used in the EC process was given by Equation (10-20). The study accounted the costs of energy, electrode, and chemicals as the major cost. The energy and electrode consumption in EC was 1.32 to 5.64 kWh/m^3 and 0.92 to 1.7 kg/m^3, respectively. The overall cost of operation was in the range of €0.91 to €1.98 /m^3 while producing 0.85 to 3.58 kg/m^3 of sludge. Although EC and CC are successful techniques for suspended particulate removal, their integration with other techniques like microfiltration has expanded the scientific horizon. Mólgora et al. (2013) performed a cost comparison between EC-MF and CC-MF techniques for arsenic removal. Their study assumed a facility with a treatment capacity of 108 m^3/h. The total energy expense was calculated in terms of hydraulic dissipation and electrolytic dissipation of metal electrodes. In EC, Equations (10-25) to (10-27) were used by assuming the motor's efficiency as 0.75.

$$\text{EEC} = \frac{[P_r + P_e] \times C_e \times 0.75}{Q} \tag{10-25}$$

$$P_r = G_t^2 \mu V_t \tag{10-26}$$

$$P_e = VI \tag{10-27}$$

where

P_r = Hydraulic power,
P_e = Electrolytic power,

C_e = Cost of electricity,
Q = Flow rate of fluid,
G_t = Velocity gradient,
μ = Dynamic viscosity,
V_t = Volume of liquid among electrodes,
I = Current, and
V = Potential difference applied.

In the case of CC, mixing constitutes the major cost for energy. The chemical energy cost (CEC) can be given by way of Equations (10-28) and (10-29).

$$\mathrm{CEC} = P_m \times C_e \times \frac{0.75}{Q} \tag{10-28}$$

$$P_m = G_t^2 \mu V_t R_T \tag{10-29}$$

All terms in CC are similar to those in EC, except that there is no energy consumption for electrolytic dissipation. R_T is the hydraulic residence time in the reactor. For energy consumptions in microfiltration (MEC), the following equation can be used [Equation (10-30)]:

$$\mathrm{MEC} = C_e V_p E_e \tag{10-30}$$

where V_p is the permeate volume, and E_e is the MF-specific energy. The value of the MF-specific energy is often taken as 0.2 kW h/m^3.

Between the two process configuration variants (CC-MF and EC-MF), CC-MF was cheaper than EC-MF. EC-MF was 1.8 times more expensive than CC-MF. Similar results were obtained by Merzouk et al. (2011) while treating red dye from aqueous solutions. The cost of operation was 0.32$/kg of dye removed for CC and around $US 0.52/kg for EC. Economic analysis was also performed by (Akbal and Camcı 2010) while studying the removal of heavy metals. The results in this study contradicted the previous perception that EC is comparatively more expensive to operate than CC. The cost of operation was $US 0.970/m^3 for EC-Al and around $US 1.176/m^3 for CC using aluminum sulfate as the coagulant. This seeds the concept of a thorough technoeconomic analysis to get a true picture of the economics of the processes (Ryan et al. 2008), in which the most optimized conditions should be used to evaluate the economics of the processes and compare to achieve a common goal. The common goal can be the same volumetric flow rate of treated water with comparable or same treatment standards. Technoeconomic analysis provides us tools to get an eagle eye view of the process and perform sensitivity analysis to understand the effect of each individual process parameter on the overall cost of the process. It is also important to understand that the nature of pollutants as well as their interaction with coagulants and their introduction system will affect process efficiency; thus, a technoeconomic analysis for one kind of pollutant system cannot be trusted or generalized for all different kinds of pollutants (Cañizares et al. 2009).

10.7 CONCLUSIONS AND FUTURE PROSPECTS

In EC, a multitude of processes are operating synergistically. The true power of this technique can be understood by the fact that by tuning the individual parameters, we can enhance pollutant removal efficiencies. It is the nature of the pollutant that decides which parameter will impact the efficiency the most. Simple dyes, viruses, heavy metals, organic loads (COD), and different types of industrial wastewater can be treated by EC. EC finds its application in the treatment of different types of water such as tannery industry wastewater, food industry wastewater, pulp and paper industry wastewater, refinery wastewater, and produced water, and also in municipal wastewater treatment plants. A comparative study of the different types of wastewater systems is important from the engineering and application points of view to decide the best available technology for pollutant removal. While considering this technology, the most important parameters that should be kept in mind are pollutant removal efficiency, cost of operation, and energy efficiency of the system. Such decisions on the choice of technology can be justified only after performing a detailed technoeconomic analysis. There are various tools available that can assist in performing real-life technoeconomic analyses using software resources. These tools can aid in finding answers to questions like should a new plant switch over to EC or continue with CC. Further, in such considerations, we should also dwell on the idea of ecological safety. An evaluation of life cycle analysis and ecological impact as well as sustainability assessment are a few important elements of analysis to actually judge the potential of a process. Such a comparison is very important to decide which technology is better for pollutant removal.

While two technologies are being compared, one should always think of the complete process chain in the whole picture. EC, CC, and adsorption cannot singly perform wastewater treatment and make the water fit enough for discharge or reuse within the permissible limits. Technoeconomic feasibility, modeling, and sustainability and ecological impact assessment of the processes should be performed for integrated systems in which EC, CC, and adsorption are part of the process chain. A justified comparison should be performed by using the most optimized conditions obtained from mathematical models and tools. So far in the literature, very few researchers have considered integrated coupled processes for wastewater treatment and juxtaposed them in an unbiased manner.

The literature reviewed so far suggests that there is still no common consensus on which technology is better in terms of energy efficiency, economics, or pollutant removal; the outcome of the process varies significantly with a change in pollutant streams and operational parameters. Thus, inevitably, researchers should perform comparative studies for different kinds of pollutants and matrix compositions (wastewater). More focus should be invested on deriving a process that can be easily scaled up and well characterized. Only then will it become feasible to justifiably choose the method of wastewater treatment.

References

Akbal, F., and S. Camcı. 2010. "Comparison of electrocoagulation and chemical coagulation for heavy metal removal." *Chem. Eng. Technol.* 33 (10): 1655–1664.

Aklil, A., M. Mouflih, and S. Sebti. 2004. "Removal of heavy metal ions from water by using calcined phosphate as a new adsorbent." *J. Hazard. Mater.* 112 (3): 183–190.

Ahluwalia, S. S., and D. Goyal. 2007. "Microbial and plant derived biomass for removal of heavy metals from wastewater." *Bioresour. Technol.* 98 (12): 2243–2257.

Ajmal, M., R. A. K. Rao, R. Ahmad, and J. Ahmad. 2000. "Adsorption studies on Citrus reticulata (fruit peel of orange): Removal and recovery of Ni(II) from electroplating wastewater." *J. Hazard. Mater.* 79 (1): 117–131.

Alvarez, M. T., C. Crespo, and B. Mattiasson. 2007. "Precipitation of Zn(II), Cu(II) and Pb(II) at bench-scale using biogenic hydrogen sulfide from the utilization of volatile fatty acids." *Chemosphere* 66 (9): 1677–1683.

Attour, A., N. Ben Grich, M. Mouldi Tlili, M. Ben Amor, F. Lapicque, and J.-P. Leclerc. 2016. "Intensification of phosphate removal using electrocoagulation treatment by continuous pH adjustment and optimal electrode connection mode." *Desalin. Water Treat.* 57 (28): 13255–13262.

Babel, S., and T. A. Kurniawan. 2003. "Low-cost adsorbents for heavy metals uptake from contaminated water: A review." *J. Hazard. Mater.* 97 (1–3): 219–243.

Bagga, A., S. Chellam, and D. A. Clifford. 2008. "Evaluation of iron chemical coagulation and electrocoagulation pretreatment for surface water microfiltration." *J. Membr. Sci.* 309 (1–2): 82–93.

Baker, F., C. Miller, A. Repik, and E. Tolles. 2003. *Activated carbon Kirk-Othmer encyclopedia of chemical technology*. New York: Wiley.

Bansode, R. R., J. N. Losso, W. E. Marshall, R. M. Rao, and R. J. Portier. 2003. "Adsorption of volatile organic compounds by pecan shell- and almond shell-based granular activated carbons." *Bioresour. Technol.* 90 (2): 175–184.

Barakat, M. A. 2011. "New trends in removing heavy metals from industrial wastewater." *Arabian J. Chem.* 4 (4): 361–377.

Bayramoglu, M., M. Eyvaz, and M. Kobya. 2007. "Treatment of the textile wastewater by electrocoagulation: economical evaluation." *Chem. Eng. J.* 128 (2–3): 155–161.

Bazrafshan, E., A. H. Mahvi, S. Naseri, and A. R. Mesdaghinia. 2008. "Performance evaluation of electrocoagulation process for removal of chromium (VI) from synthetic chromium solutions using iron and aluminum electrodes." *Turk. J. Eng. Environ. Sci.* 32 (2): 59–66.

Binnie, C., M. Kimber, and G. Smethurst. 2002. *Basic water treatment*. London: Royal Society of Chemistry.

Bishnoi, N. R., M. Bajaj, N. Sharma, and A. Gupta. 2004. "Adsorption of Cr(VI) on activated rice husk carbon and activated alumina." *Bioresour. Technol.* 91 (3): 305–307.

Blue, L. Y., M. A. Van Aelstyn, M. Matlock, and D. A. Atwood. 2008. "Low-level mercury removal from groundwater using a synthetic chelating ligand." *Water Res.* 42 (8–9): 2025–2028.

Cañizares, P., C. Jimenez, F. Martinez, M. A. Rodrigo, and C. Sáez. 2009. "The pH as a key parameter in the choice between coagulation and electrocoagulation for the treatment of wastewaters." *J. Hazard. Mater.* 163 (1): 158–164.

Cañizares, P., F. Martínez, C. Jiménez, J. Lobato, and M. A. Rodrigo. 2006. "Comparison of the aluminum speciation in chemical and electrochemical dosing processes." *Ind. Eng. Chem. Res.* 45 (26): 8749–8756.

Cañizares, P., F. Martinez, C. Jiménez, C. Sáez, and M. A. Rodrigo. 2008. "Coagulation and electrocoagulation of oil-in-water emulsions." *J. Hazard. Mater.* 151 (1): 44–51.

Chafi, M., B. Gourich, A. H. Essadki, C. Vial, and A. Fabregat. 2011. "Comparison of electrocoagulation using iron and aluminium electrodes with chemical coagulation for the removal of a highly soluble acid dye." *Desalination* 281: 285–292.

Chahbani, M.H., and D. Tondeur. 2001. "Pressure drop in fixed-bed adsorbers." *Chem. Eng. J.* 81 (1–3): 23–34.

Charerntanyarak, L. 1999. "Heavy metals removal by chemical coagulation and precipitation." *Water Sci. Technol.* 39 (10–11): 135–138.

Chen, G. 2004. "Electrochemical technologies in wastewater treatment." *Sep. Purif. Technol.* 38 (1): 11–41.

Chen, Q., Z. Luo, C. Hills, G. Xue, and M. Tyrer. 2009. "Precipitation of heavy metals from wastewater using simulated flue gas: Sequent additions of fly ash, lime and carbon dioxide." *Water Res.* 43 (10): 2605–2614.

Cooney, D., and J. Wijaya. 1987. "Effect of pH and added salts on the adsorption of ionizable organic species onto activated carbon from aqueous solution." In *Proc., 2nd Engineering Foundation Conf. Fundamentals of Adsorption*, Engineering Foundation, New York. 185–194.

Crini, G. 2005. "Recent developments in polysaccharide-based materials used as adsorbents in wastewater treatment." *Prog. Polym. Sci.* 30 (1): 38–70.

Darkrim, F., and D. Levesque. 1998. "Monte Carlo simulations of hydrogen adsorption in single-walled carbon nanotubes." *J. Chem. Phys.* 109 (12): 4981–4984.

Demirci, Y., L. C. Pekel, and M. Alpbaz. 2015. "Investigation of different electrode connections in electrocoagulation of textile wastewater treatment." *Int. J. Electrochem. Sci.* 10: 2685–2693.

Drogui, P., N. Meunier, G. Mercier, and J.-F. Blais. 2011. "Removal of Pb^{2+} and Zn^{2+} ions from acidic soil leachate: A comparative study between electrocoagulation, adsorption and chemical precipitation processes." *Int. J. Environ. Waste Manage.* 8 (3–4): 241–257.

El-Sikaily, A., A. E. Nemr, A. Khaled, and O. Abdelwehab. 2007. "Removal of toxic chromium from wastewater using green alga Ulva lactuca and its activated carbon." *J. Hazard. Mater.* 148 (1): 216–228.

Eyvaz, M., M. Kirlaroglu, T. S. Aktas, and E. Yuksel. 2009. "The effects of alternating current electrocoagulation on dye removal from aqueous solutions." *Chem. Eng. J.* 153 (1–3): 16–22.

Feng, D., and C. Aldrich. 2004. "Adsorption of heavy metals by biomaterials derived from the marine alga Ecklonia maxima." *Hydrometallurgy* 73 (1): 1–10.

Fu, F., and Q. Wang. 2011. "Removal of heavy metal ions from wastewaters: A review." *J. Environ. Manage.* 92 (3): 407–418.

Fu, F., H. Zeng, Q. Cai, R. Qiu, J. Yu, and Y. Xiong. 2007. "Effective removal of coordinated copper from wastewater using a new dithiocarbamate-type supramolecular heavy metal precipitant." *Chemosphere* 69 (11): 1783–1789.

Ghosh, P., A. N. Samanta, and S. Ray. 2011. "Reduction of COD and removal of Zn^{2+} from rayon industry wastewater by combined electro-Fenton treatment and chemical precipitation." *Desalination* 266 (1–3): 213–217.

Giesche, H. 2006. "Mercury porosimetry: A general (practical) overview." *Part. Part. Syst. Char.* 23 (1): 9–19.

Golder, A., A. Samanta, and S. Ray. 2007a. "Removal of Cr^{3+} by electrocoagulation with multiple electrodes: Bipolar and monopolar configurations." *J. Hazard. Mater.* 141 (3): 653–661.

Golder, A. K., A. K. Chanda, A. N. Samanta, and S. Ray. 2007b. "Removal of Cr(VI) from aqueous solution: Electrocoagulation vs chemical coagulation." *Sep. Sci. Technol.* 42 (10): 2177–2193.

Gregg, S. J., K. S. W. Sing, and H. W. Salzberg. 1967. "Adsorption surface area and porosity." *J. Electrochem. Soc.* 114 (11): 279C.

Guo, Z.-R., G. Zhang, J. Fang, and X. Dou. 2006. "Enhanced chromium recovery from tanning wastewater." *J. Cleaner Prod.* 14 (1): 75–79.

Gupta, V. K., and A. Rastogi. 2008. "Biosorption of lead(II) from aqueous solutions by non-living algal biomass *Oedogonium sp.* and *Nostoc sp.*—A comparative study." *Colloids Surf., B: Biointerfaces* 64 (2): 170–178.

Gupta, V., A. Rastogi, V. Saini, and N. Jain. 2006. "Biosorption of Copper(II) from Aqueous Solutions by Spirogyra Species." *J. Colloid Interface Sci.* 296: 59–63.

Hakizimana, J. N., B. Gourich, M. Chafi, Y. Stiriba, C. Vial, P. Drogui, et al. 2017. "Electrocoagulation process in water treatment: A review of electrocoagulation modeling approaches." *Desalination* 404: 1–21.

Harif, T., M. Khai, and A. Adin. 2012. "Electrocoagulation versus chemical coagulation: Coagulation/flocculation mechanisms and resulting floc characteristics." *Water Res.* 46 (10): 3177–3188.

Hegazi, H. A. 2013. "Removal of heavy metals from wastewater using agricultural and industrial wastes as adsorbents." *HBRC J.* 9 (3): 276–282.

Holt, P. K., G. W. Barton, and C. A. Mitchell. 2005. "The future for electrocoagulation as a localised water treatment technology." *Chemosphere* 59 (3): 355–367.

Holt, P. K., G. W. Barton, M. Wark, and C. A. Mitchell. 2002. "A quantitative comparison between chemical dosing and electrocoagulation." *Colloids Surf., A* 211 (2–3): 233–248.

Hudson, H. E., and J. P. Wolfner. 1967. "Design of mixing and flocculating basins." *J.– Am. Water Works Assoc.* 59 (10): 1257–1267.

Igwe, J. C., D. N. Ogunewe, and A. A. Abia. 2005. "Competitive adsorption of Zn(Ii), Cd(Ii) and Pb(Ii) ions from aqueous and non-aqueous solution by maize cob and husk." *African J. Biotechnol.* 4 (10): 1113–1116.

Ilhan, F., U. Kurt, O. Apaydin, and M. T. Gonullu. 2008. "Treatment of leachate by electrocoagulation using aluminum and iron electrodes." *J. Hazard. Mater.* 154 (1–3): 381–389.

İrdemez, Ş., N. Demircioğlu, Y. Ş. Yıldız, and Z. Bingül. 2006. "The effects of current density and phosphate concentration on phosphate removal from wastewater by electrocoagulation using aluminum and iron plate electrodes." *Sep. Purif. Technol.* 52 (2): 218–223.

Katal, R., and H. Pahlavanzadeh. 2011. "Influence of different combinations of aluminum and iron electrode on electrocoagulation efficiency: Application to the treatment of paper mill wastewater." *Desalination* 265 (1–3): 199–205.

Khandegar, V., and A. K. Saroha. 2013. "Electrocoagulation for the treatment of textile industry effluent—A review." *J. Environ. Manage.* 128: 949–963.

Kladnig, W. 1975. "Catalysis with zeolites-review." *Acta Cient. Venez.* 26 (1): 40–69.

Kobya, M., O. T. Can, and M. Bayramoglu. 2003. "Treatment of textile wastewaters by electrocoagulation using iron and aluminum electrodes." *J. Hazard. Mater.* 100 (1–3): 163–178.

Kobya, M., E. Gengec, and E. Demirbas. 2016. "Operating parameters and costs assessments of a real dyehouse wastewater effluent treated by a continuous electrocoagulation process." *Chem. Eng. Process. Process Intensif.* 101: 87–100.

Kurniawan, T. A., G. Y. S. Chan, W.-H. Lo, and S. Babel. 2006. "Physico-chemical treatment techniques for wastewater laden with heavy metals." *Chem. Eng. J.* 118 (1–2): 83–98.

Lakshmanan, D., D. A. Clifford, and G. Samanta. 2010. "Comparative study of arsenic removal by iron using electrocoagulation and chemical coagulation." *Water Res.* 44 (19): 5641–5652.

Larue, O., E. Vorobiev, C. Vu, and B. Durand. 2003. "Electrocoagulation and coagulation by iron of latex particles in aqueous suspensions." *Sep. Purif. Technol.* 31 (2): 177–192.

Le, S. T., K. C. Le, L. T. Doan, and A. T. Doan. 2017. "Effect of some effective parameters on COD removal from Nam Son landfill leachate by electrocoagulation." *Vietnam J. Sci. Technol.* 55 (5): 540.

London, F. 1937. "The general theory of molecular forces." *Trans. Faraday Soc.* 33: 8b–26.

Mamchenko, A. V., N. G. Gerasimenko, and T. A. Pakhar'. 2011. "The impact of temperature on the efficiency of the coagulation process of titanyl sulfate and aluminum sulfate." *J. Water Chem. Technol.* 33 (5): 315–322.

Martin, R., and K. O. Iwuco. 1982. "The effects of pH and suspended solids in the removal of organics from waters and wastewaters by the activated carbon adsorption process." *Water Res.* 16 (1): 73–82.

Merzouk, B., B. Gourich, K. Madani, C. Vial, and A. Sekki. 2011. "Removal of a disperse red dye from synthetic wastewater by chemical coagulation and continuous electrocoagulation. A comparative study." *Desalination* 272 (1–3): 246–253.

Missana, T., and A. Adell. 2000. "On the applicability of DLVO theory to the prediction of clay colloids stability." *J. Colloid Interface Sci.* 230 (1): 150–156.

Mohammadi, T., A. Moheb, M. Sadrzadeh, and A. Razmi. 2005. "Modeling of metal ion removal from wastewater by electrodialysis." *Sep. Purif. Technol.* 41 (1): 73–82.

Mólgora, C. C., A. M. Domínguez, E. M. Avila, P. Drogui, and G. Buelna. 2013. "Removal of arsenic from drinking water: A comparative study between electrocoagulation-microfiltration and chemical coagulation-microfiltration processes." *Sep. Purif. Technol.* 118: 645–651.

Mollah, M. Y., P. Morkovsky, J. A. Gomes, M. Kesmez, J. Parga, and D. L. Cocke. 2004. "Fundamentals, present and future perspectives of electrocoagulation." *J. Hazard. Mater.* 114 (1–3): 199–210.

Mollah, M. Y. A., R. Schennach, J. R. Parga, and D. L. Cocke. 2001. "Electrocoagulation (EC)—Science and applications." *J. Hazard. Mater.* 84 (1): 29–41.

Mouflih, M., A. Aklil, and S. Sebti. 2005. "Removal of lead from aqueous solutions by activated phosphate." *J. Hazard. Mater.* 119 (1): 183–188.

Müller, G., C. Radke, and J. Prausnitz. 1980. "Adsorption of weak organic electrolytes from aqueous solution on activated carbon. Effect of pH." *J. Phys. Chem.* 84 (4): 369–376.

Nah, I. W., K.-Y. Hwang, C. Jeon, and H. B. Choi. 2006. "Removal of Pb ion from water by magnetically modified zeolite." *Miner. Eng.* 19 (14): 1452–1455.

Nijkamp, M. G., J. E. M. J. Raaymakers, A. J. Van Dillen, and K. P. de Jong. 2001. "Hydrogen storage using physisorption—Materials demands." *Appl. Phys. A* 72 (5): 619–623.

Nouri, A., B. Shahmoradi, S. Dehestani-Athar, and A. Maleki. 2015. "Effect of temperature on pH, turbidity, and residual free chlorine in Sanandaj Water Distribution Network, Iran." *J. Adv. Environ. Health Res.* 3: 3.

Oncel, M. S., A. Muhcu, E. Demirbas, and M. Kobya. 2013. "A comparative study of chemical precipitation and electrocoagulation for treatment of coal acid drainage wastewater." *J. Environ. Chem. Eng.* 1 (4): 989–995.

Pan, B., Q. Zhang, W. Du, W. Zhang, B. Pan, Q. Zhang, Z. Xu, and Q. Zhang. 2007. "Selective heavy metals removal from waters by amorphous zirconium phosphate: Behavior and mechanism." *Water Res.* 41 (14): 3103–3111.

Papadopoulos, A., D. Fatta, K. Parperis, A. Mentzis, K.-J. Haralambous, and M. Loizidou. 2004. "Nickel uptake from a wastewater stream produced in a metal finishing industry by combination of ion-exchange and precipitation methods." *Sep. Purif. Technol.* 39 (3): 181–188.

Rahman, I. A., and V. Padavettan. 2012. "Synthesis of silica nanoparticles by sol–gel: Size-dependent properties, surface modification, and applications in silica-polymer nanocomposites—A review." *J. Nanomater.* 2012: 132424.

Randtke, S. J., and C. P. Jepsen. 1982. "Effects of salts on activated carbon adsorption of fulvic acids." *J. Am. Water Works Assoc.* 74 (2): 84–93.

Ryan, D., A. Gadd, J. Kavanagh, M. Zhou, and G. Barton. 2008. "A comparison of coagulant dosing options for the remediation of molasses process water." *Sep. Purif. Technol.* 58 (3): 347–352.

Snoeyink, V. L., W. J. Weber Jr., and H. B. Mark Jr. 1969. "Sorption of phenol and nitrophenol by active carbon." *Environ. Sci. Technol.* 3 (10): 918–926.

Şölener, M., S. Tunali, A. S. Özcan, A. Özcan, and T. Gedikbey. 2008. "Adsorption characteristics of lead(II) ions onto the clay/poly(methoxyethyl)acrylamide (PMEA) composite from aqueous solutions." *Desalination* 223 (1): 308–322.

Sundstrom, D. W., and H. E. Klei. 1979. *Wastewater treatment.* Hoboken, NJ: Prentice-Hall.

Tang, P. L., C. K. Lee, K. S. Low, and Z. Zainal. 2003. "Sorption of Cr(VI) and Cu(II) in aqueous solution by ethylenediamine modified rice hull." *Environ. Technol.* 24 (10): 1243–1251.

Tang, S., J. M. Preece, C. M. McFarlane, and Z. Zhang. 2000. "Fractal morphology and breakage of DLCA and RLCA aggregates." *J. Colloid Interface Sci.* 221 (1): 114–123.

Tran, N., P. Drogui, J.-F. Blais, and G. Mercier. 2012. "Phosphorus removal from spiked municipal wastewater using either electrochemical coagulation or chemical coagulation as tertiary treatment." *Sep. Purif. Technol.* 95: 16–25.

Tünay, O., and N. I. Kabdaşli. 1994. "Hydroxide precipitation of complexed metals." *Water Res.* 28 (10): 2117–2124.

Vasudevan, S., J. Lakshmi, and G. Sozhan. 2011. "Effects of alternating and direct current in electrocoagulation process on the removal of cadmium from water." *J. Hazard. Mater.* 192 (1): 26–34.

Vengris, T., R. Binkien, and A. Sveikauskait. 2001. "Nickel, copper and zinc removal from waste water by a modified clay sorbent." *Appl. Clay Sci.* 18 (3): 183–190.

Yilmaz, A. E., R. Boncukcuoğlu, and M. M. Kocakerim. 2007. "A quantitative comparison between electrocoagulation and chemical coagulation for boron removal from boron-containing solution." *J. Hazard. Mater.* 149 (2): 475–481.

Zhu, B., D. A. Clifford, and S. Chellam. 2005. "Comparison of electrocoagulation and chemical coagulation pretreatment for enhanced virus removal using microfiltration membranes." *Water Res.* 39 (13): 3098–3108.

CHAPTER 11

Comparative Studies between Electro-Oxidation and Other Oxidation Processes

Ouarda Yassine, Kiendrebeogo Marthe,
Ali Khosravanipour, Patrick Drogui

11.1 INTRODUCTION

Oxidation processes as biological oxidation are the cheapest and most effective processes for treating polluted effluents with an organic load of less than 1 g L^{-1} chemical oxygen demand (COD) (Fryda et al. 2003, Sirés et al. 2014). Chemical oxidation with chlorine, ozone, or hydrogen peroxide allows the treatment of recalcitrant contaminants or at least breaks them down into harmless or biodegradable products. However, in some reactions, intermediate products that are more toxic than the original products remain in the solution. Advanced oxidation processes (AOPs) are a special class of oxidation technique that allows the removal of these pollutants. They can be conveniently defined as water-phase oxidation methods based on the in situ production of reactive species (mainly but not exclusively) for the destruction of the target pollutant. Among the oxidizing species, hydroxyl radical with a standard potential of 2.80 Volts/Standard hydrogen electrode (V/SHE) is one of the powerful oxidants after fluorine (3.06 V/SHE) and positive hole electrons of TiO_2 (3.2 V/SHE). By forming dehydrogenated (aliphatic) and hydroxylated (aromatic compounds) derivatives, hydroxyl radical is able to mineralize most organic and organometallic contaminants without selection into CO_2, water, and inorganic ions. Initially, AOPs consisted of photochemical and nonphotochemical processes. In recent years, new AOPs based on electrochemistry, for example, electrochemical advanced oxidation processes (EAOPs), have been developed. Particularly, electro-oxidation (EO) treatment has received a great deal of interest in wastewater treatment and organic pollutant oxidation.

Several recent studies have already recognized that EO processes are more effective and eco-friendly than other oxidation processes for the degradation of recalcitrant pollutants and the treatment of wastewater (Martinez-Huitle et al. 2015, Rodrigo et al. 2010, Särkkä et al. 2015). A general overview of the application of EAOPs on the removal of aqueous organic pollutants has been given, first reviewing the recent works and then looking to the future (Sirés et al. 2014). The latest advances in the electrochemical processes for the oxidation of organic pollutants, discussions on the most important approaches to be adopted, and critically presenting their advantages and limitations have been summarized and dealt with in detail in several reviews (Martínez-Huitle and Panizza 2018). It becomes necessary to compare the recent progress made in EO and its limitations with other oxidation processes to have a more unbiased and modern view on the effectiveness of the EO process.

In this chapter, the EO principles of action, operating conditions, and performance factors are discussed and compared with Fenton, ozonation, photocatalytic, and sonochemical processes. The treatment of various organic pollutants by EO processes and the economic comparison between these and other processes are also reviewed.

11.2 ADVANCED OXIDATION PROCESSES

Usually, biorefractory compounds are not completely removed by wastewater treatment plants (WWTPs), and consequently, they act as an important source of toxic compounds released into the environment. For promoting high-quality effluents of treated wastewater, the implementation of sustainable technologies is considered as a possible viable option. In wastewater, the use of AOPs for reducing biologically toxic or nondegradable molecules (e.g., aromatics, pesticides, dyes, pharmaceuticals, their residues, and volatile organic compounds) has received particular attention (Borras et al. 2011, Hussain et al. 2013, Isarain-Chávez et al. 2011, Oturan et al. 2011) compared with conventional processes and options. AOPs are introduced as promising, powerful, and environmentally friendly methods for wastewater treatment. Table 11-1 shows a large number of methods classified under the broad definition of AOPs, together with many electric and electrochemical methods (Almeida et al. 2011, Brillas et al. 2009). In AOPs, different oxidant species like hydroxyl radical ($^{\bullet}OH$) and other strong oxidant species (e.g., $O_2^{\bullet}$, $HO_2^{\bullet}$, and $ROO^{\bullet}$) produced in situ destroy organic compounds. Hydroxyl radical ($^{\bullet}OH$) is the common reactive species in all AOPs. It can highly react with a wide range of organic compounds, regardless of their concentration. AOPs are able to destroy the target organic molecules and may mineralize many organics (if not all) to CO_2 and H_2O (Almeida et al. 2011).

The use of electricity in additional AOPs for water treatment was first suggested in 1889 (Chen 2004). Since then, many electrochemical technologies (the so-called EAOPs) have been used for wastewater treatment (Rajkumar and Palanivelu 2004, Robinson et al. 2001), such as anodic oxidation (AO), electro-Fenton (EF), photoelectron-Fenton (PEF), and sonoelectro-Fenton (Oturan et al. 2008), providing

Table 11-1. Main AOPs and Related Reactions Involving the Production of •OH.

AOPs		Reactions
Dark AOPs	Ozone at elevated pH	$3O_3 + OH^- + H^+ \rightarrow 2OH^\bullet + 4O_2$
	Ozone + hydrogen peroxide	$2O_3 + H_2O_2 \rightarrow 2OH^\bullet + 3O_2$
	Ozone + catalyst	$O_3 + Fe^{2+} + H_2O \rightarrow Fe^{3+} + OH^- + OH^\bullet + O_2$
	Fenton	$Fe^{2+} + H_2O_2 \rightarrow Fe^{3+} + OH^- + OH^\bullet$
Photo-assisted AOPs	Ozone/UV	$O_3 + H_2O + hv \rightarrow O_2 + H_2O_2$
	Ozone/H_2O_2/UV	The addition of H_2O_2 to the O_3/UV process accelerates the decomposition of ozone, which results in an increased rate of •OH generation
	Hydrogen peroxide/UV	$H_2O_2 + hv \rightarrow 2OH^\bullet$
	Photo-Fenton	$Fe^{2+} + H_2O_2 + hv \rightarrow Fe^{3+} + OH^- + OH^\bullet$
		$Fe(OH)^{2+} + hv \rightarrow Fe^{2+} + OH^\bullet$
		$Fe(OOCR)^{2+} + hv \rightarrow Fe^{2+} + R^\bullet + CO_2$
	Heterogeneous photocatalysis (TiO_2/UV)	$TiO_2 + hv \rightarrow TiO_2(e^- + h^+)$
		$h^+ + H_2O \rightarrow OH^\bullet + H^+$
		$e^- + O_2 \rightarrow O_2^{-\bullet}$

valuable contributions to the protection of the environment through implementation of effluent treatment and production-integrated processes. Recently, EAOPs were extensively studied along with other processes described in several publications with the aim of upscaling them to pilot plants in the near future (Dirany et al. 2010, Hammami et al. 2008, Panizza and Cerisola 2009). In EAOPs, hydroxyl radicals can be generated by direct electrochemistry; for example, in the AO case, (•OH) are generated heterogeneously by direct water discharge on the anode. Hydroxyl radicals can also be generated indirectly through the electrochemical generation of Fenton's reagent; for example, in an EF case, (•OH) are generated homogeneously via Fenton's reaction. Both processes are widely applied to the treatment of several kinds of wastewater with an almost 100% mineralization efficiency in most cases. EAOPs can be applied in a variety of media and volumes and can also eliminate pollutants in the form of gas, liquid, or solid. The nonselective character of (•OH) helps prevent the production of unwanted by-products, making EAOPs promising technologies for treatment of biorefractory compounds in water and wastewater (Barrera-Díaz et al. 2009, Sirés and Brillas 2012).

EAOPs offer several benefits for the prevention and remediation of pollution because electron is a clean reagent. Their other benefits include high energy efficiency, the ability to automate, easy handling because of the requirement of simple equipment, safety because of their operations under mild conditions (ambient temperature and pressure), and versatility because of their application in effluents with COD in the range of 0.1 to 100 g L^{-1} (Sirés et al. 2014). The main drawbacks of some EAOPs include the costs associated with the supply of electricity, and the conductivity of many wastewaters that require the addition of electrolytes.

11.3 ELECTRO-OXIDATION

For the elimination of organic pollutants in wastewater, AO or EO is the most effective of all advanced EAOPs (Brillas et al. 2009, Martínez-Huitle and Brillas 2009, Panizza and Cerisola 2009). EO is the basic process used in all these processes. As several researchers have revealed, EO involves the oxidation of pollutants in an electrochemical reactor directly on the anode by direct transfer of electrons on the anode and indirectly by chemical reaction with the electrogenerated species ($^{\bullet}OH$) by discharge of the water molecule on the anode (Comninellis and Chen 2010, Martínez-Huitle and Brillas 2009). Examples of these oxidants include the generation of chlorine, hypochlorinated acid, and hypochlorite (Chatzisymeon et al. 2006, Rajkumar et al. 2007, Sakalis et al. 2006) in accordance with what is displayed in Equations (11-1) to (11-3); the generation of hydrogen peroxide/ozone (Chu et al. 2012, Dhaouadi et al. 2009) in accordance with what is shown in Equations (11-4) to (11-7); and the generation of peroxodisulfuric acid and peroxosulfates (Michaud et al. 2000, Serrano et al. 2002) in accordance with what is exhibited in Equations (11-8) and (11-9).

$$2Cl^{-} \rightarrow Cl_2 + 2e^{-} \quad (11\text{-}1)$$

$$Cl_2 + H_2O \rightarrow HOCl + H^{+} + Cl^{-} \quad (11\text{-}2)$$

$$HOCl \rightarrow H^{+} + OCl^{-} \quad (11\text{-}3)$$

$$H_2O \rightarrow OH^{\bullet} + H^{+} + 2^{-} \quad (11\text{-}4)$$

$$2OH^{\bullet} \rightarrow H_2O_2 \quad (11\text{-}5)$$

$$H_2O_2 \rightarrow O_2 + 2H^{+} + 2e^{-} \quad (11\text{-}6)$$

$$O_2 + O^{\bullet} \rightarrow O_3 \quad (11\text{-}7)$$

$$2HSO_4^{-} \rightarrow S_2O_8^{2-} + 2H^{+} + 2^{e-} \quad (11\text{-}8)$$

$$S_2O_8^{2-} + H_2O \rightarrow 2HSO_4^{-} + ½O_2 \quad (11\text{-}9)$$

For direct oxidation, hydroxyl radicals produced on the surface of the anode are physically or chemically adsorbed through active oxygen according to the electrocatalytic activity of the anode, leading to partial or total decontamination. In addition, depending on the reaction of these hydroxyl radicals with the anode, the degradation of pollutants occurs in two ways: electrochemical conversion, in which biorefractory organic matter is selectively transformed into biodegradable compounds under the action of chemically adsorbed active oxygen; and electrochemical combustion, in which organic matter is fully mineralized into H_2O, CO_2, and inorganic ions by physically adsorbed hydroxyl radicals (Comninellis 1994). Depending on the behavior of the electrode in EO, anodes can be distinguished as nonactive anode and active anode (Comninellis and Chen 2010, Comninellis et al. 2008, Comninellis and Pulgarin 1993). The oxidation model proposed by Comninellis assumes that the reaction in both types of anodes (generically denoted M) corresponds to the oxidation of water molecules, leading to the formation of physically adsorbed hydroxyl radicals (Comninellis and Pulgarin 1993)

$$M + H_2O \rightarrow M(OH^{\bullet}) + H^+ + e^- \tag{11-10}$$

On active anodes, the surface interacts strongly with (OOH), forming a more oxidized anode or superoxide. This happens when metal oxide has a very high oxidation state compared with the standard oxygen oxidation potential. The MO/M redox couple acts as a mediator in the oxidation of organic matter R ($MO + R \rightarrow M + RO$). Unfortunately, the chemical decomposition of the superoxide in residues competes with the oxidation of pollutants (Marselli et al. 2003, Martínez-Huitle and Andrade 2011). For nonactive electrodes, where the formation of an oxide is excluded, physically adsorbed hydroxyl radicals allow nonselective oxidation of organic compounds, leading to complete CO_2 combustion. Therefore, although anodes with low oxygen surge cause a partial oxidation of organic compounds, anodes with high oxygen surge promote the mineralization of organic materials to CO_2, thus becoming ideal electrodes for wastewater treatment. In any case, the effectiveness of decontamination depends on the following several conditions: electrode passivation, anode material, current density, nature and concentration of the organic pollutant, supporting electrolyte and conductivity, pH, and temperature.

11.3.1 Electrode Material

Several researchers have reported the results of different types of organic pollutant treatment using different electrocatalytic materials (Anglada et al. 2009, Ganzenko et al. 2014, Sirés et al. 2014, Rao and Venkatarangaiah 2014). Overall, during wastewater treatment, dimensionally stable anodes (DSAs), stannic oxide (SnO_2), lead dioxide (PbO_2), graphite, and boron-doped diamond (BDD) show greater chemical resistance than other materials. Although the graphite electrode has a large area for the elimination of organic compounds, it presents an extremely high electrical potential with a low observed durability and is also expensive (Panizza and Cerisola 2009). In the case of PbO_2 anodes, the generation of the highly toxic Pb^{2+} is a matter of concern because of severe pollution (Li et al. 2011, Panizza and

Cerisola 2009). However, Ti/PbO_2 has proven to be somewhat stable, although its performance and stability depend on the preparation method. Ti/SnO_2 displays similar behavior and a limited life span (Martinez-Huitle et al. 2015). As for the Ti-Pt/PbO_2 electrode, its use resulted in higher rates for COD elimination than that with DSA electrodes (20% higher at pH 3 and 25 °C) (Martinez-Huitle et al. 2008). BDD has an inert surface with low chemisorption properties, remarkable corrosion stability even in strong acid media, and a wide range of use in both water and nonwater media (Martinez-Huitle et al. 2005, Panizza and Cerisola

Table 11-2. Comparison of Electrode Chemisorption and/or Physisorption for •OH, Oxidation Power, and performance in EO Treatment

Electrodes	*Oxidation power*	*Absorption enthalpy of M-•OH*	*Advantages*	*Disadvantages*
Ti	Lower	Chemisorption of •OH	• Stable	• Passive • Expensive
DSA			• Supports indirect oxidation • Good current efficiency • High O_2 evolution overpotential • Lower cost	• Short lifetime • Lack of electrochemical stability
Pt			• Inert • Overpotential Evolution	• Expensive
PbO_2			• Good current efficiency • Cheap high O_2 evolution overpotential	• Corrosive • May release toxic Pb^{2+}
BDD	Higher	Physisorption of •OH	• Inert • High O_2 evolution overpotential • Electrochemical stability • Good current efficiency • Good conductivity	• Very expensive

2007). They also have the highest overpotential oxygen evolution value (2.7 V for Ti/BDD) (Chen 2004), which means that more hydroxyl radicals are formed on the surface of the anode during treatment. BDD electrodes can also completely degrade refractory organic pollutants, and the nature of the pollutant does not significantly impact the efficiency of the process (Rodrigo et al. 2010). It is also known that in addition to the formation of radical hydroxyls on the surface of the electrode, diamond electrodes increase oxidation mediated by other chemically formed compounds such as persulfate, perphosphate, or hypochlorite, depending on the electrolyte used (Table 11-2).

11.3.2 Current Density

Applied current density, defined as current intensity per unit of anode surface, finds more mention in studies because it controls the reaction rate, and, therefore, defines the effectiveness of the process. It directly influences the EO process (Anglada et al. 2009, Khandegar and Saroha 2013). In general, when the EO kinetics is not limited by the massive transport of organic matter to the surface of the anode, an increase in current density leads to a greater elimination of pollutants. When the process is controlled by mass transport, an increase in density, that is, current intensity, is expected to improve the production of O_2 (Martinez-Huitle et al. 2015), leading to lower energy efficiency and higher energy costs. When the process is subject to agitation, the high-intensity values should cause an increase in the elimination of pollutants, but also a reduction in energy efficiency (Comninellis et al. 2008). Strangely, when a high current intensity is applied in the presence of an electrolyte, the electrical potential increases. It is believed that the formation of other oxidants such as chlorine, ozone, hydrogen peroxide, persulfate, and peroxophosphate seems to affect this scenario when the BDD anode is used for real effluent treatment with a flow electrolytic cell (Comninellis and Chen 2010, Panizza and Cerisola 2009). Consequently, mass transport limits are minimized by these stable oxidants. This promotes a greater elimination of organic matter. It is important to note that an increase in density does not necessarily improve the efficiency of oxidation or organic mineralization. It depends on the characteristics of the water to be treated. Certainly, variation in current density can impact the efficiency of the process, but we should not ignore the action of strong oxidants produced on the anode surface. The production of these oxidants is controlled not only by the applied current but also by other operating conditions such as dissolved O_2, pH, temperature, and inorganic ions as well as the anode properties (Sirés et al. 2014).

11.3.3 Nature and Concentration of Organic Pollutants

The nature of organic pollutants appears to have only a small influence on the rate of degradation and process efficiency in the case of nonactive anodes (Martinez-Huitle et al. 2015). However, it is relatively dependent on the oxidizing species generated on the anodic surface. As a general rule, the higher the material concentration to be removed, the greater is the energy efficiency (Panizza and Cerisola 2009,

Panizza et al. 2001, Yu et al. 2014). However, a current intensity limit of 80 should not be exceeded. When the applied intensity *I* (A) is lower than the *I* (A) limit, the electrolysis is under current control, and the energetic efficiency is 100% with a linear decrease in COD over time. When the applied *I* (A) is above the *I* (A) limit, the electrolysis is under the control of mass transport, and secondary reactions (such as oxygen evolution) will occur in a significant way, resulting in a reduction in energy efficiency. In this case, the elimination of organic matter grows exponentially because of the limitations of mass transport. The main limitation of this model is that it does not consider the existence of different phenomena on the surface of the anode involving other chemical species. Nevertheless, its application prevents a large amount of (•OH) from being wasted in nonoxidant parasitic reactions (Cañizares et al. 2002). In the case of aliphatic acids, it has been demonstrated that electrochemical processing is independent of the chemical nature of the organic compound. Nevertheless, a change in solution pH can potentially alter the chemical structure of organic pollutants (protonated or nonprotonated), favoring a decrease or increase in suppression rates (Gandini et al. 2000).

The EO of various aromatic compounds in an acid solution was achieved by varying the organic concentration and density of current. However, the nature of the organic pollutant can also impact the effectiveness of treatment (Martinez-Huitle et al. 2015).

For the EO treatment of phenol, its conversion to hydroquinone, p-benzoquinone and pyrocatechol were selectively carried out with an energy efficiency minor than 1%. With these results, two features were demonstrated: i) phenol does not undergo additional oxidation, even after a near-complete conversion of phenol with an anodic working potential attached to 1.7 V buffered by the Ti/IrO_2 anode; and ii) the high selectivity towards the formation of p-benzoquinone has ortho in a ratio of (3:1), which reveals a specific orientation of phenol on the surface of IrO_2 during EO. In fact, in the absence of interaction with the electrode, the main product is the pyrocatechol in a para/ortho ratio of 1/2.

11.3.4 pH

It is found that the effectiveness of removing pollutants by EO is maximum at some optimal pH value; it can be acidic or basic. This may be related to the chemical structure of the pollutant (i.e., protonated or not protonated, depending on the pH, thus promoting or not promoting its oxidation) (Martínez-Huitle and Brillas 2009). The pH of the solution also controls the concentration of hydroxyl radicals and other stable oxidants in the solution. The performance of the EC process in an agitated undivided cell with an Si/BDD anode and a stainless-steel cathode was investigated under the influence of current density (100 mA, 300 mA) and pH (3,10) during EO with a 220 mg dm^{-3} indigo carmine solution and 0.05 M Na_2SO_4 at 35 °C (Ammar et al. 2006). At pH 3.0, a faster TOC decay to more than 80% was obtained after 3 h with the current density increasing. However, at pH 10 and with 100 mA, the solution became colorless more quickly (120 min) than that with pH 3.0 (270 min). The first finding is caused by the greater production of active BDD(•OH) that accelerates the oxidation of organics, and the second is because of the fact electroactive species in an alkaline medium (the unprotonated form) was more easily oxidized (Martinez-Huitle et al. 2015).

11.3.5 Electro-Oxidation Removal Efficiency of Biorefractory Compounds

Several studies have demonstrated that the efficiency of EO is much greater than that of other processes (Li et al. 2011, Panizza and Cerisola 2005, Yu et al. 2014). Table 11-3 lists the degradation of different organic pollutants (e.g., pesticides, pharmaceuticals, and dyes) by EO with a BDD anode; a complete mineralization of wastewater can be obtained using proper operating conditions, thus revealing the excellent prospects for this technology. The use of graphite cathode generates H_2O_2 through the electrochemical reduction of O_2, leading to an enhancement of oxidation as reported by several researchers (Flox et al. 2006, Ma et al. 2007). Most of these studies used Na_2SO_4 as the supporting electrolyte, which was attributed to the following reasons: (1) a relatively cheap option; (2) it could be oxidized into $S_2O_8^{2-}$, which also participated in the degradation of organics; (3) it would not produce hazardous compounds during treatment; and (4) the presence of Na_2SO_4 would promote degradation (Özcan et al. 2008, Yu et al. 2014). Current density is one of the most important parameters in AO. As shown in Table 11-3, the current density is in the reported ranges (10 to 500 $mA \cdot cm^{-2}$) (Yu et al. 2014). A higher current density could lead to better treatment performance for pollutant or TOC removal, that is, removal of contaminants within the stipulated time or achieving a higher removal efficiency within the same period. All this showed that EO is capable of abating the organic load of the wastewater of almost any range of concentration. Despite this, a comparison between EO and chemical oxidation processes such as Fenton and ozonation applied in the treatment of many types of wastewater is highly necessary.

11.4 COMPARISON BETWEEN ELECTRO-OXIDATION AND CHEMICAL OXIDATION

EO is able to eliminate the organic load of industrial effluents completely; on the contrary, ozonation and Fenton oxidation may lead to the accumulation of significant amounts of oxidation-refractory compounds (Oturan et al. 2011, Rodrigo et al. 2010). Besides the formation of hydroxyl radicals (Cañizares et al. 2007, Marselli et al. 2003), EO combines other types of oxidation mechanisms such as direct EO on the conductive diamond surface and mediated oxidation by other electrochemically formed compounds such as persulfate (Cañizares et al. 2005), perphosphate (Cañizares et al. 2007) or hypochlorite (Panizza and Cerisola 2003, Polcaro et al. 2008), depending on the electrolyte. Conversely, in the Fenton process and ozonation at an alkaline pH, only the hydroxyl radical and the molecular ozone (for ozonation) and hydrogen peroxide (for Fenton) are the oxidant agents involved in the treatment. In addition, these two conventional AOPs are strongly

Table 11-3. Efficiencies of EO with the BDD Electrode for the Removal of Different Pollutants.

Compounds	*Removal (%) and electrolysis time (h)*	*Initial concentration*	*Current density*	*References*
Clofibric acid	100*** (9 h)	179 mg L^{-1}	33 mA · cm^{-2}	Sirés et al. (2006)
Biphenol A	100** (24 h)	20 mg L^{-1}	14.28 mA · cm^{-2}	Murugananthan et al. (2008)
4-Chlorophenol	100* (3 h)	1.1 mM	30 mA · cm^{-2}	Cañizares et al. (2004)
Cyanazine	68** (4 h)	55 mg L^{-1}	100 mA · cm^{-2}	Borràs et al. (2013)
2-Naphthol	100* (12 h)	5 mM	50 mA · cm^{-2}	Panizza and Cerisola (2004)
Desmetryne	68** (1.33 h)	110 mg L^{-1}	450 mA	Borras et al. (2011)
Oxalic acid	100*** (12 h)	0.12 M	100 A · m^{-2}	Martinez-Huitle et al. (2005)
Propham	94** (3 h)	0.5 mM	300 mA	Özcan et al. (2008)
Diclofenac	97** (6 h)	175 mg L^{-1}	300 mA · cm^{-2}	Brillas et al. (2010)
Ibuprofen	95** (6 h)	1.75 mM	30 mA · cm^{-2}	Ciríaco et al. (2009)
Carbamazépine	83.33*** (2 h)	10 mg L^{-1}	2 A	García-Gómez et al. (2014)
Chloranilic acid	100* (6.4 h)	450 mg L^{-1}	50 mA · cm^{-2}	Martínez-Huitle et al. (2004)
Cyanide waste	80–90* (3 h)	375 mg L^{-1}	30 mA · cm^{-2}	Cañizares et al. (2005)
Sodium dodecyl-benzenesulfonate	75** (10 h)	6.15×10^{-4} M	12 mA · cm^{-2}	Weiss et al. (2006)
p-Nitrophenol	90** (9.3 h)	1 mM	20 mA · cm^{-2}	Zhu et al. (2008)
Mecoprop	100* (3.6 h)	178 mg L^{-1}	1 A	Sirés et al. (2008)
	100*** (0.28 h)		150 mA · cm^{-2}	Flox et al. (2006)
1,2-Dichloroethane	97* (7 h)	50 mM	50 mA · cm^{-2}	Scialdone et al. (2008)
Phenol	100** (5 h)	3.9 mM	30 mA · cm^{-2}	Weiss et al. (2008)
Methamidophos	90*** (0.2 h)	50 mg L^{-1}	50 mA · cm^{-2}	Martinez-Huitle et al. (2008)

Note: * = COD; ** = TOC; *** = pollutant itself.

influenced by the composition of the waste. The mineralization rate of chemical Fenton performed at pH 3 depends on the $[H_2O_2]/[Fe^{2+}]$ ratio (Oturan et al. 2011). In organic oxidation, the instantaneous O_3 demand of ozonation has shown similarities with O_3-based AOPs, resulting in the same exposure to OH radicals (Buffle et al. 2006a). Thus, it can be pointed out that EO competes favorably with ozonation and Fenton oxidation because of the formation of more by-products of complexes in the ozonation and Fenton processes and more efficient oxidation processes associated with the electrochemical process. However, it can be stated that, in general, oxidants are less efficiently used in ozonation but more efficiently used in EO and Fenton oxidation (Cañizares et al. 2009).

In addition, the efficiency of EO is independent of the type of wastewater but increases with an increase in the pollutant concentration (COD), that is, the higher the organic load, the higher the efficiency of the process (Rodrigo et al. 2010). Thus, for a concentrated wastewater, a particular amount of electrical charge is able to remove a higher organic content than in the case of diluted waste. This means that EO is more efficient for highly polluted waste, and it also suggests that mediated oxidation should play an important role in the removal of pollutants from raw waste. It is also important to mention that the energy required to remove COD to 1,500 mg L^{-1} (a typical discharge limit in municipal sewage collectors) increases directly with an increase in the concentration of pollutants in raw wastewater (Rodrigo et al. 2010). With this knowledge, it is easy to mark the higher limit of COD concentration in which the use of EO can be recommended, by a comparison of the energy requirements of EO with the energy requirements of other technologies.

11.5 OZONATION VERSUS ELECTRO-OXIDATION

The ozonation process is an AOP involving the generation of highly free radicals, mainly hydroxyl radical ($OH^\bullet$), via chemical (O_3/OH^-) reaction as shown in the following equation:

$$3O_3 + OH^- + H^+ \rightarrow 2OH^\bullet + 4O_2 \tag{11-11}$$

Usually, ozonation is used as a wastewater pretreatment for improving the biodegradability of compounds or the low concentration of biorefractory compounds in drinking water degradation and disinfection (Balcıoğlu and Ötker 2003). However, EO is often used as a tertiary treatment for wastewater polishing. This can be explained by the formation of organic [e.g., assimilable organic carbon (AOC), aldehydes, carboxylic acids, and ketones] and inorganic (e.g., bromate) disinfection by-products (DBPs). In addition, this has been well documented in ozonation processes (Huang et al. 2005). Dissolved organic carbon (DOC) concentrations are typically higher in wastewater than in surface water, resulting in faster O_3 decomposition rates and increased consumption of hydroxyl radical ($^\bullet OH$). As a result, higher O_3 dosages are required to meet wastewater treatment goals, potentially leading to increased formation exposures of DBPs (Buffle et al. 2006b). In water disinfection, a less-efficient disinfection by O_3/H_2O_2 than by O_3 was observed because of a lack of O_3 residuals and the quantities of available OH scavengers in wastewater, specifically DOC and alkalinity (Acero and Von Gunten 2001, von Gunten 2003). Consequently, the production of hydroxyl radical is limited over time by the amount of dissolved O_3. On the contrary, EO in situ radical hydroxyl production is continuous.

In addition, ozonation is a pH-dependent process for the elimination of biorefractory compounds. At a near-neutral pH, biorefractory ozonation does not result in complete mineralization. Alachlor ozonation indicates that the benzene ring is cleaved, although complete mineralization does not occur (Somich et al. 1988). However, 2.4-dichlorophenoxy-propionic acid (2.4-DP) treated in EO with a BDD anode indicates mineralization rates of 63%, 82%, and 97% with current densities of 100, 300, and 450 $mA \cdot cm^{-2}$, respectively (Brillas et al. 2007). Also, for chlorpyrifos, mineralization by EO with a BDD anode ranges from 16% to 57% as the density of the applied current changes from 10 to 50 $mA \cdot cm^{-2}$ (Samet et al. 2010). At pH 7, herbicide compounds yield a substantial removal for 30 min (60% to 88%) in the presence of ozone (Hladik et al. 2005). This removal results from the reaction with ozone or alternatively with OH radicals, which are always present during ozonation. However, prior work with acetochlor and metolachlor has demonstrated these compounds react slowly with ozone (1.1 and 2.4 $M \cdot s^{-1}$, respectively) relative to OH radicals (6.9 to 6.3×10^{-9} $M \cdot s^{-1}$) (Acero et al. 2003).

Low pH is known to suppress the formation of hydroxyl radicals from ozone. Thus, ozone reacts directly through an electrophilic attack. Consequently, a 16% decrease in overall COD removal is obtained than that in a buffered solution at pH 7 (Balcıoğlu and Ötker 2003). Reaction products formed at an acidic pH are resistant to oxidation by ozone, whereas in the case of ozonation in a buffered solution at pH 7, both OH radicals and ozone are the oxidizing agents; hence, a significant amount of COD is removed by using the ozonation process.

Ozonation at pH greater than 7 is used to degrade biorefractory compounds with a high percentage of mineralization success. Percentages obtained in the EO/ozonation oxidation assays carried out at high concentrations of pollutants are around 88.6 to 99/67.8 to 91.9 for mineralization and 97.6 to 98.4/77.2 to 86.9 for COD elimination from the most biodegradable to the most refractory compounds, respectively (Cañizares et al. 2009). Furthermore, it can be stated that, in general, oxidants are less-efficiently used in ozonation and more-efficiently used in EO (Cañizares et al. 2009). Cañizares et al. (2009) reported that the treatment of olive oil mill wastewater gave rise to values of 1.01 and 0.16 kg COD/kg oxygen equivalent chemical oxidation capacity (OCC) for EO and ozonation at an alkaline pH, respectively. On the contrary, in the case of aliphatic pollutants, the values obtained were completely different: 0.98 and 0.11 kg COD/kg OCC, respectively, for EO and ozonation at an alkaline pH. These values represent the amount of oxidants required to achieve 50% of COD elimination in the EO and ozonation oxidation assays carried out at high concentrations of pollutants

(Cañizares et al. 2009). Consequently, the degradation of biorefractory compounds by ozonation is dependent on the nature of the compound.

11.6 PHOTOCATALYSIS PROCESS VERSUS ELECTRO-OXIDATION

Heterogeneous photocatalytic EAOPs, similar to EO processes, use electrodes and catalysts to degrade biorefractory or nonrefractory compounds into CO_2 and H_2O (Im et al. 2012, Liu et al. 2011, Oller et al. 2006). Photocatalysis involves the photoexcitement of oxides or semiconductors by absorption of light, with the passage of electrons from the valence band (VB) to the conduction band and the formation of an electron/hole pair in VB (h^+ VB), allowing the reaction with the pollutant in the adsorbed phase (Herrmann 2010). This is different from the EO process where the degradation of the pollutant on the anodic surface is mainly done by hydroxyl radicals produced there, as previously discussed. In the presence of electron traps in water, photocatalyzed charges react with water molecules to generate reactive oxygenated species ($HOO^\bullet$, $H_2O_2^\bullet$, $O_2^{\bullet-}$, $^\bullet OH$), contributing to pollutant degradation as well (Gaya and Abdullah 2008). During this reaction, electrons/holes can recombine and release heat (Herrmann 2010). Photocatalysis assisted by electrochemistry is described as a process that prevents the recombination of electrons/holes (Daghrir et al. 2013a). The external potential applying amplifies at the VB the direct oxidation reactions of the pollutant absorbed on the photoanodic surface and hydroxyl radical production by water oxidation and hydroxide ions (An and Zhou 2012, Daghrir et al. 2013b).

The effectiveness of photocatalysis or electrophotocatalysis depends on many parameters such as the nature of the photocatalyst and concentration, light intensity and wavelength, the pH of the reaction medium, temperature, and the initial pollutant concentration. In addition, electrophotocatalysis is influenced by current density and the nature of the electrolyte as in EO. Several semiconductors such as ZnO, Fe_2O_3, WO_3, CdS, ZnS, and others have been tested as photocatalysts, but TiO_2 remains the most-used photocatalyst as it is inexpensive, photostable, and biologically and chemically inert (Fujishima et al. 2000, Hoffmann et al. 1995). TiO_2 doping with metal particles such as platinum (Pt) (Katsumata et al. 2009) or chemical doping (Ahmed et al. 2011, Yang and Gao 2004) permitted, respectively, reducing the recombination of electron/hole pairs and promoting TiO_2 photoresponse in the visible light range. This photocatalyst showed that it could degrade a wide range of refractory pollutants such as pesticides, drugs, and endocrine disruptors and effectively eliminate bacteria as well (Ahmed et al. 2011, Daghrir et al. 2013a, b). The pH medium mostly influences the effectiveness of the electrophotocatalysis (EPC) process rather than that of EO. As previously discussed, the pH affected pollutant chemical structure by protonating or not protonating and, thus, promoting or not promoting its oxidation, and BDD, for example, is inert to pH change. On the contrary, in EPC, pH has an effect on adsorption and desorption properties on the photocatalyst surface. The surface

Table 11-4. Comparison of Process Advantages and Disadvantages.

	Electro-oxidation	*Electrophotocatalysis*
Advantage	• Direct and indirect action of electrical current • No chemical reagents, except electrolyte • Low pH restriction	• Requires low potential • Chemical doping for absorption in the visible range • No chemical reagents added • Photocatalyst immobilized, no filtration after treatment
Disadvantage	• High cost of electrodes with high oxygen surges • Oxygen formation to strong currents (parasitic reaction) • Relatively high energy consumption	• Cost of photoanode preparation techniques • Configuration of the reactor usually in quartz • Fouling the photoanode for high concentrations

of TiO_2 is positively charged at pH values below the zero-charge point (ZCP 6.25), and a desorption or repulsion of positively charged compounds takes place. On the contrary, at pH values higher than ZCP, the surface of TiO_2 is negatively charged, and the repulsion of negatively charged compounds is seen. However, high concentration of hydroxide ions promotes the formation of hydroxyl radicals and hence, the effectiveness of the basic pH process (Chu and Wong 2004, Wong and Chu 2003) (Table 11-4).

11.7 SONOCHEMICAL PROCESS VERSUS ELECTRO-OXIDATION

Sonochemistry applied to water treatment is considered an AOP. Any treatment involving ultrasound to degrade pollutants is considered a sonochemical treatment. During sonochemical processing, high-frequency ultrasonic waves (100 kHz to 1 MHz) are used (Serna-Galvis et al. 2016). When it passes through an aqueous medium, ultrasound generates acoustic bubbles known as *acoustic cavitation*. The microbubbles produced during acoustic cavitation will tend to increase and decrease continuously until they reach the resonance size, which is the average size of the bubble before undergoing a violent explosion. The collapse of microbubbles induces the generation of an enormous amount of heat (approximately 5,000 K), pressure (approximately 1,000 atm), and various free radicals from a dissociation of water molecules and oxygen in the aqueous medium [Equations (11-12) to (11-15)] (Serna-Galvis et al. 2019a, Serna-Galvis et al. 2016, Tran et al. 2015). Hydrogen peroxide can also be formed by a combination of hydroxyl radicals [Equation (11-16)].

$$H_2O \rightarrow OH^{\bullet} + H^{\bullet} \tag{11-12}$$

$$O_2 \rightarrow 2O^{\bullet} \tag{11-13}$$

$$H_2O \rightarrow O^{\bullet} + 2OH^{\bullet} \tag{11-14}$$

$$O_2 + H^{\bullet} \rightarrow O^{\bullet} + OH^{\bullet} \tag{11-15}$$

$$2OH^{\bullet} \rightarrow H_2O_2 \tag{11-16}$$

Sonochemical water treatment for the elimination of a variety of organic pollutants has been widely studied. In general, this treatment process is used for tertiary treatment to remove refractory and nonbiodegradable compounds. Emerging contaminants are extensively treated by using the sonochemical process. In fact, a recent study reported the degradation of 17 contaminants of emerging concern in municipal wastewater effluents by using the sonochemical process (Serna-Galvis et al. 2019a). The study was performed using effluents from the municipal WWTP of Bogotá-Colombia and also using high-frequency ultrasound (375 kHz). The performances of the process were evaluated by degradation of the following pollutants: diclofenac, carbamazepine, venlafaxine, ciprofloxacin, norfloxacin, valsartan, losartan, irbesartan, sulfamethoxazole, clarithromycin, azithromycin, erythromycin, metronidazole, trimethoprim and clindamycin, and cocaine and its major metabolite benzoylecgonine. The authors reported that all pollutants were removed after 90 min of treatment, except ciprofloxacin, norfloxacin, diclofenac, and sulfamethoxazole, in which they observed an increase in concentrations, which was attributed to the release of pollutants absorbed in suspended matter under the physical action of ultrasound (i.e., turbulence and high shear in the liquid medium by cavitation phenomena). The elimination of pharmaceutical pollutants by EO has also been studied. Hospital wastewater was treated using a sonochemical process after a conventional biological treatment process. The biological treatment removed biodegradable substances but had a limited action on the pharmaceuticals. Effluents were then submitted to the sonochemical process (375 kHz and 88 W L^{-1}, 1.5 h), which, because of its chemical (i.e., radical attacks) and physical (i.e., suspended solids' disaggregation) effects, induced a considerable degradation of pharmaceuticals (pondered removal: 58.82%) (Serna-Galvis 2019b #753). The removal of emerging contaminants, particularly pharmaceutical pollutants, has also been widely studied using EO (Sirés and Brillas 2012). The degradation of pollutants is carried out by direct electron transfer to the anode and indirect oxidation by the reactive oxygen species (ROS) formed from water discharge at the anode, such as superoxide radical ions ($O_2^{-\bullet}$), hydroperoxyl radical ($HO_2^{\bullet}$), and hydrogen peroxide (H_2O_2). The use of the nonactive electrode BDD is one of the main reasons for the great attention that had been observed recently on EO (Brillas et al. 2010). Several authors reported that BDD is powerful enough to mineralize organic pollutants and their carboxylic acids generated from wastewater and has a much higher oxidizing power than other common anodes such as iridium(IV) oxide (IrO_2) and platinum and lead dioxide (PbO_2). A summary of several studies carried out on the degradation of pharmaceutical pollutants by using EO is presented in Table 11-5.

11.8 ENERGY AND ECONOMICS COMPARISON

The oxidation processes have shown their efficiency to treat effluents contaminated by pharmaceuticals. However, few studies have been performed to analyze the cost and the feasibility of these processes for the treatment of contaminated effluents. (Lin et al. 2013) evaluated the economic feasibility of sulfamethoxazole degradation by the Ti/SnO_2-Sb/Ce-PbO_2 anode by determining the energy cost. Results showed that the residence time and energy cost of sulfamethoxazole degradation at optimal conditions from 100 to 1 $mg \cdot L^{-1}$ were 32.9 to 23.0 min and 26.3 to 46.3 $Wh \cdot L^{-1}$ ($US 1.5 to $US 2.6 m^{-3}), respectively. Brillas et al. (2009) mentioned that BDD electrodes can be competitive with traditional AOPs such as ozonation and Fenton oxidation. Ferre-Aracil et al. (2016) studied the degradation of cytostatic compounds present in hospital wastewater by ozonation. The authors have shown that for a system with a reactor volume of 500 L treating 1 $m^3 h^{-1}$ with a hydraulic retention time of 30 min and an oxygen requirement of approximately 6.7 $L \cdot min^{-1}$, the cost of treatment will be €0.3 m^{-3} (Ferre-Aracil et al. 2016). The costs reported were only the direct operating costs (electricity + O_2) of the installation for the complete elimination of cytostatic agents in wastewater.

AO was able to completely mineralize various organic compounds as compared to ozonation and Fenton oxidation, whereas the cost of oxidant generation for BDD electrodes was less than that of ozonation and comparable to that of Fenton oxidation (Brillas et al. 2009). Through the combination of direct and indirect reactions, AO is capable of oxidizing both organic and inorganic compounds. However, the major drawback of AO is the formation of intermediate products during electrolysis. This by-product can be more toxic than the parent compound itself (Dirany et al. 2011). To improve the efficiency of AO and its use as an eco-friendly method, the treatment time should be extended long enough to completely eliminate the toxicity occurring during the electrochemical treatment.

11.9 CONCLUSIONS AND FUTURE PROSPECTS

This chapter presented a comparative study among EO and other oxidation processes, namely, chemical oxidation, ozonation, photocatalysis, and the sonochemical process for the treatment of refractory compounds present in wastewater. EO has proven to be a good alternative for the removal of several persistent organic contaminants that escape the conventional treatment processes in WWTPs. Despite the efficiency of EO to remove these pollutants,

Table 11-5. Anodic Oxidation of Pharmaceuticals under Various Operating Conditions.

Compounds	*Anode material*	*Parameters*	*Performances*	*References*
Diclofenac	BDD	C_0 30 mg L^{-1}, treatment time = 4 h, NaCl = 0.1 M, bias potential = 4.0 V, pH = 8.5	Diclofenac degradation (>96%), mineralization of 72%	Zhao et al. (2009)
	Pt BDD	Diclofenac 175 mg L^{-1}, treatment time = 6 h, pH = 6.5, Na_2SO_4 = 0.05 M, current intensity = 300 mA, temperature = 35°C	Pt: TOC removal 46%, drug removal 82% BDD: TOC removal > 97, drug removal 100%	Brillas et al. (2010)
Paracetamol	Pt BDD	157 mg L^{-1} of drug (corresponding to 100 mg L^{-1} of TOC), Na_2SO_4 = 0.05 M, pH = 3, current intensity = 300 mA, t = 6 h	Pt: 19% of TOC removal BDD: 98% of TOC removal	Brillas et al. (2005)
	Ti/IrO_2 Ti/SnO_2 BDD	1 mM of drug, Na_2SO_4 = 0.025 M, pH = 7.8, current intensity = 200 mA, treatment time = 210 min TOC: 116 mg L^{-1}	Ti/IrO_2: 1% of TOC removal Ti/SnO_2: 40% of TOC removal BDD: 70% of TOC removal	Waterston et al. (2006)

(*Continued*)

Table 11-5. Anodic Oxidation of Pharmaceuticals under Various Operating Conditions. (Continued)

Compounds	*Anode material*	*Parameters*	*Performances*	*References*
Amoxicillin	BDD Ti/RuO_2-IrO_2	36.54 mg L^{-1} of amoxicillin, $NaSO_4$ = 50 mM, treatment time = 6 h, current density = 41.66 mA · cm^{-2} pH = 5.3, distance between electrodes = 3 cm, stainless cathode.	Complete mineralization with BDD (100%) and less than 25% with Ti/RuO_2-IrO_2	Sopaj et al. (2015)
Omeprazole	Pt BDD	169 mg L^{-1} of drug, treatment time = 6 h, current density = 41.66 mA · cm^{-2}, pH = 6	No mineralization with Pt 78% of mineralization was achieved with BDD.	Cavalcanti et al. (2013)
17β-Estradiol	BDD Pt	0.5 mg L^{-1} of drug, treatment time = 270 min, current intensity = 350 mA, pH = 6, Na_2SO_4 = 0.1 M	BDD: 94% of TOC removal was observed. Pt: poor efficiency of TOC removal was observed.	Yoshihara and Murugananthan (2009)
Ibuprofen	$Ti/Pt/PbO_2$ BDD	1.75 mM of drug, treatment time = 6 h, current density = 30 mA · cm^{-2}, Na_2SO_4 = 35 mM	48% of TOC and 60% of COD removal with $Ti/Pt/PbO_2$ 92% of TOC and 95% of COD removal with BDD	Ciríaco et al. (2009)

	Pt BDD	0.2 mM of drug, current intensity = 500 mA, Na_2SO_4 = 0.05 M, pH = 3	Complete removal of Ibu after 90 min with BDD and after 180 min with Pt	Ambuludi et al. (2013)
Carbamazepine	BDD PbO_2	10 mg L^{-1} of drug, min, Na_2SO_4 = 0.4 g L^{-1}, current intensity = 1.242 A, pH = 7, recycling flow rate = 162 mL min^{-1}, recycling flow rate = 232 mL min^{-1}	81.52% of CBZ removal with BDD after 120 min and with a current intensity of 1.242 A. 89.3% of CBZ removal with PbO_2 after 120 min and with a current intensity of 1.37 A.	García-Gómez et al. (2014)
Sulfamethoxazole	$Ti/Ru_{0.3}Ti_{0.7}O_2$	200 mg L^{-1} of drug, NaCl = 0.1 mol L^{-1}, pH = 3, current density = 20 mA · cm^{-2}, treatment time = 30 min	Almost complete (98%) degradation of SUL	Hussain et al. (2015)
	BDD	0.25 mM of drug, Na_2SO_4 = 0.025 M, pH = 6-7, current intensity = 10 mA, treatment time = 195 min	62% of TOC and 90% of SUL removal	Li et al. (2008)
	Ti/SnO_2-Sb/Ce-PbO_2	100 mg L^{-1} of drug, $NaClO_4$ = 10 mmol L^{-1}, pH = 3, current density = 40 mA · cm^{-2}, treatment time = 30 min	91% of TOC removal	Lin et al. (2013)

its development in the area of wastewater treatment remains limited. The main reason is that the investment costs and the operating costs are significantly higher than those of conventional techniques used in wastewater treatment. To promote large-scale development of the latter, it is necessary to adapt the EO process to existing WWTPs to increase their purification efficiency (vis-à-vis emerging and refractory pollutants) without necessarily carrying out any major construction of new infrastructure; the EO process can advantageously be combined with biological processes during the treatment of effluents containing refractory organic compounds. This type of coupling allows us to obtain benefit from the advantages of the EO processes (a shorter retention time) and biodegradation (lower operating costs). Used in synergy with biological processes, EO processes can be used to transform nonbiodegradable compounds into biodegradable products, or even as a final treatment for the complete oxidation of organic compounds to carbon dioxide.

References

Acero, J. L., and U. Von Gunten. 2001. "Characterization of oxidation processes: Ozonation and the AOP O_3/H_2O_2." *J. AWWA* 93 (10): 90–100.

Acero, J. L., F. J. Benitez, F. J. Real, and C. Maya. 2003. "Oxidation of acetamide herbicides in natural waters by ozone and by the combination of ozone/hydrogen peroxide: Kinetic study and process modeling." *Ind. Eng. Chem. Res.* 42 (23): 5762–5769.

Ahmed, S., M. G. Rasul, R. Brown, and M. A. Hashib. 2011. "Influence of parameters on the heterogeneous photocatalytic degradation of pesticides and phenolic contaminants in wastewater: A short review." *J. Environ. Manage.* 92 (3): 311–330.

Almeida, L. C., S. Garcia-Segura, N. Bocchi, and E. Brillas. 2011. "Solar photoelectro-Fenton degradation of paracetamol using a flow plant with a Pt/air-diffusion cell coupled with a compound parabolic collector: Process optimization by response surface methodology." *Appl. Catal. B* 103 (1–2): 21–30.

Ambuludi, S. L., M. Panizza, N. Oturan, A. Özcan, and M. A. Oturan. 2013. "Kinetic behavior of anti-inflammatory drug ibuprofen in aqueous medium during its degradation by electrochemical advanced oxidation." *Environ. Sci. Pollut. Res.* 20 (4): 2381–2389.

Ammar, S., R. Abdelhedi, C. Flox, C. Arias, and E. Brillas. 2006. "Electrochemical degradation of the dye indigo carmine at boron-doped diamond anode for wastewaters remediation." *Environ. Chem. Lett.* 4 (4): 229–233.

An, J., and Q. Zhou. 2012. "Degradation of some typical pharmaceuticals and personal care products with copper-plating iron doped Cu_2O under visible light irradiation." *J. Environ. Sci.* 24 (5): 827–833.

Anglada, A., A. Urtiaga, and I. Ortiz. 2009. "Contributions of electrochemical oxidation to waste-water treatment: Fundamentals and review of applications." *J. Chem. Technol. Biotechnol.* 84 (12): 1747–1755.

Balcıoğlu, I. A., and M. Ötker. 2003. "Treatment of pharmaceutical wastewater containing antibiotics by O_3 and O_3/H_2O_2 processes." *Chemosphere* 50 (1): 85–95.

Barrera-Díaz, C., I. Linares-Hernández, G. Roa-Morales, B. Bilyeu, and P. Balderas-Hernández. 2009. "Removal of biorefractory compounds in industrial wastewater by chemical and electrochemical pretreatments." *Ind. Eng. Chem. Res.* 48: 1253–1258.

Borras, N., C. Arias, R. Oliver, and E. Brillas. 2011. "Mineralization of desmetryne by electrochemical advanced oxidation processes using a boron-doped diamond anode and an oxygen-diffusion cathode." *Chemosphere* 85 (7): 1167–1175.

Borràs, N., C. Arias, R. Oliver, and E. Brillas. 2013. "Anodic oxidation, electro-Fenton and photoelectro-Fenton degradation of cyanazine using a boron-doped diamond anode and an oxygen-diffusion cathode." *J. Electroanal. Chem.* 689: 158–167.

Brillas, E., M. A. Baños, M. Skoumal, P. L. Cabot, J. A. Garrido, and R. M. Rodríguez. 2007. "Degradation of the herbicide 2,4-DP by anodic oxidation, electro-Fenton and photoelectro-Fenton using platinum and boron-doped diamond anodes." *Chemosphere* 68 (2): 199–209.

Brillas, E., S. Garcia-Segura, M. Skoumal, and C. Arias. 2010. "Electrochemical incineration of diclofenac in neutral aqueous medium by anodic oxidation using Pt and boron-doped diamond anodes." *Chemosphere* 79 (6): 605–612.

Brillas, E., I. Sirés, C. Arias, P. L. Cabot, F. Centellas, R. M. Rodríguez, et al. 2005. "Mineralization of paracetamol in aqueous medium by anodic oxidation with a boron-doped diamond electrode." *Chemosphere* 58 (4): 399–406.

Brillas, E., I. Sirés, and M. A. Oturan. 2009. "Electro-Fenton process and related electrochemical technologies based on Fenton's reaction chemistry." *Chem. Rev.* 109 (12): 6570–6631.

Buffle, M.-O., J. Schumacher, S. Meylan, M. Jekel, and U. von Gunten. 2006a. "Ozonation and advanced oxidation of wastewater: Effect of O_3 dose, pH, DOM and $HO^•$-scavengers on ozone decomposition and $HO^•$ generation." *Ozone Sci. Eng.* 28 (4): 247–259.

Buffle, M.-O., J. Schumacher, E. Salhi, M. Jekel, and U. von Gunten. 2006b. "Measurement of the initial phase of ozone decomposition in water and wastewater by means of a continuous quench-flow system: Application to disinfection and pharmaceutical oxidation." *Water Res.* 40 (9): 1884–1894.

Cañizares, P., M. Díaz, J. A. Domínguez, J. García-Gómez, and M. A. Rodrigo. 2002. "Electrochemical oxidation of aqueous phenol wastes on synthetic diamond thin-film electrodes." *Ind. Eng. Chem. Res.* 41: 4187–4194.

Cañizares, P., M. Díaz, J. A. Domínguez, J. Lobato, and M. A. Rodrigo. 2005. "Electrochemical treatment of diluted cyanide aqueous wastes." *J. Chem. Technol. Biotechnol.* 80 (5): 565–573.

Cañizares, P., J. García-Gómez, C. Saez, and M. A. Rodrigo. 2004. "Electrochemical oxidation of several chlorophenols on diamond electrodes: Part II. Influence of waste characteristics and operating conditions." *J. Appl. Electrochem.* 34 (1): 87–94.

Cañizares, P., F. Larrondo, J. Lobato, M. A. Rodrigo, and C. Sáez. 2005. "Electrochemical synthesis of peroxodiphosphate using boron-doped diamond anodes." *J. Electrochem. Soc.* 152 (11): D191.

Cañizares, P., R. Paz, C. Sáez, and M. A. Rodrigo. 2009. "Costs of the electrochemical oxidation of wastewaters: A comparison with ozonation and Fenton oxidation processes." *J. Environ. Manage.* 90 (1): 410–420.

Cañizares, P., C. Saez, J. Lobato, R. Paz, and M. A. Rodrigo. 2007. "Effect of the operating conditions on the oxidation mechanisms in conductive-diamond electrolyses." *J. Electrochem. Soc.* 154 (3): E37.

Cavalcanti, E. B., S. Garcia-Segura, F. Centellas, and E. Brillas. 2013. "Electrochemical incineration of omeprazole in neutral aqueous medium using a platinum or boron-doped diamond anode: Degradation kinetics and oxidation products." *Water Res.* 47 (5): 1803–1815.

Chatzisymeon, E., N. P. Xekoukoulotakis, A. Coz, N. Kalogerakis, and D. Mantzavinos. 2006. "Electrochemical treatment of textile dyes and dyehouse effluents." *J. Hazard. Mater.* 137 (2): 998–1007.

Chen, G. 2004. "Electrochemical technologies in wastewater treatment." *Sep. Purif. Technol.* 38 (1): 11–41.

Chu, W., and C. C. Wong. 2004. "The photocatalytic degradation of dicamba in TiO_2 suspensions with the help of hydrogen peroxide by different near UV irradiations." *Water Res.* 38 (4): 1037–1043.

Chu, Y. Y., Y. Qian, W. J. Wang, and X. L. Deng. 2012. "A dual-cathode electro-Fenton oxidation coupled with anodic oxidation system used for 4-nitrophenol degradation." *J. Hazard. Mater.* 199–200: 179–185.

Ciríaco, L., C. Anjo, J. Correia, M. J. Pacheco, and A. Lopes. 2009. "Electrochemical degradation of Ibuprofen on Ti/Pt/PbO_2 and Si/BDD electrodes." *Electrochim. Acta* 54 (5): 1464–1472.

Comninellis, C. 1994. "Electrocatalysis in the electrochemical conversion/combustion of organic pollutants for waste water treatment." *Electrochim. Acta* 39 (11/12): 1857–1862.

Comninellis, C., and G. Chen. 2010. *Electrochemistry for the environment.* New York: Springer.

Comninellis, C., A. Kapalka, S. Malato, S. A. Parsons, I. Poulios, and D. Mantzavinos. 2008. "Advanced oxidation processes for water treatment: Advances and trends for R&D." *J. Chem. Technol. Biotechnol.* 83 (6): 769–776.

Comninellis, C., and C. Pulgarin. 1993. "Electrochemical oxidation of phenol for wastewater treatment using SnO_2 anodes." *J. Appl. Electrochem.* 23: 108–112.

Daghrir, R., P. Drogui, A. Dimboukou-Mpira, and M. A. El Khakani. 2013b. "Photoelectrocatalytic degradation of carbamazepine using Ti/TiO_2 nanostructured electrodes deposited by means of a pulsed laser deposition process." *Chemosphere* 93 (11): 2756–2766.

Daghrir, R., P. Drogui, and M. A. El Khakani. 2013a. "Photoelectrocatalytic oxidation of chlortetracycline using Ti/TiO_2 photo-anode with simultaneous H_2O_2 production." *Electrochim. Acta* 87: 18–31.

Dhaouadi, A., L. Monser, and N. Adhoum. 2009. "Anodic oxidation and electro-Fenton treatment of rotenone." *Electrochim. Acta* 54 (19): 4473–4480.

Dirany, A., S. Efremova Aaron, N. Oturan, I. Sirés, M. A. Oturan, and J. J. Aaron. 2011. "Study of the toxicity of sulfamethoxazole and its degradation products in water by a bioluminescence method during application of the electro-Fenton treatment." *Anal. Bioanal. Chem.* 400 (2): 353–360.

Dirany, A., I. Sirés, N. Oturan, and M. A. Oturan. 2010. "Electrochemical abatement of the antibiotic sulfamethoxazole from water." *Chemosphere* 81 (5): 594–602.

Ferre-Aracil, J., Y. Valcárcel, N. Negreira, M. L. de Alda, D. Barceló, S. Cardona, et al. 2016. "Ozonation of hospital raw wastewaters for cytostatic compounds removal. Kinetic modelling and economic assessment of the process." *Sci. Total Environ.* 556: 70–79.

Flox, C., P. L. Cabot, F. Centellas, J. A. Garrido, R. M. Rodriguez, C. Arias, et al. 2006. "Electrochemical combustion of herbicide mecoprop in aqueous medium using a flow reactor with a boron-doped diamond anode." *Chemosphere* 64 (6): 892–902.

Fryda, M., T. Matthée, S. Mulcahy, M. Höfer, L. Schäfer, and I. Tröster. 2003. "Applications of DIACHEM® electrodes in electrolytic water treatment." *Electrochem. Soc. Interface* 12 (1): 40–44.

Fujishima, A., T. N. Rao, and D. A. Tryk. 2000. "Titanium dioxide photocatalysis." *J. Photochem. Photobiol. C* 1 (1): 1–21.

Gandini, D., E. Mahe, P. A. Michaud, W. Haenni, A. Perret, and C. Comninellis. 2000. "Oxidation of carboxylic acids at boron-doped diamond electrodes for wastewater treatment." *J. Appl. Electrochem.* 30 (12): 1345–1350.

Ganzenko, O., D. Huguenot, E. D. van Hullebusch, G. Esposito, and M. A. Oturan. 2014. "Electrochemical advanced oxidation and biological processes for wastewater treatment: A review of the combined approaches." *Environ. Sci. Pollut. Res.* 21 (14): 8493–8524.

García-Gómez, C., P. Drogui, F. Zaviska, B. Seyhi, P. Gortáres-Moroyoqui, G. Buelna, et al. 2014. "Experimental design methodology applied to electrochemical oxidation of carbamazepine using Ti/PbO_2 and Ti/BDD electrodes." *J. Electroanal. Chem.* 732: 1–10.

Gaya, U. I., and A. H. Abdullah. 2008. "Heterogeneous photocatalytic degradation of organic contaminants over titanium dioxide: A review of fundamentals, progress and problems." *J. Photochem. Photobiol. C* 9 (1): 1–12.

Hammami, S., N. Bellakhal, N. Oturan, M. A. Oturan, and M. Dachraoui. 2008. "Degradation of acid orange 7 by electrochemically generated •OH radicals in acidic aqueous medium using a boron-doped diamond or platinum anode: A mechanistic study." *Chemosphere* 73 (5): 678–684.

Herrmann, J. M. 2010. "Environmental photocatalysis: Perspectives for China." *Sci. China Chem.* 53 (9): 1831–1843.

Hladik, M. L., A. L. Roberts, and E. J. Bouwer. 2005. "Removal of neutral chloroacetamide herbicide degradates during simulated unit processes for drinking water treatment." *Water Res.* 39 (20): 5033–5044.

Hoffmann, M. R., S. T. Martin, W. Choi, and D. W. Bahnemann. 1995. "Environmental applications of semiconductor photocatalysis." *Chem. Rev.* 95 (1): 69–96.

Huang, W.-J., G.-C. Fang, and C.-C. Wang. 2005. "The determination and fate of disinfection by-products from ozonation of polluted raw water." *Sci. Total Environ.* 345 (1–3): 261–272.

Hussain, S., S. Gul, J. R. Steter, D. W. Miwa, and A. J. Motheo. 2015. "Route of electrochemical oxidation of the antibiotic sulfamethoxazole on a mixed oxide anode." *Environ. Sci. Pollut. Res.* 22 (19): 15004–15015.

Hussain, S., S. Shaikh, and M. Farooqui. 2013. "COD reduction of waste water streams of active pharmaceutical ingredient—Atenolol manufacturing unit by advanced oxidation-Fenton process." *J. Saudi Chem. Soc.* 17 (2): 199–202.

Im, J.-K., H.-S. Son, Y.-M. Kang, and K.-D. Zoh. 2012. "Carbamazepine degradation by photolysis and titanium dioxide photocatalysis." *Water Environ. Res.* 84 (7): 554–561.

Isarain-Chávez, E., R. M. Rodríguez, P. L. Cabot, F. Centellas, C. Arias, J. A. Garrido, et al. 2011. "Degradation of pharmaceutical beta-blockers by electrochemical advanced oxidation processes using a flow plant with a solar compound parabolic collector." *Water Res.* 45 (14): 4119–4130.

Katsumata, H., M. Sada, Y. Nakaoka, S. Kaneco, T. Suzuki, and K. Ohta. 2009. "Photocatalytic degradation of diuron in aqueous solution by platinized TiO_2." *J. Hazard. Mater.* 171 (1–3): 1081–1087.

Khandegar, V., and A. K. Saroha. 2013. "Electrocoagulation for the treatment of textile industry effluent—A review." *J. Environ. Manage.* 128: 949–963.

Li, S., D. Bejan, M. S. McDowell, and N. J. Bunce. 2008. "Mixed first and zero order kinetics in the electrooxidation of sulfamethoxazole at a boron-doped diamond (BDD) anode." *J. Appl. Electrochem.* 38 (2): 151–159.

Li, X., D. Pletcher, and F. C. Walsh. 2011. "Electrodeposited lead dioxide coatings." *Chem. Soc. Rev.* 40 (7): 3879–3894.

Lin, H., J. Niu, J. Xu, Y. Li, and Y. Pan. 2013. "Electrochemical mineralization of sulfamethoxazole by Ti/SnO_2-Sb/Ce-PbO_2 anode: Kinetics, reaction pathways, and energy cost evolution." *Electrochim. Acta* 97: 167–174.

Liu, R., Y. Liu, C. Liu, S. Luo, Y. Teng, L. Yang, et al. 2011. "Enhanced photoelectrocatalytic degradation of 2,4-dichlorophenoxyacetic acid by $CuInS_2$ nanoparticles deposition onto TiO_2 nanotube arrays." *J. Alloys Compd.* 509 (5): 2434–2440.

Ma, H., B. Wang, and Y. Wang. 2007. "Application of molybdenum and phosphate modified kaolin in electrochemical treatment of paper mill wastewater." *J. Hazard. Mater.* 145 (3): 417–423.

Marselli, B., J. Garcia-Gomez, P. A. Michaud, M. A. Rodrigo, and C. Comninellis. 2003. "Electrogeneration of hydroxyl radicals on boron-doped diamond electrodes." *J. Electrochem. Soc.* 150 (3): D79.

Martínez-Huitle, C. A., and L. S. Andrade. 2011. "Electrocatalysis in wastewater treatment: Recent mechanism advances." *Quim. Nova* 34 (5): 850–858.

Martínez-Huitle, C. A., and E. Brillas. 2009. "Decontamination of wastewaters containing synthetic organic dyes by electrochemical methods: A general review." *Appl. Catal. B* 87 (3–4): 105–145.

Martinez-Huitle, C. A., A. De Battisti, S. Ferro, S. Reyna, M. Cerro-López, and M. A. Quiro. 2008. "Removal of the pesticide methamidophos from aqueous solutions by electrooxidation using Pb/PbO_2, Ti/SnO_2, and Si/BDD electrodes." *Environ. Sci. Technol.* 42 (18): 6929–6935.

Martinez-Huitle, C. A., S. Ferro, and A. De Battisti. 2005. "Electrochemical incineration of oxalic acid: Reactivity and engineering parameters." *J. Appl. Electrochem.* 35 (11): 1087–1093.

Martínez-Huitle, C. A., and M. Panizza. 2018. "Electrochemical oxidation of organic pollutants for wastewater treatment." *Curr. Opin. Electrochem.* 11: 62–71.

Martínez-Huitle, C. A., M. A. Quiroz, C. Comninellis, S. Ferro, and A. D. Battisti. 2004. "Electrochemical incineration of chloranilic acid using Ti/IrO_2, Pb/PbO_2 and Si/BDD electrodes." *Electrochim. Acta* 50 (4): 949–956.

Martinez-Huitle, C. A., M. A. Rodrigo, I. Sirés, and O. Scialdone. 2015. "Single and coupled electrochemical processes and reactors for the abatement of organic water pollutants: A critical review." *Chem. Rev.* 115 (24): 13362–13407.

Michaud, P.-A., E. Mahé, W. Haenni, A. Perret, and C. Comninellisa. 2000. "Preparation of peroxodisulfuric acid using boron-doped diamond thin film electrodes." *Electrochem. Solid-State Lett.* 3 (2): 77–79.

Murugananthan, M., S. Yoshihara, T. Rakuma, and T. Shirakashi. 2008. "Mineralization of bisphenol A (BPA) by anodic oxidation with boron-doped diamond (BDD) electrode." *J. Hazard. Mater.* 154 (1–3): 213–220.

Oller, I., W. Gernjak, M. I. Maldonado, L. A. Pérez-Estrada, J. A. Sánchez-Pérez, and S. Malato. 2006. "Solar photocatalytic degradation of some hazardous water-soluble pesticides at pilot-plant scale." *J. Hazard. Mater.* 138 (3): 507–517.

Oturan, M. A., N. Oturan, M. C. Edelahi, F. I. Podvorica, and K. E. Kacemi. 2011. "Oxidative degradation of herbicide diuron in aqueous medium by Fenton's reaction based advanced oxidation processes." *Chem. Eng. J.* 171 (1): 127–135.

Oturan, M. A., I. Sirés, N. Oturan, S. Pérocheau, J.-L. Laborde, and S. Trévin. 2008. "Sonoelectro-Fenton process: A novel hybrid technique for the destruction of organic pollutants in water." *J. Electroanal. Chem.* 624 (1–2): 329–332.

Özcan, A., Y. Şahin, A. S. Koparal, and M. A. Oturan. 2008. "Propham mineralization in aqueous medium by anodic oxidation using boron-doped diamond anode: Influence of experimental parameters on degradation kinetics and mineralization efficiency." *Water Res.* 42 (12): 2889–2898.

Panizza, M., and G. Cerisola. 2003. "Electrochemical oxidation of 2-naphthol with in situ electrogenerated active chlorine." *Electrochim. Acta* 48 (11): 1515–1519.

Panizza, M., and G. Cerisola. 2004. "Influence of anode material on the electrochemical oxidation of 2-naphthol." *Electrochim. Acta* 49 (19): 3221–3226.

Panizza, M., and G. Cerisola. 2005. "Application of diamond electrodes to electrochemical processes." *Electrochim. Acta* 51 (2): 191–199.

Panizza, M., and G. Cerisola. 2007. "Electrocatalytic materials for the electrochemical oxidation of synthetic dyes." *Appl. Catal. B* 75 (1–2): 95–101.

Panizza, M., and G. Cerisola. 2009. "Direct and mediated anodic oxidation of organic pollutants." *Chem. Rev.* 109 (12): 6541–6569.

Panizza, M., P. A. Michaud, G. Cerisola, and C. Comninellis. 2001. "Electrochemical treatment of wastewaters containing organic pollutants on boron-doped diamond electrodes: Prediction of specific energy consumption and required electrode area." *Electrochem. Commun.* 3: 336–339.

Polcaro, A. M., A. Vacca, M. Mascia, and F. Ferrara. 2008. "Product and by-product formation in electrolysis of dilute chloride solutions." *J. Appl. Electrochem.* 38 (7): 979–984.

Rajkumar, D., and K. Palanivelu. 2004. "Electrochemical treatment of industrial wastewater." *J. Hazard. Mater.* 113 (1–3): 123–129.

Rajkumar, D., B. J. Song, and J. G. Kim. 2007. "Electrochemical degradation of Reactive Blue 19 in chloride medium for the treatment of textile dyeing wastewater with identification of intermediate compounds." *Dyes Pigm.* 72 (1): 1–7.

Rao, S. A. N., and V. T. Venkatarangaiah. 2014. "Metal oxide-coated anodes in wastewater treatment." *Environ. Sci. Pollut. Res.* 21 (5): 3197–3217.

Robinson, T., G. McMullan, R. Marchant, and P. Nigam. 2001. "Remediation of dyes in textile effluent: A critical review on current treatment technologies with a proposed alternative." *Bioresour. Technol.* 77 (3): 247–255.

Rodrigo, M. A., P. Cañizares, A. Sánchez-Carretero, and C. Sáez. 2010. "Use of conductive-diamond electrochemical oxidation for wastewater treatment." *Catal. Today* 151 (1–2): 173–177.

Sakalis, A., K. Fytianos, U. Nickel, and A. Voulgaropoulos. 2006. "A comparative study of platinised titanium and niobe/synthetic diamond as anodes in the electrochemical treatment of textile wastewater." *Chem. Eng. J.* 119 (2–3): 127–133.

Samet, Y., L. Agengui, and R. Abdelhédi. 2010. "Anodic oxidation of chlorpyrifos in aqueous solution at lead dioxide electrodes." *J. Electroanal. Chem.* 650 (1): 152–158.

Särkkä, H., A. Bhatnagar, and M. Sillanpää. 2015. "Recent developments of electro-oxidation in water treatment—A review." *J. Electroanal. Chem.* 754: 46–56.

Scialdone, O., A. Galia, and G. Filardo. 2008. "Electrochemical incineration of 1,2-dichloroethane: Effect of the electrode material." *Electrochim. Acta* 53 (24): 7220–7225.

Serna-Galvis, E. A., A. M. Botero-Coy, D. Martínez-Pachón, A. Moncayo-Lasso, M. Ibáñez, F. Hernández, et al. 2019a. "Degradation of seventeen contaminants of emerging concern in municipal wastewater effluents by sonochemical advanced oxidation processes." *Water Res.* 154: 349–360.

Serna-Galvis, E. A., L. Isaza-Pineda, A. Moncayo-Lasso, F. Hernández, M. Ibáñez, and R. A. Torres-Palma. 2019b. "Comparative degradation of two highly consumed antihypertensives in water by sonochemical process. Determination of the reaction zone, primary degradation products and theoretical calculations on the oxidative process." *Ultrason. Sonochem.* 58: 104635.

Serna-Galvis, E. A., J. Silva-Agredo, A. L. Giraldo-Aguirre, O. A. Flórez-Acosta, and R. A. Torres-Palma. 2016. "High frequency ultrasound as a selective advanced oxidation process to remove penicillinic antibiotics and eliminate its antimicrobial activity from water." *Ultrason. Sonochem.* 31: 276–283.

Serrano, K., P. A. Michaud, L. Comninellis, and A. Savall. 2002. "Electrochemical preparation of peroxodisulfuric acid using boron doped diamond thin film electrodes." *Electrochim. Acta* 48 (4): 431–436.

Sirés, I., and E. Brillas. 2012. "Remediation of water pollution caused by pharmaceutical residues based on electrochemical separation and degradation technologies: A review." *Environ. Int.* 40: 212–229.

Sirés, I., E. Brillas, G. Cerisola, and M. Panizza. 2008. "Comparative depollution of mecoprop aqueous solutions by electrochemical incineration using BDD and PbO_2 as high oxidation power anodes." *J. Electroanal. Chem.* 613 (2): 151–159.

Sirés, I., E. Brillas, M. A. Oturan, M. A. Rodrigo, and M. Panizza. 2014. "Electrochemical advanced oxidation processes: Today and tomorrow. A review." *Environ. Sci. Pollut. Res.* 21 (14): 8336–8367.

Sirés, I., P. L. Cabot, F. Centellas, J. A. Garrido, R. M. Rodríguez, C. Arias, et al. 2006. "Electrochemical degradation of clofibric acid in water by anodic oxidation." *Electrochim. Acta* 52 (1): 75–85.

Somich, C. J., P. C. Kearney, M. T. Muldoon, and S. Elsasser. 1988. "Enhanced soil degradation of alachlor by treatment with ultraviolet light and ozone." *J. Agric. Food Chem.* 36 (6): 1322–1326.

Sopaj, F., M. A. Rodrigo, N. Oturan, F. I. Podvorica, J. Pinson, and M. A. Oturan. 2015. "Influence of the anode materials on the electrochemical oxidation efficiency. Application to oxidative degradation of the pharmaceutical amoxicillin." *Chem. Eng. J.* 262: 286–294.

Tran, N., P. Drogui, and S. K. Brar. 2015. "Sonochemical techniques to degrade pharmaceutical organic pollutants." *Environ. Chem. Lett.* 13 (3): 251–268.

von Gunten, U. 2003. "Ozonation of drinking water: Part II. Disinfection and by-product formation in presence of bromide, iodide or chlorine." *Water Res.* 37 (7): 1469–1487.

Waterston, K., J. W. Wang, D. Bejan, and N. J. Bunce. 2006. "Electrochemical waste water treatment: Electrooxidation of acetaminophen." *J. Appl. Electrochem.* 36 (2): 227–232.

Weiss, E., K. Groenen-Serrano, and A. Savall. 2006. "Electrochemical degradation of sodium dodecylbenzene sulfonate on boron doped diamond and lead dioxide anode." *J. New Mater. Electrochem. Syst.* 9: 249–256.

Weiss, E., K. Groenen-Serrano, and A. Savall. 2008. "A comparison of electrochemical degradation of phenol on boron doped diamond and lead dioxide anodes." *J. Appl. Electrochem.* 38 (3): 329–337.

Wong, C. C., and W. Chu. 2003. "The direct photolysis and photocatalytic degradation of alachlor at different TiO_2 and UV sources." *Chemosphere* 50 (8): 981–987.

Yang, S., and L. Gao. 2004. "New method to prepare nitrogen-doped titanium dioxide and its photocatalytic activities irradiated by visible light." *J. Am. Ceram. Soc.* 87 (9): 1803–1805.

Yoshihara, S., and M. Murugananthan. 2009. "Decomposition of various endocrine-disrupting chemicals at boron-doped diamond electrode." *Electrochim. Acta* 54 (7): 2031–2038.

Yu, X., M. Zhou, Y. Hu, K. Groenen Serrano, and F. Yu. 2014. "Recent updates on electrochemical degradation of bio-refractory organic pollutants using BDD anode: A mini review." *Environ. Sci. Pollut. Res.* 21 (14): 8417–8431.

Zhao, X., Y. Hou, H. Liu, Z. Qiang, and J. Qu. 2009. "Electro-oxidation of diclofenac at boron doped diamond: Kinetics and mechanism." *Electrochim. Acta* 54 (17): 4172–4179.

Zhu, X., M. Tong, S. Shi, H. Zhao, and J. Ni. 2008. "Essential explanation of the strong mineralization performance of boron-doped diamond electrodes." *Environ. Sci. Technol.* 42 (13): 4914–4920.

CHAPTER 12

Electro-Coagulation Processes: Criteria, Considerations, and Examples for Full-Scale Applications

Ali Khosravanipour Mostafazadeh,
Patrick Drogui, R. D. Tyagi

12.1 INTRODUCTION

Scaling-up the bench-scale electro-coagulation (EC) technology is one of the most significant processes in this field. Compared with conventional treatment technologies, EC provides a relatively compact and cost-effective alternative. Because this treatment is a complex process involving a multitude of pollutant removal mechanisms operating synergistically, it is highly important to consider the reactor design criteria on a larger scale and its application on different types of wastewater. Moreover, in the last two decades, the application of EC has been extended to treatment of various effluents containing metals, foodstuff, olive oil, textile dyes, fluorine, polymeric waste, landfill leachate, turbidity, chemical and mechanical polishing waste, aqueous suspensions of ultrafine particles, nitrate, phenolic waste, and arsenic as well as municipal wastewater (Kabdaşlı et al. 2012).

In this chapter, we investigate the scale-up and design criteria of EC, economics, crucial factors, reactor types and operating conditions, current commercial plants and companies, and the types of wastewaters and pollutants and finally explore the current challenges and possible solutions.

12.2 SCALE-UP AND ECONOMICS

EC is more of a method than coagulation that avoids the introduction of soluble anions (sulfate or chloride) into the system. For instance, an operating cost study for the treatment of a textile wastewater by using EC with iron and aluminum electrodes showed that the operating cost obeys the following form:

$$C = aC_{\text{energy}} + bC_{\text{electrode}} \tag{12-1}$$

where

a = Cost of electrical energy,
b = Cost of electrode material, and
C_{energy} and $C_{\text{electrode}}$ = Consumption of energy and electrode per kilogram of eliminated chemical oxygen demand (COD), respectively.

In this regard, it is worth mentioning that a lower initial pH and a higher conductivity lead to lower energy consumption.

The best function that is tuned well to existing cost data over several orders of magnitude of system scale (including on-site systems) is a logarithmic variant of Williams' power law (Guo et al. 2014)

$$\log(y) = a(\log(x))^b + c \tag{12-2}$$

where

y = Cost (\$),
x = Capacity (m^3 day^{-1}), and
a, b, and c = Numerical correlation factors.

The operating cost must be calculated to evaluate the feasibility of EC application, because both removal efficiencies and the economic impact are significant. The total operation cost (OC) is the result of the sum of different costs related to the EC process operation as it is specified by the following equation (Garcia-Segura et al. 2017):

$$\begin{aligned} \text{OC} = {} & (\text{Price of electricity} \times \text{energy consuption}) \\ & + (\text{anode price} \times \text{electrode consuption}) \\ & + (\text{sluge production} \times \text{disposal and trasportation}) \\ & + (\text{chemical price} \times \text{chemical consuption}) \\ & + \text{maintenance} + \text{depreciation} + \text{labor cost} - \text{amortization} \end{aligned} \tag{12-3}$$

12.3 DESIGN CRITERIA

Establishing main scale-up parameters to find the relationships between laboratory and full-scale equipment is very significant in scaling up a process. In regard to EC, the active electrode surface area-to-working reactor volume (treated wastewater) ratio (S/V) can be a substantial scale-up parameter. Indeed, the electrode surface area affects current density, position and rate of cation dosing, bubble production, and bubble path length. As the S/V ratio rises, the optimum current density decreases. It has been reported in the literature that the optimum *S*/*V* values can be considered as 18.8, 42.5, and 30.8 m^2 m^{-3}, which are all of the same order of magnitude. Nevertheless, the other significant parameters must be taken into account.

Current intensity (A m^{-2}) is another important factor introduced as imposed divided by the surface of the electrode. This factor directly affects the mass of dissolved metallic coagulants. The following dimensionless scale-up parameters should be considered for an electrochemical reactor design (Holt et al. 1999):

- Reynolds number—indication of the fluid flow regime,
- Froude number—indication of buoyancy,
- Weber criteria—indication of surface tension,
- Gas saturation similarity, and
- Geometric similarity.

For designing a reactor, the major question is whether the configuration and mode of operation must be in batch or continuous mode. The latter has gained a lot of attention, which operates with a continuous feed of wastewater and under pseudo-steady–state conditions. The main benefit of this system is fixing of coagulant requirements. Alternatively, batch reactor configuration operates with a certain wastewater volume per treatment cycle. However, from a design and operational viewpoint, the operating conditions inside the reactor change as time passes and result in dynamic behavior by adding a coagulant precursor into the reactor as the anode corrodes. This phenomenon leads to a shift in both pollutant and coagulant levels over time.

The other major factor is the role of flotation. Thus, an additional separation process may be required. A low current leads to production of a low bubble density and consequently low upward momentum flux conditions that encourage sedimentation over flotation. By increasing the current, the bubble density increases, leading to a higher upward momentum flux and more probably a higher removal efficiency by flotation. EC can be combined with most common separation methods such as dissolved air flotation (DAF), electroflotation, filtration, and clarification. In general, the separation step may be combined into the reactor design, or done in downstream units. Therefore, the following can be summarized for an EC reactor design (Holt et al. 2005):

- Batch or continuous operation,
- Role played by electrolytically generated bubbles, and
- Means of separating the aggregated pollutant.

As previously mentioned, the *S*/*V* ratio is a crucial factor in an EC system design because of the injection of a right dose of metal ion coagulants into the system. The latter also depends on current density. The average current density for such a process is 27 A m^{-2}, and 10 m^2 m^{-3}/h^{-1} is the design basis for the ratio of electrode surface area to wastewater volume (Holt et al. 2005). Anything less than this value will lead to insufficient delivery of metal ions; thus, a requirement of excessive current density leads to a quick erosion of electrodes. In fact, by reducing the coagulant, the hydraulic retention time (HRT) will become too short, and by increasing the current, the heat will be lost.

The voltage of the system is varied and can be started at 3 V DC and will be increased as required depending on water conductivity (to achieve suitable current). In special cases for high conductive wastewater such as shale gas tracking flow-back water, the voltage should be reduced to 0.5 V (Eflo 2017). The space between the electrodes is the other significant parameter that is designed to be 3 mm (the thickness of the electrode is also 3 mm). The standard HRT is 1 min. By reversal polarity performance, the positive and negative particles already attached to the electrode surface are rejected (every 1 min for 1/30 s). However, periodic chemical cleaning with diluted acid is necessary. Vertical flow is appropriate for EC, because for the flow direction, the gas bubbles locate on the underside of the horizontal surface and result in insulation on the electrode and consequently a reduction in the efficiency of EC (Eflo 2017).

Although a lower voltage can reduce the power consumption of a plant, reactor configurations, applied current, water characteristics, and operation time affect the applied voltage in the system. For example, in the case of a monopolar reactor, a lower voltage is required, compared with a bipolar reactor, to obtain the same current; however, a higher current is required to achieve the same abatement percentage of contaminant.

12.4 REACTOR TYPES AND ELECTRODE ARRANGEMENT

An EC reactor configuration comprises an electrolytic cell with one anode and one cathode in the simplest form. This kind of electrode arrangement is not optimum for wastewater treatment. Because a large surface area is normally required, monopolar electrodes in parallel connections and monopolar and dipolar electrodes in series connections including different numbers of electrodes are used. In a parallel structure, the electric current is divided to every electrode. In the series arrangement, the current intensity is equal for all electrodes. Although a higher applied voltage potential is needed, it is more economical than others. Bipolar electrodes in the series connections make it possible to eliminate the interconnections between the inner electrodes, because two monopolar electrodes are connected to the power generator, leading to a simple setup, which simplifies maintenance in practice (Fernandes et al. 2015). In the other configuration, only the graphite electrodes are directly connected to the current generator (alternately positive and negative) and operate in monopolar configuration, whereas iron or aluminum electrodes are not directly connected to the generator. They are installed between two graphite electrodes and operate in bipolar configuration.

In a recent study, three types of EC reactors using iron as a sacrificial electrode were investigated for removal , namely, a flow continuous reactor, a turbulent flow reactor, and an airlift reactor. The results demonstrated that this process could remove around 98% of As with a current density of 1.2 A dm^{-2} in the flow continuous and airlift reactors; however, the removal efficiency with the turbulent flow reactor could not reach the same level, whereas the Fe/As ratio (mol/mol) in this case was lower than that of the other two reactors. It is worth noting that by increasing the current density beyond a maximum value, the performance of the reactors cannot be improved because of the passivation of the anodes (Hansen et al. 2007).

The most widely used reactor type for EC is the open batch system with vertical plate electrodes (91% compared with the other types). Among alternative reactor usage, the typical filter press cell (72%) is the most utilized configuration. Enhanced removal efficiencies have been observed by using a filter press reactor for the treatment of waste containing metals and nonmetallic inorganic and organic contaminants compared with the conventional open batch cell with plate electrodes. A cylindrical reactor with a concentrically inner electrode feature (12%) is also used as an alternative. Another type involves a metallic rod instead of an inner electrode. Otherwise, the cathode can be a rotating impeller with two blades to mix the solution and avoid precipitation in the course of operation. Finally, novel EC systems consisting of continuous reactors with rotating screw electrodes (a sacrificial anode rod with a helical cathode) have also been reported (Garcia-Segura et al. 2017).

12.5 OPERATING CONDITIONS AND PROCESS PARAMETERS

In regard to optimum operating conditions of EC, in general, the optimal treatment times of EC is in the range of 5 to 60 min (not taking into account the required sedimentation times), and more than 50% of researchers found an optimum treatment time of 30 min or less. In regard to optimal current densities, significant variations occur; however, in most studies, these are in the range of 10 to 150 A m^{-2}. A moderately narrow pH range (around neutral pH) can be applied for realizing the optimal performance of EC, although it can operate at a wide range of pH. Various electrode materials are used depending on the different types of influents. Both operating cost and electrical energy consumption (depending on different influents) are reported between €0.0047 to €6.74 m^{-3} and 0.002 to 58.0 kW · h m^{-3}, respectively; however, mostly their values are in the range of €0.1 to €1.0/m^3 and 0.4 to 4.0 kW · h m^{-3}. EC applications can be considered in various wastewater types such as tannery waste, textile wastewater; pulp and paper effluents; oily wastewater; food industry; surface water and wastewater containing heavy metals, nutrients, cyanide, and ions, and other contaminants (Kuokkanen et al. 2013). Because of a variety of electrochemical reactions taking place depending on the wastewater type and applied voltage and mixing problems in an EC system, treatment systems of this technology must be adopted to a specific application. In a patent, the inventors claimed that an EC system comprises a dosing unit that receives water from the source and injects metal ions (electrochemically generated) into the medium, a mixing unit that blends the coagulants with fluid, and a buffer tank that is able to maintain the fluid until the particles reach a suitable size [A. R. Volkel, M. H. Lean, and K. Melde "Electrocoagulation system," US Patent No. 818,235 (2012)]. In another innovative patent, the spirally wound electrode sheets were invented to avoid using hundreds of electrode plates and reduce the number of mechanical and electrical connections [C. W. Heiss, "Electrocoagulation reactor and water treatment system and method," US Patent No. 2008/069285 (2009)].

The relationship between current density (mA cm^{-2}) and the amount of dissolved metals (g of M/cm^2) can be expressed by Faraday's law as

$$W = \frac{j \cdot t \cdot M}{n \cdot F} \tag{12-4}$$

where

$W =$ Amount of electrode material dissolved (per surface area of anode electrode),
$J =$ Current density,
$t =$ Operation time (s),
$M =$ Relative molar mass of the electrode material,
$n =$ Number of electrons in the redox reaction, and
$F =$ Faraday's constant (96,500 C/mol).

The pH affects the operation and performance of EC. pH can change during the process. Actually, if the initial pH value is less than or equal to 4, the effluent pH increases; however, when the initial pH value is greater than 8, the pH declines; finally, if the initial pH value is between 6 and 8, the pH values change only marginally, which shows the different behaviors compared with conventional chemical coagulation. In fact, the equilibrium between the production and the consumption of hydroxyl ions and charge neutralization are responsible for producing a buffering solution.

Conductivity also has an influence on EC, which directly impacts the Faradic yield, cell voltage, and energy consumption. For instance, by the addition of NaCl, the conductivity increases along with the formation of chloride anions, which meaningfully reduce the opposing effects of other anions such as HCO^{-3} and SO_2^{-4}. In contrast, an extreme amount of NaCl results in an overconsumption of the aluminum electrodes (Garcia-Segura et al. 2017).

The flux (ϕ_{Fe}) of dissolved iron (mol $\cdot$ s^{-1}) into wastewater is a decisive factor in EC. It can be written as follows:

$$\phi_{Fe} = \frac{j \cdot A}{n \cdot F} = \frac{I}{n \cdot F} \tag{12-5}$$

The concentration (q) of dissolved iron (mol $\cdot$ L^{-1}) into wastewater can be written as follows:

$$q = \frac{\phi_{Fe} \cdot t}{V} = \frac{I \cdot t}{n \cdot F \cdot V} \tag{12-6}$$

The amount of dissolved iron (*g*) can be calculated as follows:

$$[Fe] = q \cdot M = \frac{I \cdot t \cdot M}{n \cdot F \cdot V} \tag{12-7}$$

where

[Fe] = Generated iron (g),
M = Molecular weight of iron (g mol^{-1}),
F = Faraday's constant (C mol^{-1}), and
n = Number of moles of electrons/moles of iron (e.g., =2).

In the research conducted by Amrose et al. (2013), it was found that for arsenic removal from groundwater, the dosage rate is a significant factor (compared to current density) for scaling up an EC reactor design. One of the main issues in that study was the slow settlement of sludge, which can be reduced by means of an Al dosage of 5 mg L^{-1}. The cost estimation resulted in $US 0.22 m^{-3} of treated water (Amrose et al. 2013).

12.6 INDUSTRIAL PLANTS OF ELECTRO-COAGULATION

Over the last few years, several industrial- and pilot-scale EC systems have been developed and are available in the market; however, the issues related to the unreliable/short lifetime of the unit and high operating costs of EC have not been resolved.

One advantage of EC is the reduction of sludge generation by 90%. Moreover, an EC unit requires only a smaller footprint because of its capability of eliminating different contaminants in a single reactor, whereas in conventional methods, a specific equipment for the distinct removal of every impurity is required (Moussa et al. 2017).

EC includes complex chemical and physical processes comprising numerous surface and interfacial phenomena; thus, understanding these phenomena from a scientific point of view is crucial because the lack of understanding limits the engineering design of an EC reactor for achieving optimum performance and prospective progress of this technique. According to previous studies, up-flow configuration is more efficient than a horizontal flow configuration. Furthermore, the use of a sonic field, a stirrer, and combined EC-EF (electro-Fenton) methods improves performance for exercising a greater control of this process (Mollah et al. 2004).

Several industrial-scale EC units have been established over the last few years; however, still some challenges related to the untrustworthiness or short life span of plants exist, and also the problem of high operating costs needs to be addressed.

Nevertheless, a reduction of sludge production by 90% has been observed by an upgraded method. Also, an EC unit requires only a smaller footprint because of its capability of eliminating various impurities in a single chamber in comparison with the conventional chemical system (Moussa et al. 2017). Some companies claim that they are able to design and build efficient EC systems. Table 12-1 summarizes the industrial EC plants along with their innovations.

- F&T water solutions is a company that fabricates EC pilots with capacities ranging from 2 to 1,000 gallons per minute (gpm) (fixed or portable units) with the option of complete automation states. Its recent technology named *Variable Electro Precipitator–VEP* is able to address the problem of shortages in the current EC units. By using an improved flow path design, maximum retention time can be obtained, and external connections are applied in place of wet electrode connections to prevent overheating and failure of the reactor. As claimed by this company, this technology was tested for different industrial applications with diverse capacities, electrode connections, and power. The capabilities (removal %) of wastewater treatment plants are as follows: BOD > 90; TSS > 99; fats/oil/grease 93 to 99; heavy metal 95 to 99; phosphate > 93; bacteria, virus, and cysts >99.99 (Ftwatersolutions 2017).
- Santa Clara Wastewater (SCWW), a company situated in Ventura County, California, is another example of a commercial corporation. Its technology has been designed to treat nonhazardous wastewater streams from a wide range of industries. Lately, it developed EC units instead of conventional chemical precipitation systems. Chemicals added to the system to physically remove pollutants (suspended and dissolved solids) through filtration cause economic consequences. In this case, SCWW was spending $100,000 on chemical additives (per month) for the removal of heavy metals alone. However, it could reduce 90% of cost by installing EC without affecting the quality of treated water.
- Genesis Water Technology (Genesiswatertech 2017) states that it can remove or remarkably reduce TSS, organics including arsenic, volatile and colloidal organic particulates including silica, emulsified oils and hydrocarbons, heavy metals including chromium (VI), fluoride, lead, and radioactive particulates, fats, bacteria, viruses, cysts, and parasites, odor-causing compounds such as hydrogen sulfide, hardness minerals such as calcium and magnesium, color, and certain radioactive compounds. The pilot can be applied for drinking water, process water, graywater recycling, and industrial and municipal wastewater. This company also claims that the integrated EC system can be used for specific applications such as energy (including mining, power generation, and oil and gas production), agriculture/aquaculture, food and beverages, textiles, pharmaceuticals, pulp and paper, potable drinking water, and municipal wastewaters. The capacity of the EC system is calculated as 10 gpm (38 lpm) and it is scalable to 18,400 GPM (100,000 m^3 day^{-1}).

Table 12-1. Industrial/Pilot Plants of EC (References are Mentioned in Section 12.6)

EC company	*Capacity*	*Technology and advantages*
F&T water solutions	From 2 to 1,000 gal. per minute (gpm)	Addresses the shortages in the current EC units
Santa Clara Wastewater (SCWW)		Reduces 90% of cost by installing EC compared with conventional coagulation
Genesis Water Technology	Available from 10 gpm and scalable to 18,400 gpm (100,000 m^3/day).	
EfloEC		Pollutants settle down as a nonleaching oxide sludge that dewaters without the requirement of polyelectrolyte chemicals
Corodex industries		
Bradley EC cell		The availability of a plurality of electrode plates capable of providing a variety of pulsable electric fields
Powell Corporation EC cell	1.5–2,500 gpm (0.34–568 m^3/h)	Electroflotation process
Gallot EC cell		Combining chemical flocculation and EC
Morkovsky and Kaspar EC cell		Continuous flow of water
Ishigaki EC cell		Electrolytic treatment of sludge
Ecoloclean EC cell		A series of devices and chemical pretreatments to separate pollutants from effluents
Electropulse cell of the Oiltrap Environmental Company		Using the electropulse system/cost reduction

- EfloEC is able to establish EC systems with a typical energy consumption of 0.9 kW · h and an electrode consumption of about 24 g/m^{-3} of treated water. Pollutants settle down as a nonleaching oxide sludge that dewaters without any requirement of polyelectrolyte chemicals. The typical contaminant removal rates of its systems are as follows: more than 99% removal rate for benzene, BOD, cadmium, iron, lindane (pesticide), magnesium, turbidity, hydrocarbon, lead, and chromium; more than 98% for mercury; around 42% for nitrate; and about 60% for ammonia. This system can be applied for a variety of wastewaters (Eflo 2017).
- Corodex industries (Corodexindustries 2017) claims that the EC process can remove over 99% of some heavy cations and can also be used to electrocute microorganisms. In such a system, colloids are charged and precipitated, significant amounts of other ions and colloids can be eliminated, and oily emulsions are broken. The reactions in the EC chamber are emulsion breaking, halogen complexing, bleaching, electron flooding, and oxidation.
- Bradley EC cell: Bradley (2004) has designed a particular EC system having a plurality of electrode plates and capable of providing a variety of pulsable electric fields. The system is configured to have at least one plate acting as a cathode and one plate acting as an anode. The inner electrodes function as an anode on one face and as a cathode on the other. The electrolytic system can be configured to favor either a simple elongated fluid pathway or an elongated fluid pathway having a spiralling flow.
- Powell Corporation EC cell: A similar EC system has been developed by Powell ["Method for electrocoagulation of liquids," US Patent No. 8133382B2 (2002)], in which an electrical field is created while applying a voltage between the electrodes. The electrolytic cell is comprised of plate electrodes vertically arranged with respect to the chamber, which induces a vertical flow liquid. The Powell System Company has developed an EC apparatus capable of treating 1.5 to 2,500 gpm (0.34 to 568 $m^3\ h^{-1}$). The electrodes used can be comprised of aluminum or iron, and the interelectrode distance is 0.32 cm. The difference potential imposed is about 3.0 V, whereas the current intensity is 0.38 A. The effluent subjected to treatment circulates in the same direction as the circulation of gas O_2 and H_2, produced, respectively, at the anode and cathode electrodes. This allows the use of the electroflotation process that takes place simultaneously in the Powell electrolytic cell. Electrolyses are carried out by means of galvanostatic methods, by applying a constant and direct current by means of a DC power supply. The Powell electrolytic cell has been used for the treatment of different kinds of wastewater such as textile wastewater, slaughterhouse wastewater, laundry wastewater, municipal effluents, metallic effluents from electroplating industries, and so on. The EC treatment cost (including installation and operation) seems to be lower than that required for chemical coagulation using alum or ferric chloride.
- Gallot EC cell: A system combining chemical flocculation and EC has been developed for purifying contaminated liquid to produce a synergistic effect

between the two processes (Gallot et al. 2005,"Contaminated liquid treatment using electrocoagulation and flotation decantation," WO2005121029A1). Chemical flocculation consists in adding specific metal salts such as $FeCl_3$, $Fe_2(SO_4)_3$, or $Al_2(SO_4)_3$, whereas EC allows delivery of the coagulant in situ by anodic dissolution. According to the authors, both processes (electrochemical and chemical processes) allow the reduction of energy and chemical consumption. Likewise, the process is more effective (in terms of cost and effectiveness) in removing pollutants, while the two processes are combined. This combined treatment method has been conceived to simultaneously remove phosphate, toxic metals, COD and biological oxygen demand (BOD), nitrogen (TKN) total suspended solids (TSSs), and microorganisms from effluents (domestic wastewater, effluents from agrofood industries, and particularly effluents from agricultural industries). For instance, the process of Gallot et al. (2005) ("Contaminated liquid treatment using electrocoagulation and flotation decantation," WO2005121029A1) has been used to purify an effluent contaminated with bacterial, organic, and inorganic pollutants. The results revealed that more than 90% of phosphorus could be removed by using this process; a COD removal of 47% was recorded, whereas more than 80% of fecal coliform (FC) was removed from the effluent. Purifying processes based on EC, followed by chemical addition, have been also developed (Morkovsky 2001) ["Process and apparatus for electrocoagulative treatment of industrial waste water," US Patent No. 6294061B1 (2001); S. Powell (2002), "Method for electrocoagulation of liquids," US Patent No. 8133382B2].

- Morkovsky and Kaspar EC cell: Morkovsky and Kaspar (2004) ("Process and apparatus for electrocoagulative treatment of industrial waste water," US Patent No. 6689271B2) have also developed a process and apparatus for EC treatment of industrial wastewater. The system is comprised of an EC reactor having charged and uncharged plates, allowing continuous flow of water. The reactor is connected to a voltage source to charge some of the plates positive and some negative, with uncharged plates between the positive and negative plates. The system allows wastewater to enter the reactor for coagulation, and the electro-coagulated effluent leaves the reactor and enters a defoam tank for agitation, which allows trapped bubbles to rise to the surface of the tank as foam. From the defoam tank, wastewater goes through a sludge thickener to allow sludge to settle at the bottom, and wastewater is drawn off from the sludge thickener to flow to a clarifier. Water is drawn off the top of the clarifier for transport to a conventional sewer system or for reuse. Kaspar Corporation (kaselco 2006) proposed an EC cell called "Kaselco" capable of treating effluents loaded with toxic metals, oily emulsions, dissolved solids, BOD, and COD. The electrodes that can be used are mild steel or aluminum electrodes.
- Ishigaki EC cell: Ishigaki (Ishigaki 1988) ("Method and apparatus for electrolytic treatment of sludge," US Patent No. 4919775) has also developed a method and apparatus for the electrolytic treatment of sludge. The electrolytic cell is equipped with a multiplicity of parallel electrode plates

(anodes and cathodes) individually connected to the power supply having a voltage of the order of 1.5 to 20 V. Iron electrode (anode) and stainless-steel electrodes (cathode) are alternately arranged side by side at a short interval in a row (15 mm). The process induces the transformation of hydrophilic substances (organic and inorganic) into hydrophobic substances by an anodic dissolution of iron. This allows improvement in the filterability of biosolids during mechanical dewatering. Inside the electrolytic cell, there is a predetermined direction close to each electrode, in which treated sludge flows upward between the electrode plates so that gas bubbles (O_2 and H_2) and floc particles that accumulate near the electrodes are regularly removed. Consequently, the fouling of the electrode is avoided, and the exchange between the electrolyte (sludge) and the electrode is favored so that biosolids are efficiently transformed by oxidation (anodic dissolution), reduction, and neutralization.

- Ecoloclean EC cell: The EC cell conceived by Ecoloclean Industries Inc. is comprised of an iron electrode built on a filter press frame (texasextension 2008). This system uses a series of devices and chemical pretreatment methods to separate pollutants from effluents (e.g., a dairy lagoon effluent). The EC unit contains positively charged electrodes that separate the negatively charged pollutants from the liquid as they flow the electrodes. In the system developed by Ecoloclean Industries Inc., an influent undergoes chemical pretreatment in a mixing tank and then passes through a centrifuge, a collection/equalization, an EC unit, a DAF unit, and then a final filter. In the study, an influent from the lagoon was pumped into a large mixing tank, where it was mixed with alum and anionic emulsion polymer to help coagulate and separate the solids. After chemical pretreatment, the influent was sent from the mixing tank to a centrifuge, which separated the solids and some metals from the liquid in the influent. From there, the liquid passed through a collection/equalization tank and on to the EC unit. In the EC unit, the liquid was passed over charged electrodes that gave off ions, causing phosphorus and the metals to coagulate and precipitate (separate). The liquid then moved to a reaction tank with a mixer that allowed sufficient time for chemical reaction to occur. A pump then moved the liquid to a DAF unit, which introduced small air bubbles that attached to the suspended solids (SSs) and floated them to the surface. Solids were then skimmed from the surface. After leaving the bottom of the DAF unit, the liquid was passed through the filter, which further removed impurities from the liquid and yielded the final treated effluent. The cost estimates for the operation of the EC system were about $0.12 per gallon of treated effluent. These costs included fees for systems setup and operation by representatives from Ecoloclean Industries Inc. This cost did not include costs for residual material removal from the dairy.
- Electropulse cell of the Oiltrap Environmental Company: The electropulse cell has been designed by the Oiltrap Environmental Company, and this cell represents a revolutionary electrical-based technology for effectively treating

complex waste streams on-site (oiltrap 2008). In this EC unit, wastewater is exposed to electropulse treatment in cells that are modular to accommodate varying flow rates. An automated backflush system ensures that the electrical plates are regularly cleaned for optimum EC effectiveness and for ensuring long life of the plate assembly. Charged wastewater flows through a series of tanks, each removing sludge and floating contaminants, thus polishing the water. According to the Oiltrap Environmental Company, EC treatment using the electropulse system has proven successful in removing a variety of contaminants that are too expensive to be removed by filtration or chemical treatment systems. Among these contaminants are emulsified oil, total petroleum hydrocarbon (TPH), SSs, and heavy metals. The cost of wastewater treatment, including electricity and consumables, is typically only about $3.0 per 1,000 gal. of treated wastewater. Likewise, significant savings in labor costs can be had because EC is fully automated.

12.7 TYPES OF WASTEWATERS AND POLLUTANTS

Various types of wastewaters can be treated and different kinds of contaminants can be removed by EC. Textile industry wastewater is an example, which contains dye-assisting chemicals such as wetting, scouring, sequestering, cleaning, anticreasing agents, and acid donors, which in some cases are bioinhibitory, toxic, and/or refractory. Color removal by EC is also done by employing several mechanisms such as adsorption, forming a complex with metal hydroxide, making ionic bonds, and reduction instantaneously with iron oxidation (Fe^{2+} to Fe^{3+}) if dissolved oxygen concentrations are at an appropriate level. Removal of organic substances such as COD and TOC can be performed by electrochemical oxidation (EO), adsorption by electrostatic attraction, and physical association.

SSs and turbidity elimination by EC have been described in some studies. The results show that iron electrodes are more effective for the treatment of some waste effluents such as textile wastewater, which is related to the settling velocity and dimensions of the settling tank. In regard to the operational cost, the estimation demonstrated $0.10/kg COD removal for iron electrodes and $0.3/kg COD elimination for aluminum electrodes. Moreover, electrode consumption expenses are approximately 50% of the total cost for iron and 80% of that for aluminum.

For leather tanning industry effluents, two mechanisms have been found for organic matter removal, namely, indirect oxidation via the formation of chlorine species and colloidal formation through adsorption/entrapment on metal hydroxide flocs for SS and turbidity removal. It is worth mentioning that organic matter removal efficiency is related to its initial concentration. For instance, in leather tanning industry effluents with high concentrations of ammonium, volatilization of ammonia occurs at pH greater than or equal to 9.5, and the decomposition of ammonia to nitrogen gas takes place via chlorine and hypochlorite by in situ formation.

Nitrate is reduced by means of electrodes such as magnesium, iron, tin, aluminum, etc.. Besides, iron or steel electrodes can be utilized for sulfide removal through the production of ferrous and ferric ions and making reactions with sulfide species by precipitations and oxidation. Pulp and paper industry wastewater contains high levels of organic matter, SSs, and color, which can be treated by EC as well.

12.7.1 Metal-Bearing Industrial Effluents

Metals (especially the free form of metals) can be efficiently eliminated by EC. In such a process, hydroxide settlement plays a vital role along with adsorption on metal hydroxide flocs. In this regard, several studies have been performed to remove zinc, copper, nickel, and cadmium from industrial effluents (from metal processing and finishing processes). Moreover, hexavalent chromium removal has been reported in several research studies.

In general, three main mechanisms are discovered for the removal of complexed heavy metals by EC using chloride. Complexed metal conversion into free metals, hydroxide precipitation of free metals, and the removal of colloidal particles through the formation of $Fe(OH)_3$ flocs via adsorption are the key mechanisms.

The total operational costs involving energy, electrode material, and sludge disposal are estimated as \$US 0.59 and \$US 0.97 m^{-3} for EC using Fe and Al electrodes, respectively. In contrast, the related costs for conventional chemical coagulation can be estimated as \$US 0.89/m^{-3} for $FeCl_3$ and \$US 1.176 m^{-3} for alum.

EC has been successfully applied for the treatment of waste containing arsenic and fluoride as well. Moreover, EC with aluminum electrodes has been used for reducing fluoride concentrations to meet the drinking water treatment standards of 0.5 to 1.5 mg L^{-1}. Fluoride can be found in industrial wastewater after glass treatment, as a fertilizer, after semiconductor manufacturing, and in metal finishing industries (Kabdaşlı et al. 2012).

12.7.2 Nonmetallic Inorganics

More than 80% of concentration abatement has been obtained for ammonia, boron, cyanide, fluoride, nitrite, nitrate, phosphate, powdered activated carbon, silica particles, sulfide, and sulfite by the use of Al, Fe, and SS electrodes. Al–SS electrodes are efficient for the elimination of ammonia. The only study for cyanide removal showed that the removal of cyanide of 300 mg concentration is practical using Al or Fe electrodes to achieve removal efficiencies from 87% to 93% in 20 min of operation (Garcia-Segura et al. 2017). In regard to fluoride removal, the use of Al–Al electrodes can improve the efficiency of elimination; however, in bipolar electrodes in a series connections (BP-S) reactor system, the removal of fluoride from synthetic waters was efficient. Actually, the performance and efficiency robustly depends on the EC reactor and supporting electrolyte. In regard to nitrate, it can be removed by EC avoiding the oxidation reduction reactions, which makes it difficult to completely remove the chemical. In the case of phosphate, sulfide, sulfite, and sulfate, EC can be a noble approach to remove higher concentrations of these species (Garcia-Segura et al. 2017).

12.7.3 Heavy Metals

Different parameters impact the removal efficiency of heavy metals (e.g., arsenate, arsenite, cadmium, manganese, and silver), such as a combination of electrodes and EC reactor configuration. In several studies, EC has been used to treat synthetic or real waste containing phenolics, dyes, pesticides, pharmaceuticals, and so on.

Although the EC method can be effective for treating real wastewater, pre- or post-treatments are required after or before EC. To do these, novel techniques such as the integration of EC with other secondary or tertiary treatment methods can be considered to meet the environmental regulations to discharge treated effluents (Garcia-Segura et al. 2017).

12.7.4 Chemical Oxygen Demand Removal

It is important to note that the efficiency of EC to remove COD can be high or extremely low depending on the characteristics of wastewater. In some cases, COD may increase even after EC treatment. When some compounds such as acetic acid react with Fe (II) to produce soluble species and remain in the solution, COD increases. This phenomenon can also be observed in the case of sequestrates or complexing agents like EDTA. In some cases, further oxidation of Fe^{2+} to Fe^{3+} in a basic pH can result in increased COD. Soluble and miscible compounds that cannot react with iron species (both types), such as glucose, lactose, phenol, and so on, can never be eliminated with EC. If the solution contains organic salts such as sodium oxalate, $(OH)^-$ preferably bounds the iron species to form insoluble iron hydroxides because iron ions are more acidic than Na^+; thus, in this case, only a very low amount of acetate and similar ions are removed (partial COD removal). However, some acids such as citric, tartaric, and so on react only with Fe^{2+} and, therefore, they can be removed more efficiently. To sum up, it is clear that SSs and liquids, FCs, fat oil and grease, and turbidity can be efficiently removed by EC (Moreno-Casillas et al. 2007).

12.8 CHALLENGES AND RECOMMENDATIONS

In general, the EC system requires simple equipment and can be operated easily without chemical addition. However, a periodic replacement of sacrificial electrodes is necessary because they are dissolved and oxidized into wastewater streams and consequently intensify the conductivity of the fluid and develop the resistant oxide film on the cathode, and finally form the sludge containing significant amounts of iron or aluminum and other recalcitrant chemicals that must be treated before disposal. Moreover, some toxic chlorinated organic compounds and trihalomethanes may be produced in situ in the presence of chloride and high humic and fulvic acids. To address these problems and improve the performance of the EC process, metal ion coagulant concentration in a medium should be monitored to thwart excess ion concentration.

Moreover, the polarity of electrodes can be changed frequently to avoid the formation of an impermeable film on the electrode and prevent reduction in efficiency. In addition, applying constant intensity instead of constant voltage can diminish the passivation of electrodes to avoid an increase in energy consumption (by applying higher overpotentials). Thus, for reducing electrode passivation, novel electrode types and arrangements can be developed. A combination of the EC technique with other treatment approaches is another alternative (as discussed in other chapters) to overcome problems like solid–liquid separation and sludge production (Fernandes et al. 2015). The cost of aluminum and iron sludge management is a crucial factor in EC. To optimize the EC process, sludge volume and its settling ability are required to be minimized to reduce costs.

EC has been empirically optimized in some studies; however, more basic research is required to improve the process from the viewpoint of engineering design and full-scale application. The EC process includes complex reaction mechanisms involving surface and interfacial interactions. To optimize the EC processes and their operating conditions, there must be an in-depth consideration of the reactor design and a fundamental understanding of the associated reactions and process control (Kabdaşlı et al. 2012).

One of the challenges in EC is developing a methodology for design and simulation of the EC reactor (e.g., predicting the behavior of the EC system in a real scale) to simplify the scaling up of a laboratory-scale EC system to an industrial one. Furthermore, more research needs to be performed in the area of continuous operation of EC instead of batch operation to meet the industry criteria. Evaluation of the economics and feasibility of EC processes is also important for future development. In this regard, the overall cost involving capital investment and operation/maintenance should be taken into account as a principal factor for industrializing the EC processes. Moreover, EC has been found effective in the treatment of produced wastewater, which is a major issue in the oil and gas industry, particularly in destabilizing tight emulsions, but more research is needed in this area (Moussa et al. 2017).

12.9 CONCLUSION

EC shows high efficiency in the removal of contaminants in different types of wastewater. EC is known as an effective method to removal toxic compounds, color, and recalcitrance. Industrialization and scaling up of EC involve some challenges such as sludge management, simulation and modeling, electrode passivation, and so on, although some industrial EC systems and some companies successfully design and build EC pilots. Several decisive factors are at play in designing a commercial EC system. For example, electrode surface-to-fluid volume, current density, residence time, reactor configuration, and electrode spacing are factors that require a careful consideration in accordance with the characteristics of water such as conductivity, pH, and so on. Cost-effectiveness analysis in terms of electricity and anode consumption is also inevitable to establish more large-scale applications of EC.

NOMENCLATURE

BOD = Biological oxygen demand
COD = Chemical oxygen demand
DC = Direct current
EC = Electro-coagulation
GPM = Gallon per minute
HTR = Hydraulic retention time
LPM = Liter per minute
OC = Operation cost
TKN = Total Kjeldahl nitrogen
TOC = Total organic carbon
TSS = Total suspended solids
WW = Wastewaters

References

Amrose, S., A. Gadgil, V. Srinivasan, K. Kowolik, M. Muller, J. Huang, et al. 2013. "Arsenic removal from groundwater using iron electrocoagulation: Effect of charge dosage rate." *J. Environ. Sci. Health Part A* 48 (9): 1019–1030.

Bradley, K. D. 2004. "Electrocoagulation system.", International patent publication number WO 2004/089832 A1.

Corodexindustries. 2017. "EFLO electro coagulation (EC)." Accessed September 20, 2017 http://www.corodexindustries.com/wastewater/industrial-wastewater/33.

Eflo. 2017. "EC electrocoagulation." Accessed September 14, 2017. http://www.eflo.com/Electrocoagulation.html.

Fernandes, A., M. J. Pacheco, L. Ciríaco, and A. Lopes. 2015. "Review on the electrochemical processes for the treatment of sanitary landfill leachates: Present and future." *Appl. Catal., B* 176–177 (Supplement C): 183–200.

Ftwatersolutions. 2017. "Electrocoagulation: General capabilities." Accessed September 28, 2017 http://www.ftwatersolutions.com/electrocoagulation.

Garcia-Segura, S., M. M. S. G. Eiband, J. V. de Melo, and C. A. Martínez-Huitle. 2017. "Electrocoagulation and advanced electrocoagulation processes: A general review about the fundamentals, emerging applications and its association with other technologies." *J. Electroanal. Chem.* 801 (Supplement C): 267–299.

Genesiswatertech. 2017. "Electrocoagulation water treatment solutions." Accessed September 14, 2017. http://genesiswatertech.com/electrocoagulation/.

Guo, T., J. Englehardt, and T. Wu. 2014. "Review of cost versus scale: Water and wastewater treatment and reuse processes." *Water Sci. Technol.* 69 (2): 223–234.

Hansen, H. K., P. Nuñez, D. Raboy, I. Schippacasse, and R. Grandon. 2007. "Electrocoagulation in wastewater containing arsenic: Comparing different process designs." *Electrochim. Acta* 52 (10): 3464–3470.

Holt, P., G. Barton, and C. Mitchell. 1999. "Electrocoagulation as a wastewater treatment." In *Proc., 3rd Annual Australian Environmental Engineering Research Event.* 23–26 November, Castlemaine, Victoria.

Holt, P. K., G. W. Barton, and C. A. Mitchell. 2005. "The future for electrocoagulation as a localised water treatment technology." *Chemosphere* 59 (3): 355–367.

Kabdaşlı, I., I. Arslan-Alaton, T. Ölmez-Hancı, and O. Tünay. 2012. "Electrocoagulation applications for industrial wastewaters: A critical review." *Environ. Technol. Rev.* 1 (1): 2–45.

kaselco. 2006. "Kaspar Corporation." www.kaselco.com. Accessed August 25, 2017

Kuokkanen, V., T. Kuokkanen, J. Rämö, and U. Lassi. 2013. "Recent applications of electrocoagulation in treatment of water and wastewater—A review." *Green Sustainable Chem.* 3: 89–121.

Mollah, M. Y. A., P. Morkovsky, J. A. G. Gomes, M. Kesmez, J. Parga, and D. L. Cocke. 2004. "Fundamentals, present and future perspectives of electrocoagulation." *J. Hazard. Mater.* 114 (1): 199–210.

Moreno-Casillas, H. A., D. L. Cocke, J. A. G. Gomes, P. Morkovsky, J. R. Parga, and E. Peterson. 2007. "Electrocoagulation mechanism for COD removal." *Sep. Purif. Technol.* 56 (2): 204–211.

Moussa, D. T., M. H. El-Naas, M. Nasser, and M. J. Al-Marri. 2017. "A comprehensive review of electrocoagulation for water treatment: Potentials and challenges." *J. Environ. Manage.* 186 (Part 1): 24–41.

oiltrap. 2008. "Electro-pulse." www.oiltrap.com. Accessed September 14, 2017

texasextension. 2008. "Texas Cooperative extension." www.texasextension.tamu.edu. Accessed September 5, 2017

CHAPTER 13

Electro-Oxidation Processes: Criteria and Considerations for Full-Scale Applications

Mahdieh Khajvand, Mitra Ebrahimi,
Ali Khosravanipour Mostafazadeh,
Patrick Drogui, R. D. Tyagi

13.1 INTRODUCTION

Choosing the best treatment methods for purifying wastewater has become one of the most environmentally challenging tasks these days. In the last few decades, electro-oxidation (EO) has gained increasing attention for the treatment of water and wastewater, mainly because of its high efficiency in the removal of biorefractory substances. Nevertheless, this area requires more investigation to overcome certain obstacles before its commercialization (Zhang et al. 2013).

EO has been employed to treat wastewater generated in various industries. The performance of EO for the treatment of synthetic and real industrial and urban effluents was reviewed by Särkkä et al. (2015) and Garcia-Segura et al. (2018). They obtained outstanding results on the abatement of chemical oxygen demand (COD) and the complete removal of persistent organic pollutants. Besides, the development trends in electrochemical technologies and their application in water and wastewater treatment based on findings in annual publications were investigated (Cong et al. 2016). They cited that published research on the oxidation process has considerably increased in number during recent years. (Ghime and Ghosh 2019) studied the effects of reactor configurations, types of electrodes, pH, temperature, and current density (CD) used in the electrochemical oxidation processes for the removal of contaminants from wastewater.

This chapter provides an overview of the fundamental aspects of EO in wastewater treatment and the scale-up of this technology considering the essential parameters in designing. Besides, the potential combination of EO with other wastewater treatment technologies to enable practical full-scale commercial applications is elucidated. This work also focuses on the wastewater treatment of various industries relying on the type of contaminants, from persistent organic pollutants to nutrients. The chapter is arranged in the following sequence. First, a summary of the EO mechanism is given, which is followed by design criteria. Next, the different coupled systems employed for EO-based wastewater treatment are brought up. This discussion is followed by the remediation of various types of wastewater by EO with a focus on the different contaminates. Finally, the critical challenges pertinent to EO and the recommendations to surmount them are debated.

13.2 MECHANISMS OF ELECTRO-OXIDATION

Two different mechanisms are available for removing impurities in wastewater by EO. Figure 13-1 illustrates the direct and indirect electrolytic treatment of pollutants. Direct anodic oxidation, where pollutants are reduced at the anode surface. Indirect oxidation where mediators such as NaCl, HClO, $H_2S_2O_8$, and so on are electrochemically formed to achieve oxidation.

It is clear that both oxidation mechanisms may occur during the EO of aqueous effluents (Chiang et al. 1995). In general, the mechanism of electrochemical degradation of wastewater pollutants is a complex phenomenon involving the coupling of the electron transfer reaction with a dissociate chemisorption step. Details about each mechanism are described as follows.

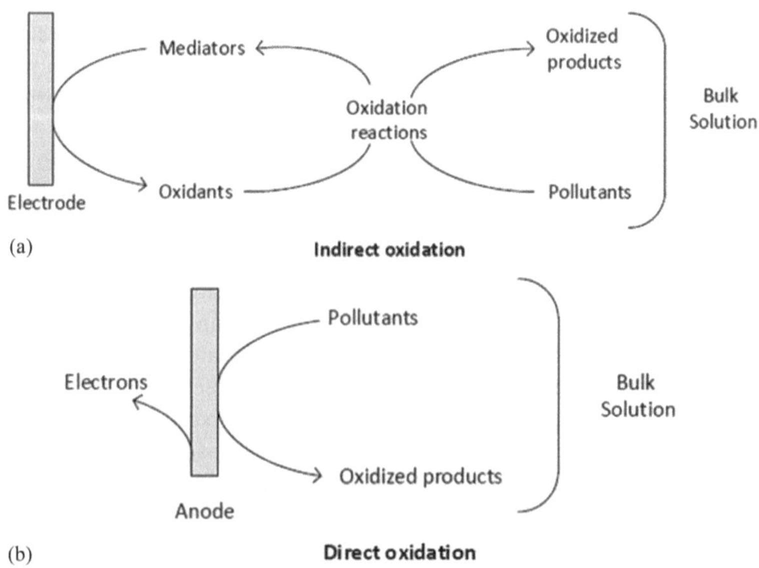

Figure 13-1. Scheme for direct and indirect electrolytic treatment of pollutants.
Source: Adapted from Chiang et al. (1995).

13.2.1 Direct Oxidation

In the direct oxidation mechanism, pollutants directly exchange electrons on the surface of the anode without using other substances. The direct oxidation of pollutants occurs in two steps: First step: The anodic oxidation of the molecule of water leads to the formation of the reactive species $OH^{\bullet}$, which is adsorbed on an active site as $M(OH^{\bullet})$. Second step: The hydroxyl radical oxidizes the organic pollutant.

The reaction between the oxidized organic compounds and the hydroxyl radicals can lead to complete oxidation.

$$H_2O + M \rightarrow M(OH^{\bullet}) + H^+ + e^- \tag{13-1}$$

In the direct oxidation reaction, oxidation is carried out by either of the two mechanisms, electrochemical conversion or electrochemical combustion. During electrochemical conversion, nonbiodegradable organic compounds (R) are partially oxidized to more biodegradable compounds Equation (13-2)], whereas in electrochemical combustion, organic pollutants are completely degraded into CO_2 and H_2O [Equation (13-3)].

$$R + M(OH^{\bullet}) \rightarrow M + RO + H^+ + e^- \tag{13-2}$$

$$R + M(OH^{\bullet}) \rightarrow M + mCO_2 + nH_2O + H^+ + e^- \tag{13-3}$$

A phenomenon of competition can take place during oxidation and, thus, reduce the efficiency of the degradation of pollutants. The formation of oxygen is an example of a parasitic reaction (Mostafazadeh et al. 2019)

$$H_2O + M(OH^{\bullet}) \rightarrow M + O_2 + 3H^+ + 3e^- \tag{13-4}$$

The correlation between the mass transfer of the substrate and the electron transfer at the electrode surface can impact the efficiency of EO. Electrode activity and CD determine the rate of electron transfer (Drogui et al. 2007).

13.2.2 Indirect Oxidation

In the indirect oxidation mechanism, pollutants do not directly exchange electrons on the surface of an anode without the presence of other substances. Some electroactive compounds exist that act as intermediators for electrons traveling between electrodes and organic compounds. This mechanism can be both a reversible and an irreversible process. The nature, structure of the electrode material, experimental condition, and electrolyte composition can influence the selection of the process.

During indirect EO, a strong oxidizing agent is electrogenerated at the anode surface and it subsequently destroys the organic compounds in the bulk solution. The most common and popular electrochemical oxidant is chlorine, which is formed by oxidizing chloride at the anode surface [Equation (13-5)]. Throughout indirect oxidation, the agents produced at the anode surface that are responsible for the oxidation of inorganic and organic compounds can be chlorine/hypochlorite, hydrogen peroxide, peroxodisulfuric acid, and ozone (Li et al. 2010, Scialdone et al. 2009). A chain of reactions that involve chlorine/hypochlorite indirect oxidation are presented in the following equations:

Anodic reactions:

$$2Cl^- \rightarrow Cl_2 + 2e^- \tag{13-5}$$

$$6HOCl + 3H_2O \rightarrow 2ClO^{3-} + 4Cl^- + 12H^+ + 1.5O_2 + 6e^- \tag{13-6}$$

$$2H_2O \rightarrow O_2 + 4H^+ + 4e^- \tag{13-7}$$

Bulk reactions:

$$Cl_2 + H_2O \rightarrow HOCl + H^+ + Cl^- \tag{13-8}$$

$$HOCl \rightarrow H^+ + OCl^- \tag{13-9}$$

Cathodic reactions:

$$2H_2O + 2e^- \rightarrow 2OH^- + H \tag{13-10}$$

$$OCl^- + H_2O + 2e^- \rightarrow Cl^- + 2OH^- \tag{13-11}$$

Hypochlorite (OCl^-) formed in bulk solution [Equations (13-8) and (13-9)] is a strong oxidizing agent to oxidize aqueous organic substances (Scialdone et al. 2009). Furthermore, for the common oxidants that can be electrochemically produced, metal catalytic mediators (Ag^{2+}, Co^{3+}, Fe^{3+}, etc.) are also employed for the generation of hydroxyl radicals, as seen in the electro-Fenton system. Nevertheless, in the treated effluent, produced ions of metal might be more toxic in comparison with the initial state. Thus, this system needs a separation process for recovering the metallic species (Anglada et al. 2009), leading to an unfavorable and intricate treatment process.

13.3 DESIGN CRITERIA

The most critical factors in designing EO systems are electrode material, cell design (configuration), operating conditions, and energy consumption. It is necessary to consider these parameters. The design criteria are summarized in Table 13-1.

13.3.1 Electrode Material

The type and extension of the production of oxidants depend on many inputs, the most relevant being the electrode material and the production of suitable raw matter for the generation of oxidants in wastewater. Their effect. The effect of electrode material on the efficiency of electrochemical advanced oxidation processes (EAOPs) is very important because of the oxidation of pollutants starting from the vicinity of the electrode surface to the bulk of the electrolyte. However, it should be noticed that these oxidants have a huge

Table 13-1. Critical Factors for Designing an EO System.

Design criteria		
Electrode material	*Cell design*	*Operating conditions*
High physical stability	Appropriate reaction engineering	Current density
High chemical stability	Simplicity and versatility	Temperature
Suitable physical shape	Uniform current density distribution	Physicochemical features
Electrical conductivity	Uniform electrode potential distribution	Energy consumption
Catalytic activity and selectivity	High mass transport rates	
Low cost/life ratio	Ability to handle solid, liquid, or gaseous products	
	The form of the product and the ease of product extraction	
	Simplicity of design, installation, and maintenance	
	Availability of electrode and membrane materials	
	Capital and running costs	
	Integration with other process needs	

effect on the oxidation mechanisms of the pollutants and occasionally can lead to the formation of unwanted intermediates or final stable products (Sirés et al. 2014).

Therefore, electrode materials can affect the selectivity and the efficiency of the process; thus, the selection of this parameter is significant. Electrode performances and a lack of ample information on this make it unfeasible to choose the optimum electrode for a given process on a theoretical basis. The first selection depends on process experience, and after that, it is tested and refined during an extensive development program. In fact, it is unrealistic to evaluate the efficiency of an electrode material or to characterize its lifetime without any extended studies under realistic operating conditions (Klamklang et al. 2012). Nevertheless, electrode materials must have the following properties (Anglada et al. 2009, Klamklang et al. 2012):

1. High physical stability: this includes good mechanical strength and good resistance to erosion and cracking;
2. High chemical stability: resistance to corrosion, the formation of unwanted oxides or hydrides, and the deposition of inhibiting organic films under all conditions;

3. Suitable physical shape: it should be viable to design the material into the required shape, to assist sound electrical connections, and to allow simple fixing of spares at a variety of scales;
4. Electrical conductivity: high conductivity throughout the electrode system, including the current feeder, electrode connections, and the entire electrode surface exposed to the electrolyte;
5. Catalytic activity and selectivity: sustaining the desired reaction, and in some cases, significant electrocatalytic properties are vital. The electrode material must encourage the desired chemical change, while inhibiting all competing chemical changes; and
6. Low cost/life ratio: The use of reasonably priced and durable electrode materials must be favored.

In the selection of an anode material, competition between organic oxidation at the anode and the side reactions of the oxygen evolution should be considered. The oxidation of water to oxygen [Equation (13-12)] happens at about 1.2 V versus the standard hydrogen electrode. However, a higher voltage is required for the EO of water to occur at the anode. For example, the potentials of oxygen evolution for different anodes such as Pt, IrO_2, graphite, SnO_2, Si/boron-doped diamond (BDD), and Ti/BDD are 1.3, 1.6, 1.7, 1.9, 2.3, and 2.7 V, respectively, at conditions of 0.5 mol L^{-1} H_2SO_4. These values are 1.9 V for PbO_2 and 2.2 V for TiO_2 at conditions of 1 mol L^{-1} H_2SO_4 (Chen 2004).

$$2H_2O \rightarrow O_2 + 4H^+ + 4e^- \qquad (13\text{-}12)$$

Some general guidelines are available to assist the selection of an electrode material. Overall, low O_2 overvoltage anodes are distinguished by a high electrochemical activity toward oxygen evolution and low chemical reactivity toward the oxidation of organic compounds. Efficient pollutant oxidation at these anodes may happen at low current densities. An important reduction of the current efficiency is expected at high current densities as a result of the production of oxygen. In contrast, at high O_2 overvoltage anodes, higher current densities may be used with minimal involvement from the oxygen evolution side reaction. Thus, in general, high O_2 overvoltage anodes are preferred. For instance, BDD anodes have been confirmed to yield higher organic oxidation rates than, and superior current efficiencies to, other commonly used metal oxides, including PbO_2 and Ti/SnO_2-Sb_2O_5 (Anglada et al. 2009).

13.3.2 Cell Design

The most crucial issue in a cell design is keeping high mass transfer rates. The main reactions in the electrochemical process take place on electrode surfaces. Gas sparging, high fluid velocity, baffles, and several types of turbulence promoters are often employed to increase the process efficiency. Also, in obtaining a high mass transfer rate, the cell structure should account for simple access to and exchange of cell parts (Wendt and Kreysa 1999). The various features that should be considered in the design of an electrochemical reactor are categorized based on cell configuration, electrode configuration, and flow type (Anglada et al. 2009):

- Cell configuration: Divided cell and undivided cell are two categorizations of an electrochemical reactor based on cell configuration;
- Flow type: Plug flow and perfect mixing are the two types of flow for the classification of electrochemical reactors; and
- Electrode configuration: The classification of electrode geometry is done in two-dimensional and three-dimensional electrodes. Each of these electrodes is of two types, static and moving. Two-dimensional and three-dimensional structures are two types of electrodes.

For gaining access to the high value of the electrode surface to cell volume ratio, the use of three-dimensional electrodes is recommended. As mentioned previously, both two-dimensional and three-dimensional electrodes can be categorized into static and moving electrodes (Anglada et al. 2009). Accordingly, the utilization of moving electrodes increases the mass-transport coefficient because of the promotion of turbulence. However, among the two-dimensional electrodes, static parallel and cylindrical electrode cells are used in major reactor designs in the latest studies. Because of the simplicity of scale-up to a larger electrode size by merely adding electrodes or an increasing number of cell stacks, parallel plate geometry is used in a filter press arrangement (Rajeshwar and Ibanez 1997).

Furthermore, cell configuration (divided and undivided) needs to be considered. In divided cells, the anolyte and catholyte are separated via a porous diaphragm or an ion-conducting membrane. The selection of the separating diaphragm or membrane in divided cells is as important as the selection of electrode materials. In general, divided cell choice should be avoided whenever possible, as separators are expensive, and tightening of a divided cell (reduction of the electrode gap) is difficult because a host of mechanical problems and those related to corrosion crop up during tightening (Wendt and Kreysa 1999).

The selection of an electrochemical reactor for a specific process is important, and it is clear that reactors for energy conversion and electrochemical synthesis will have drivers that are different from those used in the destruction of electrolyte-based contaminants. More specialized reactors are required for many applications such as metal ion removal from a dilute solution. Some studies are available on the form of the electrode, its geometry, and motion, together with the need for cell division or a thin electrolyte gap. The form of the reactants and products and the mode of operation (batch or continuous) are also important design factors. Desirable factors in reactor design (and their implications) include the following (Walsh 2001):

1. Moderate costs: such as using low-cost components, low cell voltage, and small pressure drop over the reactor;
2. Convenience and reliability in operation: designed for facile installation, maintenance, and monitoring;

3. Appropriate reaction engineering: using uniform and proper values of CD, electrode potential, mass transport, and flow; and
4. Simplicity and versatility: In an elegant design, this is attractive to end users.

Furthermore, the important parameters that should be considered in the design of reactors and to ensure their effective performance are as follows (Walsh 2001):

1. Uniform CD distribution;
2. Uniform electrode potential distribution;
3. High mass transport rates;
4. Ability to handle solid, liquid, or gaseous products;
5. The form of the product and the ease of product extraction;
6. Simplicity of design, installation, and maintenance;
7. Availability of electrode and membrane materials;
8. Capital and running costs; and
9. Integration with other process needs.

13.3.3 Operating Conditions

CD, temperature, physicochemical features, and energy consumption are the main factors that influence the efficiency of the EO process. The effects of these parameters are as follows:

Current density. Current density (CD) is among the most important factors that usually control EO processes through the reaction rate. It should be clear that an increase in CD does not necessarily result in an increase in oxidation efficiency; the effect of CD on the treatment level depends on the features of the wastewater to be treated. On the contrary, the use of a higher CD, in general, results in higher operating costs because of the increase in energy use.

Temperature. Increasing the temperature can lead to more efficient processes because of global oxidation. Although temperature cannot affect direct oxidation processes, this fact may be explained in terms of the presence of inorganic electrogenerated reagents. An enhancement in the rising temperature of the mediated oxidation processes by inorganic electrogenerated reagents (active chlorine, peroxodisulfate) has been reported. But operating at ambient temperature is preferred because electrochemical processes need lower temperatures in comparison with their nonelectrochemical counterparts (i.e., incineration, supercritical oxidation) (Cañizares et al. 2006).

Physicochemical features. The physicochemical features of wastewater, such as electrolyte nature and amount, pH value, and initial concentration of pollutants, also affect the EO process. The higher the concentration of the electrolyte, the higher the conductivity and the lower cell voltage for a given current density. Thus, treatment by EO is more suitable and cost-efficient when wastewater contains high salinity. The pH effect is similar to temperature, affecting mostly indirect oxidation processes (Anglada et al. 2009). In chloride-mediated reactions, the pH may impact the oxidation rate. During indirect oxidation, chlorine evolution occurs at the anode [Equation (13-13)] as

$$Cl^- \rightarrow 1/2Cl_2 + e^- \tag{13-13}$$

When the pH is lower than 3.3, the primary active chloro species is Cl_2, whereas at higher pH values, its diffusion away from the anode is coupled to its disproportionation reaction to form HClO at a pH < 7.5 [Equation (13-14)] and ClO^- at a pH > 7.5 [Equation (13-15)].

$$Cl_2 + H_2O \rightarrow HOCl + H^+ + Cl^- \tag{13-14}$$

$$HOCl \rightarrow OCl^- + H^+ \tag{13-15}$$

In principle, operating at acidic conditions could be the preferred option as chlorine is the strongest oxidant, followed by HClO. Accordingly, higher pH values will improve the EO of pollutants, as HClO and ClO^- are almost unaffected by the desorption of gases, and they can act as oxidizing reagents in the total volume of wastewater (Cañizares et al. 2006).

Energy consumption. Energy consumption should be reduced to minimize power costs. The total power requirement contributes to both electrolysis and movement of either the solution or the electrode. The essential parameters in reducing power are the design of the electrode and the cell. So, a very open flow-through porous electrode will have a low-pressure drop linked with it, giving rise to modest pumping costs and facilitating reactor sealing. A high-surface-area electrode, which itself is a turbulence promoter in a bed electrode, will give rise to a moderately high mass transfer coefficient and an active area without the need for high flow rates through the cell; the pumping cost will again be moderately low (Klamklang et al. 2012). The following aspects should be considered:

- Counter electrode reaction should be considered to reduce the reversible cell voltage. Thus, a suitable and stable electrode material is required;
- Overpotentials at both electrodes should be minimized by using electrocatalysts;
- Electrodes, current feeders, and connectors should be prepared from greatly conducting materials;
- Electrode and cell design should allow a small interelectrode or electrode membrane gap. The electrode may touch the membrane as in zero gap or solid polymer electrolyte cells; and
- Separator should be avoided by a suitable selection of the counter electrode chemistry, or a thin conductive membrane should be applied.

The energy efficiency of a reactor depends on the cell voltage, which can be expressed as a number of voltage components

$$E_{\text{cell}} = E_{\text{cell}}^{e} - \sum |n| - \sum IR \tag{13-16}$$

where I is the current, and R is the flow-through resistances. It is necessary to minimize the overpotential and ohmic components of voltage to maximize energy efficiency. Appropriate ways for achieving this purpose are the use of conductive, catalytic electrodes and membranes, the use of small interelectrode (or membrane-electrode) gaps, and a careful choice of the counter electrode chemistry to minimize the equilibrium cell voltage term.

13.4 INTEGRATION OF ELECTRO-OXIDATION IN WASTEWATER TREATMENT PLANTS

In recent decades, EO has attracted global attention for the treatment of wastewater because conventional biological and chemical methods fail to eliminate refractory and nonbiodegradable substances and because stricter limitations have been imposed by the new legislation. For instance, US Environmental Protection Agency (EPA) published a *White Paper Aquatic Life Criteria for Contaminants of Emerging Concern: Part I Challenges and Recommendations* in 2008, describing the technical issues and recommendations intended as a basis for revising the EPA's 1985 guideline. Through these modifications, the Agency should be better equipped to address emerging contaminants, including pharmaceuticals and personal care products, and develop criteria for ambient water quality to protect aquatic life (Workgroup 2008). EO is well known as an environmentally friendly technology, being able to mineralize diverse organic pollutants, including toxic and recalcitrant ones. However, some limitations of this method have necessitated its use in combination with other treatment technologies. The major impediments to EO are the consumption of energy, the high cost of adequate electrodes, and the risk of harmful generation of by-products. The high energy consumption in EO can be considered as a limiting factor for its full-scale commercial applications. Besides, EO is not applicable for the treatment of effluents with a high amount of suspended solids (Woisetschläger et al. 2013). Therefore, the use of this technique, in combination with other techniques, may offset the disadvantages of employing EO alone. Combined wastewater treatment systems are utilized in various applications such as industrial textile wastewater (electro-coagulation and EO in a continuous multistage reactor), landfill leachate (physicochemical pretreatment following EO), tannery wastewater (EO combined with a biological process), almond wastewater (electro-coagulation and EO), restaurant wastewater (combining electro-coagulation and EO) (Garcia-Segura et al. 2018, Mandal et al. 2017).

The coupled systems employed for EO-based wastewater treatment can be classified as pretreatment, post-treatment, and integrated treatment. Choosing the best option depends on the influent characteristics and the requirements for the final treated water. By the coupling of distinct processes, one may obtain the advantage of each process in the merged system. For instance, the biological process and EO can be employed as a treatment unit, and one may secure the twin benefits of the reduction of operating cost and shorter retention time simultaneously (Anglada et al. 2009).

13.4.1 Pretreatment

EO depends on treatment objectives, and it can be carried out either for complete mineralization of all contaminants or for partial removal of specified target pollutants and convention to intermediates. Because the total mineralization with EO might be highly expensive, employing EO as a pretreatment to convert the initial biologically recalcitrant organics to readily biodegradable intermediates sounds more reasonable (Mantzavinos and Psillakis 2004). Coupling EO with biological post-treatment improves the removal efficiency compared with each individual process by increasing the biodegradability index (BOD_5/COD). EO converts higher molecular organics to lower molecular ones, which leads to an augmentation of the BOD_5/COD ratio and consequently to an improvement in the performance of the subsequent process (Mandal et al. 2017). This procedure was applied for landfill leachate treatment. EO eliminated the higher molecular organics; thus, the EO effluent had a higher biodegradability. Then, anaerobic digestion removed the residual organic compounds (Li et al. 2007).

In addition to enhancing the biodegradability index, coupling the EO process with the biological process reduces the reaction time. EO prior to aerobic digestion was employed for the remediation of waste-activated sludge. By applying EO before aerobic treatment, the sludge retention time reduced from 23.5 days to 17.5 days. It directly brought down the costs by lowering the reactor volume and reducing utility consumption (Song et al. 2010). Fontmorin et al. (2014) employed a combined electrochemical oxidation and biological process to remove a commercial herbicide solution called U46D®. This investigation indicated a considerable improvement in the biodegradability of the treated wastewater after the EO pretreatment and a reduction in the retention time for complete mineralization by subsequent biological treatment (Fontmorin et al. 2014). Isik et al. (2020) investigated the effect of employing electrochemical pretreatment on the fungal treatment of pistachio processing wastewater. They used BDD/stainless-steel electrode pairs for EO pretreatment before fungal treatment. They reported that an improvement in the biodegradability of the treated wastewater after EO led to a significant reduction in the required time for complete mineralization by biological post-treatment. They achieved COD and total phenol removal efficiencies of 90.1% and 88.7%, respectively, and recommended this process for the treatment of wastewater with high organic pollutants and low flow (Isik et al. 2020). Studies have demonstrated the advantage of applying EO as a pretreatment to increase the biodegradability index of wastewater and, consequently, reduce the reaction time in biological post-treatment.

13.4.2 Post-Treatment

EO can be used as a post-treatment, in which the treated effluents contain a substantial amount of bioresistant and refractory pollutants. In many cases, wastewater is partially treated by other water technologies and requires further remediation to achieve the determined standard. In general, EO is employed as a final polishing step to eliminate trace organics and inorganic ions completely (Garcia-Segura et al. 2018). Some of the examples of this method are as follows:

1. Because EO can successfully eliminate organic contaminants, it has been utilized as a post-treatment of Fenton oxidation to remove residual organic substances. Besides, as the Fenton process has proved ineffective in oxidizing ammonia, EO has been employed as a polishing step for the abatement of this pollutant (Anglada et al. 2009).
2. Another hybridized process is membrane bioreactor (MBR) with the EO process as a post-treatment stage for stabilized landfill leachates. Biorefractory compounds were oxidized, and the final treated effluents met the environmental discharge standards. By the combination of these processes, the removal efficiency increased, and energy consumption reduced by more than 50% (Feki et al. 2009).
3. For the treatment of landfill leachate, EO could be employed as a pretreatment or polishing step with biological processes. In most studies, EO acted as a post-treatment technology to eliminate residual refractory organics in the effluents after biological treatment. This method led to an increase in removal efficiency, reduced energy consumption, and brought down operating costs (Mandal et al. 2017).
4. For the treatment of carwash wastewater, a two-step process of electrochemical coagulation and EO was carried out. The electro-coagulation method was able to remove COD by 75%. By applying EO as a post-treatment process, the remaining organics from the electro-coagulation unit were mineralized completely (Panizza and Cerisola 2010).
5. Szpyrkowicz et al. (2005) employed the biological aerobic sludge process followed by EO as a final polishing stage to remove COD from tannery wastewater. This process led to an 80% reduction of the plant's total volume compared with the single biological process. However, this design increased energy consumption by 20 kW·h (Szpyrkowicz et al. 2005).
6. Mostafazadeh et al. (2019) employed various combined treatments, including ultrafiltration (UF), followed by EO, by applying an anode type of BDD and a cathode type of graphite to eliminate nonylphenol ethoxylates (NPEO3-17) from real laundry wastewater. They reported that the EO treatment of UF filtrate led to an improved removal efficiency of 97% for NPEO3-17 under optimum operating conditions of a current intensity of 12 A and treatment time of 45 min (Mostafazadeh et al. 2019).

The analysis thus far indicates that EO is a promising polishing method because it can completely mineralize biorefractory compounds remaining in the effluent after pretreatment. However, the major challenges to implementing the full-scale integrated approach are reducing the operation cost (especially energy cost) and developing stable and efficient electrode material.

13.4.3 Integrated Treatment

The difference between an integrated system and pre- or post-treatment is that in an integrated design, processes are coupled, but in pre- or post-treatment, processes are operated sequentially. Examples of the integrated processes are given as follows:

1. In a study, EO with biological oxidation in different operational configurations such as individual, combined, and integrated schemes for the removal of COD and a reactive dye of Procion Scarlet in synthetic effluents was investigated. The COD and color removal of the integrated system was the same as that of the individual EO process, but the integrated method required the application of less current density (10.25 A/dm^2 for the EO process alone and 6.84 A/dm^2 for the integrated one). The less-applied current density was interpreted as the synergy effect of integrated electrochemical and biological degradation (Senthilkumar et al. 2012).
2. The coupled system of EO-ozonation was utilized for the treatment of high organic loading industrial wastewater. Ozonation alone reduced the COD of wastewater by 45%. On the contrary, by applying EO and using BDD electrodes, COD was removed by 99.9% over 2 h. However, by the coupled process, not only was COD reduced by the same 99.9% within 1 h, but also, without the addition of reagents, color and turbidity were eliminated. In the coupled system, the electrodes of the EO reactor were employed in the ozonation reactor (García-Morales et al. 2013).
3. García et al. (2018) carried out a hybrid electrochemical-granular-activated carbon system to remove pollutants from synthetic greywater. They found that the 3D reactor hybrid system was capable of eliminating color. However, adsorption resulted in a removal efficiency of 71%. Therefore, they reported that the synergy between EO and granular-activated carbon adsorption led to an improvement in the treatment performance of 3D systems up to 21% and 23% of COD and TOC removal, respectively, compared with the granular-activated carbon adsorption and 2D EO process (García et al. 2018).
4. Xu et al. (2019) also utilized novel layer-by-layer carbon nanotube (CNT)/PbO_2 anodes in a hybrid electrochemical oxidation/adsorption process. They employed this process for the low-concentration sodium pentachlorophenate treatment. They achieved a remarkable removal efficiency of 73.8% after 5 min compared with PbO_2, which was only 9.9%. The reasons for such an improved efficiency were the synergistic effect of

electrochemical oxidation/adsorption, the presence of more active sites, lower electron transfer resistance, and the higher direct oxidation capacity of the CNT/PbO_2 anode (Xu et al. 2019).

The synergy of an integrated process, as opposed to the effectiveness of the individual ones, is the main factor that attracts researchers' attention. Besides, an integrated approach can be considered as a complementary process that might overcome the limitations of individual systems.

13.5 TYPES OF WASTEWATERS AND POLLUTANTS

EO is able to nonselectively mineralize various organic pollutants to water, carbon dioxide, and inorganics, or produce innocuous substances. Several characteristics of EO lead to its use for the treatment of wastewater, such as (i) mild operation conditions, (ii) compact reactors, (iii) no additional reagents, (iv) production of no secondary waste streams, (v) easy combinability with other conventional treatment technologies, (vi) high removal efficiency, and (vii) versatility (Garcia-Segura et al. 2018).

EO is successfully applied mainly for the treatment of wastewater released by various industries, such as petrochemical, pulp and paper, and pharmaceuticals, textiles, tannery, food, agroindustry (olive oil and dairy manure), almond as well as landfill leachate, and urban wastewater (Anglada et al. 2009).

The textile industry consumes approximately 21 to 377 m^3 of water per ton of woven fabric and releases almost the same amount of wastewater. The treatment of textile wastewater is an arduous process because of (a) a high amount of total dissolved solids, (b) the presence of heavy metals, (c) the presence of nonbiodegradable organic dyes, and (d) the existence of free chlorine. The textile effluent contains high COD (150 to 10,000 mg/L) and BOD (100 to 4,000 mg/L) content (Kalra et al. 2011). EAOPs and, more specifically, EO, have been reported as powerful technologies to mineralize organic dyes in textile wastewater (Ling et al. 2016, Martínez-Huitle et al. 2012, Solano et al. 2013) compared with conventional processes that have hardly succeeded in removing these kinds of persistent pollutants.

EO has also been carried out for the treatment of upstream and downstream petroleum wastewater. Petroleum wastewater is typically characterized by high COD and turbidity and contains a mixture of hydrocarbons and various inorganic substances such as Mg^{2+}, Ca^{2+}, S^{2-}, Cl^-, and SO_4^{2-}. The most harmful pollutants of the effluent are aromatic compounds and their derivatives that are almost always highly toxic and carcinogenic. Because of the low concentration of these pollutants in effluents and their recalcitrant nature, EO can remove these contaminants more efficiently than the conventional methods (Yan et al. 2011).

The pulp and paper mill industry also consumes a large amount of water (as high as 60 m^3 per ton of paper produced). Consequently, a considerable amount of wastewater is generated from the plants, which commonly has a high COD, low biodegradability because of the presence of refractory matters, as well as various organic and inorganic substances. Such compounds comprise adsorbable organic halides, colored compounds, phenolic compounds, additives, and solvents. Consequently, the effluent might run the severe risk of environmental contamination because of an incomplete degradation of toxic chemicals. Owing to the convenience and typically greater efficiency of removing recalcitrant substances than other traditional methods, EO is one of the most useful and interesting processes for treating the wastewater of the pulp and paper mill industry (Hermosilla et al. 2015).

The tanning industry releases wastewater with highly loaded pollutants. This wastewater includes various chemicals like sodium sulfite, chromium sulfate, soda ash, ammonium chloride, sodium chloride, sulfuric acid, dyes, resins, waxes, and a wide range of solvents and additives. Because of the wide variety of chemicals, conventional methods are not able to eliminate contaminants completely. In contrast, EO succeeds in oxidizing pollutants efficiently and recently received more attention because of its robust performance in wastewater treatment at ambient conditions (Garcia-Segura et al. 2018).

Landfill leachate contains large amounts of organic contaminants and ammonia-nitrogen (as the two principal chemicals), along with inorganic salts, chlorinated organic compounds, and heavy metals. EO can be employed as a pretreatment to increase the biodegradability index of leachate and stabilize it for biological treatments. Besides, it is used as a final polishing step to eliminate trace organics and inorganic ions (Mandal et al. 2017).

As mentioned, EO is a robust technology for the removal of recalcitrant and toxic pollutants when conventional methods are not able to eliminate this kind of contamination. EO can abolish COD, color, ammonia, and microorganisms and can completely mineralize highly refractory organic pollutants such as pharmaceuticals, pesticides, persistent organic pollutants, and hydrocarbons. Also, it can eliminate a wide variety of contaminates, including anions, surfactants, heavy metals, and phenolic compounds.

13.5.1 Chemical Oxygen Demand

The most economical method for the removal of organic load characterized by COD is biological treatment. However, this method is not able to eliminate all types of organic substances. Because of the complete mineralization of pollutants, advanced oxidation processes have been employed recently. EO is a very efficient process for COD abatement. Take into account, choosing the appropriate anode required to consider both economical and efficient oxidation rate. It is essential to consider both the economic and efficient oxidation rates when choosing the appropriate anode.

Because of the amount of organic load, different technologies exist for wastewater treatment. When the amount of organic load is 0.001 to 1 g O_2/L, biological treatment is applicable. For COD in the range of 0.1 to 9 g O_2/L, advanced oxidation processes are suitable. Wet oxidation and incineration are applicable for the amount of organic load in the range of 90 to 400 and 100 to 1,000 g O_2/L, respectively. EAOPs can be employed for a wide range of COD amounts. However, the most efficient usage span is 0.1 to 25 gO_2 /L (Fryda et al. 2003).

Un et al. (2008) evaluated the influence of current density, sodium chloride concentration (as a supporting electrolyte), recirculation rate of wastewater, and temperature on the rate of pollutant abatement in the treatment of olive mill wastewater by EO. They achieved a reduction of COD from 41,000 to 167 mg/L^{-1}, which meets discharge requirements. Besides, they reported the running cost of Euro 0.88 per kg COD after seven h of electrolysis with a current density of 135 mA cm^{-2}, sodium chloride concentration of 2 M, and recirculation rate of wastewater 7.9 cm^3 s^{-1} at 40°C (Un et al. 2008). Gargouri et al. (2014) used EO for the treatment of real petroleum wastewater. They applied lead dioxide supported on tantalum (Ta/PbO_2) and BDD anodes in a batch system. They reported 85% and 96% COD removal efficiencies for PbO_2 and BDD after 11 and 7 h, respectively. Besides, they concluded that the BDD anode led to a higher removal efficiency of hydrocarbons compounds from petroleum wastewater (Gargouri et al. 2014). Azarian et al. (2018) employed a combined EC/EC process to treat tannery wastewater for reuse in the tannery industry or in the agricultural sector. The COD of the effluent was 5 mg/L, which met the standard quality for reuse in agricultural and tannery processes. Besides, the energy consumption of the coupled process decreased 70% compared with the individual EO process (Azarian et al. 2018).

13.5.2 Persistent Organic Pollutants

The presence of persistent organic pollutants (POPs) in water has emerged as a great environmental concern because of persistence against biological degradation, an inclination for bioaccumulation in fatty tissues, and high toxicity. Some POPs have been produced intentionally by industry for a broad range of applications. Others are generated unpremeditatedly as a by-product of diverse activities. These contaminants exert detrimental effects on the environment because of their toxicity, carcinogenicity, and mutagenicity, even at low concentrations. Owing to the biorefractory character of POPs, conventional water treatments are not able to eliminate them. The development of EO offers new-generation technologies for the removal of POPs to reuse treated water (Eljarrat and Barcelo 2003).

Dietrich et al. (2017) employed ultrasound and electrochemical oxidation as a hybrid system to degrade bisphenol A in synthetic wastewater. They used a low-frequency (24 kHz) ultrasound horn and two BDD electrodes. They achieved 93% degradation with a treatment time of 30 min and an initial concentration of 1 mg L^{-1} of bisphenol A (Dietrich et al. 2017). Ren et al. (2019) also used ultrasound and electrochemical oxidation to remove chlorpyrifos (pesticide) in an aqueous solution. They investigated the effect of voltage, initial electrolyte concentration, temperature, and ultrasonic power to obtain the highest removal efficiency. The optimal conditions were a gained voltage of 20 V, electrolyte concentration (Na_2SO_4) of 2 mg L^{-1}, and ultrasonic power of 200 W at 20°C. In these conditions and applying stainless steel as electrodes, 93% degradation was reported (Ren et al. 2019).

13.5.3 Dye

The effluents of a large variety of industries such as textile, leather tanning, paper, and food consist of high color and COD. The release of these colored substances into the environment causes irreparable damage because of their toxic and carcinogenic characteristics. The color structure contains two parts, a reactive group that reacts with the fiber, and a chromophore group that gives the color. Azo (R–N=N–R) is the most utilized chromophore group in industries, with approximately 65% of the world's production (Sala and Gutiérrez-Bouzán 2012). Other dyes used in industries are anthraquinone, indigoid, triphenylmethyl, sulfur, and phthalocyanine compounds. Because colors are stable against biological degradation, EO is more efficient than adsorption, coagulation, filtration, and biological treatment. The conventional methods cause sludge formation, membrane fouling, and incomplete mineralization (Gutiérrez and Crespi 1999, Martínez-Huitle and Brillas 2009).

EO has been considered as a powerful technology for mineralizing organic dyes in synthetic and real wastewater. Cotillas et al. (2018) investigated the removal of Procion Red MX-5B dye by electrochemical oxidation with BDD anodes. They evaluated the effect of current density, flow rate, initial pH, and supporting electrolyte on dye and organic matter removal. They achieved a complete degradation of dye and COD after 240 min of treatment time, regardless of the applied current density. In regard to pH, its variation did not affect the dye removal significantly. However, working at a neutral pH led to the complete removal of COD. They concluded that the process reached the highest efficiency when using low current densities, neutral pH, and high flow rates (Cotillas et al. 2018). Sartaj et al. (2020) performed two techniques, photolytic and EO, to remove Allura red and erythrosine dyes from synthetic wastewater. They reported that the photolytic process was slow as it took 1 h for achieving a removal efficiency of 95.69% for Allura red and 6 h for achieving 90.84% erythrosine removal. However, they achieved more than 99% dye removal efficiency after 5 and 10 min, respectively, for Allura red and erythrosine dye by EO using a titanium ruthenium oxide anode and a stainless-steel cathode. They concluded that EO was a more efficient approach for dye degradation than the photolytic process because of the generation of powerful oxidants (Sartaj et al. 2020).

13.5.4 Heavy Metals

Wastewater streams discharged from mining operations, electroplating industry, electronic device manufacturing, paper and pulp industry, and tanneries may contain toxic heavy metals such as lead, copper, chromium, cadmium, nickel, zinc, manganese, and arsenic in high concentrations. Heavy metals are one of the most widespread contaminants in the environment. They are nondegradable, and their accumulation inside living organisms can cause severe health problems. The conventional physicochemical treatment methods such as chemical precipitation, adsorption, and ion exchange are not able to eliminate them easily. EO can offer an efficient removal method for heavy metal remediation from wastewater. For instance, electrochemical treatment in the 3D system removes heavy metals by both adsorption and electrochemical redox reactions. It has been successfully employed for the removal of toxic metal ions for many years. The removal efficiency of some heavy metals is close to 100% with this approach, which demonstrates that this technique is a robust one for the elimination of heavy metals from wastewater (Zhang et al. 2013).

13.5.5 Pharmaceuticals

Pharmaceutical waste contains drug remnants with high BOD, COD, and pharmaceutically active compounds like hormones, antibiotics, toxic substances, and surfactants. Pharmaceuticals can enter the environment by manufacturing emissions, human consumption, excretion, and disposal of drugs into sinks. The existence of pharmaceutical residues in wastewater becomes an environmental challenge because of its destructive effects on living organisms. They can contribute to endocrine disruption, antibiotic resistance, carcinogenicity, and toxicity. Physical and physicochemical treatments, including adsorption, coagulation-flocculation, and membrane filtration and biological treatment, are not appropriate for the removal of pharmaceutical waste because of the highly resistant nature of pharmaceuticals. In light of this, EO has been broadly investigated for treating pharmaceutical wastewater because of its high versatility, efficiency, and environmental compatibility (HealthCareWithoutHarm 2018, Särkkä et al. 2015).

Thokchom et al. (2015) employed a hybrid reactor system including sonochemistry and EO to degrade ibuprofen in synthetic wastewater. The performance of the process was evaluated by making changes in electrolytes, frequency, applied voltage, ultrasonic power density, and temperature in the reactor with a platinum electrode. They achieved the highest removal efficiency of 89.32% by adding NaOH as an electrolyte, with an electrical voltage of 30 V, an ultrasound frequency of 1,000 kHz, and a power density of 100 W L^{-1} at 298 K in 1 h (Thokchom et al. 2015). Loos et al. (2018) investigated the degradation of some pharmaceutical substances like iopromide, sulfamethoxazole, 17-α-ethinylestradiol, and diclofenac in simulated wastewater and real hospital wastewater by EO via BDD electrodes. They evaluated the effect of the flow rate, applied current, and initial concentration of substances. They concluded that the degradation of iopromide was considerably slower than the others. Besides, applied current had a significant effect on degradation. However, the impact of flow rate was not remarkable and it had only a moderate effect on the removal of 17-alpha-ethinylestradiol and diclofenac (Loos et al. 2018).

13.5.6 Ammonia

Ammonia is a common pollutant found in wastewater streams because of its extensive use in fertilizers. Moreover, it may be entered into wastewater from municipal and other industrial activities. The removal or recovery of ammonia and its derivative compounds has become challenging because of its detrimental impact on the environment. As ammonia is highly toxic to aquatic animals, it is necessary to remove the excess amount of this gas present in wastewater to protect marine species. In general, biological treatment methods like air stripping, ion exchange, biological nitrogen removal systems, and breakpoint chlorination have been carried out for the removal of these pollutants with different degrees of success. These types of units are economical to operate compared with EO, but they also have their own limitations. For instance, the operation of the system is inhibited when the wastewater temperature falls below 10°C. Therefore, the EO technology for the removal of ammonia has attracted attention because it eliminates ammonia pollutants in wastewater efficiently. However, the widespread use of EO for ammonia abatement is limited because of the poor performance of electrocatalysts and the considerable costs involved in high Pt loading. The efficiency of the electrochemical removal of ammonia and operational cost ultimately depend on the type of anode material used and the electric voltage applied (Ghimire et al. 2019, Zhong et al. 2013).

13.5.7 Phenolic Compounds

Phenol and phenolic compounds are used in many industries, including dye, plastic, polymeric resin, pharmaceutical, and oil refinery. Therefore, effluents flowing from the said industries contain different concentrations of phenolic compounds. Petroleum wastewater typically comprises 500 to 1,500 mg/L phenol. Moreover, phenol concentration in coking plants is around 200 to 1,200 mg/L. These pollutants are highly dangerous because of their low biodegradability and high toxicity. Because of the high toxicity of phenol, severe restrictions are imposed for the maximum acceptable concentration in drinking and irrigation water and discharge wastewater. The removal process of phenolic compounds of wastewater can be classified into three categories: physical removal, biological treatment, and advanced oxidation processes. Many problems associated with the aforementioned methods have been reported, such as high cost, low efficiency, and generation of toxic products. Compared with conventional methods, EO has better removal efficiency because of the nature of these contaminates (El-Ashtoukhy et al. 2013). The mineralization of phenol to H_2O and CO_2 in the EO process follows the following pathway (Zhang et al. 2013):

$$\text{phenol} \xrightarrow{{}^{\bullet}\text{OH}} \text{quinone} \xrightarrow{{}^{\bullet}\text{OH}} \text{carboxylicacids} \xrightarrow{{}^{\bullet}\text{OH}} CO_2 + H_2O$$

13.6 CHALLENGES AND RECOMMENDATIONS

EO provides advantages for wastewater treatment over other approaches, such as the capability to degrade a wide range of contaminants, increase the biodegradability index, a relatively simple process to use, utilizing a clean agent, and environmental compatibility and versatility. Besides, EO can be adapted to various applications and can be easily coupled with other technologies. Although

EO has attracted attention in the last decade for the treatment of wastewater and has secured promising results to overcome the main impediments for pollutant removal, some aspects require more investigation to employ EO on a large scale.

One of the main drawbacks of EO is the relatively high operational costs, including electricity and the value of the electrode used. High energy consumption is a limiting factor for full-scale commercial applications. High current densities are often applied to obtain admitted removal efficiency but lead to increased energy consumption. Besides, the price of an effective electrode such as BDD is very high. For instance, the price of a bipolar BBD silicon-based electrode with dimensions of 100 × 100 × 2 mm is around USD 1,050 (NeoCoat company). Furthermore, electrode fouling may occur because of the deposition of material on the electrode surface. Thus, it is essential to take into account the service lifetime of electrodes. The application of electrochemical treatment has been slowed down by the relatively high costs of operation (Radjenovic and Sedlak 2015).

The next probable disadvantage is related to the formation of harmful by-products. During the chlorine-mediated oxidation process, especially for high chlorides containing wastewater, chlorates, perchlorates, and organochlorinated derivatives are generated. The generation of these highly toxic compounds, which are often more persistent and harmful than the parent compound, is a limiting factor in the scale-up process of EO and should, therefore, be addressed before industrial-scale installations (Garcia-Segura et al. 2018).

One possible approach to circumvent this limitation is utilizing this technology in combination with other methods. The EO process can be coupled with other biological/physicochemical processes to overcome the drawback of high energy consumption and consequently cut down the treatment costs. The hybridized approach improves both the purification result and the economic aspect of the process. Moreover, it is recommended to couple or integrate EO with adsorption (employing granular-activated carbon) to minimize the release of chlorinated organic by-products (Radjenovic and Sedlak 2015). Three-dimensional particle and granular electrodes can be utilized for combined adsorption-EO. It is reported that they may be suitable for the treatment of wastewater with a high potential of formation of halogenated organic by-products (Radjenovic and Sedlak 2015).

Finally, it is recommended to reduce operating costs by developing a new electric power source for the system such as solar energy or using renewable energy sources to power EO. Besides, the recent design of novel 3D electrodes allows the adoption of a more efficient EO treatment strategy. It is necessary to develop more active anodes with highly available electroactive surface areas and reasonable prices (Chaplin 2018).

13.7 CONCLUSION

By increasing social awareness on protecting the environment, purifying wastewater and reusing it in industries or as potable water is highlighted. The efficient and cost-effective removal of organic compounds, which is hardly doable by conventional methods, is the main challenge. EO is a promising technology for the treatment of wastewater, which can be easily combined with other approaches to enhance the removal percentage and achieve cost-effectiveness. EO might be applied as a pretreatment to increase the biodegradability index of wastewater containing recalcitrant organics, and thereafter by a biological post-treatment, a complete removal of pollutants can be achieved. Besides, it can be performed as a final polishing step to eliminate trace organic and inorganic ions. Finally, EO has been used in an integrated system to take advantage of the synergistic effect of various technologies to improve removal efficiency. Because EO can mineralize various organic pollutants, it has been found feasible and attractive to employ it in the treatment of different kinds of wastewater to degrade refractory organic contaminants such as pharmaceuticals and pesticides or abolish others like color, heavy metals, and ammonia. Although EO has enormous advantages, it also has drawbacks that prevent its use on an industrial scale. Therefore, more investigation must be performed before scale-up, which is essential in the future.

NOMENCLATURE

BDD = Boron-doped diamond
BOD = Biological oxygen demand
CD = Current density
CNT = Carbon nanotube
COD = Chemical oxygen demand
EC = Electro-coagulation
EO = Electro-oxidation
EPA = Environmental protection agency
POP = Persistent organic pollutant
UF = Ultrafiltratio

References

Anglada, A., A. Urtiaga, and I. Ortiz. 2009. "Contributions of electrochemical oxidation to waste-water treatment: Fundamentals and review of applications." *J. Chem. Technol. Biotechnol.* 84 (12): 1747–1755.

Azarian, G., M. Miri, and D. Nematollahi. 2018. "Combined electrocoagulation/electrooxidation process for the COD removal and recovery of tannery industry wastewater." *Environ. Prog. Sustainable Energy* 37 (2): 637–644.

Cañizares, P., R. Paz, J. Lobato, C. Sáez, and M. Rodrigo. 2006. "Electrochemical treatment of the effluent of a fine chemical manufacturing plant." *J. Hazard. Mater.* 138 (1): 173–181.

Chaplin, B. P. 2018. "Advantages, disadvantages, and future challenges of the use of electrochemical technologies for water and wastewater treatment." In *Electrochemical water and wastewater treatment*, edited by C. A. Martínez-Huitle, M. A. Rodrigo, and O. Scialdone, 451–494. Amsterdam, Netherlands: Elsevier.

Chen, G. 2004. "Electrochemical technologies in wastewater treatment." *Sep. Purif. Technol.* 38 (1): 11–41.

Chiang, L.-C., J.-E. Chang, and T.-C. Wen. 1995. "Indirect oxidation effect in electrochemical oxidation treatment of landfill leachate." *Water Res.* 29 (2): 671–678.

Cong, V. H., Y. Sakakibara, M. Komori, N. Kishimoto, T. Watanabe, I. Mishima, et al. 2016. "Recent developments in electrochemical technology for water and wastewater treatments." *J. Water Environ. Technol.* 14 (2): 25–36.

Cotillas, S., J. Llanos, P. Cañizares, D. Clematis, G. Cerisola, M. A. Rodrigo, et al. 2018. "Removal of Procion Red MX-5B dye from wastewater by conductive-diamond electrochemical oxidation." *Electrochim. Acta* 263: 1–7.

Dietrich, M., M. Franke, M. Stelter, and P. Braeutigam. 2017. "Degradation of endocrine disruptor bisphenol A by ultrasound-assisted electrochemical oxidation in water." *Ultrason. Sonochem.* 39: 741–749.

Drogui, P., J.-F. Blais, and G. Mercier. 2007. "Review of electrochemical technologies for environmental applications." *Recent Pat. Eng.* 1 (3): 257–272.

El-Ashtoukhy, E., Y. El-Taweel, O. Abdelwahab, and E. Nassef. 2013. "Treatment of petrochemical wastewater containing phenolic compounds by electrocoagulation using a fixed bed electrochemical reactor." *Int. J. Electrochem. Sci.* 8 (1): 1534–1550.

Eljarrat, E., and D. Barcelo. 2003. "Priority lists for persistent organic pollutants and emerging contaminants based on their relative toxic potency in environmental samples." *TrAC, Trends Anal. Chem.* 22 (10): 655–665.

Feki, F., F. Aloui, M. Feki, and S. Sayadi. 2009. "Electrochemical oxidation post-treatment of landfill leachates treated with membrane bioreactor." *Chemosphere* 75 (2): 256–260.

Fontmorin, J.-M., J. Siguié, F. Fourcade, F. Geneste, D. Floner, I. Soutrel, et al. 2014. "Combined electrochemical treatment/biological process for the removal of a commercial herbicide solution, U46D®." *Sep. Purif. Technol.* 132: 704–711.

Fryda, M., T. Matthée, S. Mulcahy, M. Hofer, L. Schafer, and I. Troster. 2003. "Applications of DIACHEM electrodes in electrolytic water treatment." *Interface* 12 (1): 40–44.

García, E. A., M. Agulló-Barceló, P. Bond, J. Keller, W. Gernjak, and J. Radjenovic. 2018. "Hybrid electrochemical-granular activated carbon system for the treatment of greywater." *Chem. Eng. J.* 352: 405–411.

García-Morales, M., G. Roa-Morales, C. Barrera-Díaz, B. Bilyeu, and M. Rodrigo. 2013. "Synergy of electrochemical oxidation using boron-doped diamond (BDD) electrodes and ozone (O_3) in industrial wastewater treatment." *Electrochem. Commun.* 27: 34–37.

Garcia-Segura, S., J. D. Ocon, and M. N. Chong. 2018. "Electrochemical oxidation remediation of real wastewater effluents—A review." *Process Saf. Environ. Prot.* 113: 48–67.

Gargouri, B., O. D. Gargouri, B. Gargouri, S. K. Trabelsi, R. Abdelhedi, and M. Bouaziz. 2014. "Application of electrochemical technology for removing petroleum hydrocarbons from produced water using lead dioxide and boron-doped diamond electrodes." *Chemosphere* 117: 309–315.

Ghime, D., and P. Ghosh. 2019. "Removal of organic compounds found in the wastewater through electrochemical advanced oxidation processes: A review." *Russ. J. Electrochem.* 55 (7): 591–620.

Ghimire, U., M. Jang, S. P. Jung, D. Park, S. J. Park, H. Yu, et al. 2019. "Electrochemical removal of ammonium nitrogen and cod of domestic wastewater using platinum coated titanium as an anode electrode." *Energies* 12 (5): 883.

Gutiérrez, M., and M. Crespi. 1999. "A review of electrochemical treatments for colour elimination." *Color. Technol.* 115 (11): 342–345.

HealthCareWithoutHarm. 2018. “Pharmaceutical pollution.” https://noharm-europe.org/.

Hermosilla, D., N. Merayo, A. Gascó, and Á. Blanco. 2015. “The application of advanced oxidation technologies to the treatment of effluents from the pulp and paper industry: A review.” *Environ. Sci. Pollut. Res.* 22 (1): 168–191.

Isik, Z., E. B. Arikan, Y. Ozay, H. D. Bouras, and N. Dizge. 2020. “Electrocoagulation and electrooxidation pre-treatment effect on fungal treatment of pistachio processing wastewater.” *Chemosphere* 244: 125383.

Kalra, S. S., S. Mohan, A. Sinha, and G. Singh. 2011. “Advanced oxidation processes for treatment of textile and dye wastewater: A review.” In *Proc., 2nd Int. Conf. on Environmental Science and Development*, 271–275. Singapore: IACSIT Press.

Klamklang, S., H. Vergnes, K. Pruksathorn, and S. Damronglerd. 2012. *Electrochemical incineration of organic pollutants for wastewater treatment: Past, present and prospect.* London: INTECH.

Li, T., X. Li, J. Chen, G. Zhang, and H. Wang. 2007. “Treatment of landfill leachate by electrochemical oxidation and anaerobic process.” *Water Environ. Res.* 79 (5): 514–520.

Li, W., Q. Zhou, and T. Hua. 2010. “Removal of organic matter from landfill leachate by advanced oxidation processes: A review.” *Int. J. Chem. Eng.* 2010: 270532.

Ling, Y., J. Hu, Z. Qian, L. Zhu, and X. Chen. 2016. “Continuous treatment of biologically treated textile effluent using a multi-cell electrochemical reactor.” *Chem. Eng. J.* 286: 571–577.

Loos, G., T. Scheers, K. Van Eyck, A. Van Schepdael, E. Adams, B. Van der Bruggen, et al. 2018. “Electrochemical oxidation of key pharmaceuticals using a boron doped diamond electrode.” *Sep. Purif. Technol.* 195: 184–191.

Mandal, P., B. K. Dubey, and A. K. Gupta. 2017. “Review on landfill leachate treatment by electrochemical oxidation: Drawbacks, challenges and future scope.” *Waste Manage.* 69: 250–273.

Mantzavinos, D., and E. Psillakis. 2004. “Enhancement of biodegradability of industrial wastewaters by chemical oxidation pre-treatment.” *J. Chem. Technol. Biotechnol.* 79 (5): 431–454.

Martínez-Huitle, C. A., and E. Brillas. 2009. “Decontamination of wastewaters containing synthetic organic dyes by electrochemical methods: A general review.” *Appl. Catal. B* 87 (3–4): 105–145.

Martínez-Huitle, C. A., E. V. dos Santos, D. M. de Araújo, and M. Panizza. 2012. “Applicability of diamond electrode/anode to the electrochemical treatment of a real textile effluent.” *J. Electroanal. Chem.* 674: 103–107.

Mostafazadeh, A. K., A. T. Benguit, A. Carabin, P. Drogui, and E. Brien. 2019. “Development of combined membrane filtration, electrochemical technologies, and adsorption processes for treatment and reuse of laundry wastewater and removal of nonylphenol ethoxylates as surfactants.” *J. Water Process Eng.* 28: 277–292.

Panizza, M., and G. Cerisola. 2010. “Applicability of electrochemical methods to carwash wastewaters for reuse. Part 2: Electrocoagulation and anodic oxidation integrated process.” *J. Electroanal. Chem.* 638 (2): 236–240.

Radjenovic, J., and D. L. Sedlak. 2015. “Challenges and opportunities for electrochemical processes as next-generation technologies for the treatment of contaminated water.” *Environ. Sci. Technol.* 49 (19): 11292–11302.

Rajeshwar, K., and J. G. Ibanez. 1997. *Environmental electrochemistry: Fundamentals and applications in pollution sensors and abatement.* Amsterdam, Netherlands: Elsevier.

Ren, Q., C. Yin, Z. Chen, M. Cheng, Y. Ren, X. Xie, et al. 2019. “Efficient sonoelectrochemical decomposition of chlorpyrifos in aqueous solution.” *Microchem. J.* 145: 146–153.

Sala, M., and M. C. Gutiérrez-Bouzán. 2012. "Electrochemical techniques in textile processes and wastewater treatment." *Int. J. Photoenergy* 2012: 629103.

Särkkä, H., A. Bhatnagar, and M. Sillanpää. 2015. "Recent developments of electro-oxidation in water treatment—A review." *J. Electroanal. Chem.* 754: 46–56.

Sartaj, S., N. Ali, A. Khan, S. Malik, M. Bilal, M. Khan, et al. 2020. "Performance evaluation of photolytic and electrochemical oxidation processes for enhanced degradation of food dyes laden wastewater." *Water Sci. Technol.* 81 (5): 971–984.

Scialdone, O., S. Randazzo, A. Galia, and G. Silvestri. 2009. "Electrochemical oxidation of organics in water: Role of operative parameters in the absence and in the presence of NaCl." *Water Res.* 43 (8): 2260–2272.

Senthilkumar, S., C. A. Basha, M. Perumalsamy, and H. J. Prabhu. 2012. "Electrochemical oxidation and aerobic biodegradation with isolated bacterial strains for dye wastewater: Combined and integrated approach." *Electrochim. Acta* 77: 171–178.

Sirés, I., E. Brillas, M. A. Oturan, M. A. Rodrigo, and M. Panizza. 2014. "Electrochemical advanced oxidation processes: Today and tomorrow. A review." *Environ. Sci. Pollut. Res.* 21 (14): 8336–8367.

Solano, A. M. S., C. K. C. de Araújo, J. V. de Melo, J. M. Peralta-Hernandez, D. R. da Silva, and C. A. Martínez-Huitle. 2013. "Decontamination of real textile industrial effluent by strong oxidant species electrogenerated on diamond electrode: Viability and disadvantages of this electrochemical technology." *Appl. Catal., B* 130–131: 112–120.

Song, L.-J., N.-W. Zhu, H.-P. Yuan, Y. Hong, and J. Ding. 2010. "Enhancement of waste activated sludge aerobic digestion by electrochemical pre-treatment." *Water Res.* 44 (15): 4371–4378.

Szpyrkowicz, L., S. N. Kaul, and R. N. Neti. 2005. "Tannery wastewater treatment by electro-oxidation coupled with a biological process." *J. Appl. Electrochem.* 35 (4): 381–390.

Thokchom, B., K. Kim, J. Park, and J. Khim. 2015. "Ultrasonically enhanced electrochemical oxidation of ibuprofen." *Ultrason. Sonochem.* 22: 429–436.

Un, U. T., U. Altay, A. S. Koparal, and U. B. Ogutveren. 2008. "Complete treatment of olive mill wastewaters by electrooxidation." *Chem. Eng. J.* 139 (3): 445–452.

Walsh, F. 2001. "Electrochemical technology for environmental treatment and clean energy conversion." *Pure Appl. Chem.* 73 (12): 1819–1837.

Wendt, H., and G. Kreysa. 1999. *Electrochemical engineering: Science and technology in chemical and other industries*. Berlin: Springer.

Woisetschläger, D., B. Humpl, M. Koncar, and M. Siebenhofer. 2013. "Electrochemical oxidation of wastewater —Opportunities and drawbacks." *Water Sci. Technol.* 68 (5): 1173–1179.

Workgroup, O. O. E. C. (2008). "Aquatic life criteria for contaminants of emerging concern: General challenges and recommendations." US Environmental Protection Agency Duluth, MN.

Xu, F., L. Chang, X. Duan, W. Bai, X. Sui, and X. Zhao. 2019. "A novel layer-by-layer CNT/ PbO_2 anode for high-efficiency removal of PCP-Na through combining adsorption/ electrosorption and electrocatalysis." *Electrochim. Acta* 300: 53–66.

Yan, L., H. Ma, B. Wang, Y. Wang, and Y. Chen. 2011. "Electrochemical treatment of petroleum refinery wastewater with three-dimensional multi-phase electrode." *Desalination* 276 (1–3): 397–402.

Zhang, C., Y. Jiang, Y. Li, Z. Hu, L. Zhou, and M. Zhou. 2013. "Three-dimensional electrochemical process for wastewater treatment: A general review." *Chem. Eng. J.* 228: 455–467.

Zhong, C., W. B. Hu, and Y. F. Cheng. 2013. "Recent advances in electrocatalysts for electro-oxidation of ammonia." *J. Mater. Chem. A* 1 (10): 3216–3238.

CHAPTER 14

Cost Comparison of Electro-Coagulation and Electro-Oxidation Processes with Other Clean-Up Technologies

Lalit R. Kumar, Sushil Kumar, Bharti Bhadana, Patrick Drogui, R. D. Tyagi

14.1 INTRODUCTION

Human activities in the industrial, agricultural, and domestic sectors have affected the environment in a big way, resulting in serious problems such as the generation of wastewater with a high level of pollutants. The available water on the earth is distributed in an uneven way. Salt water accounts for 97.5% of this available water, which can be used only after treatment. The remaining 2.5% water is pure and fresh, but most of it is locked up in polar ice caps and glaciers, whereas whatever is left of this pure water (0.75%) is found as soil moisture or underground (Gupta et al. 2009). The availability of useful freshwater is very little because of the rapid increase in industrialization and modern agricultural and domestic activities. The demand for water has increased tremendously and has also resulted in a large amount of wastewater containing a lot of pollutants that are harmful to both human and aquatic life. Freshwater is consumed mostly in agricultural use (70% of total consumption), followed by industrial use (22%), and domestic use (8%) (Gupta et al. 2009). Thus, it is important to treat wastewater for pollutant removal before it mixes with freshwater bodies. The waste treatment methods should not only have high pollutant removal efficiencies but also be cost-effective at the industrial scale.

Various physical, biological, and chemical processes are available to treat wastewater. Wastewater can be treated using adsorbents, but regeneration of adsorbents is difficult. Chemical coagulation needs additional chemicals to wastewater and produces a large amount of sludge. Chemical oxidation by chlorine is effective but produces toxic by-products such as organochlorine compounds.

Electro-coagulation has application in wastewater treatment of various industries such as olive mill (Ün et al. 2006), food processing (Barrera-Díaz et al. 2006), semiconductor (Hu et al. 2005), chemical mechanical polishing (Den et al. 2006, Lai and Lin 2004), restaurant (Chen et al. 2000), pulp and paper mill (Mahesh et al. 2006), and dairy and slaughterhouse (Bayramoglu et al. 2006, Kobya et al. 2006, Şengil and Özacar 2006). The electro-coagulation technique has applications mostly in the textile industry. Textile plants produce large amounts of colored wastewater, and the dyes present in wastewater are toxic, mutagenic, and carcinogenic to human and aquatic life (Can et al. 2003, Christie 2007). Because of the hydrophilic properties of reactive dyes, they are not absorbed onto biomass to any great degree. Most reactive dyes are toxic to microorganisms, recalcitrant, and resistant to biological degradation. Therefore, conventional biological treatment methods have low efficiency to treat these dyes (Christie 2007).

Electro-oxidation is a simple and effective technique for treating wastewater and disinfection of waste. Electro-oxidation has been utilized for organic pollutant degradation (olive oil, dye, and pulp and paper effluents), drinking water disinfection, and industrial circulating water and wastewater treatment (Gaied et al. 2019, Un et al. 2008). Electro-oxidation is also effective in mineralizing nonbiodegradable organic matter (Zhu et al. 2015). Electro-coagulation and electro-oxidation treatments have advantages such as a smaller footprint compared with biological treatments because of a shorter reaction time, low sludge production in comparison with chemical coagulation, low cost of equipment and operation, and easy operation (Dalvand et al. 2011). However, it is essential to evaluate the economics of the electro-coagulation and electro-oxidation processes and compare them with conventional processes so that researchers understand the industrial feasibility of these processes. Moreover, technoeconomic studies will provide a direction to where research should be conducted.

This chapter discusses the applications of electro-coagulation (EC) and electro-oxidation (EO) processes in wastewater treatment and compares these processes with other treatment methods (membrane technology, commercial adsorbents, low-cost adsorbents, and chemical coagulation) in terms of treatment efficiency and economics. The factors impacting cost in the electro-coagulation and electro-oxidation processes and other treatment processes are also discussed.

14.2 PRINCIPLES GOVERNING ELECTRO-COAGULATION AND ELECTRO-OXIDATION PROCESSES

In recent years, electrochemical treatment methods such as electro-oxidation and electro-coagulation have attracted great attention for their eco-friendly process of wastewater treatment. This section discusses the principles governing the EC and EO processes.

Electro-Coagulation. EC involves in-place generation of coagulants by anodic dissolution of an appropriate sacrificial anode (e.g., iron and aluminum) on application of direct current. The metal ions generated hydrolyze in the electro-coagulator to produce metal hydroxide ions and metal ions. Negative hydroxide ions bind with positively charged metal ions in wastewater and produce $M(OH)_3$. $M(OH)_3$ with low solubility at a pH of 6 to 7 results in the precipitation or formation of flocs.

EC is an electrolytic process consisting of dissolution of sacrificial anodes (Fe or Al) on the application of current between two electrodes for treatment of liquid wastewater containing inorganic or organic pollutants. In EC, the anodic reaction involves the dissolution of Al or Fe electrodes [Equations (14-1) to (14-3)], and the cathodic reaction [Equations (14-2) and (14-5)] involves the formation of hydrogen gas and hydroxide ions; then, hydroxide ions formed at the cathode increase the pH of the wastewater, thereby inducing the precipitation of metal ions as corresponding hydroxides and coprecipitation with iron hydroxides (Kobya et al. 2016).

The main anode and cathode reactions occurring at the Al and Fe electrodes in the EC process are as follows: anode and cathode reactions for Al electrodes:

Anode:

$$Al \rightarrow Al^{3+} + 3e^- \tag{14-1}$$

Cathode:

$$3H_2O + 3e^- \rightarrow 3/2\ H_2(g) + 3OH^- \tag{14-2}$$

Anode and cathode reactions for Fe electrodes:

Anode:

$$Fe \rightarrow Fe^{2+} + 2e^- \tag{14-3}$$

$$Fe^{2+} \rightarrow Fe^{3+} + e^- \tag{14-4}$$

Cathode:

$$2H_2O + 2e^- \rightarrow H_2(g) + 2OH^- \tag{14-5}$$

Electro-Oxidation. EO may occur either by (i) direct oxidation (anodic oxidation) or (ii) mediated oxidation. In direct EO, the oxidation of pollutants in the electrolytic cell occurs at the electrode surface or by direct electron transfer to the anode. Moreover, powerful oxidants called reactive oxygen species (hydroxyl radicals) can be formed from water discharge at the anode. Direct oxidation has an advantage in that no chemicals need to be added to the treated solution, hence producing less secondary pollution (Zhu et al. 2015). In mediated oxidation, metal ions are oxidized on an anode from a stable state to a reactive high valence state, which, in turn, attack pollutants directly and may also produce hydroxyl-free radicals to promote degradation (Zhu et al. 2015). The following are the reactions that occur at electrodes during EO:

$$2Cl^- \rightarrow Cl_2 + 2e^- \tag{14-6}$$

$$Cl_2 + H_2O \rightarrow HOCl + H^+ + Cl^- \tag{14-7}$$

$$HOCl \rightarrow H^+ + OCl^- \tag{14-8}$$

$$H_2O \rightarrow {}^*OH + H^+ + e^- \tag{14-9}$$

$$2\,{}^*OH \rightarrow H_2O_2 \tag{14-10}$$

$$O_2 + {}^*O \rightarrow O_3 \tag{14-11}$$

14.3 OPERATING COST COMPONENTS FOR ELECTRO-COAGULATION TECHNIQUE

The cost components for a wastewater treatment plant through EC can be divided into two parts: (1) fixed costs for the setting up/installation of an EC tank, and (2) operational cost—the cost required for the operation of the treatment plant. The major operating cost components for the electro-coagulation technique include the cost of electrodes and chemicals, electricity, and utility costs, and sludge transportation and disposal costs, which are accounted as direct operational costs. The other cost factors such as labor costs, equipment maintenance, and depreciation costs are accounted as indirect operational costs (Bayramoglu et al. 2007).

The major cost contributing factors in EC are electricity, labor costs, and electrode purchase cost. However, studies have reported that the operating cost of EC includes three main components besides the labor cost: (1) the cost of electrode material, (2) cost of chemical consumption, and (3) consumed electrical energy (Dalvand et al. 2011, Kobya et al. 2016). Thus,

$$\text{Operating cost} = aC_{\text{energy}} + bC_{\text{electrode}} + cC + M \tag{14-12}$$

where

a = Electricity price (\$/kWh);
C_{energy} = Energy consumption per cubic meter of wastewater (kWh/m^3);
b = Average cost of an electrode sheet in \$/kg;
$C_{\text{electrode}}$ = Consumed electrode for the treatment of cubic meter of wastewater (kg/m^3);
c = Cost of chemicals consumed (\$/kg);

C = Amount of chemicals consumed; and
M = Maintenance and labor costs.
C_{energy} and $C_{electrode}$ are calculated according to Equations (14-7) and (14-8)

$$C_{energy} = U * I * t_{EC} / V \tag{14-13}$$

where
U = cell voltage (V);
I = current (A);
t_{EC} = operating time (s); and
V = volume (m^3).

$$C_{electrode} = I * t_{EC} * M_w / (z * F * v) \tag{14-14}$$

where
M_w = molecular weight of the electrode (kg/mol);
z = number of electrons transferred; and
F = Faraday's constant (96,487 C/mol).

14.4 OPERATING COST COMPONENTS FOR ELECTRO-OXIDATION TECHNIQUE

The major operating cost component for an EO process is the cost of equipment/reactor, which is a kind of fixed cost and required only initially. The other major cost directly associated with the operation of the treatment system is the cost for power consumption to run the treatment system, cost of electrode material, and cost of regent needed for the cleaning of electrodes. Apart from these, labor cost also plays a significant role in the operation of the treatment system.

Reactor cost can be determined by using the William equation (Gaied et al. 2019, Llanos et al. 2011). During the estimation, it is assumed that no variation of money update has been considered for the next 10 years. The cost of the reactor calculated in 2010 is

$$P = 21{,}662\ \mathrm{A}^{0.7953} \tag{14-15}$$

where A = surface area of the electrode;

$$\mathrm{A} = \frac{\mathrm{Q} \times 1{,}000 \times \text{flow rate}}{J_{app} \times 24} \tag{14-16}$$

where Q is in Ah/m^3; the flow rate is in m^3/day; and J_{app} is the current density in A/m^2. Power consumption can be calculated by using Equation (14-13), and

Table 14-1. Cost Components for an Electro-Oxidation Plant with a Flow Rate of 1 m^3/day.

Item	*Cost (€)*
Fixed costs	
Reactor/equipment cost	6,022.9
Operational costs	
Indirect costs	
Maintenance (€/m^3)	0.0027
Labor costs (€/m^3)	0.053
Direct costs	
Electricity (€/kWh)	0.14
Fe electrode (€/kg)/m	1.001
Sludge treatment cost (€/m^3)	150

Source: Gaied et al. (2019), Kobya et al. (2016), Un et al. (2008).

electrode consumption can be calculated by using Equation (14-14). Operational cost can be calculated as follows:

$$\text{Operating cost} = aC_{\text{energy}} + bC_{\text{electrode}} + M \tag{14-17}$$

where

C_{energy} = the energy consumption per cubic meter of wastewater (kWh/m^3);
$C_{\text{electrode}}$ = the consumed electrode for the treatment of cubic meter of wastewater (kg/m^3);
b = the average cost of an electrode sheet in €/kg; a = the electricity price (€/kWh); and
M = the maintenance and labor cost.

Table 14-1 shows different cost components of an EO process. The initial investment cost related to the equipment is €6,022.9, and the total capital investment for the process is €34,270.5. During the process, no additional chemicals are required, and so, the cost of chemicals has not been accounted for (Gaied et al. 2019, Kobya et al. 2016).

14.5 FACTORS AFFECTING OPERATING COST OF ELECTRO-COAGULATION PROCESS

14.5.1 Effect of Time and Voltage Variations

Increasing the reaction time also increases electrical consumption and electrode consumption linearly according to Equations (14-13) and (14-14), respectively, which further increases the operating cost of EC. In one study, the effects of voltage and treatment time on electrical energy consumption was investigated. Increasing the reaction time from 5 to 90 min, the electrical energy consumption linearly increased from 0.25 to 3.31 kWh/m^3 of wastewater at 20 V and 1 cm of interelectrode distance. Electrical consumption at 75 min was 2.17 times higher than that of 30 min (Dalvand et al. 2011). Another study on the treatment of textile wastewater using EC revealed that the operating cost increased from $0.2 to $0.6/$m^3$ on increasing the reaction time from 5 to 25 min for an iron electrode, whereas it increased from $US 0.3 to $US 1.1/$m^3$ for an aluminum electrode (Bayramoglu et al. 2007).

Increasing the voltage at the same interelectrode distances also increases the electrical consumption according to Equation (14-13). At all interelectrode distances, electrical consumption at 40 V was four times higher than that of 20 V. The operating cost for treating a cubic meter colored wastewater increased from $0.256 to $0.795 on increasing the voltage from 20 to 40 V (Dalvand et al. 2011).

14.5.2 Effect of Inter-Electrode Distance

Two cases are considered while discussing the effect of the interelectrode distance. During electrolysis, the current density (or current intensity) can be maintained constant. In this case, while increasing the interelectrode distance, the resistance between the electrodes and the voltage increases so that the energy consumption also increases.

The second case consists in keeping the voltage constant during electrolysis. Increasing the interelectrode distance increases the resistance between the electrodes but decreases the current passing through the electrode (keeping the voltage same), which further decreases electrical consumption and electrode consumption according to Equations (14-7) and (14-8). As a result, with an increasing gap between electrodes, electrical energy consumption decreases, leading to reduced operational cost. However, pollutant removal efficiency decreases with increasing the interelectrode distance as the electrostatic interaction between hydroxyl ions and pollutants decreases (Dalvand et al. 2011). Hence, an optimum distance needs to be selected for efficient pollutant removal from wastewater and reduced operating cost of EC.

14.5.3 Effect of Electrolyte Concentration

Studies have reported that at constant voltage conditions, increasing the electrolyte concentration also increases the conductivity of the solution, because of which resistance decreases. Thus, the current passed increases and, hence, both electricity consumption and electrode consumption increase, leading to high operational costs. However, with increasing electrical energy consumption, the floc production rate also increases, leading to high pollutant removal efficiency. Hence, an optimum electrolyte concentration needs to be selected for high pollutant removal efficiency and economical operating costs (Dalvand et al. 2011).

14.5.4 Effect of Pollutant Concentration/Chemical Oxygen Demand

According to Faraday's law, a constant amount of metal ions (i.e., Fe^{3+} or Al^{3+}) are released into the solution at the same current, voltage, and operating time for different pollutant concentrations. Thus, the same number of flocs is produced in the solution. If the pollutant concentration is increased, current, voltage, and the time of reaction needs to be increased for the same removal efficiency, which enhances the electricity and electrode consumption cost and, thus, the operational cost of EC (Dalvand et al. 2011).

14.5.5 Effect of Electrode Connection Mode

In a recent study, EC was conducted in three different modes: monopolar parallel (mp-p) connection, monopolar series (mp-s), and bipolar. The results showed that the mp-p connection mode was the most effective connection mode as a higher current passed through the electrodes than in other connections. Dye removal efficiencies achieved for mp-p, mp-s, and bipolar were 98.59%, 91.18%, and 84.6%, respectively, for a operating time of 30 min (Dalvand et al. 2011). Consequently, mp-p connection released more metal ions and hydroxyl ions and, thus, produced more flocs and removed a higher dye concentration as compared to other configurations. Based on the aforementioned study, it can be concluded that for the same removal efficiency, mp-p connection had a lower reaction time, consuming less electricity and electrodes for the operation and, thus, proving to be comparatively more economical.

14.5.6 Material of Electrode

A comparative economic study has been performed for the treatment of textile wastewater using two different electrodes, Al and Fe. It was observed that even for the same treatment conditions, they had similar removal efficiencies, but the operating cost of the Fe electrode ($0.25/$m^3$) was lower than that of the Al electrode ($US 0.4/$m^3$) (Bayramoglu et al. 2007). Aluminum electrodes are often more expensive than iron electrodes. For instance, mild steel electrode consumption was estimated at a cost of CAN$ 228/ton, whereas a cost of CAN$ 1,596/ton was estimated for the aluminum electrode, whereas Drogui et al. (2008) studied the effectiveness of EC to remove pollutants from agroindustry wastewater.

14.5.7 Effect of Current Density

Current density is defined as the current passing per unit surface area. Its SI unit is A/m^2. As current density between electrodes increases, electrical consumption also increases, leading to high operating costs. Similar results have been obtained in one of the studies in which on increasing the current density from 30 to 60 A/m^2, the operating cost increased from $US 0.4/$m^3$ to $US 1/$m^3$ for an iron electrode and from $US 0.7/$m^3$ to $US 1.5/$m^3$ for an aluminum electrode (Bayramoglu et al. 2007), but resulting in a higher efficiency and lower treatment time for both the electrodes (Bayramoglu et al. 2007).

14.5.8 Effect of Salt Concentration

The higher the concentration of sodium chloride, the higher the current, which increases electricity and electrode consumption costs according to Equations (14-7) and (14-8), leading to increased operational and chemical consumption costs. One of the pilot plant studies reported that increasing NaCl concentration from 0.02% (w/w) to 0.08% (w/w) increased the operating cost from $0.1/m^3$ to $0.5/m^3$ but led to a higher efficiency and lower treatment time (Donini et al. 1994).

14.5.9 Effect of Feed Flow Rate

Increasing the feed flow rate reduces the residence time of the feed and, thus, energy consumption per unit volume of feed. One of the pilot plant studies reported that increasing the feed flow rate from 50 to 500 cm^3/s decreased the operating cost from \$US 0.8/m^3 to \$US 0.1/m^3 (Donini et al. 1994). According to Equation (14-14), electrode consumption is directly proportional to reaction time. Hence, when the flow rate increases, electrode and energy consumption decrease, resulting in a decrease of the operating treatment costs. However, the efficiency can reduce owing to a lower retention time.

14.5.10 Effect of Passivation

The formation of passivation occurs gradually to protect metal electrodes from corrosion. The formation of passivation layers results in an increase in resistance, and more energy is required to counter-attack the effect of passivation layers. During the electrolysis, a gradual fouling (deposition of solids on the surface of the electrodes and local enrichment in hydroxides) is observed on the electrodes, thus leading to a significant increase in the electrical resistance and passivation of the electrodes. A reduction in treatment efficiency (in terms of pollutant removal) is, therefore, recorded over time. The gradual scaling of cathode electrodes over time is mainly related to the local enrichment of hydroxide ions. Also, a deposition of the reduction products takes place on the surface of the electrodes. The thickness of the passivation layer develops fast, resulting in a rapid increase in the operating cost. The operating cost can double because of the formation of the passivation layer after a period of electrolysis (Donini et al. 1994). As time passes, an increased resistance is seen between the two electrodes, and more energy is wasted on heating the passivation layers. The energy wasted on the passivation layers can be estimated by

$$\alpha = V_L / V_{AP} \tag{14-18}$$

where

α = Ratio of energy wasted on passivation layers to total energy consumption,
V_L = Voltage drop on the passivation layers (V), and
V_{AP} = Total applied voltage (V).

During the two-month experimental period, the percentage of operating cost wasted on the passivation layers went up sharply with time, reaching 23% of the operating cost (Donini et al. 1994). Neither the current efficiency of aluminum dissolution nor the aluminum material cost showed an obvious correlation with time. Aluminum material cost accounted for most of the total cost, although the percentage of the energy cost was increasing as the passivation layers developed. Periodic inversion of polarities (once per hour) can be applied to overcome this phenomenon. This regular inversion will also allow uniform wear of the electrode plates (Al or Fe).

14.5.11 Recirculation of Feed

Proper recirculation enhances the process of mass transport in a cell, improving the floc structure for faster settling and lower cake height and leading to reduced operational costs. A better EC cell design should be able to optimize the hydrodynamic conditions inside the cell and the whole system (Donini et al. 1994). However, a higher recirculation rate leads to lower residence time inside the electrochemical cell, leading to a higher energy dissipation of the recycling pump. This higher energy dissipation increases the operating treatment costs.

14.6 FACTORS AFFECTING OPERATING COST OF ELECTRO-OXIDATION PROCESS

14.6.1 Power Consumption

Electricity is the basic need of an EO process, which has a major impact on the operational cost of the treatment system. In many studies, it has been seen that with the increase in current intensity, the treatment efficiency improved, but at the same time, more power gets consumed for the process. The more the energy consumption, the more will be the operational cost. In many countries, the cost of electricity is very high, and, therefore, in this context, it is very crucial to adopt a cost-effective, eco-friendly, and sustainable treatment technology for wastewater treatment (Gaied et al. 2019).

14.6.2 Time of Treatment

The duration of treatment affects the power consumed during the treatment process. Therefore, we can say that once the time of treatment is increased, power consumption will also increase and, as a result, the cost of operation will be high.

14.6.3 Electrode Material

Electrode material plays a significant role in the cost estimation of the EO process. During this process, large-scale consumption of electrode material takes place, and within the process time, electrodes must be replaced. The cost of an electrode depends on the type of electrode used. Iron, aluminum, and graphite electrodes are cheaper than platinum, titanium, and boron-doped diamond (BDD) electrodes.

Therefore, before the start of the treatment process, selection of the electrode is very important (Zhu et al. 2015).

14.6.4 Passivation of Electrodes

Electrode passivation is one of the major problems encountered during the treatment process. A layer forms on the electrode surface, which affects the treatment efficiency of the process. Therefore, it is quite necessary to remove this layer to achieve the desired removal efficiency. To remove this layer, the electrode needs to be cleaned, which is more often done by chemical wash. However, this type of wash requires chemical reagents, which ultimately adds to the operating cost of the process (Gaied et al. 2019).

14.6.5 Type of Wastewater

The nature of wastewater is one of the important factors that decides the adaptability of the treatment process. The selection of the electrode is also based on the type of wastewater to be treated. As explained, the electrode cost depends on the type of electrode material used. Thus, as a whole, the nature of wastewater plays a significant role in determining the cost of electrode material. In addition, the more contamination is in the wastewater, the more time it takes to oxidize the pollutants and, as a result, more energy gets consumed during the process, which impacts the total operational cost of the treatment process (Gaied et al. 2019).

14.7 COST COMPARISON OF ELECTRO-COAGULATION WITH CHEMICAL COAGULATION PROCESS

The cost comparison for the EC and chemical coagulation (CC) processes is highlighted in Table 14-2. The operating cost of CC is higher than that of EC because of high coagulant consumption in CC processes. The operating time of EC is also less than that of CC, indicating lower operational costs (electricity, coagulants, and electrode material) in EC as compared with those in CC. Besides the operating cost, CC has other environmental disadvantages; for example, the final pH of the medium is 2.9 in CC, which indicates high corrosion demanding high-cost building materials and, thus, making the process more expensive. Because of high coagulant consumption in CC, high chloride ions are present in the medium, which requires further treatment, thereby increasing the treatment cost, whereas the final pH of the medium in EC is near neutral (7.9), which needs no further treatment (Bayramoglu et al. 2007, Kobya et al. 2007). Also, COD and turbidity removal efficiencies in EC were comparable to those of CC. Moreover, EC is able to remove even the smallest colloidal particles as the applied electric field neutralizes any residual charge, thereby facilitating the coagulation.

Table 14-2. Comparison between EC and CC Processes.

	Electro-coagulation		*Chemical coagulation*			
Parameter	*Fe*	*Al*	*$FeCl_3 \cdot 6H_2O$*	*$Fe_2(SO_4)_3 \cdot 7H_2O$*	*$AlCl_3 \cdot 6H_2O$*	*$Al_2(SO_4)_3 \cdot 18H_2O$*
Operating time (min)	15	15	25	25	25	25
Final pH	7.9	7.9	2.9	3.1	4.1	4.1
COD removal (%)	65	63	71	68	68	59
Turbidity removal (%)	83	80	87	63	89	90
Coagulant consumption (kg/kg COD)	0.126	0.096	1.761	1.586	0.828	1.896
Operating cost ($US/m³)	**0.25**	**0.4**	**0.67**	**0.75**	**0.96**	**0.75**
Reference	Kobya et al. (2007)		Bayramoglu et al. (2007)			

Yuksel et al. (2012) also compared EC and CC for treating textile wastewater and dye color solution in terms of economics. It was found that EC was faster and more economical, consumed less material, and produced less sludge, and pH of the medium was more stabilized (pH 5 to 5.3) than that of CC (pH 3.3 to 3.4). Moreover, TOC, COD, and turbidity removal efficiencies were more effective in EC (80% to 85%) than in CC (71% to 75%). The operating cost of EC was 1.5 times lower than that of CC, according to Yuksel et al. (2012). EC was also economical and effective than CC in the treatment of dye solution RY 135. For RY 135, the operating cost was $US 0.26/m^3 for EC and $0.31 to $US 0.37/m^3 for CC. Because sludge production is less in EC, sludge treatment cost will also be less, thus rendering the process more economical than CC. Moreover, final pH in CC is acidic with the presence of chloride ions that can corrode building materials; therefore, CC requires more expensive corrosion-resistant building materials.

In another study by Drogui et al. (2011), three processes (chemical precipitation, electro-coagulation, and sorption) were compared for the treatment of acidic soil leachate (ASL) (Drogui et al. 2011). PAL-soil (Pointe-Aux-Lièvres-soil) leaching, followed by an EC process for metal (Pb^{2+} and Zn^{2+}) recovery from ASL, contributes a total cost of $35/t, whereas chemical precipitation using $Ca(OH)_2$ and NaOH leads to a total cost of $38/t and $51/t, respectively. The treatment using the adsorption process involves a total cost of $39/t. The major cost contributor in the sorption process is waste disposal cost, which includes transportation and disposal of hazardous waste, but such cost is very low in the EC and chemical precipitation processes. The major cost contributor in the chemical precipitation process is the cost of chemicals $Ca(OH)_2$ and NaOH besides HCl cost, whereas that in the EC process is the cost of the electrode and drying cost of wet residues besides HCl. Thus, at the industrial scale, EC for metal recovery is more economical and the final pH of the solution is around 7, which does not need further treatment. Also, the EC process (45 min) is faster than the chemical precipitation (60 min) and the adsorption processes (300 min). Moreover, the EC process is as effective as the traditional CC process in removing metal ions.

14.8 COST COMPARISON OF ELECTRO-OXIDATION WITH CHEMICAL OXIDATION PROCESSES

The major cost in the EO process is the cost of electricity and cost of consumed electrodes. Minor costs include the type of cleaning reagent used for cleaning the electrodes and preventing passivation. The cost of the process is highly dependent on the type of electrodes used for it. When iron, stainless steel, and graphite are used, the process is normally cheap, but at the same time, it also depends on the time of treatment. However, when platinum or titanium are used, the cost of electrodes is more. However, in the case of chemical oxidation, the major cost is the cost of chemicals consumed for the treatment of wastewater. The chemicals used have a high oxidizing nature, which enables oxidation of the contaminants present in wastewater.

Table 14-3 shows the cost comparison between the EO and chemical oxidation processes for the treatment of different types of wastewater. Gonzalez-Olmos et al. (2018) performed a study for the pretreatment of wastewater to minimize membrane fouling in a water reuse process. During the process, BDD electrodes were used as the anode and cathode materials at different current intensities and for different operational time durations. At a current density of 7 mA/cm^2 for a duration of 180 min and at a precleaning operation time of 8 min, the total operation cost worked out as €13.3/m^3 of wastewater. When the current intensity changed to 21 mA/cm^2 for a duration of 180 min and when the operation time before cleaning was 40 min, the total operation cost was €3.31/m^3 of wastewater, whereas for a treatment duration of 10 min and for a precleaning operational time of 14 min, the operational cost reduced and worked out as €0.36/m^3 (Gonzalez-Olmos et al. 2018).

In another study, El-Dein et al. (2006) estimated the treatment cost for wastewater having a high dye concentration using H_2O_2/UV and O_3. The cost of treatment when H_2O_2/UV was used was lower than that when the ozonation oxidation process was used. The estimated operational cost per m^3 of wastewater worked out to be €74 for H_2O_2/UV and €54 for ozonation oxidation (El-Dein et al. 2006).

Table 14-3. Cost Comparison Between Electro-Oxidation and Chemical Oxidation.

Process	*Type of wastewater*	*Reagent*	*Cost/m³*	*References*
Electro-oxidation	Municipal wastewater	Iron electrodes	17.3 €	Gaied et al. (2019)
	Municipal wastewater	Platinum electrodes	21.8 €	Gaied et al. (2019)
	Landfill leachate	Boron-doped diamond and stainless-steel electrodes	9.3 €	Anglada et al. (2010)
	Municipal wastewater	Boron-doped diamond electrodes	13.3 €	Gonzalez-Olmos et al. (2018)
	Synthetic textile wastewater (Containing RV-2 dye)	Graphite and stainless steel	1.54 $	Hamad et al. (2018)
	Synthetic textile wastewater (Containing AV-14 dye)	Graphite and stainless steel	1.29 $	Hamad et al. (2018)
Electrochemical oxidation	Textile wastewater	Mild steel and graphite electrodes and NaCl as electrolyte	5.83 $	Pillai and Gupta (2017)
	Coke oven wastewater	Mild steel and graphite electrodes and NaCl as electrolyte	21.4 $	Pillai and Gupta (2016)
Chemical oxidation	Textile dye wastewater	H_2O_2/UV	74 €	El-Dein et al. (2006)
	Textile dye wastewater	O_3	54 €	El-Dein et al. (2006)

14.9 ECONOMICS OF ADSORBENTS IN WATER TREATMENT

For an adsorbent to be effective, it should have a porous structure with a high surface area. Also, the time taken for the adsorption equilibrium should be as less as possible so that it can be used to remove contaminants in less time. Table 14-4 lists the adsorbents commonly used in industries, their applications, surface area, and the price. Activated carbon for water treatment can be divided into two forms: powdered activated carbon (PAC) and granular activated carbon (GAC). GAC is widely used in wastewater treatment plants as the granular form is more adaptable to continuous contacting and there is no need to separate the carbon from the bulk. However, PAC is also used in wastewater treatment plants because of its low capital cost and lesser contact time. Regeneration of activated carbon has been reported using NaOH, acetic, and formic acid. However, because of the lower adsorption capacity after regeneration, 10% of virgin carbon (paid by the end user) is added to regenerate carbon to maintain the product within specification. Although the operational costs for adsorption to treat wastewater are low, the cost of adsorbent and the additional cost of regeneration can increase the overall cost (Gupta et al. 2009). Commercial activated carbon costs around $US 1.37 to 20/kg,

Table 14-4. List of Commercial Adsorbents Used for Wastewater Treatment.

Adsorbent	*Application*	*Surface area (m^2/g)*	*Price ($/kg)*	*Reference*
Alumina	Decolorization, refining of petroleum oils and waxes, removal of water from gas streams	200–300		Do (1998)
Bauxite	Removal of aerobic and anaerobic bacteria	25–250		Akhurst et al. (2006)
Silica gel	Purification of hydrocarbons and drying of gases and liquids	250–900		Allingham et al. (1958)
Zeolites	First ion exchangers for water softener	Up to 700		Kesraoui-Ouki et al. (1994)
Activated carbon	Wastewater treatment; adsorb metal ions, phenols, dyes, pesticides, chlorinated hydrocarbons, humic substances, polychlorinated biphenyls	500–2,000	1.37–20	Nicolet and Rott (1999)

whereas commercial GAC costs around $US 3.3/kg, which are expensive than the iron and aluminum electrodes ($US 1.8 and 0.3/kg, respectively) used in EC. Hence, attempts have been made to use low-cost adsorbents (LCAs), which are natural minerals and industrial by-products having good adsorption properties.

LCAs can be divided into two categories based on their availability: natural materials such as wood, peat, coal, lignite, and so on and industrial/agricultural/domestic waste or by-products such as slag, sludge, fly ash, bagasse, red mud, and so on. The development of low-cost adsorbents involves the following steps: collection of raw LCA material, activation using chemical or physical methods, washing and drying, sieving and storage, and characterization of adsorbents (Gupta et al. 2009). Table 14-5 lists the advantages and applications of different low-cost adsorbents with their price.

Although several economic advantages can be had from the use of LCAs, such as their cost-effectiveness and their versatility in removing organic and inorganic contaminants, there are also disadvantages such as the following:

- In many cases, the surface area of the new adsorbents is low, leading to poor adsorptive capacity;
- They are usually nonregenerable;
- Relationship between the constituents and the functional groups present on them is not clearly defined;
- Research using LCAs is limited to batch studies, and LCAs have not been tested using columns for their large-scale use;
- Performance of LCAs has been reported only for specific adsorbates but not for all contaminants;
- Performance of LCAs has not been compared with activated carbon; and
- Their activation requires high temperature followed by washing and drying, thus increasing the pretreatment costs.

14.10 ECONOMICS OF MEMBRANE FILTRATION TECHNOLOGY

Mechanical filtration is mostly used in primary settlement tanks for removal of suspended particles after the settling process. Filtration is also used for treatment of textile wastewater (Vergili et al. 2012). The major cost contributing factors in membrane technology are membrane cost, membrane replacement frequency, and power required for operation. Other disadvantages of using membrane technology in primary settling tanks exist. For example, large suspended particles can damage the membrane and increase the membrane replacement cost; emulsified oil in wastewater can lead to a fouling of membrane, in, turn, leading to frequent requirements of back-flushing and chemical treatment, further increasing the maintenance costs. The other main disadvantage of membrane technology is that a single type of membrane cannot

Table 14-5. Low-Cost Adsorbents Used for Wastewater Treatment with Their Advantages and Costs.

LCA	*Advantage/ Applications*	*Cost ($/kg)*	*Reference*
Natural			
Wood	1. Low-cost, no need for regeneration after use 2. No pretreatment 3. After use, burning of wood can generate steam.	—	Poots et al. (1976b)
Coal-based adsorbents (lignite, char fines, bituminous coal)	1. Equilibrium achieved in a short time (40 min) because of the presence of acidic groups in coal-based adsorbents 2. Effective removal of metal ions at a pH of 3.8–5.5	—	Mittal and Venkobachar (1993)
Peat	1. No pretreatment before use 2. Displayed adsorption capacity of 16.3 mg/g with a contact time of 2 h for the removal of telon blue	0.018–0.069	Poots et al. (1976a)
Chitin and chitosan	1. Effective in removing metal ions and color 2. High adsorption capacity for Cr (273 mg/g)	Chitosan – 16	Babel and Kurniawan (2003)
Biomass	1. Good adsorptive capacity for methylene blue and Victoria blue 2. Can be used for 10 cycles after amine modification	—	Low et al. (1995)
Clay (Fuller's earth)	Effective in adsorbing positively charged species	0.04 0.1	Crini (2006)
Manganese oxide		0.05–0.06	Babel and Kurniawan (2003)
Bentonite		0.05	Gupta et al. (2009)

(*Continued*)

Table 14-5. Low-Cost Adsorbents Used for Wastewater Treatment with Their Advantages and Costs. (Continued)

LCA	*Advantage/ Applications*	*Cost ($/kg)*	*Reference*
Industrial/domestic/agricultural waste			
Blast furnace slag	High removal efficiencies of metal ions	0.04	Srivastava et al. (1997)
Spheroidal cellulose	Strong affinity for cations such as metal ions and organic bases	1.07	Babel and Kurniawan (2003)
Spahganum moss peat		0.02	Sharma and Forster (1993)
Bagasse fly ash	High removal efficiencies for phenol	0.009	Srivastava et al. (1995)
Starch xanthate		1	Tare et al. (1992)
Carbonaceous adsorbent	Effective in the removal of copper ion, Phenol, and dyes	0.1	Jain et al. (2003)
Perlite		<1.5	Mathialagan and Viraraghavan (2002)
Coconut shell charcoal	Removal of metal ions and phenols	0.34	Babel and Kurniawan (2004)
Activated carbon (almond shell)		1.54–2.93	Toles et al. (2000)

Table 14-6. Economic Comparison Between Membrane Technology and EC for Treatment of Textile Wastewater.

Parameter	*Membrane technology*	*EC*
Total capital cost ($)	682,000	10,500
Unit operating cost ($/m^3)	1.21	0.25–0.4
Operating time (min)	1,505	15
Reference	Vergili et al. (2012)	Bayramoglu et al. (2007)

treat wastewater completely as it removes only certain impurities but not all. In one of the studies, a combination of ultrafiltration (UF) and nanofiltration (NF) was used to treat textile wastewater. The UF membrane yielded removal rates of 14.8%, 99.3%, and 41.1% for COD, dye, and alkalinity, respectively, whereas the NF membrane yielded rates of 59.4%, 25.2%, and 60% for COD, dye, and alkalinity, respectively (Vergili et al. 2012). Table 14-6 compares the capital costs and the operating costs of EC and membrane technology (UF/NF). It can be easily concluded that membrane technology requires more capital investment because of high membrane costs and more operating time, which lead to increased operating costs. In membrane technology, 31.7% of the total operating cost comes from facility-dependent cost, which results from membrane replacement and maintenance costs. Hence, it cannot be said that membrane technology is always cheaper than EC.

The choice of membrane material also has an effect on the overall cost, with the cellulose ester (CE) membrane being the cheapest and ceramic and polysulfone membranes being the most expensive. Large-diameter fibers are, in general, more expensive for the same membrane module, whereas small-diameter fibers have a cost advantage (Owen et al. 1995). The process of membrane filtration becomes advantageous and effective only if certain steps are taken that include having a better understanding of membrane fouling mechanisms, devising more effective fouling control strategies, employing high-quality membrane materials and module designs, and implementing new membrane integrity management technologies. Also, it should be noted that membrane technology is effective only at the desalination stage of wastewater treatment.

14.11 COMPARISON BETWEEN DIFFERENT WASTEWATER TREATMENT PROCESSES

Table 14-7 compares different wastewater treatment processes, EC, CC, membrane filtration, commercial adsorbents, and low-cost adsorbents. The advantages and disadvantages of each process and their ultimate impact on process economics are discussed in Table 14-7.

Table 14-7. Comparison of Different Wastewater Treatment Technologies, Their Merits and Demerits, and Their Effects on Economics and Downstream Processing.

Technology	*Advantage/Disadvantage*	*Effect on economics and downstream processing*
Electro-coagulation	Less sludge production	Because sludge treatment is 40% of wastewater plant operating costs, lesser costs will be spent on sludge treatment/disposal, and less environmental pollution is created.
	Lesser operating time as compared with other techniques	The process is more economical as the operating cost is directly proportional to the time of operation
	Flocs formed by EC are larger, more stable, and more acid-resistant	Can be easily separated by filtration
	Can remove the smallest of colloidal particles, leaving no residual colloidal particle in the sludge	No need for coagulation at later stages, thus saving costs of coagulation at later stages
	Chemical consumption is very low and the final pH is between 5 and 7.	Less requirement to neutralize chemicals, thereby saving costs on excessive chemicals, and less possibility of pollution caused by an excessive use of chemicals. The final pH can be easily maintained between 5 and 7 by regulating the current intensity during hydrolysis.
	Capable of removing suspended solids and membrane fouling species	No fouling of membrane during filtration after primary treatment, thus saving costs of back-flushing and membrane replacement

Chemical coagulation	More sludge production than in EC	The more the sludge treatment/disposal costs, the more the environmental pollution
	The final pH is 3–3.5.	1. Use of excessive NaOH to neutralize pH, increasing the costs on chemicals and also increasing secondary pollution 2. High use of chemicals can kill microbes, which are essential for aerobic and anaerobic digestion
	Contains chloride ions after treatment	Expensive corrosion-resistant building materials need to be used, thereby increasing capital costs
	Slower than EC	Operating time is directly proportional to operating costs
	High coagulant consumption	Increased operational costs
Commercial adsorbents	High purchasing cost of adsorbents	Increased costs of consumables, leading to increased operational costs
	High cost of regeneration	Increased operational costs
Low-cost adsorbents	Adsorbs specific impurities	For other impurities, further treatment needs to be performed, thereby increasing the treatment cost
	Surface area is not so high.	Ineffective removal of COD and other impurities is seen, and, thus, further treatment is required, which increases treatment costs
	Physical or chemical activation is followed by washing and drying before use as an adsorbent	Increased pretreatment costs
Membrane Technology	High capital cost	Initial investments are high
	Longer operating time	Increased electricity and operating costs
	Impurities present in wastewater can clog the membrane.	Increased costs resulting from back-flushing, chemical treatment; Increased maintenance costs
	Membrane replacement frequency is high.	High costs of consumables (membrane) in the operating cost
Chlorination	Production of toxic waste products	Increases further treatment costs

14.12 CONCLUDING REMARKS

The following conclusions can be drawn in light of the aforementioned studies of EC and other clean-up technologies:

1. Electro-coagulation is a faster process than chemical coagulation. It consumes less chemicals and produces less sludge than CC, proving that EC is more economical than CC. Moreover, in EC, the final pH after treatment is 5 to 7, suggesting that no chemicals should be added for neutralization, whereas in CC, there is the requirement for neutralization as the final pH after treatment is 3 to 3.5. The excessive use of chemicals in chemical coagulation further increases the treatment costs in downstream processing and also leads to secondary pollution. Moreover, chloride ions are present in the treated water after CC, which can corrode the walls of the tank, indicating that corrosion-resistant building materials should be used in the construction of the tank. But such materials further increase the capital cost in CC.
2. Electro-oxidation is one of the eco-friendly processes with minimal (or no) sludge generation during the treatment of wastewater. This process does not require any chemical coagulants during the treatment process.
3. Commercial adsorbents like activated carbon are used in water treatment processes, which have a high purchase cost. Moreover, their regeneration requires 10% of virgin carbon, which increases the operational cost of the process. In this regard, low-cost adsorbents obtained from industrial and/or domestic waste with high adsorption capacities for metal ions, dyes, and so on can also be used in water treatment processes. However, they have not been reported in large-scale applications, and their use as adsorbents requires physical or chemical activation, which leads to increased operational costs. Moreover, they are usually specific to adsorbing only certain types of impurities, and not all sorts of impurities are present in wastewater, thus making their use in wastewater treatment uneconomical.
4. use of membrane filtration has also been reported for wastewater treatment. However, high membrane costs, membrane fouling, membrane replacement frequency, and longer process time are some of the challenges in membrane filtration. Likewise, high membrane replacement frequency and back-flushing increase the maintenance costs. Membrane filtration might not be suitable for wastewater with high suspended solid particles (as they can damage the membrane and the filter) and, therefore, has to be used in combination with other processes to avoid premature damage of the membrane. Its use is effective at the later stages of wastewater treatment when suspended solid particles have already been removed.

ACKNOWLEDGMENTS

The authors would like to thank Dr. Pratik Kumar (Assistant Professor, IIT Jammu) for providing his valuable suggestions for writing the chapter.

References

Akhurst, D. J., G. B. Jones, M. Clark, and D. McConchie. 2006. "Phosphate removal from aqueous solutions using neutralised bauxite refinery residues (Bauxsol™)." *Environ. Chem.* 3 (1): 65–74.

Allingham, M. M., J. Cullen, C. Giles, S. Jain, and J. Woods. 1958. "Adsorption at inorganic surfaces. II. Adsorption of dyes and related compounds by silica." *J. Chem. Technol. Biotechnol.* 8 (2): 108–116.

Anglada, A., D. Ortiz, A. Urtiaga, and I. Ortiz. 2010. "Electrochemical oxidation of landfill leachates at pilot scale: Evaluation of energy needs." *Water Sci. Technol.* 61 (9): 2211–2217.

Babel, S., and T. A. Kurniawan. 2003. "Low-cost adsorbents for heavy metals uptake from contaminated water: A review." *J. Hazard. Mater.* 97 (1–3): 219–243.

Babel, S., and T. A. Kurniawan. 2004. "Cr(VI) removal from synthetic wastewater using coconut shell charcoal and commercial activated carbon modified with oxidizing agents and/or chitosan." *Chemosphere* 54 (7): 951–967.

Barrera-Díaz, C., G. Roa-Morales, L. Avila-Córdoba, T. Pavón-Silva, and B. Bilyeu. 2006. "Electrochemical treatment applied to food-processing industrial wastewater." *Ind. Eng. Chem. Res.* 45 (1): 34–38.

Bayramoglu, M., M. Eyvaz, and M. Kobya. 2007. "Treatment of the textile wastewater by electrocoagulation: economical evaluation." *Chem. Eng. J.* 128 (2–3): 155–161.

Bayramoglu, M., M. Kobya, M. Eyvaz, and E. Senturk. 2006. "Technical and economic analysis of electrocoagulation for the treatment of poultry slaughterhouse wastewater." *Sep. Purif. Technol.* 51 (3): 404–408.

Can, O. T., M. Bayramoglu, and M. Kobya. 2003. "Decolorization of reactive dye solutions by electrocoagulation using aluminum electrodes." *Ind. Eng. Chem. Res.* 42 (14): 3391–3396.

Chen, X., G. Chen, and P. L. Yue. 2000. "Separation of pollutants from restaurant wastewater by electrocoagulation." *Sep. Purif. Technol.* 19 (1–2): 65–76.

Christie, R. M. 2007. *Environmental aspects of textile dyeing.* 1st ed. Sawton, Cambridge: Woodhead Publishing.

Crini, G. 2006. "Non-conventional low-cost adsorbents for dye removal: A review." *Bioresour. Technol.* 97 (9): 1061–1085.

Dalvand, A., M. Gholami, A. Joneidi, and N. M. Mahmoodi. 2011. "Dye removal, energy consumption and operating cost of electrocoagulation of textile wastewater as a clean process." *Clean-Soil Air Water* 39 (7): 665–672.

Den, W., C. Huang, and H.-C. Ke. 2006. "Mechanistic study on the continuous flow electrocoagulation of silica nanoparticles from polishing wastewater." *Ind. Eng. Chem. Res.* 45 (10): 3644–3651.

Do, D. D. 1998. *Adsorption analysis: Equilibria and kinetics: (With CD containing computer Matlab programs).* Singapore: World Scientific Publishing.

Donini, J., J. Kan, J. Szynkarczuk, T. A. Hassan, and K. L. Kar. 1994. "The operating cost of electrocoagulation." *Can. J. Chem. Eng.* 72 (6): 1007–1012.

Drogui, P., M. Asselin, S. K. Brar, H. Benmoussa, and J.-F. Blais. 2008. "Electrochemical removal of pollutants from agro-industry wastewaters." *Sep. Purif. Technol.* 61 (3): 301–310.

Drogui, P., N. Meunier, G. Mercier, and J.-F. Blais. 2011. "Removal of Pb^{2+} and Zn^{2+} ions from Acidic Soil Leachate: a comparative study between electrocoagulation, adsorption and chemical precipitation processes." *Int. J. Environ. Waste Manage.* 8 (3–4): 241–257.

El-Dein, A. M., J. Libra, and U. Wiesmann. 2006. "Cost analysis for the degradation of highly concentrated textile dye wastewater with chemical oxidation H_2O_2/UV and biological treatment." *J. Chem. Technol. Biotechnol.* 81 (7): 1239–1245.

Gaied, F., B. Louhichi, M. Bali, and M. R. Jeday. 2019. "Tertiary treatment of wastewater by electro-coagulation, electro-Fenton and advanced electro-oxidation processes: Comparative and economic study." *Songklanakarin J. Sci. Technol.* 41 (5): 1084–1092.

Gonzalez-Olmos, R., A. Penadés, and G. Garcia. 2018. "Electro-oxidation as efficient pretreatment to minimize the membrane fouling in water reuse processes." *J. Membr. Sci.* 552: 124–131.

Gupta, V. K., P. J. M. Carrott, M. M. L. Ribeiro Carrott, and J. Suhas . 2009. "Low-cost adsorbents: Growing approach to wastewater treatment—A review." *Crit. Rev. Environ. Sci. Technol.* 39 (10): 783–842.

Hamad, H., D. Bassyouni, E.-S. El-Ashtoukhy, N. Amin, and M. Abd El-Latif. 2018. "Electrocatalytic degradation and minimization of specific energy consumption of synthetic azo dye from wastewater by anodic oxidation process with an emphasis on enhancing economic efficiency and reaction mechanism." *Ecotoxicol. Environ. Saf.* 148: 501–512.

Hu, C. Y., S. L. Lo, W. H. Kuan, and Y. D. Lee. 2005. "Removal of fluoride from semiconductor wastewater by electrocoagulation–flotation." *Water Res.* 39 (5): 895–901.

Jain, A. K., V. K. Gupta, and A. Bhatnagar and Suhas. 2003. "Utilization of industrial waste products as adsorbents for the removal of dyes." *J. Hazard. Mater.* 101 (1): 31–42.

Kesraoui-Ouki, S., C. R. Cheeseman, and R. Perry. 1994. "Natural zeolite utilisation in pollution control: A review of applications to metals' effluents." *J. Chem. Technol. Biotechnol.* 59 (2): 121–126.

Kobya, M., M. Bayramoglu, and M. Eyvaz. 2007. "Techno-economical evaluation of electrocoagulation for the textile wastewater using different electrode connections." *J. Hazard. Mater.* 148 (1–2): 311–318.

Kobya, M., E. Gengec, and E. Demirbas. 2016. "Operating parameters and costs assessments of a real dyehouse wastewater effluent treated by a continuous electrocoagulation process." *Chem. Eng. Process. Process Intensif.* 101: 87–100.

Kobya, M., E. Senturk, and M. Bayramoglu. 2006. "Treatment of poultry slaughterhouse wastewaters by electrocoagulation." *J. Hazard. Mater.* 133 (1–3): 172–176.

Lai, C. L., and S. H. Lin. 2004. "Treatment of chemical mechanical polishing wastewater by electrocoagulation: System performances and sludge settling characteristics." *Chemosphere* 54 (3): 235–242.

Llanos, J., R. Camarillo, A. Pérez, P. Cañizares, and M. A. Rodrigo. 2011. "Costs estimation of an integrated process for the treatment of heavy-metal loaded aqueous effluents." *J. Appl. Electrochem.* 41 (9): 1099–1107.

Low, K. S., C. K. Lee, and K. K. Tan. 1995. "Biosorption of basic dyes by water hyacinth roots." *Bioresour. Technol.* 52 (1): 79–83.

Mahesh, S., B. Prasad, I. D. Mall, and I. M. Mishra. 2006. "Electrochemical degradation of pulp and paper mill wastewater. Part 1. COD and color removal." *Ind. Eng. Chem. Res.* 45 (8): 2830–2839.

Mathialagan, T., and T. Viraraghavan. 2002. "Adsorption of cadmium from aqueous solutions by perlite." *J. Hazard. Mater.* 94 (3): 291–303.

Mittal, A. K., and C. Venkobachar. 1993. "Sorption and desorption of dyes by sulfonated coal." *J. Environ. Eng.* 119 (2): 366–368.

Nicolet, L., and U. Rott. 1999. "Recirculation of powdered activated carbon for the adsorption of dyes in municipal wastewater treatment plants." *Water Sci. Technol.* 40 (1): 191–198.

Owen, G., M. Bandi, J. A. Howell, and S. J. Churchouse. 1995. "Economic assessment of membrane processes for water and waste water treatment." *J. Membr. Sci.* 102: 77–91.

Pillai, S. I. M., and A. K. Gupta. 2016. "Anodic oxidation of coke oven wastewater: Multiparameter optimization for simultaneous removal of cyanide, COD and phenol." *J. Environ. Manage.* 176: 45–53.

Pillai, I. M. S., and A. K. Gupta. 2017. "Performance analysis of a continuous serpentine flow reactor for electrochemical oxidation of synthetic and real textile wastewater: Energy consumption, mass transfer coefficient and economic analysis." *J. Environ. Manage.* 193: 524–531.

Poots, V. J. P., G. McKay, and J. J. Healy. 1976a. "The removal of acid dye from effluent using natural adsorbents—I. Peat." *Water Res.* 10 (12): 1061–1066.

Poots, V. J. P., G. McKay, and J. J. Healy. 1976b. "The removal of acid dye from effluent using natural adsorbents—II. wood." *Water Res.* 10 (12): 1067–1070.

Şengil, İ. A., and M. Özacar. 2006. "Treatment of dairy wastewaters by electrocoagulation using mild steel electrodes." *J. Hazard. Mater.* 137 (2): 1197–1205.

Sharma, D. C., and C. F. Forster. 1993. "Removal of hexavalent chromium using sphagnum moss peat." *Water Res.* 27 (7): 1201–1208.

Srivastava, S. K., V. K. Gupta, and D. Mohan. 1997. "Removal of lead and chromium by activated slag—A blast-furnace waste." *J. Environ. Eng.* 123 (5): 461–468.

Srivastava, S., V. Gupta, I. Yadav, and D. Mohan. 1995. "Removal of 2, 4-dinitrophenol using bagasse fly ash- A sugar industry waste material." *Fresenius Environ. Bull.* 4 (9): 550–557.

Tare, V., S. Chaudhari, and M. Jawed. 1992. "Comparative evaluation of soluble and insoluble xanthate process for heavy metal removal from wastewaters." *Water Sci. Technol.* 26 (1–2): 237–246.

Toles, C. A., W. E. Marshall, L. H. Wartelle, and A. McAloon. 2000. "Steam- or carbon dioxide-activated carbons from almond shells: physical, chemical and adsorptive properties and estimated cost of production." *Bioresour. Technol.* 75 (3): 197–203.

Un, U. T., U. Altay, A. S. Koparal, and U. B. Ogutveren. 2008. "Complete treatment of olive mill wastewaters by electrooxidation." *Chem. Eng. J.* 139 (3): 445–452.

Ün, Ü. T., S. Uğur, A. S. Koparal, and Ü. B. Öğütveren. 2006. "Electrocoagulation of olive mill wastewaters." *Sep. Purif. Technol.* 52 (1): 136–141.

Vergili, I., Y. Kaya, U. Sen, Z. B. Gönder, and C. Aydiner. 2012. "Techno-economic analysis of textile dye bath wastewater treatment by integrated membrane processes under the zero liquid discharge approach." *Resour. Conserv. Recycl.* 58: 25–35.

Yuksel, E., E. Gurbulak, and M. Eyvaz. 2012. "Decolorization of a reactive dye solution and treatment of a textile wastewater by electrocoagulation and chemical coagulation: Techno-economic comparison." *Environ. Prog. Sustainable Energy* 31 (4): 524–535.

Zhu, R., C. Yang, M. Zhou, and J. Wang. 2015. "Industrial park wastewater deeply treated and reused by a novel electrochemical oxidation reactor." *Chem. Eng. J.* 260: 427–433.

CHAPTER 15

Challenges and Future Perspectives of Electro-Coagulation and Electro-Oxidation Processes

Song Yan, M. R. Karimi Estahbanati, Patrick Drogui, R. D. Tyagi, Rao Y. Surampalli, Tian C. Zhang

15.1 INTRODUCTION

Electro-Coagulation (EC) has been proven to be a low-cost treatment technology for its ability to effectively remove an extremely wide range of pollutants in various types of water and wastewater, together with its inherent simplicity of design and operation. The available literature (Garcia-Segura et al. 2017, Hakizimana et al. 2017, Holt et al. 2005, Khandegar and Saroha 2013, Maitlo et al. 2019, Mollah et al. 2004, Moussa et al. 2017, Nariyan et al. 2017, Reilly et al. 2019, Song et al. 2017) has shown that EC is a viable approach for the elimination of different pollutants from wastewater, at the lab- and pilot scale, both in batch and in continuous applications.

EC is an efficient technique because the in situ adsorption of hydroxide on mineral surfaces is hundred times greater than that on precipitated hydroxides when metal hydroxides are used as coagulants. EC needs simple equipment, designable for virtually any size, supplied as a skid-mounted unit, and fully assembled for inlet/outlet and backwash connections. EC is cost-effective and easily operable. The start-up and operating costs as well as power requirements of EC processes are relatively low. EC requires low maintenance cost with no moving parts. Because no chemical addition is required in this process, EC brings minimum chance of secondary pollution. EC needs low current and can be run via green power sources (e.g., solar, windmills, and fuel cells). Compared with the conventional chemical and biological techniques, EC can effectively destabilize the smallest colloidal particles because these particles have a greater probability of being coagulated by the electric field that sets them in motion. Gas bubbles produced during electrolysis can enhance flotation, and the EC process has a minimal sludge production. EC has the potential to be the distinct economic and environmental choice for wastewater treatment/management. Although EC has numerous advantages, it has some drawbacks such as the need to replace *sacrificial anodes* periodically, the lack of any systematic evaluation of existing applications, an almost complete lack of detailed experimental data on the operation of batch EC reactors, no generic solution to the problem of electrode passivation, and so on.

Electro-oxidation (EO) is an application of electrochemical properties to remove many types of organic, inorganic, and microbial pollutants from water via direct and/or indirect oxidation inside an electrolysis cell. Insoluble electrodes having high overvoltage of oxygen are currently used (Deng and Englehardt 2007, Gandini et al. 1998, Rajkumar et al. 2005). At the anode, these electrodes allow oxidizing many types of organic molecules contained in water and wastewater (Morão et al. 2004, Wang et al. 2004). The oxidation of these pollutants can exclusively take place at the anode surface (direct effect of electrolysis) and/or by *in situ* generation of oxidants in solution (indirect effect of electrolysis).

EO has been applied to treat actual effluents and has shown noticeable results on the abatement of chemical oxygen demand (COD) and dissolved organic carbon (DOC). Moreover, a combination of this electrochemical method with other treatment techniques in pre-, post-, or integrated treatment opens a wide range of novel and enhanced processes. More research on electrocatalytic anodic materials is needed to reduce the initial investments on the acquisition of electrodes. One of the critical points is the control and evaluation of the possible generation of toxic derivatives such as organohalides. The future application of EO will be directly related to the design of EO reactors design and the scale-up process to ensure the effective and economical exploitation of this technology (Garcia-Segura et al. 2018).

To support future applications of EC and EO treatments, it is necessary to address limitations and new approaches to minimize their formation of toxic by-products and the loss of efficiency, their improvements in design and operation, minimization of their power consumption; and their future development direction. Therefore, in this chapter, the drawbacks of the EC and EO processes were discussed, and the future research was addressed to make the EC and EO be a more distinct economic and environmental choice for treatment of water and wastewater.

15.2 CHALLENGES OF ELECTRO-COAGULATION

15.2.1 Electro-Coagulation Reactor Design and Operation

There is a growing awareness that localized plants (each tailored to a particular region and application) have a significant role to play in the formulation of the overall solution strategy. Holt et al. (2005) addressed EC's surprising lack of acceptance as a mainstream treatment option and put forward the following reasons for this failure by identifying deficiencies in a number of key areas:

- Lack of any systematic evaluation of existing applications leading to an agreed set of guidelines suitable for the design/operation of new applications;
- Almost complete lack of detailed experimental data on the operation of batch EC reactors, one of the more suitable options for (modest size scale) localized water treatment;
- No widely applicable mechanistically based approach to the mathematical modeling of EC reactors;
- Failure to fully appreciate that the performance of an EC reactor is largely determined by the interactions that occur between the three foundation technologies of electrochemistry, coagulation, and flotation; and
- No generic solution to the problem of electrode passivation.

It has attempted to demystify EC by showing that it is possible to classify a wide diversity of reactor systems on a simple basis, to obtain and interpret detailed dynamic data from a batch reactor system, and to show that the complexity of EC is a natural (and understandable) consequence of the interactions between a number of processes occurring in parallel. It is hoped that this evaluation will play its part in focusing attention on EC as a viable localized water treatment technology in the near future. Apart from a technical focus on quantifying interactions between the foundation technologies and an economic focus on the relative cost of EC, future research also needs to examine reliable means of reducing electrode passivation (Holt et al. 2005).

The full potential of the EC technique as a wastewater treatment alternative has yet to be fully realized. It is still an empirically optimized process that requires more fundamental understanding for improved engineering design. EC comprises complex chemical and physical processes involving many surface and interfacial phenomena. However, there is a paucity of scientific understanding of these phenomena, which limits the engineering design of the EC reactor for achieving optimum performance and future progress of this novel and innovative technology. A literature survey indicates that up-flow configuration of the electro-coagulator is better than a horizontal flow configuration. In addition, the use of a sonic field, stirrer, and hybrid EC–electroflotation (EC–EF) technique increases the efficiency and provides better control of this process (Mollah et al. 2004).

Considering the EC process, a number of major deficiencies have been identified. The *first* is that the literature does not reveal any systematic approach to EC reactor design and operation. Reported designs range through laboratory-, pilot-, and field scales as both stand-alone reactors and fully integrated units within a wastewater purification system. Certainly, no dominant reactor design is in use today, whereas very little is available by way of comparative reactor performance. This article, thus, begins by reviewing the literature on EC, seeking to introduce a structured approach to highlight commonalities and differences between the works reported. The *second* deficiency is the almost complete lack of published data on the operation of batch EC reactors, seen as the most likely means of providing a localized water treatment facility. One section contains detailed experimental results from a laboratory-scale batch reactor, whereas the following section outlines a simple modeling approach that is consistent with the observed data. The *third* deficiency is intimately linked with the previous two, in that little guidance is available for and a priori reactor design or performance prediction. Thus, in Section 15.5, a framework is proposed as an aid to further understanding of the myriad interactions that can occur among the fundamental processes (i.e., electrochemistry, coagulation, sedimentation, and flotation) taking place within an EC reactor. The *final* deficiency relates to the fact that an EC reactor is an electrochemical cell, and, thus, its performance is directly related to the operational state of its electrodes. These vary widely in terms of the material used and the design and mode of operation from simple vertical plate arrangements to packed-bed-style reactors containing various metallic packings (Ögütveren et al. 1992, Barkley et al. 1993). Electrode passivation, specifically of aluminum electrodes, has been widely observed and recognized as detrimental to reactor performance (Osipenko and Pogorelyi 1977, Novikova et al. 1982). The formation of an inhibiting layer, usually an oxide, on the electrode surface prevents metal dissolution and electron transfer, thereby limiting coagulant addition to the solution. Over time, the thickness of this layer increases, reducing the efficacy of the EC process as a whole. Therefore, it is imperative to explore new materials, different electrode types and arrangements (Pretorius et al. 1991, Mameri et al. 1998), and more sophisticated reactor operational strategies (such as periodic polarity reversal of the electrodes).

EC has been successfully applied for handling specific problematic factors (such as color, recalcitrance, and toxicity), which, otherwise, are difficult to deal with via conventional treatment methods. However, many researchers have argued that the full potential of the EC technique as a wastewater treatment alternative has yet to be fully realized. A number of possible future studies using the same experimental setup are apparent. For achieving optimal performance and future progress in the application of this innovative technology, considerably more work needs to be done in the area of reactor design and understanding of process control. This technology will continue to make inroads into the wastewater treatment arena because of the numerous advantages it offers and the nature of the changing strategic needs for wastewater in the world (Samir and Abbas 2013). EC technology is not yet fully tuned and optimized, and extensive research is required to introduce this technology worldwide as a reliable water treatment technology (Moussa et al. 2017).

The current research activities are pollutant oriented, focusing only on the optimization of operating parameters to achieve optimum removal of a specific contaminant in a bench-scale experiment. On the contrary, future researchers need to establish new areas and develop a more systematic approach for EC reactor design, operation, and modeling for predicting the performance prior to actual operation (Holt et al. 2005). Having such models will facilitate the scaling up of bench-scale reactors to meet industrial capacity. Another issue that must be addressed in future research is assessment of the economic feasibility of an EC operation. The overall cost of wastewater treatment using EC (operating and capital investment costs) must be estimated, as it is a major criterion for industry to choose the desired treatment technology. Moreover, the majority of EC studies have been performed using small-scale batch reactors, whereas most industrial applications require continuous flow mode. Hence, more research is required to evaluate the performance of EC units operating in continuous flow mode with a more effective reactor design. Future research should also consider new potential areas of EC application such as produced water treatment, which is a major challenge for the oil and gas industry. EC can play a key role in produced water treatment, especially in destabilizing tight emulsions (Moussa et al. 2017).

15.2.2 Sacrificial Electrodes and Other Challenges

Sacrificial electrodes are dissolved into wastewater streams as a result of oxidation and need to be regularly (or periodically) replaced. EC requires a minimum level of conductivity depending on the reactor design, thus limiting its use in water containing low dissolved solids. In the case of the removal of organic compounds, some toxic chlorinated organic compounds may be formed in situ if chlorides are also present. Wastewater with a high humic and fluvic acid content may be amenable to the formation of trihalomethanes. If phenols and algal metabolic and decomposition products are present, chlorine may lead to bad taste and odor.

There are other drawbacks of EC technology, such as an impermeable oxide film may be formed on the cathode leading to loss of efficiency of the EC unit; the high conductivity of the wastewater suspension is required; gelatinous hydroxide may tend to solubilize in some cases.

15.2.3 Cost of Electro-Coagulation

Operating cost calculations have been made in a few articles (Ashraf et al. 2019, Barzegar et al. 2019, Çırak 2018, Garcia-Segura et al. 2017, Hakizimana et al. 2017, Hashim et al. 2019, Maitlo et al. 2019, Priya and Jeyanthi 2019, Zazou et al. 2019). The calculations typically include the cost of chemicals, electrodes, and energy. It should be noted that the price of materials and energy changes over the course of time, and, therefore, operating costs are only rough estimates. Cost calculations do not typically include investment costs, which may be significant, including, for example, power supplies, electrochemical cell vessels, and sludge separation systems. Dalvand et al. (2011) investigated the operational cost at optimum conditions; electrode consumption, energy consumption, and operating cost were \$0.052 kg/m^3, \$1.303 kW · h/m^3, and \$0.256/m^3, respectively. One study (Sridhar et al. 2011) analyzed the operating cost of EC treatment for pulp and paper industry bleaching effluents to be in the range of \$1.52 to \$1.72/m^3. Kobya et al. (2010) found that the treatment of cadmium and nickel from electroplating rinse water could achieve a removal efficiency of 99.4% for cadmium, 99.1% for nickel, and 99.7% for cyanide. The cost for treatment was \$1.05/m^3 for cadmium and \$2.45/m^3 for nickel and cyanide, provided that the treatment maintained optimum conditions. Orori et al. (2010) took samples at five locations from a Kraft pulp and paper mill effluent primary settling tank, two aerated lagoons, a stabilization pond, and at the discharge port to compare treatment efficiency using graphite electrodes and aluminum electrodes with wood ash (Orori et al. 2010). The overall treatment with aluminum electrodes was better (60% BOD and 58.8% COD) but was more expensive than that with graphite (\$0.0006 to \$0.0008/m^3 of wastewater) versus (\$8.34 to \$31.74/m^3 of wastewater). Kobya et al. (2010) also studied Remazol Red 3B decolorization using iron electrodes and found that 99% decolorization was possible under optimum conditions. The authors found that energy consumption could achieve 3.3 kW · h/kg dye at a cost of E0.6/m^3. Meas et al. (2010) concluded that aluminum electrodes are capable of treating fluorescent penetrant liquid for nondestructing testing part of aircraft industry. Having used EC, the removal was found to be 95% of COD, 99% color, and 99% turbidity. With this high level of treatment, the cost was able to have a return period of 17 weeks (Samir and Abbas 2013).

Many process options for wastewater treatment are available, and these are necessary for EO to be cost-effective (Butler et al. 2011). Asselin et al. (2008) analyzed oil bilge wastewater (OBW) at the laboratory scale using bipolar (BP) and monopolar configuration with iron and aluminum electrodes. Using optimum conditions, treatment of OBW by EC could remove 93% BOD, 95.6% oil and grease, 99.8% total suspended solids, and 98.4% turbidity; the cost was \$0.46/m^3 for energy and electrode consumption, chemicals, and sludge disposal. Ghosh et al. (2008) used electrochemicals with aluminum electrodes for the purpose of removing iron [Fe(II)] from tap water having considered amorphous aluminum hydroxides, current densities, and electrode density. When treating a concentration of 15 mg Fe(II)/L, it would cost \$6.05/m^3 of tap water. Drogui et al. (2008) used EC with mild steel electrodes to treat agroindustry (meat processing, cereal, and food beverages) wastewater. An 82% removal of COD was achieved with treatment costs between \$0.95 and \$4.93/m^3, where the costs included electrical power, chemical, and electrode consumption.

The application can be extended to the shipping industry. When using EC-flocculation in this industry, a study (Dia et al. 2018) showed the capability of removing 80% turbidity, 56% COD, 90% oil and grease, and 89% BOD having used BP electrode arrangements. The costs, including energy and electrode consumption and sludge disposal, ranged between CAN\$ 1.54 and 2.40/m^3 of ship waste effluent. Khansorthong and Hunsom (2009) used EC to treat wastewater from the pulp and paper mill industry using parallel iron electrode comparing current density, pH, and flow rate. When using the optimum conditions, it was found that to remove color (97%) and COD (77%), it would cost \$0.29/m^3 of wastewater.

As a possibly the most relevant drawback, the high operating costs is due to the high energy consumption since electricity is basically the only consumable in EC. To overcome this problem, there are two possible ways: (1) use of this technology in combination with other techniques, as a pre-treatment or a polishing step, and (2) use of renewable energy sources to power the EO units. There are several reports describing the application of combined methods involving EO to treat sanitary landfill leachates. The encouraging results obtained so far set the basis for future work. In fact, similar operating costs for conventional treatments and for EO, applied as a polishing step, have been already observed. In regard to the use of renewable energy sources, the integration of photovoltaic modules in electrochemical wastewater treatment has been already reported. Within this field, there is still a wide range of possibilities that can be explored (Fernandes et al. 2015).

The applied electrical current was proved as a major factor affecting the EC performance. However, on the other hand, the increase in the applied electrical current directly affected the operational cost of the process, and, therefore, it should be meticulously optimized considering the whole parametric system (Çırak 2018). The cost of the electricity consumed during the process effectively diminished to $0.003. Therefore, it can be concluded that EC can be effectively implemented for the separation of colloidal mineral substances from heated aqueous systems because the process becomes easier, cheaper, and more effective at higher temperatures (Çırak 2018).

15.3 FUTURE PERSPECTIVES OF ELECTRO-COAGULATION

15.3.1 Improvement of Electro-Coagulation Systems for Scale-Up/ Commercialization

EC has achieved higher removal efficiencies by using principally Fe or Al anodes and has been considered as an economical technology. This electrochemical technology is able to reduce high pollutant contents considerably, although it does not achieve complete removal in some cases. Therefore, the use of alternate materials as sacrificial anodes has been reported for EC, such as Zn and Cu, to propose the most economical methodologies. In spite of the very good results obtained with electrochemical techniques on pollution abatement, EC processes are often seen as only interfacial reactions and therefore difficult (if not impossible) to be applied to large volumes of pollution. The *interfacial problem* is, in fact, a problem shared by several unit operations already established at the industrial level. However, it is a sort of handicap that, to be efficiently overcome, needs pilot- and semi-industrial-scale studies (Fernandes et al. 2015).

EC has demonstrated to be an effective and practical technology to treat leachates from sanitary landfills (Fernandes et al. 2015). Its application requires simple equipment, provides ease of operation, and avoids the need for chemical addition. Nevertheless, some aforementioned drawbacks are associated with the application of this process. To increase the performance of the EC process and to reduce the increase in conductivity of the remaining effluents, attention has to be paid to the metal ion concentration in solution, avoiding unnecessary high metal contents. Complementary studies like the one conducted by Fernandes et al. (2015)

should also be performed using aluminum to establish the ideal concentration of iron or aluminum ions in solution that allows high EC efficiency without unnecessary waste of the sacrificial anodes. In fact, EC performance depends on the availability to produce metal cations that act as coagulant. The production of metal cations shows an ability to bind together to form aggregates of flocs composed of a combination of glyphosate and metal oxide. Hence, by removing the glyphosate, the toxicity in the water is eliminated, meaning that the water has been rendered safe for consumption. Formation of new polymeric compounds confirms the mechanism and the formation of a new equation (Danial et al. 2019).

Changing the electrodes' polarity may help reducing the efficiency loss caused by the impermeable oxide film formed on the cathode. Also, constant applied intensity instead of constant applied potential reduces electrode passivation, although it increases energy consumption, because, to keep the intensity when a passivating film is formed, higher overpotentials have to be applied. This increase in overpotential presents an extra feature, that is, the possibility of the direct oxidation of the organic matter present in the effluent.

It is also possible to find reliable ways of reducing electrode passivation and the consequent inactivation of the EC treatment unit by the development of different electrode types and arrangements as well as more sophisticated reactor operational strategies. Dura and Breslin (2019) reported that AISI 420, a stainless-steel alloy with 13% to 14% Cr as an alloying component, dissolves to give relatively high concentrations of Fe^{2+}. It has good efficiencies for the removal of phosphates, an organic dye molecule, and Zn^{2+} ions, and comes close to the efficiencies of pure iron, while limiting the formation of rust (Dura and Breslin 2019). However, on increasing the chromium content with the AISI 310 alloy, the removal efficiency reduces considerably, and the anode adopts higher potentials to give the required current, which, in turn, gives rise to a trans-passive dissolution and secondary pollution by Cr(VI). This comparison clearly illustrates that the composition of steel alloys and, in particular, the chromium concentrations are highly relevant in EC. Although AISI 420 compares well with pure iron, it is more expensive than iron. However, it is widely used in the oil and gas production industry and is considerably cheaper than the highly corrosion-resistant steel alloys. Given that it is stable and not consumed when the EC unit is not in use, it should function longer than iron, bringing the final costs closer. This simple schematic can be easily scaled up to include multiple anodes/cathodes that facilitate the flow of water through the unit, making it suitable for small-scale water treatment facilities (Dura and Breslin 2019).

Integration of the EC process with other treatment methods can also be a possible solution to overcome some of the drawbacks, namely the solid-liquid separation and the sludge elimination. The EC technology can be designed to include membrane separation, reverse osmosis, electrofiltration, sludge dewatering, thermo-oxidation, and other conventional technologies to enhance the recovery of fine particles and metal ions from wastewater. This is an area that urges for intensive work in the future. Otherwise, EC will become just another way of diverting - the problem, with the addition of metal ions to the final solution.

Integration of EC or EO processes, respectively, besides enhancing the treatment efficacy also has the potential to eliminate possible toxic species, such as chromium, zinc, and part of the iron introduced during the EC process. However, this kind of studies need to be further conducted to include many other metal ions that were not included before. Research on the best experimental conditions are also welcome that will lead to a more efficient recovery of metals from solution and the possibility of the selective recovery of the most dangerous metal ions or those that are present in higher concentration (Fernandes et al. 2015).

More research efforts are needed to clarify the operating conditions that can be applied to scale up these methods to the industrial level, even when these parameters depend on each specific kind of effluents (Garcia-Segura et al. 2017). Nevertheless, the EC approach can prove more effective than other conventional technologies, even when the mechanism to eliminate pollutants depends strongly on the nature of the contaminant (heavy metal, organic matter, inorganic species, etc.) as well as the other species present in the effluent.

To solve the problems such as passivation encountered in the operation of the EC process, the configuration of the EC reactor needs to be optimized. In EC systems, a range of physical phenomena are related to the interaction of multiphysics field, including flow field, electric field as well as temperature field, which can be mathematically described by partial differential equations. Future developments will rely on the close collaboration of analytical chemists, engineers, and electrochemists to ensure effective application and exploitation of these electrochemical technologies for environmental protection.

Although the concept of employing electrical current to dissolve coagulants and generate oxidants in situ has a long history, recently, much progress has been made to elucidate various performance aspects, including pH profiles between electrodes, precipitated solid phases, electrode passivation, and other factors influencing EC performance. However, more work needs to be done since a universal understanding of the nature and composition of precipitated phases based on water chemistry and electrolysis conditions continues to remain elusive as well as the final fate of the sludge produced during the treatment of wastewater (e.g., ceramic materials, asphalt, construction applications). In addition, preliminary cost estimates suggest that EC is competitive to conventional coagulation, especially for smaller installations. Importantly, electrochemical treatment does not consume buffering capacity unlike hydrolyzing metal salt coagulants that behave as bronze acids.

More work remains to be done to predict precipitate morphology and composition based on background water chemistry and EC operating conditions. Such knowledge would aid our understanding of contaminant removal and fouling control by EC and future design of hybrid systems. Further, long-term pilot-scale on-site evaluations for individual applications are necessary to build an empirical database for integrated treatment approaches in the parallel and/or serial mode to developing a more detailed understanding of underlying mechanisms and best design practices (Garcia-Segura et al. 2017).

EC may be suitable for a wide range of water chemistries and eliminates the need to add base to maintain pH during coagulation (leading to reduced chemical handling and increased process simplicity). Promising results have been obtained when EC has been coupled with other water treatment technologies. However, electrode passivation and inconsistent coagulant dosing over long periods of operation continue to limit the real-world application of EC. The laboratory- and field-scale experiments are sufficiently encouraging to expand the EC processes to the pilot- or field-plant scale. To do so, many process variables require to be optimized in terms of electricity consumption, water aging times (for silica polymerization), water pH, EC cell design (including electrode geometry and separation), residence times in EC cells, residence times in settling tanks, floc recirculation, and EC cell dynamics.

15.3.2 Role of Nanotechnology in Electro-Coagulation

Coagulation is a chemical process that aggregates the particles via neutralization of their charges and then separates them from the liquid phase. Traditionally, chemicals with positive charge (like alum) are added to neutralize negative charges of the particles (Bratby 2016). The recovery factor (R_f) and the strength factor (S_f) of a coagulation process can be defined as follows (Hui et al. 2016):

$$R_f = \frac{d_3 - d_2}{d_1 - d_2} \times 100 \tag{15-1}$$

$$S_f = \frac{d_2}{d_1} \times 100 \tag{15-2}$$

where d_1, d_2, and d_3 are the average sizes of flocs before breakage, after breakage, and after regrowth, respectively. Recently, it has been reported that incorporation of nanoparticles as adsorbents or connectors in coagulation can enhance the aggregation of particles. EC is another method for treatment of wastewater which incorporates the consumable electrode to produce nanoions in the water. These ions immediately covert to polymeric hydroxides (like aluminum hydroxide) as coagulation agent (Comninellis and Chen 2010). Using nanoparticles in the process of EC has synergistic effect to neutralize the particles' charge.

Incorporation of Fe-TiO_2 and Al_2O_3 nanoparticles in coagulation successfully raised the floc size and the recovery factor (because of bridging effects between nanoparticles) and decreased the strength factor (Xu et al. 2014a, c). By analyzing the effect of operating parameters, it can be concluded that (1) higher nanoparticle concentrations enhance the rate of particle formation at an acidic or basic condition; and (2) at neutral pH, the size of floc increases because of the lack of charge on the surface of nanoparticles (Xu et al. 2014b, c). It is shown that using carbon nanotube (CNT) could efficiently enhance the removal of humic acid during the coagulation process (Hui et al. 2016). The influence of CNT incorporation is affected by two pathways: (1) the interaction between humic acid molecules and CNT particles, and (2) the coagulation process in humic acid removal. Nano-AL_{13} showed a higher charge neutralization ability in comparison with polyaluminum chloride and aluminum sulfate, thereby illustrating the highest efficiency in the treatment of TiO_2–humic acid water (Xu et al. 2014b). In addition, a stable structure of nano-AL_{13} was observed in different pH, leading to the generation of strong flocs with better regrowth ability (Xu et al. 2014b). Nanoclays, which are accessible in the nature can be added to the coagulant for improving the efficiency because of its high surface area and adsorption capacity. Incorporation of different nano clays for removal of polycyclic aromatic hydrocarbon in coagulation was compared; the removal effectiveness was found to be in the order of halloysite nanoclay < normal bentonite < hydrophilic nanobentonite < nanomontmorillonite (Shabeer et al. 2014). A comparison of different nanoclays, nanobentonites, nanohalloysites, and organically modified nanomontmorillonites for removing various pesticides by coagulation showed that organically modified nanomontmorillonite exhibits the best result (Shabeer et al. 2015). Investigation of the effect of silica-modified polyacrylamide nanomaterial as coagulant in separation of high density waste drilling mud revealed that this coagulant has more efficiency than normal polyacrylamide (Wang et al. 2011).

Incorporation of metallic nanoparticles such as magnetite and zerovalent iron (ZVI) in the EC process can enhance its efficiency. ZVI, as an unstable and effective nanomaterial with a specific surface area of 20 to 25 m^2/g, can enhance the reduction in the EC process (Sukopová et al. 2013). Nunez et al. (2011) showed that the addition of ZVI into the EC process could enhance arsenic removal efficiency at an acidic pH (in the range of 2 to 4) by precipitation of arsenic. Analysis of the operating parameters of this process showed the concentration of ZVI did not increase the removal efficiency. Incorporation of ZVI not only can enhance the efficiency of the EC process, but also can help in removing heavy metals such as Cu, Pb, and Zn (Hansen and Ottosen 2010, Nunez et al. 2011). This process can also be combined with an electromagnetic separator for removing heavy metals (Núñez and Hansen 2013).

Other than the addition of nanomaterials to water, they can be produced during the process or can be used in construction of electrodes for EC processes. As the combination of ultrasound and H_2O_2 leads to hydroxyl radicals generation, their incorporation in an EC process with an Fe electrode can assist in the formation of magnetite nanoparticles by oxidizing $Fe(OH)_2$ (Darvishi Cheshmeh Soltani et al. 2020). The formed magnetite nanoparticle can play two notable roles: (1) as an efficient coagulant (because of its high surface area), and (2) in production of $OH^{\bullet}$ using H_2O_2 and ultrasound (Darvishi Cheshmeh Soltani et al. 2020). CNT can be used as an electrode material for EC as a well-liked carbonaceous nanomaterial with unique properties such as high electrical conductivity, high chemical stability, appropriate surface area, and remarkable mechanical strength (Garcia-Segura et al. 2017, Karimi Estahbanati et al. 2019, Zarei et al. 2009). Using CNT-polytetrafluoroethylene (PTFE) electrode as a cathode in proxy coagulation improved the degradation efficiency of dyes (Zarei et al. 2010). This electrode can generate high amounts of hydrogen peroxide because of high surface area of CNT. The generation of H_2O_2 with the CNT-PTFE electrode was found to be three times more than that of the carbon-PTFE electrode (Zarei et al. 2010). The peroxicoagulation process by CNT-PTFE had more degradation efficiency in comparison with electro-Fenton and peroxicoagulation by carbon-PTFE (Zarei et al. 2009).

Metal oxide nanoparticles exhibit better properties than their microparticles in terms of particle size and specific surface area (Chen and Deng 2012, Shon et al. 2010). Controlling the size of formed metal oxide nanoparticles during EC can enhance the process efficiency. Current intensity is an important parameter that can influence the size of nanoparticles formed during the EC process,, thereby affecting the rate of formation and size of the flocs. It is shown that the removal efficiency is increased by raising the current intensity (Chen and Deng 2012, Darvishi Cheshmeh Soltani et al. 2020, Khandegar and Saroha 2013). The type of electrolyte is another important factor that affects the size of generated nanoparticles. It is shown that using $(NH_4)_2SO_4$ electrode in EC generated smaller nanoparticle than NH_4Cl and $(NH_4)HCO_2$ (Tchomgui-Kamga et al. 2013). Water temperature can also affect the EC process by changing the pore size of nanoparticles (Kamaraj et al. 2016).

The operating conditions of EC can lead to various pathways for formation of nanoparticles with different sizes. Dubrawski et al. (2015) investigated the production of different mixed-valent iron nanoparticles in EC and found that the absence/presence of oxygen or ions and also the duration of the experiment affect the type of generated iron nanoparticles in forms of magnetite, green rust, goethite, and lepidocrocite (Figure 15-1). They proposed two main pathways for the generation of nanoparticles via Fe(0): (1) as a primary product (formation within seconds in the absence of dissolved oxygen), and (2) as a transformation product of green rust (from minutes to days depending on pH, electrolyte composition, and aging conditions). Each nanoparticle has specific surface area

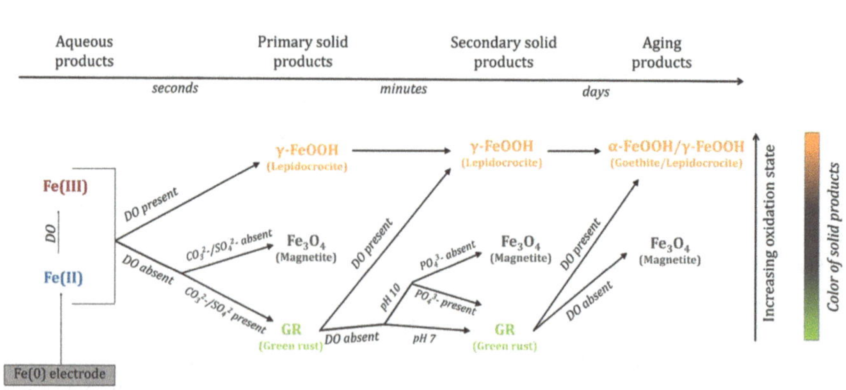

Figure 15-1. Schematic diagram of the formation pathways of mixed-valent Fe with different nanoparticles during EC.

Source: Adapted with permission from Dubrawski et al. (2015). © 2015 American Chemical Society.

affecting the efficiency of EC; for example, green rust was specified by smaller size than magnetite.

15.3.3 Combination with Other Treatment Processes

The full potential of the EC technique as a wastewater treatment alternative has yet to be fully realized. It is still an empirically optimized process that requires a more fundamental understanding for achieving improved engineering design. The EC comprises complex chemical and physical processes involving many surface and interfacial phenomena. There is, however, a paucity of scientific understanding of these phenomena, which limits the engineering design of the EC reactor for achieving optimum performance and for future progress of this innovative technology. A literature survey indicates that up-flow configuration of the EC unit (reactor) is better than a horizontal flow configuration. Selecting an EC configuration should be based on the company's priority for cost (electrical consumption) or removal efficiency. For alleviating electrical consumption and obtaining high removal efficiency the combination of both EC configurations could be suggested.

While when EC alone would not be effective for pollutants removal, EC process can be also hybridized or integrated with other wastewater treatment processes, for example, advanced oxidation process (AOP, membrane filtration technology as in membrane bioreactor systems. EC/ozone as a hybrid process exhibited high efficiency in COD and TOC removal from the greywater. EC/ozone was well operated in neutral conditions in which the pH of greywater was close to neutral condition. 85% of COD and 70% of TOC were removed by EC/ozone under conditions of pH = 7.0, 15 mA/cm^2, 47.4 mg/L ozone, and 60 min electrolysis time. EC with iron electrode had a synergistic effect for the EC/ozone process whereas EC with aluminium electrode did not indicate such effect. Ozone compared to other chemical oxidants (H_2O_2, PMS and PDS) was more efficacious in combination with the EC process. In contrast with US irradiation, UV could significantly affect the EC/ozone process in terms of organic compounds removal and inactivation of bacteria. UV improved disinfection performance of EC/ozone/UV with 4 logs removal of total coliform and 96% of *Escherichia coli*. The costs of EC/ozone and EC/ozone/UV were estimated 1.9 and 4.03 ($\$/m^3$), respectively. The results showed that EC/ozone/UV can be a practicable method for treatment of wastewater and disinfection of polluted water with high efficiency (Barzegar et al. 2019). In addition, the use of a sonic field, stirrer, and hybrid EC–EF technique increases efficiency and provides better control of this process. However, additional test series concerning continuous operations are mandatory to generalize this concept for future large-scale applications (Mohammadi et al. 2019).

15.3.4 Fuel Cell and Use of Renewable Energies

Its large electric consumption of EC is considered to be the main limiting factor. The possibility of improved or alternative options (e.g., fuel cells) has been widely sought to overcome such limitations. Therefore, many studies have been carried out recently (Garcia-Segura et al. 2017, Hakizimana et al. 2017, Maitlo et al. 2019, Moussa et al. 2017, Reilly et al. 2019, Song et al. 2017) to develop various types of fuel cell (e.g., AMD-FC, AEFC, air-cathode FC, and MFC) technologies that have proven to be efficient both in the removal of pollutants and in power generation.

The performance of fuel cells changes greatly in terms of power generation depending on the kind of fuel cell used. As such, their performance is considerably affected by the type and concentration of the supporting electrolyte. Conventional fuel cell systems are, however, found to have very low treatment capacity (e.g., 12 and 150 mL) at the expense of extended operating times (e.g., 10 and 74 h). In addition, the operating costs of different methods were compared, for example between metal-air fuel cell EC (MAFCEC) and conventional methods. Accordingly, it was confirmed that the MAFCEC is the most cost-effective treatment option for the treatment of pollutants like As (arsenic) in water. The previously mentioned limitations are the most important challenges to overcome to expand the practical applications of these kinds of fuel cells. Among all the available fuel cell technologies, MAFCEC has been investigated most extensively to pursue treatment of As along with power generation. The basic configuration of MAFCEC systems has been divided into a variety of forms (e.g., single-chambered batch reactor, double chambered with continuous recirculation, single chambered with continuous recirculation, and a single-chambered stack with continuous circulation) with various treatment capacities (e.g., 90 mL to 20 L). Similarly, the effects of numerous parameters like the supporting electrolyte (e.g., NaCl, Na_2SO_4, and $NaHCO_3$), initial pH of electrolyte, type of membrane (e.g., anionic and cationic membranes), and type of catholyte (e.g., HCl and NaCl) have been investigated for treatment of As in water to assess the performance of MAFCEC systems. Accordingly, it was confirmed that the stacked IAFCEC would be the best option for practical water treatment technology. This system is less dependent on the concentration of the supporting electrolyte and its initial pH. Therefore, based on this type of performance evaluation, it is concluded that the MAFCEC system is the most energy-effective treatment method for the treatment of As in water. The MAFCEC system is a novel form of EC process for the treatment of As contaminated water without an external electricity supply. Therefore, this innovate type of fuel cell technology can be applicable to secure drinking water in remote areas where commercial electricity supply is limited. To date, the real-world applicability of this technology has been investigated for the removal of As along with electricity energy generation with a small-scale MAFCEC and liter-scale MAFCEC stack. However, further research on MAFCEC with a gravitational water flow system will be required to enhance its real-world applicability as the electrical energy produced by the MAFCEC would not be sufficient to operate the pump for water circulation. In addition, future work on the application of MAFCEC systems as a versatile treatment process will be required to examine effects of a wide range of water pollutants (e.g., different types of heavy metals, natural organic matter, fluoride, oil and greases, algae, viruses, turbidity, colors, and nutrients) on the performance of MAFCEC systems. The development of this technology would help achieve a new milestone for the establishment of a sustainable, cost-effective, and efficient technology for removal of pollutants.

Among the factors limiting widespread application of EC, electricity consumption, as one of the major operating costs in the EC process, should be given priority. Such prioritization is necessary because a large-scale and continuous operation of the EC process for arsenic removal from wastewater requires considerable power. With the rapid development of green energy, application of renewable energy sources such as solar energy, biochemical fuel cells, and wind energy for continuous EC operation is considered to be cost-effective and environmentally friendly. As a result, the technology of green energy coupling with EC seems to be promising for large scale arsenic removal from wastewaters.

Also, to lower the energetic costs, these technologies should be applied to partial streams or as integrated processes with biological solutions. The energetic consumption can also be solved with green energy, like wind energy, tidal or solar, or using biogas from an anaerobic digestion reactor to produce energy. For example, incorporating renewable energy sources in the electrochemical treatment of sanitary landfill leachate (SLL)s closes with a flourish future prospects in this field (Fernandes et al. 2015). A solar-powered EC system (Nawarkar and Salkar 2019) is applied successfully for the municipal wastewater (MWW) in Sangli, Maharashtra, India, using aluminium electrodes and the results showed that (1) lab-scale continuous solar-powered EC (SPEC) process is found significantly effective by using batteries, (2) removal efficiencies for current density and hydraulic detention time of 48 A/m^2 and 16 min, respectively, are 90% for COD, 94.56% for turbidity and 49.78% for TDS, (3) the effluent pH is within the range of 7.4 to 8.5, and, therefore, further treatment for pH correction is not required; and 4) The SPEC reactor configuration tried is feasible for the treatment of MWW and is also cost effective (Nawarkar and Salkar 2019).

Another major area for future research is to improve the electrocatalytic activity and electrochemical stability of new electrode materials and to develop different electrode types and arrangements and more sophisticated reactor operational strategies, to reduce operational costs and to overcome some other identified drawbacks. With the same goal, integration of these methods with each other or with other treatment methods should also be explored further.

For the electrochemists and chemical engineers, this work is extremely significant because future advances will include more real applications of EC in the industry that will motivate the substitution of traditional treatment processes by new alternatives. The future for EC applicability is bright. Some engineers are starting to make important contributions to design industrial plants and to start the commercialization of this technology to more industries because it is important to remark that a significant amount of industrial producers of EC systems already exist; thus the commercialization of this technology is already effective, some years ago.

15.3.5 Cost Estimation of Electro-Coagulation Treatment Processes

Its large electric consumption of EC is considered to be the main limiting factor. There is a study (Reilly et al. 2019) highlighting the range in EC power consumption (kW · h/m^3) required by various investigators to process meat and dairy waste (Reilly et al. 2019). The construction of detailed life-cycle analyses is crucial to quantify the capex, opex, and environmental impacts before widespread adoption of the EC technology in the slaughterhouse meat and dairy processing industries. EC has good potential for use in the treatment of dairy processing and slaughterhouse meat industry wastewater effluents. Typically, investigators have been able to remove over 75% of COD from both dairy and meat waste effluents when using Al (aluminum) and Fe (iron) electrodes in batches for less than 1 h. The application of EC technology for treatment of dairy and meat effluents remains in its infancy and information about larger-scale operation of EC for the treatment of these waste effluents is still lacking. Therefore, more pilot scale studies would be beneficial for supporting the transition of the technology from laboratory to industrial scale.

Furthermore, the vast majority of studies have applied batch EC methodologies but there may be cost benefits associated with implementing continuous EC treatment. Note that the energy analysis does not include the mechanical energy required for the electrode oscillation, and, therefore, further work is needed to explore this aspect.

The operating costs of EC (laboratory scale) were calculated with the optimum treatment conditions (Al electrodes, 200 mA and 45 min versus Fe electrodes, 400 mA, 60 min for EC). Al electrodes may present a good alternative to Fe electrodes for Fe-rich acid mine drainage (AMD) treatment at a lower current intensity and reaction time (lower investment and operating costs), but they require pH adjustment of final effluent, which could entail additional treatment costs (\$0.03/m^3 treated effluent because of the cost of NaOH). The results of the calculations showed a slight cost variation between the two electrodes. For a complete Fe removal, the operating cost (\$/m^3) for treating Fe-rich AMD with Fe electrodes was around 0.56 versus 0.81 with Al electrodes (a working volume of 250 mL). To estimate the real cost of microbial fuel cells (MFCs) and EC at the field scale, additional data must be considered, such as direct costs related to labor (operating and supervision wages), sludge dewatering and disposal as well as indirect costs related to administrative staff, overhead, insurance and taxes, R&D, reimbursement of capital costs including all equipment required.

Comparative efficiency for Fe removal from synthetic Fe-rich AMD (pH = 2.4 to 2.5 and Fe around 500 mg/L) using MFCs and EC was evaluated in batch testing. The parameters were optimized, including sodium acetate concentration in the MFC (200 to 1,000 mg/L), electrode materials (Fe and Al), reaction times (15 to 90 min), and current intensities (100 to 500 mA) in the EC cell. Both technologies were able to produce a treated effluent complying with the Canadian discharge standards (pH 6.0 to 9.5, Fe <3 mg/L). Yet, EC in its optimal conditions (Fe electrodes, 400 mA and 60 min) was found to be more cost-effective than MFC. The EC process also showed more suitability for large-scale use, because the operating time in EC is comparatively shorter than that of the MFC (minutes versus days). Further studies should be conducted by using different types of AMD with more complex compositions (e.g., polymetallic contaminated AMD at higher concentrations and real AMD).

15.3.6 Mathematical Model of Electro-Coagulation

The study of the electro-dissolution of the aluminum electrodes was performed experimentally in a batch EC reactor (Graça et al. 2019). Both total and dissolved aluminum concentrations were determined during the EC operation. The total aluminum concentration was fitted with Faraday's equation, and a current efficiency value of 1.6 was obtained, indicating a super-faradaic behavior of the electrochemical process. The determination of the dissolved aluminum was performed for cases when the electrical current is on and off. The results obtained indicated that different coagulation mechanisms should be present in each case. A mathematical model was developed. The simulated results showed that the model reasonably predicts the evolution of both concentrations of dissolved aluminum and pH. In addition, the mathematical model developed was able to reasonably predict the EC operation performance for As removal. Moreover, the presented model intends to be a base for the development of more complex models that may be required to predict the performance of the EC reactor to treat a specific effluent (Graça et al. 2019).

Continual research using this technology will not only improve its efficiency, but new modeling techniques can be used to predict many factors and develop equations that will predict the effectiveness of treatment (Butler et al. 2011). For future work, the empirical model developed here could be enhanced by including the influences of other operating parameters (such as water temperature) and/or by studying wider ranges of the parameters already included. The sludge produced could be characterized using scanning electron microscope technology to provide information on the development of better sludge handling methods (Karimi Estahbanati et al. 2019).

As a result, future research is to overcome the drawbacks mainly on high energy consumption in large scales and difficulty of EC reactor design, and to enlarge the practical application and deeply explore the EC performance.

15.4 CHALLENGES OF ELECTRO-COAGULATION

Electrolytic cells offer multiple technical benefits, including a lack of need for storage and handling of chemicals, easily automated operation, compact and modular reactor design, and an ability to adjust to variable organic loads in wastewater. Under optimum experimental conditions, EO has been reported to remove almost all COD and ammonium nitrogen in wastewater. Electrochemical systems are particularly well suited for the treatment of hazardous organic waste. For example, anodes made of materials such as boron-doped diamond (BDD) and Ti/SnO_2 can be used for the treatment of aqueous film-forming foams that contain high concentrations of fluorinated surfactants. In the case of wastewater with a high concentration of halides, EO may be coupled or integrated with granular activated carbon to minimize the release of chlorinated organic by-products. Nevertheless, for some waste streams, electrochemical treatment may not be technically and economically feasible because of increased toxicity of the treated effluent. Efforts to overcome these fundamental obstacles that have thus far limited the practical use of traditional EO processes have been focused on reducing their chemical and energy input demands, targeting the utilization of heterogeneous catalysts, and, specifically, investigating engineered nanomaterial (ENM) applications to the EO process.

15.4.1 Electro-Oxidation Reactor Design and Operation

One of the major criteria that should be considered for recommending a sustainable treatment technology or a combination of treatment technologies is treatment efficiencies. EO as a pretreatment step may solve the problem as it is capable of breaking down higher molecular compounds to lower molecular compounds and making the EO treated wastewater suitable for biological treatment. EO can be employed as a complementary treatment system with biological process for wastewater treatment or as an auxiliary unit in case of any emergency situations (Deng and Englehardt 2007). For example, EO processes are effective for treatment of bioreactor landfill leachate, particularly for ammonium nitrogen removal (Anglada et al. 2010b). However, the recommendation of any general treatment system or combination of treatment systems for wastewater treatment is difficult due to the varying composition of wastewater from site to site and time to time. Above all, the treatment system must be simple, universally adaptable, and able to eliminate negative environmental impacts of wastewater (Mandal et al. 2017).

Significant hurdles must be overcome before bench-scale ENM discoveries can be translated to water treatment practices. A critical issue frequently neglected in research focused on ENM development is diminished performance of the ENM in complex water matrices. Another issue that needs to be carefully considered during the ENM development stage pertains to reactor design and systems engineering. Further, many ENMs with state-of-the-art nanoarchitectures have not yet significantly infiltrated the field of water treatment despite their possessing unique properties that are particularly noteworthy for environmental applications (Hodges et al. 2018).

The feasibility of the scale-up of these approaches for commercial purposes and the economics of the processes involved for large-scale application have not been investigated. Many of the recent technologies focusing on the improvement of wettability of surfaces (or reduction of interfacial problems) have been carried out using facile methods and synthetic oily wastewater, and, therefore, more research efforts still need to be directed toward the development of practical solutions for real-world applications. Innovative techniques for membrane fouling reduction, such as the use of sponge balls to wipe the membrane surface and clean the membrane of accumulated debris, should be enhanced. Doing this will improve membrane efficiency, prolong membrane life, and reduce operating costs. Guo et al. (2008) achieved a 100% increase in sustainable membrane flux, 95% DOC removal, and 97% COD removal through the addition of sponge with a volume fraction of 10% in a membrane bioreactor (Jamaly et al. 2015).

15.4.2 Cost and Environmental Impact of Electro-Oxidation

One critical challenge to wider adoption of electrochemical treatment of industrial and environmental applications is a need to reduce energy consumption and cost. In conventional wastewater treatment plants (WWTPs) that process 100,000 to 500,000 m^3 of wastewater daily, the cost would be about \$0.21/m^3 (Tang et al. 2019). The energy consumption for electrochemical technologies is 1.0 to 3.43 kW · h/m^3 for sea water and industry wastewater (Tang et al. 2019), which is \$0.06 to 0.21/m^3 if assuming 1 kW · h costs \$0.06. For large-scale municipal WWTPs with a high treatment capacity, the energy consumption of 1.0 kW · h/m^3 by the EO process is still very high, which indicates that the electrochemical method is not economical at present (Tang et al. 2019).

High operational cost for electricity as needed for all electrochemical methods and probable formation of chlorinated organic compounds during the chlorine mediated oxidation process especially for high chloride containing wastewater treatment are identified as major drawbacks of the system (Anglada et al. 2010a). However, the EO process can be applied in combination with other biological/physicochemical processes to overcome the drawback of high energy consumption (Mandal et al. 2017).

Another challenge for electrochemical treatment is a need to control the formation of toxic by-products when treating halide-containing wastewater. When electrochemical treatment systems are developed, bench-scale studies should be performed with authentic water samples and, when oxidation occurs, measurements of adsorbable organic halogen, toxicity, chlorate, and perchlorate are needed. Given that most contaminants are present at low concentrations, their electrochemical reactivity will be subject to mass transfer limitations. Thus, research on electrochemical treatment systems should be redirected toward the actual conditions that occur in wastewater (e.g., concentrations of contaminants that are lower than those of dissolved salts and organic matter) (Radjenovic and Sedlak 2015).

Pressures including environmental pollution, stricter wastewater effluent discharge standards, and so on need to decrease carbon emissions in WWTPs (Rothausen and Conway 2011). In general, implementation of energy optimization measures is accompanied by a reduction in greenhouse gas (GHG) emissions. The amount of GHG emissions saved cannot be generalized as the reduction of GHG depends on the particular situation of electricity produced in the country where the WWTP is located (Nowak et al. 2011). Environmental pollutants show a remarkable decrease after anaerobic digestion (Dong et al. 2014). However, at present, three quarters of total methane emission originated from WWTPs were avoided by utilizing the biogas. Methane emissions can be significantly reduced by better handling of the ventilation air of sludge handling facilities. Methane present in the wastewater was for a large part aerobically oxidized in the activated sludge tanks. This could be exploited as a means to further reduce methane emissions from wastewater treatment (Daelman et al. 2012, Tang et al. 2019).

15.5 FUTURE PERSPECTIVES OF ELECTRO-OXIDATION

The main obstacles to be overcome before the full-scale implementation of electrochemical oxidation is the reduction of operating costs and the development of efficient and stable electrode materials. Another major area for future research is the improvement of the electrocatalytic activity and electrochemical stability of the electrode materials for the development of better and more cost-effective electrodes, which will result in lower operational and capital costs (Anglada et al. 2009).

15.5.1 Improvement in Electro-Oxidation Systems for Scale-Up/Commercialization

Despite successes in laboratory-scale tests, a practical large-scale implementation of EO for water and wastewater treatment has yet to be achieved. Progress in the development of consistent mathematical models will undoubtedly contribute to this field; these models can describe the complex scenarios and reactions involved in the EO of multicomponent mixtures of various pollutants as well as integrate knowledge about reactor hydrodynamics with the design of EO systems. In 2007, a BDD EO pilot plant was installed in the context of a project focusing on the treatment of landfill leachate using AOPs. Raw leachate contains about $2.8\ g \cdot L^{-1}$ of TOC and $1.2\ g \cdot L^{-1}$ of ammonia. On-site tests were performed combining aerobic treatment, the Fenton process, and EO. The EO consists of an electrochemical reactor with $1.05\ m^2$ of BDD anodes. The overall efficiency in the combined system was 99% organic matter mineralization, that is, 50% of the initial TOC was degraded in the aerobic treatment, 35% in the Fenton process, and the remaining 15% in the EO step. The ammonia removal efficiency was more than 90%, of which 50% was attributed to EO treatment, because the Fenton process was unable to reduce the ammonia concentration. The results were consistent with the predictions obtained from tests previously performed at the laboratory scale, showing the robustness of the system (Anglada et al. 2009). Using flow-through reactors with three-dimensional electrodes has been shown to be an approach that can significantly enhance EO performance compared with the conventional two-dimensional reactor design at the bench scale. Also, improving the long-term stability and electrocatalytic performance of electrode materials and thereby lowering the cost of these materials is a critical area of research where new developments in materials science and nanotechnology may be important (Radjenovic and Sedlak 2015).

15.5.2 Role of Nanotechnology in Electro-Oxidation

Recent advances in nanotechnology have stimulated extensive research investigating ENM applications to AOPs. In this perspective, we critically evaluate previously studied ENM catalysts and the next-generation treatment technologies they seek to enable. Opportunities for improvement exist at the intersection of materials science and treatment process engineering, as future research should aim to enhance catalyst properties while considering the unique roadblocks to practical ENM implementation in water treatment (Hodges et al. 2018).

There is an urgent need for future research to rationally design ENMs for the intended purpose of water treatment. In addition to optimizing novel engineered systems from a technical standpoint, much remains to be done to accurately assess their worth with respect to cost and life-cycle impacts, which are largely variable and uncertain at these early stages of development. Provided such advancements, the demonstrated properties of these materials suggest significant opportunities for ENM catalysts in the development of modular decentralized water treatment systems, which can ultimately transform how water is treated, reused, and safely discharged to the environment (Hodges et al. 2018).

15.5.3 Combination of Electro-Oxidation and Other Treatment Processes

In general, a single technology cannot satisfy all reuse and disposal requirements for different types of waste/wastewater. Combining biological methods with EO is by far the most suitable alternative found in the literature. For example, the use of microbial decomposition to remove pollutants from oily wastewater may have its drawbacks, because most of the biological processes are time-consuming and require highly expensive enzymes. In the future, therefore, an integrated approach seems plausible to remove the many components with dissimilar chemical and biological compositions and properties in oily wastewater. This can ensure that the different regulatory limits in many localities or municipalities are met (Jamaly et al. 2015). Although biological-EO coupled system performances are evaluated in lab scale (Lei et al. 2007, Aloui et al. 2009, Feki et al. 2009, Del Moro et al. 2016), more pilot and real field studies are required for detecting real constraints related to application of this technology. Therefore, more studies should focus on the implementation of full-scale biological-EO coupled technology with detailed economic analysis. Improvement of electrochemical properties of the electrode materials is another important area of research in this field (Deng and Englehardt 2007) (Mandal et al. 2017).

In general, there is no sequence of process for integrated treatment whereas two or more water treatment technologies are combined into a reaction vessel or single treatment train to serve the purpose. In a study, Senthilkumar et al. (2012) assessed the EO performance in different operational configurations such as individual, combined, and integrated schemes for the removal of a reactive dye of Procion Scarlet in synthetic effluent. Interestingly, they found that the EO performance in individual and integrated treatments was almost identical. The maximum COD and color removal efficiencies of 80% and 97.9%, respectively, were achieved with the individual EO treatment at the optimum current density of 102.5 mA/cm^2, whereas 80% and 96.4% COD and color removal efficiencies were achieved at a lower current density of 68.4 mA/cm^2 for the integrated treatment. Out of the three operational configurations, the combined treatment (i.e., biological followed by EO post-treatment) resulted in the maximum COD removal and color removal rates of 90% and 98.5% at the current densities of 76.0 and 102.0 mA/cm^2, respectively. However, the integrated treatment required lesser energy than the combined treatment.

Urtiaga et al. (2013) assessed a pilot-scale integrated treatment consisting of activated sludge, ultrafiltration, RO, and EO for the removal of 77 emerging contaminants from municipal WWTP effluent. The final EO treatment step using BDD anode was able to reduce the micropollutants content from 149 to less than 10 μg/L. This study has demonstrated the good performance of integrated treatment as well as the ability of EO as the final polishing step to reduce the trace emerging pollutants from the treated effluent.

Another interesting process-hybridization study by Moreira et al. (2016) investigated the feasibility of incorporating electrochemical advanced oxidation processes (EAOPs) in a multistage combined treatment system for the remediation of SLL. They combined biological treatment, coagulation/aeration, EOAP, and biological treatment for removing the biodegradable organic matter, ammonium ions, alkalinity, humic acids, and suspended solid particles. They found that the biodegradable organic carbon was almost completely removed after 42 to 47 h of biological system, whereas BOD_5, DOC, and COD abatements were 90% to 95%, 13% to 33%, and 9% to 31%, respectively. In this regard, the biological treatment was capable of removing the nonbioresistant fraction (i.e., high BOD_5 removal), whereas the bioresistant fraction such as humic acids (i.e., relatively low DOC and COD abatements) was not removed. Subsequent pretreatment using coagulation/aeration resulted in large decolorization of the treated effluent owing to the removal of humic acids, total suspended solids, and volatile suspended solids that altogether contributed to 63% to 65% DOC abatement. However, they still found that the presence of nitrite in the influent of EAOP presents a negative effect on the additional H_2O_2 consumption and high pH drop. A subsequent study (Moreira et al. 2016) on the similar biological and coagulation pretreated spent pot liner revealed the important role of BDD EO treatment on the organics degradation under potent solar radiation. They also evaluated different operational variables of EO processes, such as initial total dissolved iron concentration ([TDI]0), pH, Fe(III)-to-oxalate molar ratio, operating temperature, and UV lamp sources for the optimum performance of EO post-treatment (Garcia-Segura et al. 2018). They concluded that the most outstanding results were the complete removal of persistent organic pollutants and the wide range of application of this technology.

15.5.4 Cost Estimation and Environmental Impact of Electro-Oxidation Processes

Cost and environmental impacts of the treatment system are also major criteria that should be considered for application in a sustainable treatment technology or in a combination of treatment technologies. How to minimize power consumption of electrochemical technology furthest is the development direction in the future. Without any doubt, newer electrode materials that will ensure lower energy consumption and more deposition of pollutants are required in the future. To reduce the environmental stress resulting from the high energy requirements of EO treatment, future research can also be directed toward the use of renewable or clean energy sources such as solar and wind energy. Doing this will ensure a reduction of depletion of energy resources and improvement in the carbon footprint for wastewater treatment via EO methods (Jamaly et al. 2015).

A challenge for municipal wastewater treatment is the protection of human health and the environment by reducing pollution, while reducing energy consumption. A paradigm shift is underway toward making wastewater treatment facilities more sustainable by recovering energy and materials. Although EO technology is an environmentally friendly method for advanced treatment in WWTPs, it is not presently economically suitable for isolated or large-scale treatment of wastewater. This indicates that general conventional activated sludge (CAS), combined with biological technologies (sludge digestion and anaerobic digestion) and bioelectrochemical treatments (BETs), can not only achieve energy-neutral operations but also yield an energy surplus. For CAS combined with advanced AC adsorption treatment (PAC-sand filter), biological technologies (sludge digestion and anaerobic digestion), BETs, and electrochemical technology for AC regeneration to achieve an energy-neutral system, GHG emissions would be greatly reduced after reclaiming biogas and using BETs in WWTPs. With continued research, coupling WWTPs with biological technologies, BETs, and electrochemical technologies could increase electricity recovery enough to achieve energy self-sufficiency and decrease GHG emissions (Tang et al. 2019).

It is prudent to evaluate a proposed technology for its environmental footprint prior to its recommendation for widespread applications. In recent times, life-cycle assessment (LCA) has gained popularity as a technique to assess potential environmental impacts corresponding to all the life-cycle stages (from cradle to grave) of products/processes. It has been found that the International Organization for Standardization–based LCA method is the most reliable and transparent method among all the methods to verify potential impacts and evaluate the environmental footprint of any industrial process, including waste treatment methods.

15.6 SUMMARY

Electro-coagulation has a long history as a water and wastewater treatment process and has been employed to remove various types and a wide range of pollutants.

However, the scarce of a systematic approach to EC reactor design/operation has limited its implementation. There is a growing need for applying EC processes to water and wastewater treatment. Further studies can promote successive passing from laboratory-scale models to the large-scale or even industrial-scale applications.

Electro-oxidation is a competitive and promising technology in terms of treatment efficiencies and is one of the eco-friendly processes in which there is minimal or no sludge generation during the treatment of wastewater. The process does not require any chemical coagulants during the treatment process. To solve the problem of high energy consumption, further research on EO process should be focused on developing cost effective and stable electrodes and EO combined with other efficient technologies for full scale implementation. Pilot scale studies should be conducted for detecting actual constrains in the field, simulating real-world conditions using actual wastewater. Furthermore, LCA should also be incorporated into the decision-making process.

References

Aloui, F., F. Fki, S. Loukil, and S. Sayadi. 2009. "Application of combined membrane biological reactor and electro-oxidation processes for the treatment of landfill leachates." *Water Sci. Technol.* 60 (3): 605–614.

Anglada, A., R. Ibanez, A. Urtiaga, and I. Ortiz. 2010a. "Electrochemical oxidation of saline industrial wastewaters using boron-doped diamond anodes." *Catal. Today* 151 (1–2): 178–184.

Anglada, A., D. Ortiz, A. Urtiaga, and I. Ortiz. 2010b. "Electrochemical oxidation of landfill leachates at pilot scale: Evaluation of energy needs." *Water Sci. Technol.* 61 (9): 2211–2217.

Anglada, Á., A. Urtiaga, and I. Ortiz. 2009. "Contributions of electrochemical oxidation to waste-water treatment: Fundamentals and review of applications." *J. Chem. Technol. Biotechnol.* 84 (12): 1747–1755.

Ashraf, S. N., J. Rajapakse, L. A. Dawes, and G. J. Millar. 2019. "Electrocoagulation for the purification of highly concentrated brine produced from reverse osmosis desalination of coal seam gas associated water." *J. Water Process Eng.* 28: 300–310.

Asselin, M., P. Drogui, S. K. Brar, H. Benmoussa, and J.-F. Blais. 2008. "Organics removal in oily bilgewater by electrocoagulation process." *J. Hazard. Mater.* 151 (2–3): 446–455.

Barkley, N., C. Farrell, and T. Gardner-Clayson. 1993. "Alternating current electrocoagulation for superfund site remediation." *Air & Waste* 43 (5): 784–789.

Barzegar, G., J. Wu, and F. Ghanbari. 2019. "Enhanced treatment of greywater using electrocoagulation/ozonation: Investigation of process parameters." *Process Saf. Environ. Prot.* 121: 125–132.

Bratby, J. 2016. *Coagulation and flocculation in water and wastewater treatment.* London: IWA.

Butler, E., Y.-T. Hung, R. Y.-L. Yeh, and M. Suleiman Al Ahmad. 2011. "Electrocoagulation in wastewater treatment." *Water* 3 (2): 495–525.

Chen, X., and H. Deng. 2012. "Removal of humic acids from water by hybrid titanium-based electrocoagulation with ultrafiltration membrane processes." *Desalination* 300: 51–57.

Çırak, M. 2018. "High-temperature electrocoagulation of colloidal calcareo-argillaceous suspension." *Powder Technol.* 328: 13–25.

Comninellis, C., and G. Chen. 2010. *Electrochemistry for the environment*. Berlin: Springer.

Daelman, M. R., E. M. van Voorthuizen, U. G. van Dongen, E. I. Volcke, and M. C. van Loosdrecht. 2012. "Methane emission during municipal wastewater treatment." *Water Res.* 46 (11): 3657–3670.

Dalvand, A., M. Gholami, A. Joneidi, and N. M. Mahmoodi. 2011. "Dye removal, energy consumption and operating cost of electrocoagulation of textile wastewater as a clean process." *Clean–Soil, Air, Water* 39 (7): 665–672.

Danial, R., S. Sobri, L. C. Abdullah, and M. N. Mobarekeh. 2019. "FTIR, CHNS and XRD analyses define mechanism of glyphosate herbicide removal by electrocoagulation." *Chemosphere* 233: 559–569.

Darvishi Cheshmeh Soltani, R., S. Jorfi, S. Alavi, P. Astereki, and F. Momeni. 2020. "Electrocoagulation of textile wastewater in the presence of electro-synthesized magnetite nanoparticles: Simultaneous peroxi- and ultrasonic-electrocoagulation." *Sep. Sci. Technol.* 55 (5): 945–954.

Del Moro, G., L. Prieto-Rodríguez, M. De Sanctis, C. Di Iaconi, S. Malato, and G. Mascolo. 2016. "Landfill leachate treatment: Comparison of standalone electrochemical degradation and combined with a novel biofilter." *Chem. Eng. J.*, 288: 87–98.

Deng, Y., and J. D. Englehardt. 2007. "Electrochemical oxidation for landfill leachate treatment." *Waste Manage.* 27 (3): 380–388.

Dia, O., P. Drogui, G. Buelna, and R. Dubé. 2018. "Hybrid process, electrocoagulation-biofiltration for landfill leachate treatment." *Waster Manage.* 75: 391–399.

Dong, J., Y. Chi, Y. Tang, F. Wang, and Q. Huang. 2014. "Combined life cycle environmental and exergetic assessment of four typical sewage sludge treatment techniques in China." *Energy Fuels* 28 (3): 2114–2122.

Drogui, P., M. Asselin, S. K. Brar, H. Benmoussa, and J.-F. Blais. 2008. "Electrochemical removal of pollutants from agro-industry wastewaters." *Sep. Purif. Technol.* 61 (3): 301–310.

Dubrawski, K. L., C. M. van Genuchten, C. Delaire, S. E. Amrose, A. J. Gadgil, and M. Mohseni. 2015. "Production and transformation of mixed-valent nanoparticles generated by Fe(0) electrocoagulation." *Environ. Sci. Technol.* 49 (4): 2171–2179.

Dura, A., and C. B. Breslin. 2019. "Electrocoagulation using stainless steel anodes: Simultaneous removal of phosphates, Orange II and zinc ions." *J. Hazard. Mater.* 374: 152–158.

Feki, F., F. Aloui, M. Feki, and S. Sayadi. 2009. "Electrochemical oxidation post-treatment of landfill leachates treated with membrane bioreactor." *Chemosphere* 75 (2): 256–260.

Fernandes, A., M. J. Pacheco, L. Ciríaco, and A. Lopes. 2015. "Review on the electrochemical processes for the treatment of sanitary landfill leachates: Present and future." *Appl. Catal., B* 176–177: 183–200.

Gandini, D., C. Comninellis, N. B. Tahar, and A. Savall. 1998. "Électrodépollution: Traitement électrochimique des eaux résiduaires chargées en matières organiques toxiques." *Actual. Chim.* 10: 68–73.

Garcia-Segura, S., M. M. S. Eiband, J. V. de Melo, and C. A. Martínez-Huitle. 2017. "Electrocoagulation and advanced electrocoagulation processes: A general review about the fundamentals, emerging applications and its association with other technologies." *J. Electroanal. Chem.* 801: 267–299.

Garcia-Segura, S., J. D. Ocon, and M. N. Chong. 2018. "Electrochemical oxidation remediation of real wastewater effluents—A review." *Process Saf. Environ. Prot.* 113: 48–67.

Ghosh, D., H. Solanki, and M. K. Purkait. 2008. "Removal of Fe(II) from tap water by electrocoagulation technique." *J. Hazard. Mater.* 155 (1–2): 135–143.

Graça, N. S., A. M. Ribeiro, and A. E. Rodrigues. 2019. "Modeling the electrocoagulation process for the treatment of contaminated water." *Chem. Eng. Sci.* 197: 379–385.
Guo, W., S. Vigneswaran, H. Ngo, J. Kandasamy, and S. Yoon. 2008. "The role of a membrane performance enhancer in a membrane bioreactor: A comparison with other submerged membrane hybrid systems." *Desalination* 231 (1–3): 305–313.
Hakizimana, J. N., B. Gourich, M. Chafi, Y. Stiriba, C. Vial, P. Drogui, et al. 2017. "Electrocoagulation process in water treatment: A review of electrocoagulation modeling approaches." *Desalination* 404: 1–21.
Hansen, H. K., and L. M. Ottosen. 2010. "Removal of arsenic from wastewaters by airlift electrocoagulation: Part 3: Copper smelter wastewater treatment." *Sep. Sci. Technol.* 45 (9): 1326–1330.
Hashim, K. S., R. Al Khaddar, N. Jasim, A. Shaw, D. Phipps, P. Kot, et al. 2019. "Electrocoagulation as a green technology for phosphate removal from river water." *Sep. Purif. Technol.* 210: 135–144.
Hodges, B. C., E. L. Cates, and J.-H. Kim. 2018. "Challenges and prospects of advanced oxidation water treatment processes using catalytic nanomaterials." *Nat. Nanotechnol.* 13 (8): 642–650.
Holt, P. K., G. W. Barton, and C. A. Mitchell. 2005. "The future for electrocoagulation as a localised water treatment technology." *Chemosphere* 59 (3): 355–367.
Hui, X., R. Jiao, F. Xiao, and D. Wang. 2016. "Enhanced removal for humic-acid (HA) and coagulation process using carbon nanotubes (CNTs)/polyalumium chloride (PACl) composites coagulants." *Colloids Surf., A* 490: 189–199.
Jamaly, S., A. Giwa, and S. W. Hasan. 2015. "Recent improvements in oily wastewater treatment: Progress, challenges, and future opportunities." *J. Environ. Sci.* 37: 15–30.
Kamaraj, R., A. Pandiarajan, S. Jayakiruba, M. Naushad, and S. Vasudevan. 2016. "Kinetics, thermodynamics and isotherm modeling for removal of nitrate from liquids by facile one-pot electrosynthesized nano zinc hydroxide." *J. Mol. Liq.* 215: 204–211.
Karimi Estahbanati, M. R., M. Feilizadeh, M. Shokrollahi Yancheshmeh, and M. C. Iliuta. 2019. "Effects of carbon nanotube and carbon sphere templates in TiO_2 composites for photocatalytic hydrogen production." *Ind. Eng. Chem. Res.* 58 (8): 2770–2783.
Khandegar, V., and A. K. Saroha. 2013. "Electrocoagulation for the treatment of textile industry effluent—A review." *J. Environ. Manage.* 128: 949–963.
Khansorthong, S., and M. Hunsom. 2009. "Remediation of wastewater from pulp and paper mill industry by the electrochemical technique." *Chem. Eng. J.* 151 (1–3): 228–234.
Kobya, M., E. Demirbas, A. Dedeli, and M. T. Sensoy. 2010. "Treatment of rinse water from zinc phosphate coating by batch and continuous electrocoagulation processes." *J. Hazard. Mater.* 173 (1–3): 326–334.
Maitlo, H. A., J. H. Kim, K.-H. Kim, J. Y. Park, and A. Khan. 2019. "Metal–air fuel cell electrocoagulation techniques for the treatment of arsenic in water." *J. Cleaner Prod.* 207: 67–84.
Mameri, N., A. Yeddou, H. Lounici, D. Belhocine, H. Grib, and B. Bariou. 1998. "Defluoridation of septentrional Sahara water of North Africa by electrocoagulation process using bipolar aluminium electrodes." *Water Res.* 32 (5): 1604–1612.
Mandal, P., B. K. Dubey, and A. K. Gupta. 2017. "Review on landfill leachate treatment by electrochemical oxidation: Drawbacks, challenges and future scope." *Waste Manage.* 69: 250–273.
Meas, Y., J. A. Ramirez, M. A. Villalon, and T. W. Chapman. 2010. "Industrial wastewaters treated by electrocoagulation." *Electrochim. Acta* 55 (27): 8165–8171.

Mohammadi, A., A. Khadir, and R. M. A. Tehrani. 2019. "Optimization of nitrogen removal from an anaerobic digester effluent by electrocoagulation process." *J. Environ. Chem. Eng.* 7 (3): 103195.

Mollah, M. Y. A., P. Morkovsky, J. A. G. Gomes, M. Kesmez, J. Parga, and D. L. Cocke. 2004. "Fundamentals, present and future perspectives of electrocoagulation." *J. Hazard. Mater.* 114 (1–3): 199–210.

Morão, A., A. Lopes, M. T. Pessoa de Amorim, and I. C. Gonçalves. 2004. "Degradation of mixtures of phenols using boron doped diamond electrodes for wastewater treatment." *Electrochim. Acta* 49 (9–10): 1587–1595.

Moreira, F. C., J. Soler, A. Fonseca, I. Saraiva, R. A. R. Boaventura, E. Brillas, et al. 2016. "Electrochemical advanced oxidation processes for sanitary landfill leachate remediation: Evaluation of operational variables." *Appl. Catal., B* 182: 161–171.

Moussa, D. T., M. H. El-Naas, M. Nasser, and M. J. Al-Marri. 2017. "A comprehensive review of electrocoagulation for water treatment: Potentials and challenges." *J. Environ. Manage.* 186: 24–41.

Nariyan, E., A. Aghababaei, and M. Sillanpää. 2017. "Removal of pharmaceutical from water with an electrocoagulation process; effect of various parameters and studies of isotherm and kinetic." *Sep. Purif. Technol.* 188: 266–281.

Nawarkar, C. J., and V. D. Salkar. 2019. "Solar powered electrocoagulation system for municipal wastewater treatment." *Fuel* 237: 222–226.

Novikova, S., T. Shkorbatova, and E. Y. Sokol. 1982. "Purification of effluents from the production of synthetic detergents by electrocoagulation." *Soviet J. Water Chem. Technol.* 4 (4): 353–357.

Nowak, O., S. Keil, and C. Fimml. 2011. "Examples of energy self-sufficient municipal nutrient removal plants." *Water Sci. Technol.* 64 (1): 1–6.

Núñez, P., and H. K. Hansen. 2013. "Zero valent iron nanoparticle assisted electrocoagulation of arsenic with electromagnetic separation of solids." In *Proc., 16th Int. Conf. on Heavy Metals in the Environment.* Les Ulis, France: *E3S Web of Conferences, EDP Sciences.* https://www.e3s-conferences.org/component/issues/?task=all&Itemid=121.

Nunez, P., H. K. Hansen, S. Aguirre, and C. Maureira. 2011. "Electrocoagulation of arsenic using iron nanoparticles to treat copper mineral processing wastewater." *Sep. Purif. Technol.* 79 (2): 285–290.

Ögütveren, Ü. B., N. Gönen, and S. Koparal. 1992. "Removal of dye stuffs from waste water: Electrocoagulation of Acilan Blau using soluble anode." *J. Environ. Sci. Health., Part A* 27 (5): 1237–1247.

Orori, O. B., L. Etiegni, K. Senelwa, M. M. Mwamburi, K. B. Balozi, G. K. Barisa, et al. 2010. "Electro-coagulation treatment efficiency of graphite, iron and aluminum electrodes using alum and wood ash electrolytes on a Kraft pulp and paper mill effluent." *Water Sci. Technol.* 62 (7): 1526–1535.

Osipenko, V., and P. Pogorelyi 1977. "Electrocoagulation neutralization of chromium containing effluent." *Metallurgist* 21 (9): 628–630.

Pretorius, W., W. Johannes, and G. Lempert. 1991. "Electrolytic iron flocculant production with a bipolar electrode in series arrangement." *Water Sa* 17 (2): 133–138.

Priya, M., and J. Jeyanthi. 2019. "Removal of cod, oil and grease from automobile wash water effluent using electrocoagulation technique." *Microchem. J.* 150: 104070.

Radjenovic, J., and D. L. Sedlak. 2015. "Challenges and opportunities for electrochemical processes as next-generation technologies for the treatment of contaminated water." *Environ. Sci. Technol.* 49 (19): 11292–11302.

Rajkumar, D., J. Guk Kim, and K. Palanivelu. 2005. "Indirect electrochemical oxidation of phenol in the presence of chloride for wastewater treatment." *Chem. Eng. Technol.* 28 (1): 98–105.

Reilly, M., A. P. Cooley, D. Tito, S. A. Tassou, and M. K. Theodorou. 2019. "Electrocoagulation treatment of dairy processing and slaughterhouse wastewaters." *Energy Procedia* 161: 343–351.

Rothausen, S. G., and D. Conway. 2011. "Greenhouse-gas emissions from energy use in the water sector." *Nat. Clim. Chang.* 1 (4): 210–219.

Samir, A., and S. A. Abbas. 2013. "Electrocoagulation technology in wastewater treatment: a review of methods and applications." *Civ. Environ. Res.* 3 (11): 29–43.

Senthilkumar, S., C. A. Basha, M. Perumalsamy, and H. Prabhu. 2012. "Electrochemical oxidation and aerobic biodegradation with isolated bacterial strains for dye wastewater: Combined and integrated approach." *Electrochim. Acta* 77: 171–178.

Shabeer, T. A., A. Saha, V. T. Gajbhiye, S. Gupta, K. M. Manjaiah, and E. Varghese. 2014. "Removal of poly aromatic hydrocarbons (PAHs) from water: Effect of nano and modified nano-clays as a flocculation aid and adsorbent in coagulation-flocculation process." *Polycyclic Aromat. Compd.* 34 (4): 452–467.

Shabeer, T. P. A., A. Saha, V. T. Gajbhiye, S. Gupta, K. M. Manjaiah, and E. Varghese. 2015. "Exploitation of nano-bentonite, nano-halloysite and organically modified nano-montmorillonite as an adsorbent and coagulation aid for the removal of multi-pesticides from water: A sorption modelling approach." *Water Air Soil Pollut.* 226 (3): 41.

Shon, H. K., S. Phuntsho, S. Vigneswaran, J. Kandasamy, L. D. Nghiem, G. J. Kim, et al.2010. "Preparation of titanium dioxide nanoparticles from electrocoagulated sludge using sacrificial titanium electrodes." *Environ. Sci. Technol.* 44 (14): 5553–5557.

Song, P., Z. Yang, G. Zeng, X. Yang, H. Xu, L. Wang, et al. 2017. "Electrocoagulation treatment of arsenic in wastewaters: A comprehensive review." *Chem. Eng. J.* 317: 707–725.

Sridhar, R., V. Sivakumar, V. P. Immanuel, and J. P. Maran. 2011. "Treatment of pulp and paper industry bleaching effluent by electrocoagulant process." *J. Hazard. Mater.* 186 (2–3): 1495–1502.

Sukopová, M., J. Matysíková, O. Škorvan, and M. Holba. 2013. "Application of iron nanoparticles for industrial wastewater treatment." In *Proc., 5th Int. Conf. NANOCON.* 1–5. Brno, Czech Republic. http://nanocon2014.tanger.cz/files/proceedings/14/reports/2016.pdf.

Tang, J., C. Zhang, X. Shi, J. Sun, and J. A. Cunningham. 2019. "Municipal wastewater treatment plants coupled with electrochemical, biological and bio-electrochemical technologies: Opportunities and challenge toward energy self-sufficiency." *J. Environ. Manage.* 234: 396–403.

Tchomgui-Kamga, E., N. Audebrand, and A. Darchen. 2013. "Effect of co-existing ions during the preparation of alumina by electrolysis with aluminum soluble electrodes: Structure and defluoridation activity of electro-synthesized adsorbents." *J. Hazard. Mater.* 254–255: 125–133.

Urtiaga, A., G. Pérez, R. Ibáñez, and I. Ortiz. 2013. "Removal of pharmaceuticals from a WWTP secondary effluent by ultrafiltration/reverse osmosis followed by electrochemical oxidation of the RO concentrate." *Desalination* 331, 26–34.

Wang, A., J. Qu, H. Liu, and J. Ge. 2004. "Degradation of azo dye Acid Red 14 in aqueous solution by electrokinetic and electrooxidation process." *Chemosphere* 55 (9): 1189–1196.

Wang, F., J. Fan, H. Zhu, K. Han, J. Zou, and H. Sun. 2011. "Preparation of nano-modified polyacrylamide and its application on solid–liquid separation in waste drilling mud." *Adv. Chem. Eng. Sci.* 1 (2): 33–36.

Xu, H., R. Jiao, F. Xiao, and D. Wang. 2014a. "Relative importance of hydrolyzed Al species (Ala, Alb, Alc) on residual Al and effects of nano-particles (Fe-surface modified TiO_2 and Al_2O_3) on coagulation process." *Colloids Surf., A* 446: 139–150.

Xu, H., R. Jiao, F. Xiao, and D.. Wang. 2014b. "Effects of different coagulants in treatment of TiO_2–humic acid (HA) water and the aggregate characterization in different coagulation conditions." *Colloids Surf., A* 446: 213–223.

Xu, H., F. Xiao, D. Wang, and C. Ye. 2014c. "Survey of treatment process in water treatment plant and the characteristics of flocs formed by two new coagulants." *Colloids Surf., A* 456: 211–221.

Zarei, M., A. Niaei, D. Salari, and A. R. Khataee. 2010. "Removal of four dyes from aqueous medium by the peroxi-coagulation method using carbon nanotube–PTFE cathode and neural network modeling." *J. Electroanal. Chem.* 639 (1–2): 167–174.

Zarei, M., D. Salari, A. Niaei, and A. Khataee. 2009. "Peroxi-coagulation degradation of CI Basic Yellow 2 based on carbon-PTFE and carbon nanotube-PTFE electrodes as cathode." *Electrochim. Acta* 54 (26): 6651–6660.

Zazou, H., H. Afanga, S. Akhouairi, H. Ouchtak, A. A. Addi, R. A. Akbour, et al. 2019. "Treatment of textile industry wastewater by electrocoagulation coupled with electrochemical advanced oxidation process." *J. Water Process Eng.* 28: 214–221.

About the Editors

Patrick Drogui, Ph.D., is a professor and researcher at the Institut national de la recherche scientifique's Centre Eau Terre Environnement. He is an internationally recognized expert in electrochemical and membrane technologies and oxidative processes for different applications in water treatment ranging from industrial wastewater to potable water. The originality of his research work lies in the development of advanced electrolytic processes for water and wastewater treatment requiring no or minimum addition of chemical products. A coinventor on eight patents, Dr. Drogui is currently working on new electrolytic and/or membrane, photocatalytic, and electrophotocatalytic processes for water treatment and the removal of phosphate and emerging contaminants. He is the author of more than 250 papers published in scientific refereed journals. He is the author of 1 book, 20 book chapters, and 61 industrial and research reports. He received 4 fellowships and 1 award for the best research project from Pollutec and delivered 16 invited conferences and 122 communications and conferences. He is also the 2019 recipient of the Eddy Wastewater Principles/Processes Medal Award of the Water Environment Federation. Dr. Drogui is the director of Collaborative Research and Training Experience Program (CREATE)-Program in Environmental Decontamination Technologies and Integrated Water and Wastewater Management (TEDGIEER) from National Science Engineering Research Council (NSERC) of Canada. He is also the director of the Innovation Centre for Advanced Technologies and Decentralized Sanitation of Liquid Effluents (CITADEL).

R. D. Tyagi, Ph.D., is an internationally recognized professor of Biochemical Engineering and chief scientific officer of BOSK Bioproducts. Prior to joining BOSK, Dr. Tyagi served as a professor for 36 years and as senior Canada Research chair on value-added products from wastes for 18 years at the University of Quebec, Canada. He is also adjunct professor at the University of Missouri, Columbia (USA), a member of the Hall of Excellence of University of Québec, and a member of the European Academy of Sciences and Arts; Academician, European Academy of Sciences & Arts. He conducts research on hazardous/solids waste management, water/wastewater treatment, sludge treatment/disposal, and bioconversion of wastewater and wastewater sludge into value-added products. He has developed novel technologies of simultaneous sewage sludge digestion and metal leaching and bioconversion of wastewater sludge (biosolids) into *Bacillus thuringiensis*–based biopesticides, bioplastics, biofertilizers, and biocontrol agents. He is the recipient of the Outstanding Scientist Award and lifetime achievement of The International Bioprocessing Association (an International Forum on Industrial Bioprocesses) for making significant contributions in the area of environmental biotechnology. He is also a recipient of the ASCE State-of-the-Art of Civil Engineering Award, ASCE Rudolph Hering Medal, ASCE Wesley W. Horner Medal, and ASCE Best Practice-Oriented Paper Award. He received the Superior Achievement Award, Excellence in Environmental Engineering Award of the American Academy of Environmental Engineers in 2015. He also received the 2010 and 2018 Global Honor Award (applied research) of the International Water Association (IWA), 2010 Grand Prize (university research) of the American Academy of Environmental Engineering, Life-Time Achievement Award (Specialist medal, 2007) for outstanding research contributions in wastewater sludge management by the IWA and Excellence, Award of the Natural Sciences and Engineering Research Council of Canada for industry-university collaborative research, and the lifetime achievement award of the International Society of Waste Management, Air and Water in recognition of his significant contributions in the fields of waste management and environment protection. He is a Fellow of the International Water Association. He has served on the boards of several professional organizations and on the panels of several granting agencies, including the International Water Association, where he worked as a specialist on "Appropriate Technology for Developing Countries". Currently, he is on the Executive Committee of the Canadian Association on Water Quality. Dr. Tyagi has either published or presented more than 600 papers in refereed journals and conference proceedings and is the author of 22 books, 145 book chapters, and 10 research reports, and holds 9 patents.

Rao Y. Surampalli, Ph.D., P.E., Dist.M.ASCE, NAC, received his M.S. and Ph.D. degrees in environmental engineering from Oklahoma State and Iowa State Universities, respectively. He is a registered professional engineer in the branches of Civil and Environmental Engineering and also a Board-Certified Environmental and Water Resources Engineer (BCEE, D.WRE) of the American Academy of Environmental Engineers (AAEE) and the American Academy of Water Resources Engineers. His private practice and his career in government, university, and applied research have given him the opportunity to experience and appreciate the varied interests and challenges of the environmental engineering profession; he has conducted applied research on more than 50 environmental science and engineering topics. He is adjunct professor in five universities and distinguished/honorary visiting professor in eight universities. He serves or has served on more than 75 national and international committees, review panels, or advisory boards, including the ASCE National Committee on Energy, Environment and Water Policy. A distinguished engineering alumnus of both the Oklahoma State and Iowa State Universities, Dr. Surampalli has received more than 30 national awards and honors from ASCE, WEF, IWA, AAEE, NSPE, and AAES. He is an elected Fellow of the American Association for the Advancement of Science (F.AAAS), an elected member of the European Academy of Sciences and Arts (EASA), an elected member of the Russian Academy of Engineering (RAE), an elected member of the U.S. National Academy of Construction (NAC), an elected Fellow of the Water Environment Federation (F.WEF) and the International Water Association (F.IWA), and recognized as a Distinguished Member of the American Society of Civil Engineers (Dist.M.ASCE). He is also editor-in-chief of the ASCE *Journal of Hazardous, Toxic, and Radioactive Waste* and past vice chair of the editorial board of the journal *Water Environment Research*. He has authored more than 600 technical publications in journals and conference proceedings, including more than 360 refereed journal articles, 20 patents, 27 books, and 171 book chapters.

Tian C. Zhang, Ph.D., P.E., is a professor in the Department of Civil Engineering at the University of Nebraska–Lincoln (UNL). He received his Ph.D. in environmental engineering from the University of Cincinnati in 1994 and joined the UNL faculty in August 1994. Professor Zhang teaches courses related to water/wastewater treatment, remediation of hazardous waste, and nonpoint pollution control. Professor Zhang's research involves the fundamentals and applications of nanotechnology and conventional technologies for water, wastewater, and stormwater treatment and management, remediation of contaminated environments, and detection/control of emerging contaminants in the environment. Professor Zhang has published more than 185 peer-reviewed journal papers, 82 book chapters, and 20 books since 1994. He is diplomate of Water Resources Engineer (D.WRE) of the American Academy of Water Resources Engineers, a board-certified environmental engineer (BCEE) of the American Academy of Environmental Engineers, an elected Fellow of ASCE (F.ASCE), an elected Fellow of the American Association for the Advancement of Science (F.AAAS), and an elected member of the European Academy of Sciences and Arts (EASA). He has been an associate editor of the *Journal of Environmental Engineering* since 2007, *Journal of Hazardous, Toxic, and Radioactive Waste* since 2006, and *Water Environment Research* since 2008. He has been a registered professional engineer in Nebraska since 2000.

Song Yan, Ph.D., is currently working as a research associate at the Institut National de la Recherche Scientifique-Eau, Terre, et Environnement (INRS-ETE), Université du Québec, Canada. She received her Ph.D. in water science from the INRS, University of Quebec. Dr. Yan's research focuses on the production of bioplastics, biodiesel, value-added products using wastewater and sludge (biosolids), greenhouse gas emissions from wastewater and sludge treatment processes, and wastewater and sludge treatment process control. Dr. Yan has published 1 patent, 4 books, 36 book chapters, and 78 papers in refereed journals and many other scientific reports. She has presented 60 papers in national and international conferences. Dr. Yan was part of the team that received the Superior Achievement Award for Excellence in Environmental Engineering and Science in 2010 and 2015 from the American Academy of Environmental Engineers and Scientists (AAEES) for university research work. She was selected as an outstanding reviewer of ASCE's *Journal of Environmental Engineering* in 2012 and holds a membership in the Canadian Association for Water Quality (CAWQ) and the International Water Association (IWA).

Xiaolei Zhang, Ph.D., received her doctorate degree in water science from INRS University of Québec in 2013. Currently, she is an assistant professor at the Department of Civil and Environmental Engineering of the Harbin Institute of Technology (Shenzhen). Her research work focuses on water and wastewater treatment, energy recovery from wastewater, bioconversion of municipal waste and industrial organic waste to bioenergy, green technology for bioenergy generation, and technoeconomic assessment of bioenergy production. She was part of the team that received the Superior Achievement Award for Excellence in Environmental Engineering and Science (the highest award of the 2015 competition) from the American Academy of Environmental Engineers and Scientists (AAEES) for university research work on "Bioconversion of Wastewater Sludge and Glycerol to Biodiesel." He has published more than 40 peer-reviewed journal papers, 10 book chapters, holds 3 patents, and made many conference presentations.

Index

Note: Page numbers followed by *f* and *t* indicate figures and tables.